f_t — tensile stress

f_v — shearing stress

F'_{nt} — nominal bolt tensile strength (stress) in the presence of shear

F_a — allowable axial compressive stress

F_{BM} — shear strength (stress) of base metal in a welded connection

F_{cr} — critical compressive or bending stress used to determine nominal strength

F_{cry} — flexural buckling strength corresponding to the axis of symmetry in a structural tee or double-angle compression member

F_{crz} — stress used in computing torsional or flexural-torsional buckling strength of a structural tee or double-angle compression member

F_e — Euler buckling stress, critical elastic buckling stress in an unsymmetrical compression member (torsional or flexural-torsional buckling stress)

F_{ex}, F_{ey}, F_{ez} — stresses used in computing torsional or flexural-torsional buckling strength

F_n — nominal bolt shear or tensile strength (stress)

F_{nt} — nominal bolt tensile strength (stress)

F_{nv} — nominal bolt shear strength (stress)

F_{pl} — stress at proportional limit

F_t — allowable member tensile stress, ultimate tensile stress of a bolt, allowable bolt tensile stress

F_u — ultimate tensile stress

F_v — allowable member shear stress

F_W — ultimate shearing stress of weld electrode

F_y — yield stress

F_{yf}, F_{yw} — yield stresses of flange and web

F_{yr} — yield stress of reinforcing steel

F_{yst} — yield stress of a stiffener

F_{yt} — yield strength of tension flange

g — gage distance for bolts (transverse spacing)

G — shear modulus of elasticity = 11,200 ksi for structural steel

G_A, G_B — factors for use in nomographs for effective length factor K

h — width of web from toe of flange fillet to toe of flange fillet for a rolled shape, width of web from inside of flange to inside of flange for a welded shape, bolt hole diameter

h_c — twice the distance from the elastic neutral axis to the inside face of the compression flange of a built-up flexural member (same as h for girders with equal flanges)

h_o — distance between W-shape flange centroids

h_p — twice the distance from the plastic neutral axis to the inside face of the compression flange of a built-up flexural member (same as h for girders with equal flanges)

h_{sc} — hole factor for slip-critical bolts

H — ...tion of flexural-...sion members, ...loads, flange force in a moment connection

I — moment of inertia (second moment of area)

$\bar{I}$ — moment of inertia of component area about its centroidal axis

I_c — moment of inertia of a column cross section

I_{eff} — effective transformed moment of inertia of a partially composite beam

I_g — moment of inertia of a girder cross section

I_{LB} — lower-bound moment of inertia of a composite beam

I_s — moment of inertia of steel section

I_{st} — moment of inertia of a stiffener cross section

I_{tr} — moment of inertia of transformed section

I_x, I_y — moments of inertia about x and y axes

j — constant used in computing required moment of inertia of a plate girder stiffener

J — torsional constant, polar moment of inertia

k — distance from outer face of flange to toe of fillet in the web of a rolled shape

k_c — factor used in computing the flexural strength of a plate girder

k_s — multiplier for bolt slip-critical strength when tension is present

k_v — factor used in computing shear strength

K — effective length factor for compression members

K_x, K_y, K_z — effective length factors for x, y, and z axes

K_xL, K_yL, K_zL — effective lengths for buckling about x, y, and z axes

ℓ — length of a connection, length of end welds, factor for computing column base plate thickness, largest unbraced length of the flange of a plate girder

L — service live load effect to be used in computation of factored load combinations, member length, story height, length of a weld segment

L_b — unbraced beam length, unbraced length of a column in the equation for required bracing stiffness

L_c — column length, distance from edge of bolt hole to edge of connected part or to edge of adjacent hole

L_g — length of girder

L_p — largest unbraced beam length for which lateral-torsional buckling will not occur

L_{pd} — largest unbraced beam length for which plastic analysis can be used.

L_r — unbraced beam length at which elastic lateral-torsional buckling will occur, service roof live load effect to be used in computation of factored load combinations

m — length of unit width of plate in bending (for beam bearing plate and column base plate design)

M — bending moment

Steel Design

Fourth Edition

William T. Segui
The University of Memphis

Australia • Brazil • Canada • Mexico • Singapore • Spain
United Kingdom • United States

THOMSON
™

Steel Design, Fourth Edition
by William T. Segui

Associate Vice President and Editorial Director:
Evelyn Veitch

Publisher:
Chris Carson

Developmental Editors:
Kamilah Reid Burrell/Hilda Gowans

Permissions Coordinator:
Vicki Gould

Production Services:
RPK Editorial Services

Copy Editor:
Harlan James

Proofreader:
Erin Wagner

Indexer:
Shelly Gerger-Knechtl

Production Manager:
Renate McCloy

Creative Director:
Angela Cluer

Interior Design:
Carmela Pereira

Cover Design:
Andrew Adams

Compositor:
International Typesetting and Composition

Printer:
R. R. Donnelley

Cover Image Credit:
© Getty Images/Terry Vine

COPYRIGHT © 2007 by Nelson, a division of Thomson Canada Limited.

Printed and bound in United States
2 3 4 07

Fore more information contact Nelson, 1120 Birchmount Road, Toronto, Ontario, Canada, MIK 5G4. Or you can visit our Internet site at
http://www.nelson.com

Library Congress Control Number: 2006908598

ISBN 10: 0-495-24471-6

ISBN 13: 0-978-0-495-24471-4

North America
Nelson 1120 Birchmount Road
Toronto, Ontario MIK 5G4
Canada

Asia
Thomson Learning
5 Shenton Way #01-01
UIC Building
Singapore 068808

Australia/New Zealand
Thomson Learning
102 Dodds Street
Southbank, Victoria
Australia 3006

Europe/Middle East/Africa
Thomson Learning
High Holborn House
50/51 Bedford Row
London WCIR 4LR
United Kingdom

Latin America
Thomson Learning
Seneca, 53
Colonia Polanco
11560 Mexico D.F.
Mexico

Spain
Paraninfo
Calle/Magallanes, 25
28015 Madrid, Spain

Contents

Preface

Steel Design, Fourth Edition covers the fundamentals of structural steel design. The emphasis is on the design of members and their connections rather than the integrated design of buildings. This book is intended for junior- and senior-level engineering students, although some of the later chapters can be used in graduate courses. Practicing civil engineers who need a review of current practice and the current AISC Specification and Manual will find the book useful as a reference. Students should have a background in mechanics of materials and analysis of statically determinate structures.

Steel Design is a revision of *LRFD Steel Design,* but because of the nature of the 2005 Specification and Manual of the American Institute of Steel Construction (AISC), it is more than a new edition of *LRFD Steel Design,* hence the change in title. Prior to the 2005 AISC documents, load and resistance factor design (LRFD) was covered by the 1999 AISC Specification and *LRFD Manual of Steel Construction, Third Edition.* Allowable stress design (ASD) was covered by the 1978 AISC Specification and *Manual of Steel Construction, Ninth Edition.* In 2005, the two approaches were unified in a single specification and a single manual, the thirteenth edition of the *Steel Construction Manual.* In addition, changes were made to many provisions of the specification, both in form and substance.

Both LRFD and ASD are covered in this textbook, but the emphasis is on LRFD. In most examples, both LRFD and ASD solutions are given. In those examples, the LRFD solution is given first. In some cases, the ASD solution is abbreviated but complete and independent of the LRFD solution. This usually involves some duplication, but is necessary if a reader is interested in only the ASD solution. In some ASD solutions where there would be a lengthy duplication, the reader is referred to the LRFD solution for that portion. In some of the examples, particularly in the later chapters, only an LRFD solution is given.

This book is designed so that an instructor can easily teach either LRFD or ASD. If time permits, both can be covered. One possibility is to cover the requirements for both but use mostly LRFD examples as the course progresses. As will be seen, the differences in the two approaches are mainly conceptual, and there is very little difference in the computations.

It is essential that students have a copy of the *Steel Construction Manual.* In order to promote familiarity with it, material from the *Manual* is not reproduced in this book so that the reader will be required to refer to the *Manual.* All notation in *Steel Design* is consistent with that in the *Manual* and AISC equation numbers are used along with sequential numbering of other equations according to the textbook chapter.

U.S. customary units are used throughout with no introduction of SI units. Although the *AISC Specification* now uses a dual system of units, the steel construction industry is still in a period of transition.

As far as design procedures are concerned, the application of fundamental principles is encouraged. Although this book is oriented toward practical design, sufficient theory is included to avoid a "cookbook" approach. Direct design methods are used where feasible, but no complicated design formulas have been developed. Instead, trial and error, with "educated guesses," is the rule. Tables, curves, and other design aids from the *Manual* are used, but they have a role that is subordinate to the use of basic equations. Assigned problems provide practice with both approaches and, where appropriate, the required approach is specified in the statement of the problem. In keeping with the objective of providing a basic textbook, a large number of assigned problems are given at the end of each chapter. Answers to selected problems are given at the back of the book, and an Instructor's Manual with solutions is available.

I would like to express my appreciation to Christopher Carson, General Manager of Thomson Engineering; Hilda Gowans, Developmental Editor for Thompson Learning; and Rose Kernan of RPK Editorial Services for their help during the production of this book. In addition, Christopher Hewitt of the American Institute of Steel Construction was very helpful in providing updates on the AISC Specification and Manual revisions as well as other assistance. Finally, I want to thank my wife, Angela, for her encouragement and for her valuable suggestions and assistance in proofreading the manuscript of this book.

I would appreciate learning of any errors that users of this book discover. I can be contacted at wsegui@memphis.edu.

William T. Segui

1 Introduction

1.1 STRUCTURAL DESIGN

The structural design of buildings, whether of structural steel or reinforced concrete, requires the determination of the overall proportions and dimensions of the supporting framework and the selection of the cross sections of individual members. In most cases the functional design, including the establishment of the number of stories and the floor plan, will have been done by an architect, and the structural engineer must work within the constraints imposed by this design. Ideally, the engineer and architect will collaborate throughout the design process to complete the project in an efficient manner. In effect, however, the design can be summed up as follows: The architect decides how the building should look; the engineer must make sure that it doesn't fall down. Although this distinction is an oversimplification, it affirms the first priority of the structural engineer: safety. Other important considerations include serviceability (how well the structure performs in terms of appearance and deflection) and economy. An economical structure requires an efficient use of materials and construction labor. Although this objective can usually be accomplished by a design that requires a minimum amount of material, savings can often be realized by using more material if it results in a simpler, more easily constructed project. In fact, materials account for a relatively small portion of the cost of a typical steel structure as compared with labor and other costs (Cross, 2005).

A good design requires the evaluation of several framing plans — that is, different arrangements of members and their connections. In other words, several alternative designs should be prepared and their costs compared. For each framing plan investigated, the individual components must be designed. To do so requires the structural analysis of the building frames and the computation of forces and bending moments in the individual members. Armed with this information, the structural designer can then select the appropriate cross section. Before any analysis, however, a decision must be made on the primary building material to be used; it will usually be reinforced concrete, structural steel, or both. Ideally, alternative designs should be prepared with each.

The emphasis in this book will be on the design of individual structural steel members and their connections. The structural engineer must select and evaluate the overall structural system in order to produce an efficient and economical design but cannot do so without a thorough understanding of the design of the components (the "building blocks") of the structure. Thus component design is the focus of this book.

Before discussing structural steel, we need to examine various types of structural members. Figure 1.1 shows a truss with vertical concentrated forces applied at the joints along the top chord. In keeping with the usual assumptions of truss analysis — pinned connections and loads applied only at the joints — each component of the truss will be a two-force member, subject to either axial compression or tension. For simply supported trusses loaded as shown — a typical loading condition — each of the top chord members will be in compression, and the bottom chord members will be in tension. The web members will either be in tension or compression, depending on their location and orientation and on the location of the loads.

Other types of members can be illustrated with the rigid frame of Figure 1.2a. The members of this frame are rigidly connected by welding and can be assumed to form a continuous structure. At the supports, the members are welded to a rectangular plate that is bolted to a concrete footing. Placing several of these frames in parallel and connecting them with additional members that are then covered with roofing material and walls produces a typical building system. Many important details have not been mentioned, but many small commercial buildings are constructed essentially in this manner. The design and analysis of each frame in the system begins with the idealization of the frame as a two-dimensional structure, as shown in Figure 1.2b. Because the frame has a plane of symmetry parallel to the page, we are able to treat the frame as two-dimensional and represent the frame members by their centerlines. (Although it is not shown in Figure 1.1, this same idealization is made with trusses, and the members are usually represented by their centerlines.) Note that the supports are represented as hinges (pins), not as fixed supports. If there is a possibility that the footing will undergo a slight rotation, or if the connection is flexible enough to allow a slight rotation, the support must be considered to be pinned. One assumption made in the usual methods of structural analysis is that deformations are very small, which means that only a slight rotation of the support is needed to qualify it as a pinned connection.

Once the geometry and support conditions of the idealized frame have been established, the loading must be determined. This determination usually involves apportioning a share of the total load to each frame. If the hypothetical structure under consideration is subjected to a uniformly distributed roof load, the portion carried by one frame will be a uniformly distributed line load measured in force per unit length, as shown in Figure 1.2b. Typical units would be kips per foot.

FIGURE 1.1

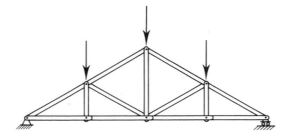

For the loading shown in Figure 1.2b, the frame will deform as indicated by the dashed line (drawn to a greatly exaggerated scale). The individual members of the frame can be classified according to the type of behavior represented by this deformed shape. The horizontal members *AB* and *BC* are subjected primarily to bending, or flexure, and are called *beams*. The vertical member *BD* is subjected to couples transferred from each beam, but for the symmetrical frame shown, they are equal and opposite, thereby canceling each other. Thus member *BD* is subjected only to axial compression arising from the vertical loads. In buildings, vertical compression members such as these are referred to as *columns*. The other two vertical members, *AE* and *CF,* must resist not only axial compression from the vertical loads but also a significant amount of bending. Such members are called *beam-columns*. In reality, all members, even those classified as beams or columns, will be subjected to both bending and axial load, but in many cases, the effects are minor and can be neglected.

FIGURE 1.2

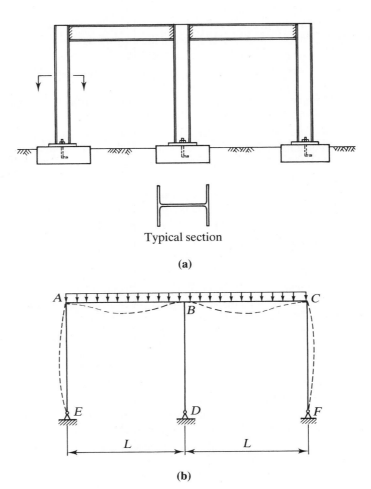

Typical section

(a)

(b)

In addition to the members described, this book covers the design of connections and the following special members: composite beams, composite columns, and plate girders.

1.2   LOADS

The forces that act on a structure are called *loads*. They belong to one of two broad categories: *dead load* and *live load*. Dead loads are those that are permanent, including the weight of the structure itself, which is sometimes called the *self-weight*. In addition to the weight of the structure, dead loads in a building include the weight of nonstructural components such as floor coverings, partitions, and suspended ceilings (with light fixtures, mechanical equipment, and plumbing). All of the loads mentioned thus far are forces resulting from gravity and are referred to as *gravity loads*. Live loads, which can also be gravity loads, are those that are not as permanent as dead loads. They may or may not be acting on the structure at any given time, and the location may not be fixed. Examples of live loads include furniture, equipment, and occupants of buildings. In general, the magnitude of a live load is not as well defined as that of a dead load, and it usually must be estimated. In many cases, a structural member must be investigated for various positions of a live load so that a potential failure condition is not overlooked.

If a live load is applied slowly and is not removed and reapplied an excessive number of times, the structure can be analyzed as if the load were static. If the load is applied suddenly, as would be the case when the structure supports a moving crane, the effects of impact must be accounted for. If the load is applied and removed many times over the life of the structure, fatigue stress becomes a problem, and its effects must be accounted for. Impact loading occurs in relatively few buildings, notably industrial buildings, and fatigue loading is rare, with thousands of load cycles over the life of the structure required before fatigue becomes a problem. For these reasons, all loading conditions in this book will be treated as static, and fatigue will not be considered.

Wind exerts a pressure or suction on the exterior surfaces of a building, and because of its transient nature, it properly belongs in the category of live loads. Because of the relative complexity of determining wind loads, however, wind is usually considered a separate category of loading. Because lateral loads are most detrimental to tall structures, wind loads are usually not as important for low buildings, but uplift on light roof systems can be critical. Although wind is present most of the time, wind loads of the magnitude considered in design are infrequent and are not considered to be fatigue loads.

Earthquake loads are another special category and need to be considered only in those geographic locations where there is a reasonable probability of occurrence. A structural analysis of the effects of an earthquake requires an analysis of the structure's response to the ground motion produced by the earthquake. Simpler methods are sometimes used in which the effects of the earthquake are simulated by a system of horizontal loads, similar to those resulting from wind pressure, acting at each floor level of the building.

Snow is another live load that is treated as a separate category. Adding to the uncertainty of this load is the complication of drift, which can cause much of the load to accumulate over a relatively small area.

Other types of live load are often treated as separate categories, such as hydrostatic pressure and soil pressure, but the cases we have enumerated are the ones ordinarily encountered in the design of structural steel building frames and their members.

1.3   BUILDING CODES

Buildings must be designed and constructed according to the provisions of a building code, which is a legal document containing requirements related to such things as structural safety, fire safety, plumbing, ventilation, and accessibility to the physically disabled. A building code has the force of law and is administered by a governmental entity such as a city, a county, or, for some large metropolitan areas, a consolidated government. Building codes do not give design procedures, but they do specify the design requirements and constraints that must be satisfied. Of particular importance to the structural engineer is the prescription of minimum live loads for buildings. Although the engineer is encouraged to investigate the actual loading conditions and attempt to determine realistic values, the structure must be able to support these specified minimum loads.

Although some large cities have their own building codes, many municipalities will modify a "model" building code to suit their particular needs and adopt it as modified. Model codes are written by various nonprofit organizations in a form that can be easily adopted by a governmental unit. Three national code organizations have developed model building codes: the *Uniform Building Code* (International Conference of Building Officials, 1999), the *Standard Building Code* (Southern Building Code Congress International, 1999), and the *BOCA National Building Code* (BOCA, 1999) (BOCA is an acronym for Building Officials and Code Administrators.) These codes have generally been used in different regions of the United States. The *Uniform Building Code* has been essentially the only one used west of the Mississippi, the *Standard Building Code* has been used in the southeastern states, and the *BOCA National Building Code* has been used in the northeastern part of the country.

A unified building code, the *International Building Code* (International Code Council, 2003), has been developed to eliminate some of the inconsistencies among the three national building codes. This was a joint effort by the three code organizations (ICBO, BOCA, and SBCCI). These organizations have merged into the International Code Council, and the new code has replaced the three regional codes.

Although it is not a building code, ASCE 7, *Minimum Design Loads for Buildings and Other Structures* (American Society of Civil Engineers, 2002) is similar in form to a building code. This standard provides load requirements in a format suitable for adoption as part of a code. The *International Building Code* incorporates much of ASCE 7 in its load provisions.

1.4 DESIGN SPECIFICATIONS

In contrast to building codes, design specifications give more specific guidance for the design of structural members and their connections. They present the guidelines and criteria that enable a structural engineer to achieve the objectives mandated by a building code. Design specifications represent what is considered to be good engineering practice based on the latest research. They are periodically revised and updated by the issuance of supplements or completely new editions. As with model building codes, design specifications are written in a legal format by nonprofit organizations. They have no legal standing on their own, but by presenting design criteria and limits in the form of legal mandates and prohibitions, they can easily be adopted, by reference, as part of a building code.

The specifications of most interest to the structural steel designer are those published by the following organizations.

1. **American Institute of Steel Construction (AISC):** This specification provides for the design of structural steel buildings and their connections. It is the one of primary concern in this book, and we discuss it in detail (AISC, 2005a).
2. **American Association of State Highway and Transportation Officials (AASHTO):** This specification covers the design of highway bridges and related structures. It provides for all structural materials normally used in bridges, including steel, reinforced concrete, and timber (AASHTO, 2002, 2004).
3. **American Railway Engineering and Maintenance-of-Way Association (AREMA):** The AREMA *Manual of Railway Engineering* covers the design of railway bridges and related structures (AREMA, 2005). This organization was formerly known as the American Railway Engineering Association (AREA).
4. **American Iron and Steel Institute (AISI):** This specification deals with cold-formed steel, which we discuss in Section 1.6 of this book (AISI, 2001).

1.5 STRUCTURAL STEEL

The earliest use of iron, the chief component of steel, was for small tools, in approximately 4000 B.C. (Murphy, 1957). This material was in the form of wrought iron, produced by heating ore in a charcoal fire. In the latter part of the eighteenth century and in the early nineteenth century, cast iron and wrought iron were used in various types of bridges. Steel, an alloy of primarily iron and carbon, with fewer impurities and less carbon than cast iron, was first used in heavy construction in the nineteenth century. With the advent of the Bessemer converter in 1855, steel began to displace wrought iron and cast iron in construction. In the United States, the first structural steel railroad bridge was the Eads bridge, constructed in 1874 in St. Louis, Missouri (Tall, 1964). In 1884, the first building with a steel frame was completed in Chicago.

The characteristics of steel that are of the most interest to structural engineers can be examined by plotting the results of a tensile test. If a test specimen is subjected to an axial load P, as shown in Figure 1.3a, the stress and strain can be computed as follows:

$$f = \frac{P}{A} \quad \text{and} \quad \varepsilon = \frac{\Delta L}{L}$$

where

f = axial tensile stress
A = cross-sectional area
ε = axial strain
L = length of specimen
ΔL = change in length

If the load is increased in increments from zero to the point of fracture, and stress and strain are computed at each step, a stress–strain curve such as the one shown in Figure 1.3b can be plotted. This curve is typical of a class of steel known as *ductile,* or *mild, steel*. The relationship between stress and strain is linear up to the proportional limit; the material is said to follow *Hooke's law*. A peak value, the upper yield point, is quickly reached after that, followed by a leveling off at the lower yield point.

FIGURE 1.3

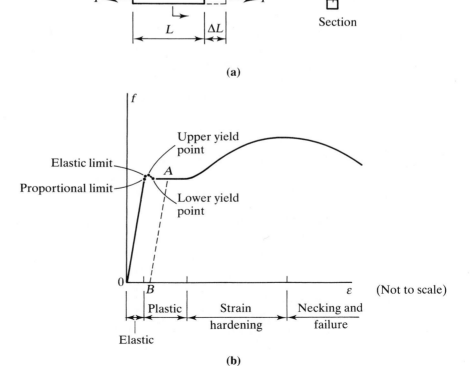

(a)

(b)

The stress then remains constant, even though the strain continues to increase. At this stage of loading, the test specimen continues to elongate as long as the load is not removed, even though the load cannot be increased. This constant stress region is called the *yield plateau,* or *plastic range*. At a strain of approximately 12 times the strain at yield, strain hardening begins, and additional load (and stress) is required to cause additional elongation (and strain). A maximum value of stress is reached, after which the specimen begins to "neck down" as the stress decreases with increasing strain, and fracture occurs. Although the cross section is reduced during loading (the Poisson effect), the original cross-sectional area is used to compute all stresses. Stress computed in this way is known as *engineering stress*. If the original length is used to compute the strain, it is called *engineering strain*.

Steel exhibiting the behavior shown in Figure 1.3b is called *ductile* because of its ability to undergo large deformations before fracturing. Ductility can be measured by the elongation, defined as

$$e = \frac{L_f - L_0}{L_0} \times 100 \tag{1.1}$$

where

e = elongation (expressed as a percent)
L_f = length of the specimen at fracture
L_0 = original length

The elastic limit of the material is a stress that lies between the proportional limit and the upper yield point. Up to this stress, the specimen can be unloaded without permanent deformation; the unloading will be along the linear portion of the diagram, the same path followed during loading. This part of the stress–strain diagram is called the *elastic range*. Beyond the elastic limit, unloading will be along a straight line parallel to the initial linear part of the loading path, and there will be a permanent strain. For example, if the load is removed at point *A* in Figure 1.3b, the unloading will be along line *AB,* resulting in the permanent strain *OB*.

Figure 1.4 shows an idealized version of this stress–strain curve. The proportional limit, elastic limit, and the upper and lower yield points are all very close to one another and are treated as a single point called the *yield point,* defined by the stress F_y. The other point of interest to the structural engineer is the maximum value of stress that can be attained, called the *ultimate tensile strength, F_u*. The shape of this curve is typical of mild structural steels, which are different from one another primarily in the values of F_y and F_u. The ratio of stress to strain within the elastic range, denoted E and called *Young's modulus,* or *modulus of elasticity,* is the same for all structural steels and has a value of 29,000,000 psi (pounds per square inch) or 29,000 ksi (kips per square inch).

Figure 1.5 shows a typical stress–strain curve for high-strength steels, which are less ductile than the mild steels discussed thus far. Although there is a linear elastic portion and a distinct tensile strength, there is no well-defined yield point or yield plateau. To use these higher-strength steels in a manner consistent with the use of ductile steels, some value of stress must be chosen as a value for F_y so that the same procedures and formulas can be used with all structural steels. Although there is no yield

FIGURE 1.4

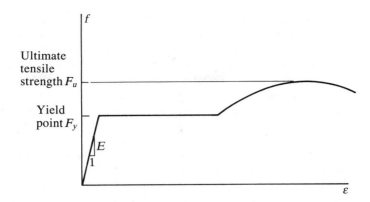

point, one needs to be defined. As previously shown, when a steel is stressed beyond its elastic limit and then unloaded, the path followed to zero stress will not be the original path from zero stress; it will be along a line having the slope of the linear portion of the path followed during loading — that is, a slope equal to E, the modulus of elasticity. Thus there will be a residual strain, or permanent set, after unloading. The yield stress for steel with a stress–strain curve of the type shown in Figure 1.5 is called the *yield strength* and is defined as the stress at the point of unloading that corresponds to a permanent strain of some arbitrarily defined amount. A strain of 0.002 is usually selected, and this method of determining the yield strength is called the *0.2% offset method*. As previously mentioned, the two properties usually needed in structural steel design are F_u and F_y, regardless of the shape of the stress–strain curve and regardless of how F_y was obtained. For this reason, the generic term *yield stress* is used, and it can mean either yield point or yield strength.

The various properties of structural steel, including strength and ductility, are determined by its chemical composition. Steel is an alloy, its principal component being iron. Another component of all structural steels, although in much smaller amounts, is carbon, which contributes to strength but reduces ductility. Other components of some grades of steel include copper, manganese, nickel, chromium, molybdenum, and silicon. Structural steels can be grouped according to their composition as follows.

FIGURE 1.5

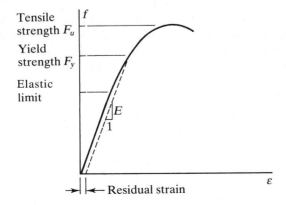

1. **Plain carbon steels:** mostly iron and carbon, with less than 1% carbon.
2. **Low-alloy steels:** iron and carbon plus other components (usually less than 5%). The additional components are primarily for increasing strength, which is accomplished at the expense of a reduction in ductility.
3. **High-alloy or specialty steels:** similar in composition to the low-alloy steels but with a higher percentage of the components added to iron and carbon. These steels are higher in strength than the plain carbon steels and also have some special quality, such as resistance to corrosion.

Different grades of structural steel are identified by the designation assigned them by the American Society for Testing and Materials (ASTM). This organization develops standards for defining materials in terms of their composition, properties, and performance, and it prescribes specific tests for measuring these attributes (ASTM, 2005a). One of the most commonly used structural steels is a mild steel designated as ASTM A36, or A36 for short. It has a stress–strain curve of the type shown in Figures 1.3b and 1.4 and has the following tensile properties.

Yield stress: $F_y = 36{,}000$ psi (36 ksi)

Tensile strength: $F_u = 58{,}000$ psi to 80,000 psi (58 ksi to 80 ksi)

A36 steel is classified as a plain carbon steel, and it has the following components (other than iron).

Carbon: 0.26% (maximum)

Phosphorous: 0.04% (maximum)

Sulfur: 0.05% (maximum)

These percentages are approximate, the exact values depending on the form of the finished steel product. A36 is a ductile steel, with an elongation as defined by Equation 1.1 of 20% based on an undeformed original length of 8 inches.

Steel producers who provide A36 steel must certify that it meets the ASTM standard. The values for yield stress and tensile strength shown are minimum requirements; they may be exceeded and usually are to a certain extent. The tensile strength is given as a range of values because for A36 steel, this property cannot be achieved to the same degree of precision as the yield stress.

Other commonly used structural steels are ASTM A572 Grade 50 and ASTM A992. These two steels are very similar in both tensile properties and chemical composition, with a maximum carbon content of 0.23%. A comparison of the tensile properties of A36, A572 Grade 50, and A992 is given in Table 1.1.

TABLE 1.1

Property	A36	A572 Gr. 50	A992
Yield point, min.	36 ksi	50 ksi	50 ksi
Tensile strength, min.	58 to 80 ksi	65 ksi	65 ksi
Yield to tensile ratio, max.	—	—	0.85
Elongation in 8 in., min.	20%	18%	18%

1.6   STANDARD CROSS-SECTIONAL SHAPES

In the design process outlined earlier, one of the objectives — and the primary emphasis of this book — is the selection of the appropriate cross sections for the individual members of the structure being designed. Most often, this selection will entail choosing a standard cross-sectional shape that is widely available rather than requiring the fabrication of a shape with unique dimensions and properties. The selection of an "off-the-shelf" item will almost always be the most economical choice, even if it means using slightly more material. The largest category of standard shapes includes those produced by *hot-rolling*. In this manufacturing process, which takes place in a mill, molten steel is taken from the furnace and poured into a *continuous casting* system where the steel solidifies but is never allowed to cool completely. The hot steel passes through a series of rollers that squeeze the material into the desired cross-sectional shape. Rolling the steel while it is still hot allows it to be deformed with no resulting loss in ductility, as would be the case with cold-working. During the rolling process, the member increases in length and is cut to standard lengths, usually a maximum of 65 to 75 feet, which are subsequently cut (in a fabricating shop) to the lengths required for a particular structure.

Cross sections of some of the more commonly used hot-rolled shapes are shown in Figure 1.6. The dimensions and designations of the standard available shapes are defined in the ASTM standards (ASTM, 2005b). The *W-shape,* also called a *wide-flange shape,* consists of two parallel flanges separated by a single web. The orientation of these elements is such that the cross section has two axes of symmetry. A typical designation would be "W18 × 50," where W indicates the type of shape, 18 is the nominal depth parallel to the web, and 50 is the weight in pounds per foot of length. The nominal depth is the approximate depth expressed in whole inches. For some of the lighter shapes, it is equal to the depth to the nearest inch, but this is not a general rule for the W-shapes. All of the W-shapes of a given nominal size can be grouped into families that have the same depth from inside-of-flange to inside-of-flange but with different flange thicknesses.

The *American Standard,* or *S-shape,* is similar to the W-shape in having two parallel flanges, a single web, and two axes of symmetry. The difference is in the proportions: The flanges of the W are wider in relation to the web than are the flanges of the S. In addition, the outside and inside faces of the flanges of the W-shape are parallel, whereas the inside faces of the flanges of the S-shape slope with respect to the outside faces. An example of the designation of an S-shape is "S18 × 70," with the S indicating the type of shape, and the two numbers giving the depth in inches and the weight in pounds per foot. This shape was formerly called an *I-beam.*

The angle shapes are available in either equal-leg or unequal-leg versions. A typical designation would be "L6 × 6 × ¾" or "L6 × 4 × ⅝." The three numbers are the lengths of each of the two legs as measured from the corner, or heel, to the toe at the other end of the leg, and the thickness, which is the same for both legs. In the case of the unequal-leg angle, the longer leg dimension is always given first. Although this designation provides all of the dimensions, it does not provide the weight per foot.

The *American Standard Channel,* or *C-shape,* has two flanges and a web, with only one axis of symmetry; it carries a designation such as "C9 × 20." This notation is

FIGURE 1.6

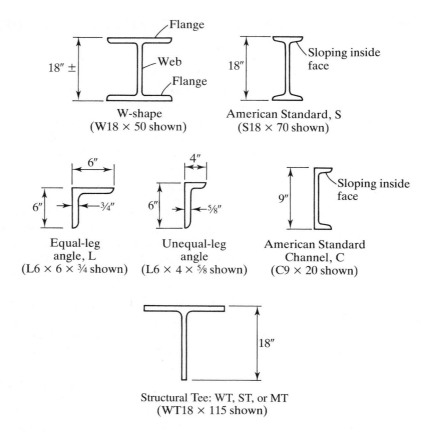

W-shape
(W18 × 50 shown)

American Standard, S
(S18 × 70 shown)

Equal-leg
angle, L
(L6 × 6 × ¾ shown)

Unequal-leg
angle
(L6 × 4 × ⅝ shown)

American Standard
Channel, C
(C9 × 20 shown)

Structural Tee: WT, ST, or MT
(WT18 × 115 shown)

similar to that for W- and S-shapes, with the first number giving the total depth in inches parallel to the web and the second number the weight in pounds per linear foot. For the channel, however, the depth is exact rather than nominal. The inside faces of the flanges are sloping, just as with the American Standard shape. Miscellaneous Channels — for example, the MC10 × 25 — are similar to American Standard Channels.

The *Structural Tee* is produced by splitting an I-shaped member at middepth. This shape is sometimes referred to as a *split-tee*. The prefix of the designation is either WT, ST, or MT, depending on which shape is the "parent." For example, a WT18 × 105 has a nominal depth of 18 inches and a weight of 105 pounds per foot, and is cut from a W36 × 210. Similarly, an ST10 × 33 is cut from an S20 × 66, and an MT5 × 4 is cut from an M10 × 8. The "M" is for "miscellaneous." The M-shape has two parallel flanges and a web, but it does not fit exactly into either the W or S categories. The HP shape, used for bearing piles, has parallel flange surfaces, approximately the same width and depth, and equal flange and web thicknesses. HP-shapes are designated in the same manner as the W-shape; for example, HP14 × 117.

Other frequently used cross-sectional shapes are shown in Figure 1.7. *Bars* can have circular, square, or rectangular cross sections. If the width of a rectangular shape is 8 inches or less, it is classified as a bar. If the width is more than 8 inches, the shape

FIGURE 1.7

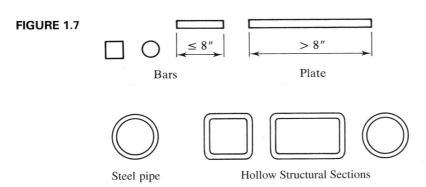

is classified as a *plate*. The usual designation for both is the abbreviation PL (for plate, even though it could actually be a bar) followed by the thickness in inches, the width in inches, and the length in feet and inches; for example, PL $\frac{3}{8} \times 5 \times 3'$-$2\frac{1}{2}''$. Although plates and bars are available in increments of $\frac{1}{16}$ inch, it is customary to specify dimensions to the nearest $\frac{1}{8}$ inch. Bars and plates are formed by hot-rolling.

Also shown in Figure 1.7 are hollow shapes, which can be produced either by bending plate material into the desired shape and welding the seam or by hot-working to produce a seamless shape. Most hollow structural sections available in the United States today are produced by cold-forming and welding (Sherman, 1997). The shapes are categorized as steel pipe, round HSS, and square and rectangular HSS. The designation HSS is for "Hollow Structural Sections."

Steel pipe is available as standard, extra-strong, or double-extra-strong, with designations such as Pipe 5 Std., Pipe 5 x-strong, or Pipe 5 xx-strong, where 5 is the nominal outer diameter in inches. The different strengths correspond to different wall thicknesses for the same outer diameter. For nominal outer diameters greater than 12 inches, the designation is the outer diameter and wall thickness in inches, expressed to three decimal places; for example, Pipe 14.000×0.375.

Round HSS are designated by outer diameter and wall thickness, expressed to three decimal places; for example, HSS 8.625×0.250. Square and rectangular HSS are designated by nominal outside dimensions and wall thickness, expressed in rational numbers; for example, HSS $7 \times 5 \times \frac{3}{8}$.

Other shapes are available, but those just described are the ones most frequently used. In most cases, one of these standard shapes will satisfy design requirements. If the requirements are especially severe, then a built-up section, such as one of those shown in Figure 1.8, may be needed. Sometimes a standard shape is augmented by additional cross-sectional elements, as when a cover plate is welded to one or both flanges of a W-shape. Building up sections is an effective way of strengthening an existing structure that is being rehabilitated or modified for some use other than the one for which it was designed. Sometimes a built-up shape must be used because none of the standard rolled shapes are large enough; that is, the cross section does not have enough area or moment of inertia. In such cases, plate girders can be used. These can be I-shaped sections, with two flanges and a web, or box sections, with two flanges and two webs. The components can be welded together and can be designed to have exactly the properties needed. Built-up shapes can also be created by

FIGURE 1.8

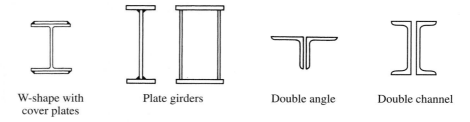

W-shape with cover plates   Plate girders   Double angle   Double channel

attaching two or more standard rolled shapes to each other. A widely used combination is a pair of angles placed back-to-back and connected at intervals along their length. This is called a *double-angle shape*. Another combination is the double-channel shape (either American Standard or Miscellaneous Channel). There are many other possibilities, some of which we illustrate throughout this book.

The most commonly used steels for rolled shapes and plate material are ASTM A36, A572, and A992. ASTM A36 is usually specified for angles and plates; A36 or A572 Grade 50 for S, M, and channel shapes; A572 Grade 50 for HP shapes; and A992 for W shapes. (These three steels were compared in Table 1.1 in Section 1.5.) Steel pipe is available in ASTM A53 Grade B only. ASTM A500 is usually specified for hollow structural sections (HSS). These recommendations are summarized in Table 1.2. Other steels can be used for these shapes, but the ones listed in Table 1.2 are the most common (Carter, 2004).

Another category of steel products for structural applications is cold-formed steel. Structural shapes of this type are created by bending thin material such as sheet steel or plate into the desired shape without heating. Typical cross sections are shown in Figure 1.9. Only relatively thin material can be used, and the resulting shapes are suitable only for light applications. An advantage of this product is its versatility, since almost any conceivable cross-sectional shape can easily be formed. In addition, cold-working will increase the yield point of the steel, and under certain conditions it may be accounted for in design (AISI, 2001). This increase comes at the expense of a reduction in ductility, however. Because of the thinness of the cross-sectional elements, the problem of instability (discussed in Chapters 4 and 5) is a particularly important factor in the design of cold-formed steel structures.

TABLE 1.2

Shape	Preferred Steel
Angles	A36
Plates	A36
S, M, C, MC	A36 or A572 Grade 50
HP	A572 Grade 50
W	A992
Pipe	A53 Grade B (only choice)
HSS	A500 Grade B or C

FIGURE 1.9

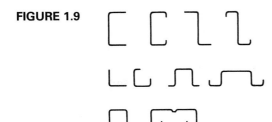

Problems

Note
The following problems illustrate the concepts of stress and strain covered in Section 1.5. The materials cited in these problems are not necessarily steel.

1.5-1 A tensile test was performed on a metal specimen with a circular cross section. The diameter was measured to be 0.550 inch. Two marks were mode along the length of the specimen and were measured to be 2.030 inches apart. This distance is defined as the *gage length,* and all length measurements are made between the two marks. The specimen was loaded to failure. Fracture occurred at a load of 28,500 pounds. The specimen was then reassembled, and the diameter and gage length were measured to be 0.430 inch and 2.300 inches. Determine the

a. Ultimate tensile stress in ksi.

b. Elongation as a percentage.

c. Reduction in cross-sectional area as a percentage.

1.5-2 A tensile test was performed on a metal specimen having a circular cross section with a diameter of $\frac{1}{2}$ inch. The *gage length* (the length over which the elongation is measured) is 2 inches. For a load 13.5 kips, the elongation was 4.66×10^{-3} inches. If the load is assumed to be within the linear elastic range of the material, determine the modulus of elasticity.

1.5-3 A tensile test was performed on a metal specimen having a circular cross section with a diameter of 0.510 inch. For each increment of load applied, the strain was directly determined by means of a *strain gage* attached to the specimen. The results are shown in Table 1.5.1.

a. Prepare a table of stress and strain.

b. Plot these data to obtain a stress–strain curve. Do not connect the data points; draw a *best-fit* straight line through them.

c. Determine the modulus of elasticity as the slope of the best-fit line.

Load (lb)	Strain $\times 10^6$ (in./in.)
0	0
250	37.1
500	70.3
1000	129.1
1500	230.1
2000	259.4
2500	372.4
3000	457.7
3500	586.5

TABLE 1.5.1

1.5-4   A tensile test was performed on a metal specimen with a diameter of $\frac{1}{2}$ inch and a gage length (the length over which the elongation is measured) of 4 inches. The data were plotted on a load-displacement graph, P vs. ΔL. A best-fit line was drawn through the points, and the slope of the straight-line portion was calculated to be $P/\Delta L = 1392$ kips/in. What is the modulus of elasticity?

1.5-5   The results of a tensile test are shown in Table 1.5.2. The test was performed on a metal specimen with a circular cross section. The diameter was $\frac{3}{8}$ inch and the gage length (the length over which the elongation is measured) was 2 inches.

a. Use the data in Table 1.5.2 to produce a table of stress and strain values.

b. Plot the stress–strain data and draw a best-fit curve.

c. Compute the modulus of elasticity from the initial slope of the curve.

d. Estimate the yield stress.

Load (lb)	Elongation $\times 10^6$ (in.)
0	0
550	350
1100	700
1700	900
2200	1350
2800	1760
3300	2200
3900	2460
4400	2860
4900	3800
4970	5300
5025	7800

TABLE 1.5.2

1.5-6   The data in Table 1.5.3 were obtained from a tensile test of a metal specimen with a rectangular cross section of 0.2011 in.² in area and a gage length (the length over which the elongation is measured) of 2.000 inches. The specimen was not loaded to failure.

a. Generate a table of stress and strain values.
b. Plot these values and draw a best-fit line to obtain a stress–strain curve.
c. Determine the modulus of elasticity from the slope of the linear portion of the curve.
d. Estimate the value of the proportional limit.
e. Use the 0.2% offset method to determine the yield stress.

Load (kips)	Elongation $\times$ 10^3 (in.)
0	0
1	0.160
2	0.352
3	0.706
4	1.012
5	1.434
6	1.712
7	1.986
8	2.286
9	2.612
10	2.938
11	3.274
12	3.632
13	3.976
14	4.386
15	4.640
16	4.988
17	5.432
18	5.862
19	6.362
20	7.304
21	8.072
22	9.044
23	11.310
24	14.120
25	20.044
26	29.106

TABLE 1.5.3

2

Concepts in Structural Steel Design

2.1 DESIGN PHILOSOPHIES

As discussed earlier, the design of a structural member entails the selection of a cross section that will safely and economically resist the applied loads. Economy usually means minimum weight — that is, the minimum amount of steel. This amount corresponds to the cross section with the smallest weight per foot, which is the one with the smallest cross-sectional area. Although other considerations, such as ease of construction, may ultimately affect the choice of member size, the process begins with the selection of the lightest cross-sectional shape that will do the job. Having established this objective, the engineer must decide how to do it safely, which is where different approaches to design come into play. The fundamental requirement of structural design is that the required strength not exceed the available strength; that is,

required strength ≤ available strength

In *allowable strength design* (ASD), a member is selected that has cross-sectional properties such as area and moment of inertia that are large enough to prevent the maximum applied axial force, shear, or bending moment from exceeding an allowable, or permissible, value. This allowable value is obtained by dividing the nominal, or theoretical, strength by a factor of safety. This can be expressed as

required strength ≤ allowable strength   (2.1)

where

$$\text{allowable strength} = \frac{\text{nominal strength}}{\text{safety factor}}$$

Strength can be an axial force strength (as in tension or compression members), a flexural strength (moment strength), or a shear strength.

If stresses are used instead of forces or moments, the relationship of Equation 2.1 becomes

maximum applied stress ≤ allowable stress   (2.2)

This approach is called *allowable stress design*. The allowable stress will be in the elastic range of the material (see Figure 1.3). This approach to design is also called *elastic design* or *working stress design*. Working stresses are those resulting from the working loads, which are the applied loads. Working loads are also known as *service loads.*

Plastic design is based on a consideration of failure conditions rather than working load conditions. A member is selected by using the criterion that the structure will fail at a load substantially higher than the working load. Failure in this context means either collapse or extremely large deformations. The term *plastic* is used because, at failure, parts of the member will be subjected to very large strains — large enough to put the member into the plastic range (see Figure 1.3b). When the entire cross section becomes plastic at enough locations, "plastic hinges" will form at those locations, creating a *collapse mechanism.* As the actual loads will be less than the failure loads by a factor of safety known as the *load factor,* members designed this way are not unsafe, despite being designed based on what happens at failure. This design procedure is roughly as follows.

1. Multiply the working loads (service loads) by the load factor to obtain the failure loads.
2. Determine the cross-sectional properties needed to resist failure under these loads. (A member with these properties is said to have sufficient strength and would be at the verge of failure when subjected to the factored loads.)
3. Select the lightest cross-sectional shape that has these properties.

Members designed by plastic theory would reach the point of failure under the factored loads but are safe under actual working loads.

Load and resistance factor design (LRFD) is similar to plastic design in that strength, or the failure condition, is considered. Load factors are applied to the service loads, and a member is selected that will have enough strength to resist the factored loads. In addition, the theoretical strength of the member is reduced by the application of a resistance factor. The criterion that must be satisfied in the selection of a member is

$$\text{Factored load} \leq \text{factored strength} \tag{2.3}$$

In this expression, the factored load is actually the sum of all service loads to be resisted by the member, each multiplied by its own load factor. For example, dead loads will have load factors that are different from those for live loads. The factored strength is the theoretical strength multiplied by a resistance factor. Equation 2.3 can therefore be written as

$$\sum (\text{Loads} \times \text{load factors}) \leq \text{resistance} \times \text{resistance factor} \tag{2.4}$$

The factored load is a failure load greater than the total actual service load, so the load factors are usually greater than unity. However, the factored strength is a reduced, usable strength, and the resistance factor is usually less than unity. The factored loads are the loads that bring the structure or member to its limit. In terms of safety, this *limit state* can be fracture, yielding, or buckling, and the factored resistance is the useful strength of the member, reduced from the theoretical value by the resistance factor. The limit state can also be one of serviceability, such as a maximum acceptable deflection.

2.2 AMERICAN INSTITUTE OF STEEL CONSTRUCTION SPECIFICATION

Because the emphasis of this book is on the design of structural steel building members and their connections, the Specification of the American Institute of Steel Construction is the design specification of most importance here. It is written and kept current by an AISC committee comprising structural engineering practitioners, educators, steel producers, and fabricators. New editions are published periodically, and supplements are issued when interim revisions are needed. Allowable stress design has been the primary method used for structural steel buildings since the first AISC Specification was issued in 1923, although plastic design was made part of the Specification in 1963. In 1986, AISC issued the first specification for load and resistance factor design along with a companion *Manual of Steel Construction.* The purpose of these two documents was to provide an alternative to allowable stress design, much as plastic design is an alternative. The current specification (AISC, 2005a) incorporates both LRFD and ASD.

The LRFD provisions are based on research reported in eight papers published in 1978 in the *Structural Journal of the American Society of Civil Engineers* (Ravindra and Galambos; Yura, Galambos, and Ravindra; Bjorhovde, Galambos, and Ravindra; Cooper, Galambos, and Ravindra; Hansell et al.; Fisher et al.; Ravindra, Cornell, and Galambos; Galambos and Ravindra, 1978).

Although load and resistance factor design was not introduced into the AISC Specification until 1986, it is not a recent concept; since 1974, it has been used in Canada, where it is known as *limit states design.* It is also the basis of most European building codes. In the United States, LRFD has been an accepted method of design for reinforced concrete for years and is the primary method authorized in the American Concrete Institute's Building Code, where it is known as *strength design* (ACI, 2005). Highway bridge design standards provide for both allowable stress design (AASHTO, 2002) and load and resistance factor design (AASHTO, 2004).

The AISC Specification is published as a stand-alone document, but it is also part of the *Steel Construction Manual,* which we discuss in the next section. Except for such specialized steel products as cold-formed steel, which is covered by a different specification (AISI, 2001), the AISC Specification is the standard by which virtually all structural steel buildings in this country are designed and constructed. Hence the student of structural steel design must have ready access to his document. The details of the Specification will be covered in the chapters that follow, but we discuss the overall organization here.

The Specification consists of three parts: the main body, the appendixes, and the Commentary. The body is alphabetically organized into Chapters A through M. Within each chapter, major headings are labeled with the chapter designation followed by a number. Further subdivisions are numerically labeled. For example, the types of structural steel authorized are listed in Chapter A, "General Provisions," under Section A3. Material, and, under it, Section 1. Structural Steel Materials. The main body of the Specification is followed by appendixes 1–7. The Appendix section is followed by the Commentary, which gives background and elaboration on many of the provisions of the Specification. Its organizational scheme is the same as that of the Specification, so material applicable to a particular section can be easily located.

The Specification incorporates both U.S. customary and metric (SI) units. Where possible, equations and expressions are expressed in non-dimensional form by leaving quantities such as yield stress and modulus of elasticity in symbolic form, thereby avoiding giving units. When this is not possible, U.S. customary units are given, followed by SI units in parentheses. Although there is a strong move to metrication in the steel industry, most structural design in the United States is still done in U.S. customary units, and this textbook uses only U.S. customary units.

2.3   LOAD FACTORS, RESISTANCE FACTORS, AND LOAD COMBINATIONS FOR LRFD

Equation 2.4 can be written more precisely as

$$\Sigma \gamma_i Q_i \leq \phi R_n \tag{2.5}$$

where
   Q_i = a load effect (a force or a moment)
   γ_i = a load factor
   R_n = the nominal resistance, or strength, of the component under consideration
   ϕ = resistance factor

The factored resistance ϕR_n is called the *design strength*. The summation on the left side of Equation 2.5 is over the total number of load effects (including, but not limited to, dead load and live load), where each load effect can be associated with a different load factor. Not only can each load effect have a different load factor but also the value of the load factor for a particular load effect will depend on the combination of loads under consideration. Equation 2.5 can also be written in the form

$$R_u \leq \phi R_n \tag{2.6}$$

where

   R_u = required strength = sum of factored load effects (forces or moments)

Section B2 of the AISC Specification requires that the load factors and load combinations given in ASCE 7 (ASCE 2002) be used. These load factors and load combinations are based on extensive statistical studies. The seven combinations are as follows:

1. $1.4(D + F)$
2. $1.2(D + F + T) + 1.6(L + H) + 0.5(L_r$ or S or $R)$
3. $1.2D + 1.6(L_r$ or S or $R) + (0.5L$ or $0.8W)$
4. $1.2D + 1.6W + 0.5L + 0.5(L_r$ or S or $R)$
5. $1.2D + 1.0E + 0.5L + 0.2S$
6. $0.9D + 1.6W + 1.6H$
7. $0.9D + 1.0E + 1.6H$

where

D = dead load
E = earthquake load
F = load due to fluids with well-defined pressures and maximum heights
H = load due to lateral earth pressure, groundwater pressure, or pressure of bulk materials
L = live load
L_r = roof live load
R = rain load*
S = snow load
T = self-straining force
W = wind load

(If a governing building code specifies other load combinations, then they should be used.)

Normally, fluid pressure F, earth pressure H, and self-straining force T are not applicable to the design of structural steel members, and we will omit them from this point forward. In addition, combinations 6 and 7 can be combined. With these and one other slight modification, the list of required load combinations becomes

$$1.4D \tag{1}$$

$$1.2D + 1.6L + 0.5(L_r \text{ or } S \text{ or } R) \tag{2}$$

$$1.2D + 1.6(L_r \text{ or } S \text{ or } R) + (0.5L \text{ or } 0.8W) \tag{3}$$

$$1.2D + 1.6W + 0.5L + 0.5(L_r \text{ or } S \text{ or } R)^\dagger \tag{4}$$

$$1.2D \pm 1.0E + 0.5L + 0.2S^\dagger \tag{5}$$

$$0.9D \pm (1.6W \text{ or } 1.0E) \tag{6}$$

Combinations 5 and 6 account for the possibility of the dead load and wind or earthquake load counteracting each other; for example, in combination 6, the net load effect could be the difference between $0.9D$ and $1.6W$ or between $0.9D$ and $1.0E$. (Wind or earthquake load may tend to overturn a structure, but the dead load will have a stabilizing effect.)

As previously mentioned, the load factor for a particular load effect is not the same in all load combinations. For example, in combination 2 the load factor for the live load L is 1.6, whereas in combination 3, it is 0.5. The reason is that the live load is being taken as the dominant effect in combination 2, and one of the three effects,

---

*This load does not include *ponding*, a phenomenon that we discuss in Chapter 5.

†For garages, areas of public assembly, and where the live load exceeds 100 psf, the load factor for the live load L in combinations 3, 4, and 5 should be 1.0 instead of 0.5; that is,

$$1.2D + 1.6(L_r \text{ or } S \text{ or } R) + (1.0L + 0.8W) \tag{3}$$

$$1.2D + 1.6W + 1.0L + 0.5(L_r \text{ or } S \text{ or } R) \tag{4}$$

$$1.2D \pm 1.0E + 1.0L + 0.2S \tag{5}$$

L_r, S, or R, will be dominant in combination 3. In each combination, one of the effects is considered to be at its "lifetime maximum" value and the others at their "arbitrary point in time" values.

The resistance factor ϕ for each type of resistance is given by AISC in the Specification chapter dealing with that resistance, but in most cases, one of two values will be used: 0.90 for limit states involving yielding or compression buckling and 0.75 for limit states involving rupture (fracture).

2.4   SAFETY FACTORS AND LOAD COMBINATIONS FOR ASD

For allowable strength design, the relationship between loads and strength (Equation 2.1) can be expressed as

$$R_a \leq \frac{R_n}{\Omega} \tag{2.7}$$

where

R_a = required strength
R_n = nominal strength (same as for LRFD)
Ω = safety factor
R_n/Ω = allowable strength

The required strength R_a is the sum of the service loads or load effects. As with LRFD, specific combinations of loads must be considered. Load combinations for ASD are also given in ASCE 7. As with the LRFD combinations, we will omit fluid pressure F, earth pressure H, and self-straining force T. With these omissions, the combinations are

D	(1)
$D + L$	(2)
$D + (L_r \text{ or } S \text{ or } R)$	(3)
$D + 0.75L + 0.75(L_r \text{ or } S \text{ or } R)$	(4)
$D \pm (W \text{ or } 0.7E)$	(5)
$D + 0.75(W \text{ or } 0.7E) + 0.75L + 0.75(L_r \text{ or } S \text{ or } R)$	(6)
$0.6D \pm (W \text{ or } 0.7E)$	(7)

The factors shown in these combinations are not load factors. The 0.75 factor in some of the combinations accounts for the unlikelihood that all loads in the combination will be at their lifetime maximum values simultaneously. The 0.7 factor applied to the seismic load effect E is used because ASCE 7 uses a strength approach (i.e., LRFD) for computing seismic loads, and the factor is an attempt to equalize the effect for ASD.

Corresponding to the two most common values of resistance factors in LRFD are the following values of the safety factor Ω in ASD: For limit states involving yielding or compression buckling, $\Omega = 1.67^*$. For limit states involving rupture, $\Omega = 2.00$. The relationship between resistance factors and safety factors is given by

$$\Omega = \frac{1.5}{\phi} \tag{2.8}$$

For reasons that will be discussed later, this relationship will produce similar designs for LRFD and ASD, under certain loading conditions.

If both sides of Equation 2.7 are divided by area (in the case of axial load) or section modulus (in the case of bending moment), then the relationship becomes

$$f \leq F$$

where
    f = applied stress
    F = allowable stress

This formulation is called *allowable stress design.*

Example 2.1   A column (compression member) in the upper story of a building is subject to the following loads:

    Dead load: 109 kips compression

    Floor live load: 46 kips compression

    Roof live load: 19 kips compression

    Snow: 20 kips compression

  a. Determine the controlling load combination for LRFD and the corresponding factored load.

  b. If the resistance factor ϕ is 0.90, what is the required *nominal* strength?

  c. Determine the controlling load combination for ASD and the corresponding required service load strength.

  d. If the safety factor Ω is 1.67, what is the required nominal strength based on the required service load strength?

Solution   Even though a load may not be acting directly on a member, it can still cause a load effect in the member. This is true of both snow and roof live load in this example. Although this building is subjected to wind, the resulting forces on the structure are resisted by members other than this particular column.

  a. The controlling load combination is the one that produces the largest factored load. We evaluate each expression that involves dead load, *D,* live load resulting from equipment and occupancy, *L,* roof live load, *Lr,* and snow, *S.*

---

*The value of Ω is actually $1\frac{2}{3} = 5/3$ but has been rounded to 1.67 in the AISC specification.

Combination 1: $1.4D = 1.4(109) = 152.6$ kips

Combination 2: $1.2D + 1.6L + 0.5(L_r$ or S or $R)$. Because S is larger than L_r and $R = 0$, we need to evaluate this combination only once, using S.

$1.2D + 1.6L + 0.5S = 1.2(109) + 1.6(46) + 0.5(20) = 214.4$ kips

Combination 3: $1.2D + 1.6(L_r$ or S or $R) + (0.5L$ or $0.8W)$. In this combination, we use S instead of L_r, and both R and W are zero.

$1.2D + 1.6S + 0.5L = 1.2(109) + 1.6(20) + 0.5(46) = 185.8$ kips

Combination 4: $1.2D + 1.6W + 0.5L + 0.5(L_r$ or S or $R)$. This expression reduces to $1.2D + 0.5L + 0.5S$, and by inspection, we can see that it produces a smaller result than combination 3.

Combination 5: $1.2D \pm 1.0E + 0.5L + 0.2S$. As $E = 0$, this expression reduces to $1.2D + 0.5L + 0.2S$, which produces a smaller result than combination 4.

Combination 6: $0.9D \pm (1.6W$ or $1.0E)$. This expression reduces to $0.9D$, which is smaller than any of the other combinations.

Answer   Combination 2 controls, and the factored load is 214.4 kips.

b. If the factored load obtained in part (a) is substituted into the fundamental LRFD relationship, Equation 2.6, we obtain

$$R_u \leq \phi R_n$$
$$214.4 \leq 0.90R_n$$
$$R_n \geq 238 \text{ kips}$$

Answer   The required nominal strength is 238 kips.

c. As with the combinations for LRFD, we will evaluate the expressions involving D, L, L_r, and S for ASD.

Combination 1: $D = 109$ kips. (Obviously this case will never control when live load is present.)

Combination 2: $D + L = 109 + 46 = 155$ kips

Combination 3: $D + (L_r$ or S or $R)$. Since S is larger than L_r, and $R = 0$, this combination reduces to $D + S = 109 + 20 = 129$ kips

Combination 4: $D + 0.75L + 0.75(L_r$ or S or $R)$. This expression reduces to $D + 0.75L + 0.75S = 109 + 0.75(46) + 0.75(20) = 158.5$ kips

Combination 5: $D \pm (W$ or $0.7E)$. Because W and E are zero, this expression reduces to combination 1.

Combination 6: $D + 0.75(W$ or $0.7E) + 0.75L + 0.75 (L_r$ or S or $R)$. Because W and E are zero, this expression reduces to combination 4.

Combination 7: $0.6D \pm (W$ or $0.7E)$. Because W and E are zero, this expression reduces to $0.6D$, which is smaller than combination 1.

Answer   Combination 4 controls, and the required service load strength is 158.5 kips.

d. From the ASD relationship, Equation 2.7,

$$R_a \leq \frac{R_n}{\Omega}$$

$$158.5 \leq \frac{R_n}{1.67}$$

$$R_n \geq 265 \text{ kips}$$

Answer   The required nominal strength is 265 kips.

Example 2.1 illustrates that the controlling load combination for LRFD may not control for ASD.

When LRFD was introduced into the AISC Specification in 1986, the load factors were determined in such a way as to give the same results for LRFD and ASD when the loads consisted of dead load and a live load equal to three times the dead load. The resulting relationship between the resistance factor ϕ and the safety factor Ω, as expressed in Equation 2.8, can be derived as follows. Let R_n from Equations 2.6 and 2.7 be the same when $L = 3D$. That is,

$$\frac{R_u}{\phi} = R_a \Omega$$

$$\frac{1.2D + 1.6L}{\phi} = (D + L)\Omega$$

or

$$\frac{1.2D + 1.6(3D)}{\phi} = (D + 3D)\Omega$$

$$\Omega = \frac{1.5}{\phi}$$

2.5   PROBABILISTIC BASIS OF LOAD AND RESISTANCE FACTORS

Both the load and the resistance factors specified by AISC are based on probabilistic concepts. The resistance factors account for uncertainties in material properties, design theory, and fabrication and construction practices. Although a complete treatment of probability theory is beyond the scope of this book, we present a brief summary of the basic concepts here.

FIGURE 2.1

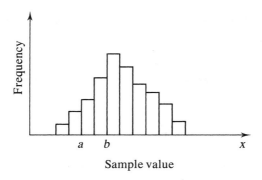

Sample value

Experimental data can be represented in the form of a histogram, or bar graph, as shown in Figure 2.1, with the abscissa representing sample values, or events, and the ordinate representing either the number of samples having a certain value or the frequency of occurrence of a certain value. Each bar can represent a single sample value or a range of values. If the ordinate is the percentage of values rather than the actual number of values, the graph is referred to as a *relative* frequency distribution. In such a case the sum of the ordinates will be 100%. If the abscissa values are random events, and enough samples are used, each ordinate can be interpreted as the probability, expressed as a percentage, of that sample value or event occurring. The relative frequency can also be expressed in decimal form, with values between 0 and 1.0. Thus the sum of the ordinates will be unity, and if each bar has a unit width, the total area of the diagram will also be unity. This result implies a probability of 1.0 that an event will fall within the boundaries of the diagram. Furthermore, the probability that a certain value or something smaller will occur is equal to the area of the diagram to the left of that value. The probability of an event having a value falling between a and b in Figure 2.1 equals the area of the diagram between a and b.

Before proceeding, some definitions are in order. The *mean*, $\bar{x}$, of a set of sample values, or *population*, is the arithmetic average, or

$$\bar{x} = \frac{1}{n} \sum_{i=1}^{n} x_i$$

where x_i is a sample value and n is the number of values. The *median* is the middle value of x, and the *mode* is the most frequently occurring value. The *variance, v,* is a measure of the overall variation of the data from the mean and is defined as

$$v = \frac{1}{n} \sum_{i-1}^{n} (x_i - \bar{x})^2$$

The *standard deviation s* is the square root of the variance, or

$$s = \sqrt{\frac{1}{n} \sum_{i=1}^{n} (x_i - \bar{x})^2}$$

Like the variance, the standard deviation is a measure of the overall variation, but it has the same units and the same order of magnitude as the data. The *coefficient of variation, V,* is the standard deviation divided by the mean, or

$$V = \frac{s}{\overline{x}}$$

If the actual frequency distribution is replaced by a theoretical continuous function that closely approximates the data, it is called a *probability density function.* Such a function is illustrated in Figure 2.2. Probability functions are designed so that the total area under the curve is unity. That is, for a function $f(x)$,

$$\int_{-\infty}^{+\infty} f(x)\, dx = 1.0$$

which means that the probability that one of the sample values or events will occur is 1.0. The probability of one of the events between a and b in Figure 2.2 equals the area under the curve between a and b, or

$$\int_{a}^{b} f(x)\, dx$$

When a theoretical probability density function is used, the following notation is conventional:

$\mu =$ mean

$\sigma =$ standard deviation

The probabilistic basis of the load and resistance factors used by AISC is presented in the ASCE structural journal and is summarized here (Ravindra and Galambos, 1978). Load effects, Q, and resistances, R, are random variables and depend on many factors. Loads can be estimated or obtained from measurements and inventories of actual structures, and resistances can be computed or determined experimentally. Discrete values of Q and R from observations can be plotted as frequency distribution histograms or represented by theoretical probability density functions. We use this latter representation in the material that follows.

FIGURE 2.2

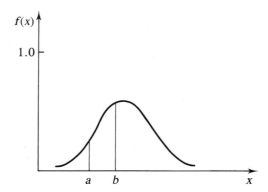

If the distributions of Q and R are combined into one function, $R - Q$, positive values of $R - Q$ correspond to survival. Equivalently, if a probability density function of R/Q, the factor of safety, is used, survival is represented by values of R/Q greater than 1.0. The corresponding probability of failure is the probability that R/Q is less than 1; that is,

$$P_F = P\left[\left(\frac{R}{Q}\right) < 1\right]$$

Taking the natural logarithm of both sides of the inequality, we have

$$P_F = P\left[\ln\left(\frac{R}{Q}\right) < \ln 1\right] = P\left[\ln\left(\frac{R}{Q}\right) < 0\right]$$

The frequency distribution curve of $\ln(R/Q)$ is shown in Figure 2.3. The *standardized* form of the variable $\ln(R/Q)$ can be defined as

$$U = \frac{\ln\left(\frac{R}{Q}\right) - \left[\ln\left(\frac{R}{Q}\right)\right]_m}{\sigma_{\ln(R/Q)}}$$

where

$$\left[\ln\left(\frac{R}{Q}\right)\right]_m = \text{ the mean value of } \ln\left(\frac{R}{Q}\right)$$

$$\sigma_{\ln(R/Q)} = \text{ standard deviation of } \ln\left(\frac{R}{Q}\right)$$

FIGURE 2.3

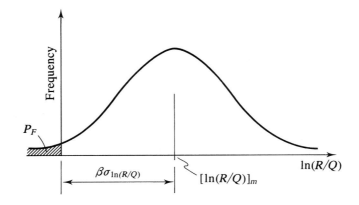

This transformation converts the abscissa U to multiples of standard deviations and places the mean of U at $U = 0$. The probability of failure can then be written as

$$P_F = P\left[\ln\left(\frac{R}{Q}\right) < 0\right] = P\left(\left\{U\sigma_{\ln(R/Q)} + \left[\ln\left(\frac{R}{Q}\right)\right]_m\right\} < 0\right)$$

$$= P\left\{U < -\frac{\left[\ln\left(\frac{R}{Q}\right)\right]_m}{\sigma_{\ln(R/Q)}}\right\} = F_u\left\{-\frac{\left[\ln\left(\frac{R}{Q}\right)\right]_m}{\sigma_{\ln(R/Q)}}\right\}$$

where F_u is the *cumulative distribution function* of U, or the probability that U will not exceed the argument of the function. If we let

$$\beta = \frac{\left[\ln\left(\frac{R}{Q}\right)\right]_m}{\sigma_{\ln(R/Q)}}$$

then

$$\left[\ln\left(\frac{R}{Q}\right)\right]_m = \beta\sigma_{\ln(R/Q)}$$

The variable β can be interpreted as the number of standard deviations from the origin that the mean value of $\ln(R/Q)$ is. For safety, the mean value *must* be more than zero, and as a consequence, β is called the *safety index* or *reliability index*. The larger this value, the larger will be the margin of safety. This means that the probability of failure, represented by the shaded area in Figure 2.3 labeled P_F, will be smaller. The reliability index is a function of both the load effect Q and the resistance R. Use of the same reliability index for all types of members subjected to the same type of loading gives the members relatively uniform strength. The "target" values of β shown in Table 2.1, selected and used in computing both load and resistance factors for the AISC Specification, were based on the recommendations of Ravindra and Galambos (1978), who also showed that

$$\phi = \frac{R_m}{R_n}e^{-0.55\beta V_R}$$

TABLE 2.1
Target Values
of β

Type of Component	Loading Condition		
	$D + (L \text{ or } S)$	$D + L + W$	$D + L + E$
Members	3.0	2.5	1.75
Connections	4.5	4.5	4.5

where

R_m = mean value of the resistance R
R_n = nominal or theoretical resistance
V_R = coefficient of variation of R

2.6   STEEL CONSTRUCTION MANUAL

Anyone engaged in structural steel design in the United States must have access to AISC's *Steel Construction Manual* (AISC, 2005b). This publication contains the AISC Specification and numerous design aids in the form of tables and graphs, as well as a "catalog" of the most widely available structural shapes.

The first nine editions of the *Manual* and the accompanying specifications were based on ASD. The ninth edition was followed by editions one through three of the LRFD-based manuals. The current version, which incorporates both ASD and LRFD, is therefore the thirteenth edition.

This textbook was written under the assumption that you would have access to the *Manual* at all times. To encourage use of the *Manual,* we did not reproduce its tables and graphs in this book. The *Manual* is divided into 17 parts as follows:

Part 1. Dimensions and Properties. This part contains details on standard hot-rolled shapes, pipe, and hollow structural sections, including all necessary cross-sectional dimensions and properties such as area and moment of inertia.

Part 2. General Design Considerations. This part includes a brief overview of various specifications (including a detailed discussion of the AISC Specification), codes and standards, some fundamental design and fabrication principles, and a discussion of the proper selection of materials.

Part 3. Design of Flexural Members. This part contains a discussion of Specification requirements and design aids for beams, including composite beams (in which a steel shape acts in combination with a reinforced concrete floor or roof slab) and plate girders. Composite beams are covered in Chapter 9 of this textbook, "Composite Construction," and plate girders are covered in Chapter 10, "Plate Girders."

Part 4. Design of Compression Members. Part 4 includes a discussion of the Specification requirements for compression members and numerous design aids. Design aids for composite columns, consisting of hollow structural sections or pipe filled with plain (unreinforced) concrete, are also included. Composite columns are covered in Chapter 9 of this textbook.

Part 5. Design of Tension Members. This part includes design aids for tension members and a summary of the Specification requirements for tension members.

Part 6. Design of Members Subject to Combined Loading. Part 6 covers members subject to combined axial tension and flexure, combined axial compression and flexure, and combined torsion, flexure, shear, and/or axial force. Of particular interest is the material on combined axial compression and flexure, which is the subject of Chapter 6 of this textbook, "Beam–Columns."

Parts 7–15 cover connections:

Part 7. Design Considerations for Bolts.

Part 8. Design Considerations for Welds.

Part 9. Design of Connecting Elements.

Part 10. Design of Simple Shear Connections.

Part 11. Design of Flexible Moment Connections.

Part 12. Design of Fully Restrained (FR) Moment Connections.

Part 13. Design of Bracing Connections and Truss Connections.

Part 14. Design of Beam Bearing Plates, Column Base Plates, Anchor Rods, and Column Splices.

Part 15. Design of Hanger Connections, Bracket Plates, and Crane–Rail Connections.

Part 16. Specifications and Codes. This part contains the AISC Specification and Commentary, a specification for high-strength bolts (RCSC, 2004), and the AISC Code of Standard Practice (AISC, 2005c).

Part 17. Miscellaneous Data and Mathematical Information. This part includes properties of standard steel shapes in SI units, conversion factors and other information on SI units, weights and other properties of building materials, mathematical formulas, and properties of geometric shapes.

All design aids in the *Manual* give values for both allowable strength design (ASD) and load and resistance factor design (LRFD). The *Manual* uses a color-coding scheme for these values: ASD allowable strength values (R_n/Ω) are shown as black numbers on a green background, and LRFD design strength values (ϕR_n) are shown as blue numbers on a white background.

The AISC Specification is only a small part of the *Manual*. Many of the terms and constants used in other parts of the *Manual* are presented to facilitate the design process and are not necessarily part of the Specification. In some instances, the recommendations are only "rules of thumb" based on common practice, not requirements of the Specification. Although such information is not in conflict with the Specification, it is important to recognize what is a *requirement* (when adopted by a building code) and what is not.

The *Manual* is accompanied by a compact disk that contains worked-out examples that illustrate Specification requirements and the use of the design aids in the *Manual*. This companion disk also contains the Specification and Commentary, a database of standard hot-rolled section properties, and web links to other AISC resources.

2.7   DESIGN COMPUTATIONS AND PRECISION

The computations required in engineering design and analysis are done with either a digital computer or an electronic calculator. When doing manual computations with the aid of an electronic calculator, an engineer must make a decision regarding the degree of precision needed. The problem of how many significant figures to use in engineering computations has no simple solution. Recording too many significant digits is misleading and can imply an unrealistic degree of precision. Conversely, recording too few figures can lead to meaningless results. The question of precision was mostly academic before the early 1970s, when the chief calculating tool was the slide rule. The guiding principle at that time was to read and record numbers as accurately as possible, which meant three or four significant figures.

There are many inherent inaccuracies and uncertainties in structural design, including variations in material properties and loads; load estimates sometimes border on educated guesses. It hardly makes sense to perform computations with 12 significant figures and record the answer to that degree of precision when the yield stress is known only to the nearest 10 kips per square inch (two significant figures). Furthermore, data given in the *Steel Construction Manual* has been rounded to three significant figures. To avoid results that are even less precise, however, it is reasonable to assume that the given parameters of a problem, such as the yield stress, are exact and then decide on the degree of precision required in subsequent calculations.

A further complication arises when electronic calculators are used. If all of the computations for a problem are done in one continuous series of operations on a calculator, the number of significant figures used is whatever the calculator uses, perhaps 10 or 12. But if intermediate values are rounded, recorded, and used in subsequent computations, then a consistent number of significant figures will not have been used. Furthermore, the manner in which the computations are grouped will influence the final result. In general, the result will be no more accurate than the least accurate number used in the computation—and sometimes less because of round-off error. For example, consider a number calculated on a 12-digit calculator and recorded to four significant figures. If this number is multiplied by a number expressed to five significant figures, the product will be precise to four significant figures at most, regardless of the number of digits displayed on the calculator. Consequently, it is not reasonable to record this number to more than four significant figures.

It is also unreasonable to record the results of every calculator multiplication or division to a predetermined number of significant figures in order to have a consistent degree of precision throughout. A reasonable approach is to perform operations on the calculator in any convenient manner and record intermediate values to whatever degree of precision is deemed adequate (without clearing the intermediate value from the calculator if it can be used in the next computation). The final results should then be expressed to a precision consistent with this procedure, usually to one significant figure less than the intermediate results, to account for round-off error.

It is difficult to determine what the degree of precision should be for the typical structural steel design problem. Using more than three or four significant figures is probably unrealistic in most cases, and results based on less than three may be too approximate to be of any value. In this book we record intermediate values to three or four digits (usually four), depending on the circumstances, and record final results to three digits. For multiplication and division, each number used in an intermediate calculation should be expressed to four significant figures, and the result should be recorded to four significant figures. For addition and subtraction, determining the location of the right-most significant digit in a column of numbers is done as follows: from the left-most significant digit of all numbers involved, move to the right a number of digits corresponding to the number of significant digits desired. For example, to add 12.34 and 2.234 (both numbers have four significant figures) and round to four significant figures,

$$
\begin{array}{r}
12.34 \\
+\ 2.234 \\
\hline
14.574
\end{array}
$$

and the result should be recorded as 14.57, even though the fifth digit of the result was significant in the second number. As another example, consider the addition of the following numbers, both accurate to four significant figures:

$$36,000 + 1.240 = 36,001.24$$

The result should be recorded as 36,000 (four significant figures). When subtracting numbers of almost equal value, significant digits can be lost. For example, in the operation

$$12,458.62 - 12,462.86 = -4.24$$

four significant figures are lost. To avoid this problem, when subtracting, start with additional significant figures if possible.

When rounding numbers where the first digit to be dropped is a 5 with no digits following, two options are possible. The first is to add 1 to the last digit retained. The other is to use the "odd-add" rule, in which we leave the last digit to be retained unchanged if it is an even number, and add 1 if it is an odd number, making it even. In this book, we follow the first practice. The "odd-add" rule tends to average out the rounding process when many numerical operations are involved, as in statistical methods, but that is not the case in most structural design problems. In addition, most calculators, spreadsheet programs, and other software use the first method, and our results will be consistent with those tools; therefore, we will round up when the first digit dropped is a 5 with no digits following.

Problems

Note   All given loads are service loads.

2-1   A column in a building is subjected to the following load effects:

9 kips compression from dead load

5 kips compression from roof live load

6 kips compression from snow

7 kips compression from 3 inches of rain accumulated on the roof

8 kips compression from wind

a. If load and resistance factor design is used, determine the factored load (required strength) to be used in the design of the column. Which AISC load combination controls?

b. What is the required *design* strength of the column?

c. What is the required *nominal* strength of the column for a resistance factor ϕ of 0.90?

d. If allowable strength design is used, determine the required load capacity (required strength) to be used in the design of the column. Which AISC load combination controls?

e. What is the required *nominal* strength of the column for a safety factor Ω of 1.67?

2-2 Repeat Problem 2-1 without the possibility of rain accumulation on the proof.

2-3 A beam is part of the framing system for the floor of an office building. The floor is subjected to both dead loads and live loads. The maximum moment caused by the service dead load is 45 ft-kips, and the maximum moment for the service live load is 63 ft-kips (these moments occur at the same location on the beam and can therefore be combined).

a. If load and resistance factor design is used, determine the maximum factored bending moment (required moment strength). What is the controlling AISC load combination?

b. What is the required *nominal* moment strength for a resistance factor ϕ of 0.90?

c. If allowable strength design is used, determine the required moment strength. What is the controlling AISC load combination?

d. What is the required *nominal* moment strength for a safety factor Ω of 1.67?

2-4 A tension member must be designed for a service dead load of 18 kips and a service live load of 2 kips.

a. If load and resistance factor design is used, determine the maximum factored load (required strength) and the controlling AISC load combination.

b. If allowable strength design is used, determine the maximum load (required strength) and the controlling AISC load combination.

2-5 A flat roof is subject to the following uniformly distributed loads: a dead load of 21 psf (pounds per square foot of roof surface), a roof live load of 12 psf, a snow load of 13.5 psf, and a wind load of 22 psf *upward*. (Although the wind itself is in a horizontal direction, the force that it exerts on this roof is upward. It will be upward regardless of wind direction. The dead, live and snow loads are *gravity loads* and *act downward.*)

a. If load and resistance factor design is used, compute the factored load (required strength) in pounds per square foot. Which AISC load combination controls?

b. If allowable strength design is used, compute the required load capacity (required strength) in pounds per square foot. Which AISC load combination controls?

3

Tension Members

3.1 INTRODUCTION

Tension members are structural elements that are subjected to axial tensile forces. They are used in various types of structures and include truss members, bracing for buildings and bridges, cables in suspended roof systems, and cables in suspension and cable-stayed bridges. Any cross-sectional configuration may be used, because for any given material, the only determinant of the strength of a tension member is the cross-sectional area. Circular rods and rolled angle shapes are frequently used. Built-up shapes, either from plates, rolled shapes, or a combination of plates and rolled shapes, are sometimes used when large loads must be resisted. The most common built-up configuration is probably the double-angle section, shown in Figure 3.1, along with other typical cross sections. Because the use of this section is so widespread, tables of properties of various combinations of angles are included in the AISC *Steel Construction Manual*.

The stress in an axially loaded tension member is given by

$$f = \frac{P}{A}$$

where P is the magnitude of the load and A is the cross-sectional area (the area normal to the load). The stress as given by this equation is exact, provided that the cross section under consideration is not adjacent to the point of application of the load, where the distribution of stress is not uniform.

If the cross-sectional area of a tension member varies along its length, the stress is a function of the particular section under consideration. The presence of holes in a member will influence the stress at a cross section through the hole or holes. At these locations, the cross-sectional area will be reduced by an amount equal to the area removed by the holes. Tension members are frequently connected at their ends with bolts, as illustrated in Figure 3.2. The tension member shown, a $\frac{1}{2} \times 8$ plate, is connected to a *gusset plate*, which is a connection element whose purpose is to transfer the load from the member to a support or to another member. The area of the bar at section a–a is $(\frac{1}{2})(8) = 4$ in.^{2}, but the area at section b–b is only $4 - (2)(\frac{1}{2})(\frac{7}{8}) = 3.13$ in.^{2} and will be more highly stressed. This reduced area is referred to as the *net area,* or *net section*, and the unreduced area is the *gross area*.

The typical design problem is to select a member with sufficient cross-sectional area to resist the loads. A closely related problem is that of analysis, or review, of a

FIGURE 3.1

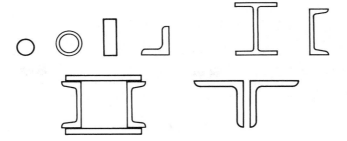

FIGURE 3.2

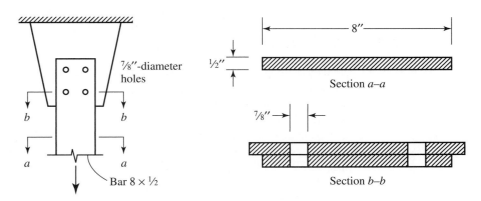

given member, where in the strength is computed and compared with the load. In general, analysis is a direct procedure, but design is an iterative process and may require some trial and error.

Tension members are covered in Chapter D of the Specification. Requirements that are common with other types of members are covered in Chapter B, "Design Requirements."

3.2 TENSILE STRENGTH

A tension member can fail by reaching one of two limit states: excessive deformation or fracture. To prevent excessive deformation, initiated by yielding, the load on the gross section must be small enough that the stress on the gross section is less than the yield stress F_y. To prevent fracture, the stress on the net section must be less than the tensile strength F_u. In each case, the stress P/A must be less than a limiting stress F or

$$\frac{P}{A} < F$$

Thus the load P must be less than FA, or

$$P < FA$$

The *nominal* strength in yielding is

$$P_n = F_y A_g$$

and the nominal strength in fracture is

$$P_n = F_u A_e$$

where A_e is the *effective* net area, which may be equal to either the net area or, in some cases, a smaller area. We discuss effective net area in Section 3.3.

Although yielding will first occur on the net cross section, the deformation within the length of the connection will generally be smaller than the deformation in the remainder of the tension member. The reason is that the net section exists over a relatively small length of the member, and the total elongation is a product of the length and the strain (a function of the stress). Most of the member will have an unreduced cross section, so attainment of the yield stress on the gross area will result in larger total elongation. It is this larger deformation, not the first yield, that is the limit state.

LRFD: In load and resistance factor design, the factored tensile load is compared to the design strength. The design strength is the resistance factor times the nominal strength. Equation 2.6,

$$R_u = \phi R_n$$

can be written for tension members as

$$P_u \le \phi_t P_n$$

where P_u is the governing combination of factored loads. The resistance factor ϕ_t is smaller for fracture than for yielding, reflecting the more serious nature of fracture.

For yielding, $\phi_t = 0.90$

For fracture, $\phi_t = 0.75$

Because there are two limit states, both of the following conditions must be satisfied:

$$P_u \le 0.90 F_y A_g$$
$$P_u \le 0.75 F_u A_e$$

The smaller of these is the design strength of the member.

ASD: In allowable strength design, the total service load is compared to the allowable strength (allowable load):

$$P_a \le \frac{P_n}{\Omega_t}$$

where P_a is the required strength (applied load), and P_n / Ω_t is the allowable strength. The subscript "a" indicates that the required strength is for "allowable strength design," but you can think of it as standing for "applied" load.

For yielding of the gross section, the safety factor Ω_t is 1.67, and the allowable load is

$$\frac{P_n}{\Omega_t} = \frac{F_y A_g}{1.67} = 0.6 F_y A_g$$

(The factor 0.6 appears to be a rounded value, but recall that 1.67 is a rounded value. If $\Omega_t = \frac{5}{3}$ is used, the allowable load is exactly $0.6\,F_y A_g$.)

For fracture of the net section, the safety factor is 2.00 and the allowable load is

$$\frac{P_n}{\Omega_t} = \frac{F_u A_e}{2.00} = 0.5 F_u A_e$$

Alternatively, the service load stress can be compared to the allowable stress. This can be expressed as

$$f_t \le F_t$$

where f_t is the applied stress and F_t is the allowable stress. For yielding of the gross section,

$$f_t = \frac{P_a}{A_g} \quad \text{and} \quad F_t = \frac{P_n/\Omega_t}{A_g} = \frac{0.6 F_y A_g}{A_g} = 0.6 F_y$$

For fracture of the net section,

$$f_t = \frac{P_a}{A_e} \quad \text{and} \quad F_t = \frac{P_n/\Omega_t}{A_e} = \frac{0.5 F_u A_e}{A_e} = 0.5 F_u$$

You can find values of F_y and F_u for various structural steels in Table 2-3 in the *Manual.* All of the steels that are available for various hot-rolled shapes are indicated by shaded areas. The black areas correspond to preferred materials, and the gray areas represent other steels that are available. Under the W heading, we see that A992 is the preferred material for W shapes, but other materials are available, usually at a higher cost. For some steels, there is more than one grade, with each grade having different values of F_y and F_u. In these cases, the grade must be specified along with the ASTM designation—for example, A572 Grade 50. For A242 steel, F_y and F_u depend on the thickness of the flange of the cross-sectional shape. This relationship is given in footnotes in the table. For example, to determine the properties of a W33 × 221 of ASTM A242 steel, first refer to the dimensions and properties table in Part 1 of the *Manual* and determine that the flange thickness t_f is equal to 1.28 inches. This matches the thickness range indicated in footnote 1; therefore, $F_y = 50$ ksi and $F_u = 70$ ksi. Values of F_y and F_u for plates and bars are given in Table 2-4, and information on structural fasteners, including bolts and rods, can be found in Table 2-5.

The exact amount of area to be deducted from the gross area to account for the presence of bolt holes depends on the fabrication procedure. The usual practice is to drill or punch standard holes (i.e., not oversized) with a diameter $\frac{1}{16}$ inch larger than the fastener diameter. To account for possible roughness around the edges of the hole, Section D3 of the AISC Specification (in the remainder of this book, references to the Specification will usually be in the form AISC D3) requires the addition of $\frac{1}{16}$ inch to the actual hole diameter. This amounts to using an effective hole diameter $\frac{1}{8}$ inch larger than the fastener diameter. In the case of slotted holes, $\frac{1}{16}$ inch should be added to the actual *width* of the hole. You can find details related to standard, oversized, and slotted holes in AISC J3.2, "Size and Use of Holes" (in Chapter J, "Design of Connections").

Example 3.1   A$\frac{1}{2}$×5 plate of A36 steel is used as a tension member. It is connected to a gusset plate with four $\frac{5}{8}$-inch-diameter bolts as shown in Figure 3.3. Assume that the effective net area A_e equals the actual net area A_n (we cover computation of effective net area in Section 3.3).

a) What is the design strength for LRFD?

b) What is the allowable strength for ASD?

FIGURE 3.3

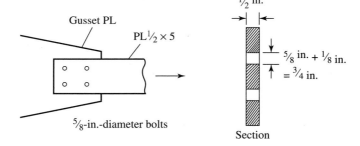

Solution   For yielding of the gross section,

$$A_g = 5(1/2) = 2.5 \text{ in.}^2$$

and the nominal strength is

$$P_n = F_y A_g = 36(2.5) = 90.0 \text{ kips}$$

For fracture of the net section,

$$A_n = A_g - A_{holes} = 2.5 - (\tfrac{1}{2})(\tfrac{3}{4}) \times 2 \text{ holes}$$
$$= 2.5 - 0.75 = 1.75 \text{ in.}^2$$
$$A_e = A_n = 1.75 \text{ in.}^2 \text{ (This is true for this example, but } A_e \text{ does not always equal } A_n.)$$

The nominal strength is

$$P_n = F_u A_e = 58(1.75) = 101.5 \text{ kips}$$

a) The design strength based on yielding is

$$\phi_t P_n = 0.90(90) = 81.0 \text{ kips}$$

The design strength based on fracture is

$$\phi_t P_n = 0.75(101.5) = 76.1 \text{ kips}$$

Answer   The design strength for LRFD is the smaller value: $\phi_t P_n = 76.1$ kips.

b) The allowable strength based on yielding is

$$\frac{P_n}{\Omega_t} = \frac{90}{1.67} = 53.9 \text{ kips}$$

The allowable strength based on fracture is

$$\frac{P_n}{\Omega_t} = \frac{101.5}{2.00} = 50.8 \text{ kips}$$

Answer   The allowable service load is the smaller value = 50.8 kips.

Alternative Solution Using Allowable Stress: For yielding,

$$F_t = 0.6F_y = 0.6(36) = 21.6 \text{ ksi}$$

and the allowable load is

$$F_t A_g = 21.6(2.5) = 54.0 \text{ kips}$$

(The slight difference between this value and the one based on allowable strength is because the value of Ω in the allowable strength approach has been rounded from 5/3 to 1.67; the value based on the allowable stress is the more accurate one.)
For fracture,

$$F_t = 0.5F_u = 0.5(58) = 29.0 \text{ ksi}$$

and the allowable load is

$$F_t A_e = 29.0(1.75) = 50.8 \text{ kips}$$

Answer   The allowable service load is the smaller value = 50.8 kips.

Because of the relationship given by Equation 2.8, the allowable strength will always be equal to the design strength divided by 1.5. In this book, however, we will do the complete computation of allowable strength even when the design strength is available.

The effects of stress concentrations at holes appear to have been overlooked. In reality, stresses at holes can be as high as three times the average stress on the net section, and at fillets of rolled shapes they can be more than twice the average (McGuire, 1968). Because of the ductile nature of structural steel, the usual design practice is to neglect such localized overstress. After yielding begins at a point of stress concentration, additional stress is transferred to adjacent areas of the cross section. This stress redistribution is responsible for the "forgiving" nature of structural steel. Its ductility permits the initially yielded zone to deform without fracture as the stress on the remainder of the cross section continues to increase. Under certain conditions, however, steel may lose its ductility and stress concentrations can precipitate brittle fracture. These situations include fatigue loading and extremely low temperature.

Example 3.2   A single-angle tension member, an L$3\frac{1}{2} \times 3\frac{1}{2} \times \frac{3}{8}$, is connected to a gusset plate with $\frac{7}{8}$-inch-diameter bolts as shown in Figure 3.4. A36 steel is used. The service loads are 35 kips dead load and 15 kips live load. Investigate this member for compliance with the AISC Specification. Assume that the effective net area is 85% of the computed net area.
a) Use LRFD.
b) Use ASD.

FIGURE 3.4

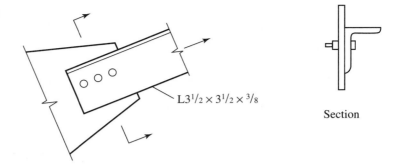

L3$\frac{1}{2} \times 3\frac{1}{2} \times \frac{3}{8}$

Section

Solution First, compute the nominal strengths.

Gross section:

$$A_g = 2.48 \text{ in.}^2 \quad \text{(from Part 1 of the } Manual\text{)}$$
$$P_n = F_y A_g = 36(2.48) = 89.28 \text{ kips}$$

Net section:

$$A_n = 2.48 - \left(\frac{3}{8}\right)\left(\frac{7}{8} + \frac{1}{8}\right) = 2.105 \text{ in.}^2$$

$$A_e = 0.85 A_n = 0.85(2.105) = 1.789 \text{ in.}^2 \quad \text{(in } this \text{ example)}$$
$$P_n = F_u A_e = 58(1.789) = 103.8 \text{ kips}$$

a). The design strength based on yielding is

$$\phi_t P_n = 0.90(89.28) = 80.4 \text{ kips}$$

The design strength based on fracture is

$$\phi_t P_n = 0.75(103.8) = 77.9 \text{ kips}$$

The design strength is the smaller value: $\phi_t P_n = 77.9$ kips

Factored load:

When only dead load and live load are present, the only load combinations with a chance of controlling are combinations 1 and 2.

Combination 1: $1.4D = 1.4(35) = 49$ kips

Combination 2: $1.2D + 1.6L = 1.2(35) + 1.6(15) = 66$ kips

The second combination controls; $P_u = 66$ kips.

(When only dead load and live load are present, combination 2 will always control when the dead load is less than eight times the live load. In future examples, we will not check combination 1 [1.4D] when it obviously does not control.)

Answer Since $P_u < \phi_t P_n$, (66 kips < 77.9 kips), the member is satisfactory.

b) For the gross section, The allowable strength is

$$\frac{P_n}{\Omega_t} = \frac{89.28}{1.67} = 53.5 \text{ kips}$$

For the net section, the allowable strength is

$$\frac{P_n}{\Omega_t} = \frac{103.8}{2.00} = 51.9 \text{ kips}$$

The smaller value controls; the allowable strength is 51.9 kips. When the only loads are dead load and live load, ASD load combination 2 will always control:

$$P_a = D + L = 35 + 15 = 50 \text{ kips}$$

Answer   Since 50 kips $<$ 51.9 kips, the member is satisfactory.

Alternative Solution Using Allowable Stress

For the gross area, the applied stress is

$$f_t = \frac{P_a}{A_g} = \frac{50}{2.48} = 20.2 \text{ ksi}$$

and the allowable stress is

$$F_t = 0.6F_y = 0.6(36) = 21.6 \text{ ksi}$$

For this limit state, $f_t < F_t$ (OK)

For the net section,

$$f_t = \frac{P_a}{A_e} = \frac{50}{1.789} = 28.0 \text{ ksi}$$

$$F_t = 0.5F_u = 0.5(58) = 29.0 \text{ ksi} > 28.0 \text{ ksi} \quad \text{(OK)}$$

Answer   Since $f_t < F_t$ for both limit states, the member is satisfactory. ∎

What is the difference in computational effort for the two different approaches? Regardless of the method used, the two nominal strengths must be computed (if a stress approach is used with ASD, an equivalent computation must be made). With LRFD, the nominal strengths are multiplied by resistance factors. With ASD, the nominal strengths are divided by load factors. Up to this point, the number of steps is the same. The difference in effort between the two methods involves the load side of the relationships. In LRFD, the loads are factored before adding. In ASD, in most cases the loads are simply added. Therefore, for tension members LRFD requires slightly more computation.

Example 3.3

A double-angle shape is shown in Figure 3.5. The steel is A36, and the holes are for $\frac{1}{2}$-inch-diameter bolts. Assume that $A_e = 0.75A_n$.

a. Determine the design tensile strength for LRFD.

b. Determine the allowable strength for ASD.

FIGURE 3.5

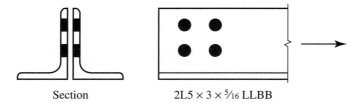

Section   2L5 × 3 × ⁵⁄₁₆ LLBB

Solution

Figure 3.5 illustrates the notation for unequal-leg double-angle shapes. The notation LLBB means "long-legs back-to-back," and SLBB indicates "short-legs back-to-back."

When a double-shape section is used, two approaches are possible: (1) consider a single shape and double everything, or (2) consider two shapes from the outset. (Properties of the double-angle shape are given in Part 1 of the *Manual.*) In this example, we consider one angle and double the result. For one angle, the nominal strength based on the gross area is

$$P_n = F_y A_g = 36(2.41) = 86.76 \text{ kips}$$

(The gross area has been taken from the table for double-angle shapes.)

There are two holes in each angle, so the net area of one angle is

$$A_n = 2.41 - \left(\frac{5}{16}\right)\left(\frac{1}{2} + \frac{1}{8}\right) \times 2 = 2.019 \text{ in.}^2$$

The effective net area is

$$A_e = 0.75(2.019) = 1.514 \text{ in.}^2$$

The nominal strength based on the net area is

$$P_n = F_u A_e = 58(1.514) = 87.81 \text{ kips}$$

a. The design strength based on yielding of the gross area is

$$\phi_t P_n = 0.90(86.76) = 78.08 \text{ kips}$$

The design strength based on fracture of the net area is

$$\phi_t P_n = 0.75(87.81) = 65.86 \text{ kips}$$

Answer

Because 65.86 kips < 78.08 kips, fracture of the net section controls, and the design strength for the two angles is $2 \times 65.86 = 132$ kips.

b. The allowable stress approach will be used. For the gross section,

$$F_t = 0.6F_y = 0.6(36) = 21.6 \text{ ksi}$$

The corresponding allowable load is

$$F_t A_g = 21.6(2.41) = 52.06 \text{ kips}$$

For the net section,

$$F_t = 0.5F_u = 0.5(58) = 29 \text{ ksi}$$

The corresponding allowable load is

$$F_t A_e = 29(1.514) = 43.91 \text{ kips}$$

Answer Because 43.91 kips < 52.06 kips, fracture of the net section controls, and the allowable strength for the two angles is $2 \times 43.91 = 87.8$ kips.

3.3 EFFECTIVE AREA

Of the several factors influencing the performance of a tension member, the manner in which it is connected is the most important. A connection almost always weakens the member, and the measure of its influence is called the *joint efficiency*. This factor is a function of the ductility of the material, fastener spacing, stress concentrations at holes, fabrication procedure, and a phenomenon known as *shear lag*. All contribute to reducing the effectiveness of the member, but shear lag is the most important.

Shear lag occurs when some elements of the cross section are not connected, as when only one leg of an angle is bolted to a gusset plate, as shown in Figure 3.6. The consequence of this partial connection is that the connected element becomes overloaded and the unconnected part is not fully stressed. Lengthening the connected region will reduce this effect. Research reported by Munse and Chesson (1963) suggests that shear lag be accounted for by using a reduced, or effective, net area. Because shear lag affects both bolted and welded connections, the effective net area concept applies to both types of connections.

For bolted connections, the effective net area is

$$A_e = A_n U \qquad \text{(AISC Equation D3-1)}$$

For welded connections, we refer to this reduced area as the *effective area* (rather than the effective *net* area), and it is given by

$$A_e = A_g U$$

FIGURE 3.6

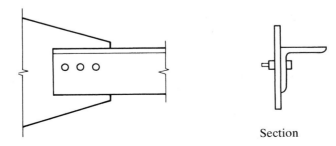

Section

where the reduction factor U is given in AISC D3.3, Table D3.1. The table gives a general equation that will cover most situations as well as alternative numerical values for specific cases. These definitions of U will be presented here in a different format from that in the Specification. The rules for determining U fall into five categories:

1. A general category for any type of tension member except plates and round HSS with $\ell \geq 1.3D$ (See Figure 3.7e.)
2. Plates
3. Round HSS with $\ell \geq 1.3\,D$
4. Alternative values for single angles
5. Alternative values for W, M, S, and HP shapes

1. For any type of tension member except plates and round HSS with $\ell \geq 1.3D$

$$U = 1 - \frac{\overline{x}}{\ell} \tag{3.1}$$

where

$\overline{x}$ = distance from centroid of connected area to the plane of the connection
ℓ = length of the connection

This definition of $\overline{x}$ was formulated by Munse and Chesson (1963). If a member has two symmetrically located planes of connection, $\overline{x}$ is measured from the centroid of the nearest one-half of the area. Figure 3.7 illustrates $\overline{x}$ for various types of connections.

The length ℓ in Equation 3.1 is the length of the connection in the direction of the load, as shown in Figure 3.8. For bolted connections, it is measured from the center of the bolt at one end of the connection to the center of the bolt at the other end. For welds, it is measured from one end of the weld to the other. If there are segments of different lengths in the direction of the load, the longest segment is used.

The Commentary of the AISC Specification further illustrates $\overline{x}$ and ℓ.

2. Plates

In general, $U = 1.0$ for plates, since the cross section has only one element and it is connected. There are two special cases for welded plates:

a. Connected with longitudinal welds on each side and no transverse weld (see Figure 3.9):

- For $\ell \geq 2w$ $U = 1.0$
- For $1.5w \leq \ell < 2w,$ $U = 0.87$
- For $w \leq \ell < 1.5w,$ $U = 0.75$

b. Connected with transverse welds only: $U = 1.0$ and A_n = area of connected element. Figure 3.10 illustrates the difference between transverse and longitudinal welds. Connections by transverse welds alone are uncommon.

3. Round HSS with $\ell \geq 1.3D$ (see Figure 3.7d):

$U = 1.0$

FIGURE 3.7

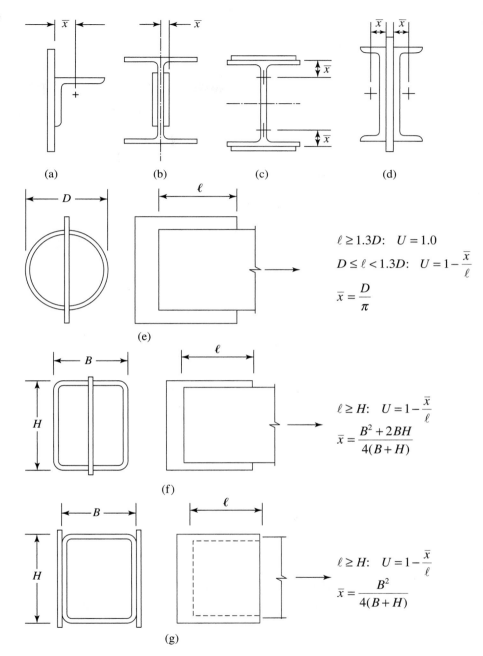

$$\ell \geq 1.3D: \quad U = 1.0$$

$$D \leq \ell < 1.3D: \quad U = 1 - \frac{\bar{x}}{\ell}$$

$$\bar{x} = \frac{D}{\pi}$$

$$\ell \geq H: \quad U = 1 - \frac{\bar{x}}{\ell}$$

$$\bar{x} = \frac{B^2 + 2BH}{4(B+H)}$$

$$\ell \geq H: \quad U = 1 - \frac{\bar{x}}{\ell}$$

$$\bar{x} = \frac{B^2}{4(B+H)}$$

4. Alternatives to Equation 3.1 for Single Angles:

The following values may be used in lieu of Equation 3.1

- For four or more fasteners in the direction of loading, $U = 0.80$
- For two or three fasteners in the direction of loading, $U = 0.60$

FIGURE 3.8

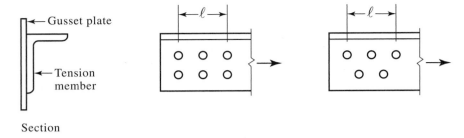

(a) Bolted

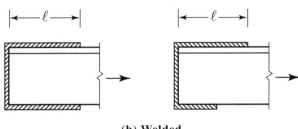

(b) Welded

FIGURE 3.9

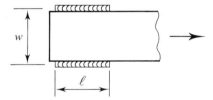

FIGURE 3.10

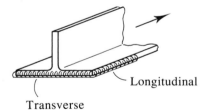

5. Alternatives to Equation 3.1 for W, M, S, HP, or Tees Cut from These Shapes:

If the following conditions are satisfied, the corresponding values may be used in lieu of Equation 3.1.

- Connected through the flange with three or more fasteners in the direction of loading, with a width at least $\frac{2}{3}$ of the depth: $U = 0.90$.
- Connected through the flange with three or more fasteners in the direction of loading, with a width less than $\frac{2}{3}$ of the depth: $U = 0.85$.
- Connected through the web with four or more fasteners in the direction of loading: $U = 0.70$

FIGURE 3.11

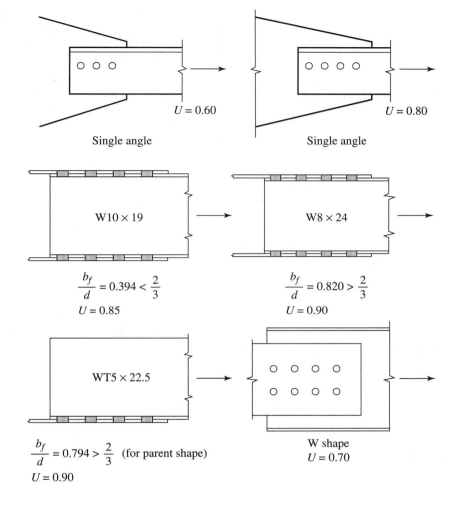

$U = 0.60$

Single angle

$U = 0.80$

Single angle

W10 × 19

$\dfrac{b_f}{d} = 0.394 < \dfrac{2}{3}$

$U = 0.85$

W8 × 24

$\dfrac{b_f}{d} = 0.820 > \dfrac{2}{3}$

$U = 0.90$

WT5 × 22.5

$\dfrac{b_f}{d} = 0.794 > \dfrac{2}{3}$ (for parent shape)

$U = 0.90$

W shape
$U = 0.70$

Figure 3.11 illustrates the alternative values of U for various connections.

AISC D3.3 mandates that for shapes such as angles, double angles, and WT shapes, the value of U should not be less than 0.60.

Example 3.4   Determine the effective net area for the tension member shown in Figure 3.12.

Solution $A_n = A_g - A_{holes}$

$$= 5.77 - \frac{1}{2}\left(\frac{5}{8} + \frac{1}{8}\right)(2) = 5.02 \text{ in.}^2$$

FIGURE 3.12

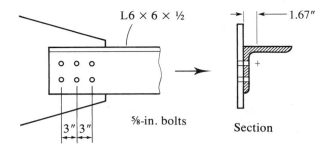

L6 × 6 × ½

1.67″

⅝-in. bolts

Section

3″ 3″

Only one element (one leg) of the cross section is connected, so the net area must be reduced. From the properties tables in Part 1 of the *Manual*, the distance from the centroid to the outside face of the leg of an L6 ×6 ×½ is

$$\bar{x} = 1.67 \text{ in.}$$

The length of the connection is

$$\ell = 3 + 3 = 6 \text{ in.}$$

$$\therefore U = 1 - \left(\frac{\bar{x}}{\ell}\right) = 1 - \left(\frac{1.67}{6}\right) = 0.7217$$

$$A_e = A_n U = 5.02(0.7217) = 3.623 \text{ in.}^2$$

The alternative value of U could also be used. Because this angle has three bolts in the direction of the load, the reduction factor U can be taken as 0.60, and

$$A_e = A_n U = 5.02(0.60) = 3.012 \text{ in.}^2$$

Either U value is acceptable, and the Specification permits the larger one to be used. However, the value obtained from Equation 3.1 is more accurate. The alternative values of U can be useful during preliminary design, when actual section properties and connection details are not known. ■

Example 3.5   If the tension member of Example 3.4 is welded as shown in Figure 3.13, determine the effective area.

Solution   As in Example 3.4, only part of the cross section is connected and a reduced effective area must be used.

$$U = 1 - \left(\frac{\bar{x}}{\ell}\right) = 1 - \left(\frac{1.67}{5.5}\right) = 0.6964$$

Answer $A_e = A_g U = 5.77(0.6964) = 4.02 \text{ in.}^2$ ■

FIGURE 3.13

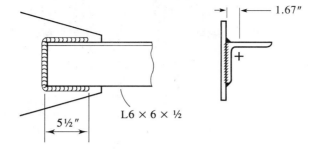

L6 × 6 × ½

5½"

1.67"

3.4   STAGGERED FASTENERS

If a tension member connection is made with bolts, the net area will be maximized if the fasteners are placed in a single line. Sometimes space limitations, such as a limit on dimension *a* in Figure 3.14a, necessitate using more than one line. If so, the reduction in cross-sectional area is minimized if the fasteners are arranged in a staggered pattern, as shown. Sometimes staggered fasteners are required by the geometry of a connection, such as the one shown in Figure 3.14b. In either case, any cross section passing through holes will pass through fewer holes than if the fasteners are not staggered.

If the amount of stagger is small enough, the influence of an offset hole may be felt by a nearby cross section, and fracture along an inclined path such as *abcd* in Figure 3.14c is possible. In such a case, the relationship $f = P/A$ does not apply, and stresses on the inclined portion *b*–*c* are a combination of tensile and shearing stresses. Several approximate methods have been proposed to account for the effects

FIGURE 3.14

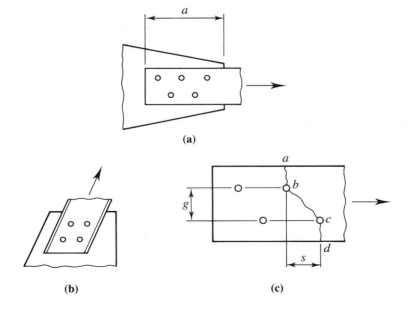

(a)

(b)   (c)

of staggered holes. Cochrane (1922) proposed that when deducting the area corresponding to a staggered hole, use a reduced diameter, given by

$$d' = d - \frac{s^2}{4g} \tag{3.2}$$

where d is the hole diameter, s is the stagger, or pitch, of the bolts (spacing in the direction of the load), and g is the gage (transverse spacing). This means that in a failure pattern consisting of both staggered and unstaggered holes, use d for holes at the end of a transverse line between holes ($s = 0$) and use d' for holes at the end of an inclined line between holes.

The AISC Specification, in Section D3, uses this approach, but in a modified form. If the net area is treated as the product of a thickness times a net width, and the diameter from Equation 3.2 is used for all holes (since $d' = d$ when the stagger $s = 0$), the net width in a failure line consisting of both staggered and unstaggered holes is

$$w_n = w_g - \Sigma d'$$
$$= w_g - \Sigma \left(d - \frac{s^2}{4g} \right)$$
$$= w_g - \Sigma d + \Sigma \frac{s^2}{4g}$$

where w_n is the net width and w_g is the gross width. The second term is the sum of all hole diameters, and the third term is the sum of $s^2/4g$ for all inclined lines in the failure pattern.

When more than one failure pattern is conceivable, all possibilities should be investigated, and the one corresponding to the smallest load capacity should be used. Note that this method will not accommodate failure patterns with lines parallel to the applied load.

Example 3.6   Compute the smallest net area for the plate shown in Figure 3.15. The holes are for 1-inch-diameter bolts.

FIGURE 3.15

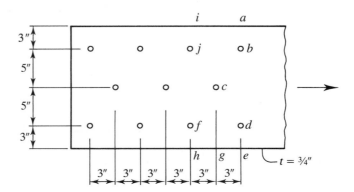

Solution   The effective hole diameter is $1 + \frac{1}{8} = 1\frac{1}{8}$ in. For line *abde,*

$$w_n = 16 - 2(1.125) = 13.75 \text{ in.}$$

For line *abcde,*

$$w_n = 16 - 3(1.125) + \frac{2(3)^2}{4(5)} = 13.52 \text{ in.}$$

The second condition will give the smallest net area:

Answer $A_n = tw_n = 0.75(13.52) = 10.1 \text{ in.}^2$

Equation 3.2 can be used directly when staggered holes are present. In the computation of the net area for line *abcde* in Example 3.6,

$$A_n = A_n - \sum t \times (d \text{ or } d')$$

$$= 0.75(16) - 0.75(1.125) - 0.75\left[1.125 - \frac{(3)^2}{4(5)}\right] \times 2 = 10.1 \text{ in.}^2$$

As each fastener resists an equal share of the load (an assumption used in the design of simple connections; see Chapter 7), different potential failure lines may be subjected to different loads. For example, line *abcde* in Figure 3.15 must resist the full load, whereas *ijfh* will be subjected to $\frac{8}{11}$ of the applied load. The reason is that $\frac{3}{11}$ of the load will have been transferred from the member before *ijfh* receives any load.

When lines of bolts are present in more than one element of the cross section of a rolled shape, and the bolts in these lines are staggered with respect to one another, the use of areas and Equation 3.2 is preferable to the net-width approach of the AISC Specification. If the shape is an angle, it can be visualized as a plate formed by "unfolding" the legs to more clearly identify the pitch and gage distances. AISC B2 specifies that any gage line crossing the heel of the angle be reduced by an amount that equals the angle thickness. Thus the distance *g* in Figure 3.16, to be used in the $s^2/4g$ term, would be $3 + 2 - \frac{1}{2} = 4\frac{1}{2}$ inches.

FIGURE 3.16

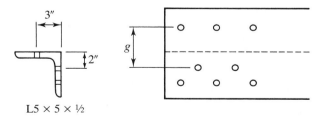

L5 × 5 × ½

Example 3.7   An angle with staggered fasteners in each leg is shown in Figure 3.17. A36 steel is used, and holes are for $\frac{7}{8}$-inch-diameter bolts.
(a) Determine the design strength for LRFD.
(b) Determine the allowable strength for ASD.

FIGURE 3.17

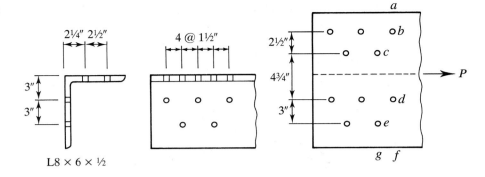

L8 × 6 × ½

Solution

From the dimensions and properties tables, the gross area is $A_g = 6.75$ in.^{2}. The effective hole diameter is $\frac{7}{8} + \frac{1}{8} = 1$ in.

For line *abdf*, the net area is

$$A_n = A_n - \Sigma t_w \times (d \text{ or } d')$$
$$= 6.75 - 0.5(1.0) \times 2 = 5.75 \text{ in.}^2$$

For line *abceg*,

$$A_n = 6.75 - 0.5(1.0) - 0.5\left[1.0 - \frac{(1.5)^2}{4(2.5)}\right] - 0.5(1.0) = 5.363 \text{ in.}^2$$

Because $\frac{1}{10}$ of the load has been transferred from the member by the fastener at *d*, this potential failure line must resist only $\frac{9}{10}$ of the load. Therefore, the net area of 5.363 in.^{2} should be multiplied by $\frac{10}{9}$ to obtain a net area that can be compared with those lines that resist the full load. Use $A_n = 5.363(\frac{10}{9}) = 5.959$ in.^{2} For line *abcdeg*,

$$g_{cd} = 3 + 2.25 - 0.5 = 4.75 \text{ in.}$$
$$A_n = 6.75 - 0.5(1.0) - 0.5\left[1.0 - \frac{(1.5)^2}{4(2.5)}\right] - 0.5\left[1.0 - \frac{(1.5)^2}{4(4.75)}\right] - 0.5\left[1.0 - \frac{(1.5)^2}{4(3)}\right]$$
$$= 5.015 \text{ in.}^2$$

The last case controls; use

$$A_n = 5.015 \text{ in.}^2$$

Both legs of the angle are connected, so

$$A_e = A_n = 5.015 \text{ in.}^2$$

The nominal strength based on fracture is

$$P_n = F_u A_e = 58(5.015) = 290.9 \text{ kips}$$

The nominal strength based on yielding is

$$P_n = F_y A_g = 36(6.75) = 243.0 \text{ kips}$$

a) The design strength based on fracture is

$$\phi_t P_n = 0.75(290.9) = 218 \text{ kips}$$

The design strength based on yielding is

$$\phi_t P_n = 0.90(243.0) = 219 \text{ kips}$$

Answer   Design strength = 218 kips.

b). For the limit state of fracture, the allowable stress is

$$F_t = 0.5F_u = 0.5(58) = 29.0 \text{ ksi}$$

and the allowable strength is

$$F_t A_e = 29.0(5.015) = 145 \text{ kips}$$

For yielding,

$$F_t = 0.6F_y = 0.6(36) = 21.6 \text{ ksi}$$
$$F_t A_g = 21.6(6.75) = 146 \text{ kips}$$

Answer   Allowable strength = 145 kips.

Example 3.8   Determine the smallest net area for the American Standard Channel shown in Figure 3.18. The holes are for ⅝-inch-diameter bolts.

Solution
$$A_n = A_g - \sum t_w \times (d \text{ or } d')$$

$$d = \text{bolt diameter} + \frac{1}{8} = \frac{5}{8} + \frac{1}{8} = \frac{3}{4} \text{ in.}$$

Line *abe:*

$$A_n = A_g - t_w d = 3.81 - 0.437\left(\frac{3}{4}\right) = 3.48 \text{ in.}^2$$

FIGURE 3.18

4 @ 2″

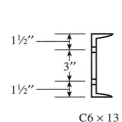

C6 × 13

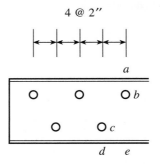

Line *abcd:*

$$A_n = A_g - t_w (d \text{ for hole at } b) - t_w (d' \text{ for hole at } c)$$

$$= 3.81 - 0.437 \left(\frac{3}{4} \right) - 0.437 \left[\frac{3}{4} - \frac{(2)^2}{4(3)} \right] = 3.30 \text{ in.}^2$$

Answer Smallest net area = 3.30 in.^{2}

When staggered holes are present in shapes other than angles, and the holes are in different elements of the cross section, the shape can still be visualized as a plate, even if it is an I-shape. The AISC Specification furnishes no guidance for gage lines crossing a "fold" when the different elements have different thicknesses. A method for handling this case is illustrated in Figure 3.19. In Example 3.8, all of the holes are in one element of the cross section, so this difficulty does not arise. Example 3.9 illustrates the case of staggered holes in different elements of an S-shape.

FIGURE 3.19

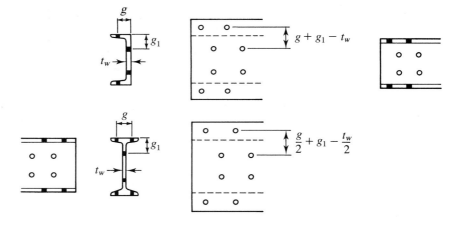

Example 3.9 Find the available strength of the S-shape shown in Figure 3.20. The holes are for ¾-inch-diameter bolts. Use A36 steel.

FIGURE 3.20

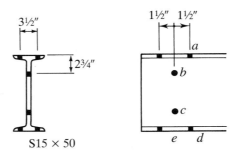

S15 × 50

Solution   Compute the net area:

$$A_n = A_g - \sum t \times (d \text{ or } d')$$

$$\text{Effective hole diameter} = \frac{3}{4} + \frac{1}{8} = \frac{7}{8}$$

For line *ad*,

$$A_n = 14.7 - 4\left(\frac{7}{8}\right)(0.622) = 12.52 \text{ in.}^2$$

For line *abcd*, the gage distance for use in the $s^2/4g$ term is

$$\frac{g}{2} + g_1 - \frac{t_w}{2} = \frac{3.5}{2} + 2.75 - \frac{0.550}{2} = 4.225 \text{ in.}$$

Starting at *a* and treating the holes at *b* and *d* as the staggered holes gives

$$A_n = A_g - \sum t \times (d \text{ or } d')$$

$$= 14.7 - 2(0.622)\left(\frac{7}{8}\right) - (0.550)\left[\frac{7}{8} - \frac{(1.5)^2}{4(4.225)}\right]$$

$$- (0.550)\left(\frac{7}{8}\right) - 2(0.622)\left[\frac{7}{8} - \frac{(1.5)^2}{4(4.225)}\right] = 11.73 \text{ in.}^2$$

Line *abcd* controls. As all elements of the cross section are connected,

$$A_e = A_n = 11.73 \text{ in.}^2$$

For the net section, the nominal strength is

$$P_n = F_u A_e = 58(11.73) = 680.3 \text{ kips}$$

For the gross section,

$$P_n = F_y A_g = 36(14.7) = 529.2 \text{ kips}$$

LRFD Solution   The design strength based on fracture is

$$\phi_t P_n = 0.75(680.3) = 510 \text{ kips}$$

The design strength based on yielding is

$$\phi_t P_n = 0.90(529.2) = 476 \text{ kips}$$

Yielding of the gross section controls.

Answer   Design strength = 476 kips.

ASD Solution   The allowable stress based on fracture is

$$F_t = 0.5 F_u = 0.5(58) = 29.0 \text{ ksi}$$

and the corresponding allowable strength is $F_t A_e = 29.0(11.73) = 340 \text{ kips}$

The allowable stress based on yielding is

$$F_t = 0.6 F_y = 0.6(36) = 21.6 \text{ ksi}$$

and the corresponding allowable strength is $F_t A_g = 21.6(14.7) = 318$ kips

Yielding of the gross section controls.

Answer   Allowable strength = 318 kips.   ■

3.5   BLOCK SHEAR

For certain connection configurations, a segment or "block" of material at the end of the member can tear out. For example, the connection of the single-angle tension member shown in Figure 3.21 is susceptible to this phenomenon, called *block shear*. For the case illustrated, the shaded block would tend to fail by shear along the longitudinal section *ab* and by tension on the transverse section *bc*.

For certain arrangements of bolts, block shear can also occur in gusset plates. Figure 3.22 shows a plate tension member connected to a gusset plate. In this connection, block shear could occur in both the gusset plate and the tension member. For the gusset plate, tension failure would be along the transverse section *df,* and shear failure would occur on two longitudinal surfaces, *de* and *fg*. Block shear failure in the plate tension member would be tension on *ik* and shear on both *hi* and *jk*. This topic is not covered explicitly in AISC Chapter D ("Design of Members for Tension"), but the introductory user note directs you to Chapter J ("Design of Connections"), Section J4.3, "Block Shear Strength."

The model used in the AISC Specification assumes that failure occurs by rupture (fracture) on the shear area and rupture on the tension area. Both surfaces contribute to the total strength, and the resistance to block shear will be the sum of the strengths of the two surfaces. The shear rupture stress is taken as 60% of the tensile ultimate stress, so the nominal strength in shear is $0.6F_u A_{nv}$ and the nominal strength in tension is $F_u A_{nt}$,

where
A_{nv} = net area along the shear surface or surfaces
A_{nt} = net area along the tension surface

This gives a nominal strength of

$$R_n = 0.6F_u A_{nv} + F_u A_{nt} \tag{3.3}$$

FIGURE 3.21

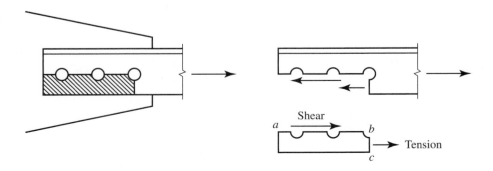

FIGURE 3.22

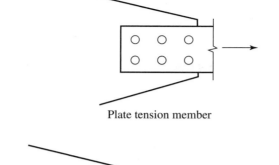

Plate tension member

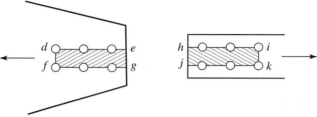

Block shear in gusset plate   Block shear in tension member

The AISC Specification uses Equation 3.3 for angles and gusset plates, but for certain types of coped beam connections (to be covered in Chapter 5), the second term is reduced to account for nonuniform tensile stress. The tensile stress is nonuniform when some rotation of the block is required for failure to occur. For these cases,

$$R_n = 0.6F_u A_{nv} + 0.5F_u A_{nt} \tag{3.4}$$

The AISC Specification limits the $0.6F_u A_{nv}$ term to $0.6F_y A_{gv}$, where

$0.6F_y$ = shear yield stress

A_{gv} = gross area along the shear surface or surfaces

and gives one equation to cover all cases as follows:

$$R_n = 0.6F_u A_{nv} + U_{bs}F_u A_{nt} \le 0.6F_y A_{gv} + U_{bs}F_u A_{nt} \qquad \text{(AISC Equation J4-5)}$$

where $U_{bs} = 1.0$ when the tension stress is uniform (angles, gusset plates, and most coped beams) and $U_{bs} = 0.5$ when the tension stress is nonuniform. A nonuniform case is illustrated in the Commentary to the Specification.

For LRFD, the resistance factor ϕ is 0.75, and for ASD, the safety factor Ω is 2.00. Recall that these are the factors used for the fracture—or rupture—limit state, and block shear is a rupture limit state.

Although AISC Equation J4-5 is expressed in terms of bolted connections, block shear can also occur in welded connections, especially in gusset plates.

Example 3.10   Compute the block shear strength of the tension member shown in Figure 3.23. The holes are for ⅞-inch-diameter bolts, and A36 steel is used.

a) Use LRFD.   b) Use ASD.

FIGURE 3.23

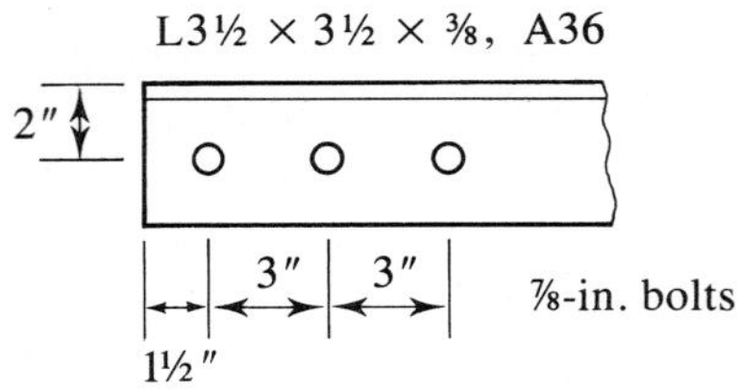

L3½ × 3½ × ⅜, A36

2"

3" 3"

⅞-in. bolts

1½"

Solution The shear areas are

$$A_{gv} = \frac{3}{8}(7.5) = 2.813 \text{ in.}^2$$

and, since there are 2.5 hole diameters,

$$A_{nv} = \frac{3}{8}\left[7.5 - 2.5\left(\frac{7}{8} + \frac{1}{8}\right)\right] = 1.875 \text{ in.}^2$$

The tension area is

$$A_{nt} = \frac{3}{8}\left[1.5 - 0.5\left(\frac{7}{8} + \frac{1}{8}\right)\right] = 0.3750 \text{ in.}^2$$

(The factor of 0.5 is used because there is one-half of a hole diameter in the tension section.)

Since the block shear will occur in an angle, $U_{bs} = 1.0$, and from AISC Equation J4-5,

$$R_n = 0.6F_u A_{nv} + U_{bs}F_u A_{nt}$$
$$= 0.6(58)(1.875) + 1.0(58)(0.3750) = 87.00 \text{ kips}$$

with an upper limit of

$$0.6F_y A_{gv} + U_{bs}F_u A_{nt} = 0.6(36)(2.813) + 1.0(58)(0.3750) = 82.51 \text{ kips}$$

The nominal block shear strength is therefore 82.51 kips.

a) Answer The design strength for LRFD is $\phi R_n = 0.75(82.51) = 61.9$ kips.

b) Answer The allowable strength for ASD is $\dfrac{R_n}{\Omega} = \dfrac{82.51}{2.00} = 41.3$ kips.

■

3.6 DESIGN OF TENSION MEMBERS

The design of a tension member involves finding a member with adequate gross and net areas. If the member has a bolted connection, the selection of a suitable cross section requires an accounting for the area lost because of holes. For a member with a rectangular cross section, the calculations are relatively straightforward. If a rolled shape is to be used, however, the area to be deducted cannot be predicted in advance because the member's thickness at the location of the holes is not known.

A secondary consideration in the design of tension members is slenderness. If a structural member has a small cross section in relation to its length, it is said to be *slender*. A more precise measure is the slenderness ratio, L/r, where L is the member length and r is the minimum radius of gyration of the cross-sectional area. The minimum radius of gyration is the one corresponding to the minor principal axis of the cross section. This value is tabulated for all rolled shapes in the properties tables in Part 1 of the *Manual*.

Although slenderness is critical to the strength of a compression member, it is inconsequential for a tension member. In many situations, however, it is good practice to limit the slenderness of tension members. If the axial load in a slender tension member is removed and small transverse loads are applied, undesirable vibrations or deflections might occur. These conditions could occur, for example, in a slack bracing rod subjected to wind loads. For this reason, the user note in AISC D1 suggests a maximum slenderness ratio of 300. It is only a recommended value because slenderness has no structural significance for tension members, and the limit may be exceeded when special circumstances warrant it. This limit does not apply to cables, and the user note explicitly excludes rods.

The central problem of all member design, including tension member design, is to find a cross section for which the required strength does not exceed the available strength. For tension members designed by LRFD, the requirement is

$$P_u \le \phi_t P_n \quad \text{or} \quad \phi_t P_n \ge P_u$$

where P_u is the sum of the factored loads. To prevent yielding,

$$0.90 F_y A_g \ge P_u \quad \text{or} \quad A_g \ge \frac{P_u}{0.90 F_y}$$

To avoid fracture,

$$0.75 F_u A_e \ge P_u \quad \text{or} \quad A_e \ge \frac{P_u}{0.75 F_u}$$

For allowable strength design, if we use the allowable *stress* form, the requirement corresponding to yielding is

$$P_a \le F_t A_g$$

and the required gross area is

$$A_g \ge \frac{P_a}{F_t} \quad \text{or} \quad A_g \ge \frac{P_a}{0.6 F_y}$$

For the limit state of fracture, the required effective area is

$$A_e \ge \frac{P_a}{F_t} \quad \text{or} \quad A_e \ge \frac{P_a}{0.5 F_u}$$

The slenderness ratio limitation will be satisfied if

$$r \ge \frac{L}{300}$$

where r is the minimum radius of gyration of the cross section and L is the member length.

Example 3.11   A tension member with a length of 5 feet 9 inches must resist a service dead load of 18 kips and a service live load of 52 kips. Select a member with a rectangular cross section. Use A36 steel and assume a connection with one line of $\frac{7}{8}$-inch-diameter bolts.

LRFD Solution $P_u = 1.2D + 1.6L = 1.2(18) + 1.6(52) = 104.8$ kips

Required $A_g = \dfrac{P_u}{\phi_t F_y} = \dfrac{P_u}{0.90 F_y} = \dfrac{104.8}{0.90(36)} = 3.235$ in.^{2}

Required $A_e = \dfrac{P_u}{\phi_t F_u} = \dfrac{P_u}{0.75 F_u} = \dfrac{104.8}{0.75(58)} = 2.409$ in.^{2}

Try $t = 1$ in.

Required $w_g = \dfrac{\text{required } A_g}{t} = \dfrac{3.235}{1} = 3.235$ in.

Try a $1 \times 3\frac{1}{2}$ cross section.

$A_e = A_n = A_g - A_{hole}$

$= (1 \times 3.5) - \left(\dfrac{7}{8} + \dfrac{1}{8} \right)(1) = 2.5$ in.^{2} > 2.409 in.^{2} (OK)

Check the slenderness ratio:

$I_{min} = \dfrac{3.5(1)^3}{12} = 0.2917$ in.^{4}

$A = 1(3.5) = 3.5$ in.^{2}

From $I = Ar^2$ we obtain

$r_{min} = \sqrt{\dfrac{I_{min}}{A}} = \sqrt{\dfrac{0.2917}{3.5}} = 0.2887$ in.^{2}

Maximum $\dfrac{L}{r} = \dfrac{5.75(12)}{0.2887} = 239 < 300$ (OK)

Answer   Use a PL $1 \times 3\frac{1}{2}$

ASD Solution $P_a = D + L = 18 + 52 = 70.0$ kips

For yielding, $F_t = 0.6 F_y = 0.6(36) = 21.6$ ksi, and

Required $A_g = \dfrac{P_a}{F_t} = \dfrac{70}{21.6} = 3.24$ in.^{2}

For fracture, $F_t = 0.5 F_u = 0.5(58) = 29.0$ ksi, and

Required $A_e = \dfrac{P_a}{F_t} = \dfrac{70}{29.0} = 2.414$ in.^{2}

(The rest of the design *procedure* is the same as for LRFD. The numerical results may be different)

Try $t = 1$ in.

$$\text{Required } w_g = \frac{\text{required } A_g}{t} = \frac{3.241}{1} = 3.241 \text{ in.}$$

Try a $1 \times 3\,\tfrac{1}{2}$ cross section.

$$A_e = A_n = A_g - A_{hole}$$

$$= (1 \times 3.5) - \left(\frac{7}{8} + \frac{1}{8}\right)(1) = 2.5 \text{ in.}^2 > 2.414 \text{ in.}^2 \qquad \text{(OK)}$$

Check the slenderness ratio:

$$I_{min} = \frac{3.5(1)^3}{12} = 0.2917 \text{ in.}^4$$

$$A = 1(3.5) = 3.5 \text{ in.}^2$$

From $I = Ar^2$, we obtain

$$r_{min} = \sqrt{\frac{I_{min}}{A}} = \sqrt{\frac{0.2917}{3.5}} = 0.2887 \text{ in.}^2$$

$$\text{Maximum } \frac{L}{r} = \frac{5.75(12)}{0.2887} = 239 < 300 \qquad \text{(OK)}$$

Answer   Use a PL $1 \times 3\tfrac{1}{2}$.

Example 3.11 illustrates that once the required area has been determined, the procedure is the same for both LRFD and ASD. Note also that in this example, the required areas are virtually the same for LRFD and ASD. This is because the ratio of live load to dead load is approximately 3, and the two approaches will give the same results for this ratio.

The member in Example 3.11 is less than 8 inches wide and thus is classified as a bar rather than a plate. Bars should be specified to the nearest ¼ inch in width and to the nearest ⅛ inch in thickness (the precise classification system is given in Part 1 of the *Manual* under the heading "Plate Products"). It is common practice to use the PL (Plate) designation for both bars and plates.

If an angle shape is used as a tension member and the connection is made by bolting, there must be enough room for the bolts. Space will be a problem only when there are two lines of bolts in a leg. The usual fabrication practice is to punch or drill holes in standard locations in angle legs. These hole locations are given in a table in Part 1 of the *Manual*. This table is located at the end of the dimensions and properties table for angles. Figure 3.24 presents this same information. Gage distance g applies when there is one line of bolts, and g_1 and g_2 apply when there are two lines. Figure 3.24 shows that an angle leg must be at least 5 inches long to accommodate two lines of bolts.

FIGURE 3.24

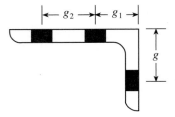

Usual Gages for Angles (inches)

Leg	8	7	6	5	4	3½	3	2½	2	1¾	1½	1⅜	1¼	1
g	4½	4	3½	3	2½	2	1¾	1⅜	1⅛	1	⅞	⅞	¾	⅝
g_1	3	2½	2¼	2										
g_2	3	3	2½	1¾										

Example 3.12   Select an unequal-leg angle tension member 15 feet long to resist a service dead load of 35 kips and a service live load of 70 kips. Use A36 steel. The connection is shown in Figure 3.25.

LRFD Solution   The factored load is

$$P_u = 1.2D + 1.6L = 1.2(35) + 1.6(70) = 154 \text{ kips}$$

$$\text{Required } A_g = \frac{P_u}{\phi_t F_y} = \frac{154}{0.90(36)} = 4.75 \text{ in.}^2$$

$$\text{Required } A_e = \frac{P_u}{\phi_t F_u} = \frac{154}{0.75(58)} = 3.54 \text{ in.}^2$$

The radius of gyration should be at least

$$\frac{L}{300} = \frac{15(12)}{300} = 0.6 \text{ in.}$$

To find the lightest shape that satisfies these criteria, we search the dimensions and properties table for the unequal-leg angle that has the smallest acceptable gross area and

FIGURE 3.25

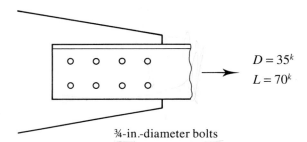

$D = 35^k$
$L = 70^k$

¾-in.-diameter bolts

then check the effective net area. The radius of gyration can be checked by inspection. There are two lines of bolts, so the connected leg must be at least 5 inches long (see the usual gages for angles in Figure 3.24). Starting at either end of the table, we find that the shape with the smallest area that is at least equal to 4.75 in.^{2} is an L6 × 4 × $\frac{1}{2}$ with an area of 4.75 in.^{2} and a minimum radius of gyration of 0.864 in.

Try L6 × 4 × $\frac{1}{2}$.

$$A_n = A_g - A_{holes} = 4.75 - 2\left(\frac{3}{4}+\frac{1}{8}\right)\left(\frac{1}{2}\right) = 3.875 \text{ in.}^2$$

Because the length of the connection is not known, Equation 3.1 cannot be used to compute the shear lag factor U. Since there are four bolts in the direction of the load, we will use the alternative value of $U = 0.80$.

$$A_e = A_n U = 3.875(0.80) = 3.10 \text{ in.}^2 < 3.54 \text{ in.}^2 \qquad (\text{N.G.})^*$$

Try the next larger shape from the dimensions and properties tables.

Try L5 × 3$\frac{1}{2}$ × $\frac{5}{8}$ ($A_g = 4.92$ in.^{2} and $r_{min} = 0.746$ in.)

$$A_n = A_g - A_{holes} = 4.92 - 2\left(\frac{3}{4}+\frac{1}{8}\right)\left(\frac{5}{8}\right) = 3.826 \text{ in.}^2$$
$$A_e = A_n U = 3.826(0.80) = 3.06 \text{ in.}^2 < 3.54 \text{ in.}^2 \qquad (\text{N.G.})$$

(Note that this shape has slightly more gross area than that produced by the previous trial shape, but because of the greater leg thickness, slightly more are is deducted for the holes.) Passing over the next few heavier shapes,

Try L8 × 4 × $\frac{1}{2}$ ($A_g = 5.75$ in.^{2} and $r_{min} = 0.863$ in.)

$$A_n = A_g - A_{holes} = 5.75 - 2\left(\frac{3}{4}+\frac{1}{8}\right)\left(\frac{1}{2}\right) = 4.875 \text{ in.}^2$$
$$A_e = A_n U = 4.875(0.80) = 3.90 \text{ in.}^2 < 3.54 \text{ in.}^2 \qquad (\text{OK})$$

Answer   This shape satisfies all requirements, so use an L8 × 4 × $\frac{1}{2}$.

ASD Solution   The total service load is

$$P_a = D + L = 35 + 70 = 105 \text{ kips}$$

$$\text{Required } A_g = \frac{P_a}{F_t} = \frac{P_a}{0.6F_y} = \frac{105}{0.6(36)} = 4.86 \text{ in.}^2$$

$$\text{Required } A_e = \frac{P_a}{0.5F_u} = \frac{105}{0.5(58)} = 3.62 \text{ in.}^2$$

$$\text{Required } r_{min} = \frac{L}{300} = \frac{15(12)}{300} = 0.6 \text{ in.}$$

---

*The notation N.G. means "No Good."

Try L8 × 4 × ½ (A_g = 5.75 in.^{2} and r_{min} = 0.863 in.). For a shear lag factor U of 0.80,

$$A_n = A_g - A_{holes} = 5.75 - 2\left(\frac{3}{4} + \frac{1}{8}\right)\left(\frac{1}{2}\right) = 4.875 \text{ in.}^2$$

$$A_e = A_n U = 4.875(0.80) = 3.90 \text{ in.}^2 > 3.62 \text{ in.}^2 \qquad \text{(OK)}$$

Answer   This shape satisfies all requirements, so use an L8 ×4 × ½.   ■

The ASD solution in Example 3.12 is somewhat condensed, in that some of the discussion in the LRFD solution is not repeated and only the final trial is shown. All essential computations are included, however.

Tables for the Design of Tension Members

Part 5 of the *Manual* contains tables to assist in the design of tension members of various cross-sectional shapes, including Table 5-2 for angles. The use of these tables will be illustrated in the following example.

Example 3.13   Design the tension member of Example 3.12 with the aid of the tables in Part 5 of the *Manual*.

LRFD Solution   From Example 3.12,

$$P_u = 154 \text{ kips}$$
$$r_{min} \geq 0.600 \text{ in.}$$

The tables for design of tension members give values of A_g and A_e for various shapes based on the assumption that $A_e = 0.75 A_g$. In addition, the corresponding available for angles strengths based on yielding and rupture (fracture) are given. All values available for angles are for A36 steel. Starting with the lighter shapes (the ones with the smaller gross area), we find that an L6 × 4 × ½, with $\phi_t P_n$ = 154 kips based on the gross section and $\phi_t P_n$ = 155 kips based on the net section, is a possibility. From the dimensions and properties tables in Part 1 of the *Manual*, r_{min} = 0.980 in. To check this selection, we must compute the actual net area. If we assume that U = 0.80,

$$A_n = A_g - A_{holes} = 4.75 - 2\left(\frac{3}{4} + \frac{1}{8}\right)\left(\frac{1}{2}\right) = 3.875 \text{ in.}^2$$

$$A_e = A_n U = 3.875(0.80) = 3.10 \text{ in.}^2$$

$$\phi_t P_n = \phi_t F_u A_e = 0.75(58)(3.10) = 135 \text{ kips} < 154 \text{ kips} \qquad \text{(N.G.)}$$

This shape did not work because the ratio of actual effective net area A_e to gross area A_g is not equal to 0.75. The ratio is closer to

$$\frac{3.10}{4.75} = 0.6526$$

This corresponds to a required $\phi_t P_n$ (based on rupture) of

$$\frac{0.75}{\text{actual ratio}} \times P_u = \frac{0.75}{0.6526}(154) = 177 \text{ kips}$$

Try an L8 × 4 × ½, with $\phi_t P_n = 186$ kips (based on yielding) and $\phi_t P_n = 187$ (based on rupture strength). From the dimensions and properties tables in Part 1 of the *Manual*, $r_{min} = 0.863$ in. The actual effective net area and rupture strength are computed as follows:

$$A_n = A_g - A_{holes} = 5.75 - 2\left(\frac{3}{4} + \frac{1}{8}\right)\left(\frac{1}{2}\right) = 4.875 \text{ in.}^2$$

$$A_e = A_n U = 4.875(0.80) = 3.90 \text{ in.}^2$$

$$\phi_t P_n = \phi_t F_u A_e = 0.75(58)(3.90) = 170 > 154 \text{ kips} \qquad \text{(OK)}$$

Answer Use an L8 × 4 × ½, connected through the 8-inch leg.

ASD Solution From Example 3.12,

$$P_a = 105 \text{ kips}$$

Required $r_{min} = 0.600$ in.

From *Manual* Table 5-2, try an L5 × 3½ × ⅝, with $P_n/\Omega_t = 106$ kips based on yielding of the gross section and $P_n/\Omega_t = 107$ kips based on rupture of the net section. From the dimensions and properties tables in Part 1 of the *Manual*, $r_{min} = 0.746$ in. Using a shear lag factor U of 0.80, the actual effective net area is computed as follows:

$$A_n = A_g - A_{holes} = 4.92 - 2\left(\frac{3}{4} + \frac{1}{8}\right)\left(\frac{5}{8}\right) = 3.826 \text{ in.}^2$$

$$A_e = A_n U = 3.826(0.80) = 3.061 \text{ in.}^2$$

and the allowable strength based on rupture of the net section is

$$\frac{P_n}{\Omega_t} = \frac{F_u A_e}{\Omega_t} = \frac{58(3.061)}{2.00} = 88.8 \text{ kips} < 105 \text{ kips} \qquad \text{(N.G.)}$$

This shape did not work because the ratio of actual effective net area A_e to gross area A_g is not equal to 0.75. The ratio is closer to

$$\frac{3.061}{4.92} = 0.6222$$

This corresponds to a required P_n/Ω_t (based on rupture), for purposes of using Table 5-2, of

$$\frac{0.75}{0.6222}(105) = 127 \text{ kips}$$

Using this as a guide, try L6×4×⅝, with $P_n/\Omega_t = 126$ kips based on yielding of the gross section and $P_n/\Omega_t = 127$ kips based on rupture of the net section. From the dimensions and properties tables in Part 1 of the *Manual*, $r_{min} = 0.859$ in.

$$A_n = A_g - A_{holes} = 5.86 - 2\left(\frac{3}{4} + \frac{1}{8}\right)\left(\frac{5}{8}\right) = 4.766 \text{ in.}^2$$

$$A_e = A_n U = 4.766(0.80) = 3.81 \text{ in.}^2$$

$$\frac{P_n}{\Omega_t} = \frac{F_u A_e}{\Omega_t} = \frac{58(3.81)}{2.00} = 111 \text{ kips} > 105 \text{ kips} \qquad \text{(OK)}$$

Answer   Use an L6 × 4 × ⅝, connected through the 8-inch leg. ∎

Note that if the effective net area must be computed, the tables do not save much effort. In addition, you must still refer to the dimensions and properties tables to find the radius of gyration. The tables for design do, however, provide all other information in a compact form, and the search may go more quickly.

When structural shapes or plates are connected to form a built-up shape, they must be connected not only at the ends of the member but also at intervals along its length. A continuous connection is not required. This type of connection is called *stitching,* and the fasteners used are termed *stitch bolts.* The usual practice is to locate the points of stitching so that L/r for any component part does not exceed L/r for the built-up member. The user note in AISC D4 recommends that built-up shapes whose component parts are separated by intermittent fillers be connected at intervals such that the maximum L/r for any component does not exceed 300. Built-up shapes consisting of plates or a combination of plates and shapes are addressed in AISC Section J3.5 of Chapter J ("Design of Connections"). In general, the spacing of fasteners or welds should not exceed 24 times the thickness of the thinner plate, or 12 inches. If the member is of "weathering" steel subject to atmospheric corrosion, the maximum spacing is 14 times the thickness, or 7 inches.

3.7   THREADED RODS AND CABLES

When slenderness is not a consideration, rods with circular cross sections and cables are often used as tension members. The distinction between the two is that rods are solid and cables are made from individual strands wound together in ropelike fashion. Rods and cables are frequently used in suspended roof systems and as hangers or suspension members in bridges. Rods are also used in bracing systems; in some cases, they are pretensioned to prevent them from going slack when external loads are removed. Figure 3.26 illustrates typical rod and cable connection methods.

When the end of a rod is to be threaded, an upset end is sometimes used. This is an enlargement of the end in which the threads are to be cut. Threads reduce the cross-sectional area, and upsetting the end produces a larger gross area to start with. Standard upset ends with threads will actually have more net area in the threaded portion than in the unthreaded part. Upset ends are relatively expensive, however, and in most cases unnecessary.

FIGURE 3.26

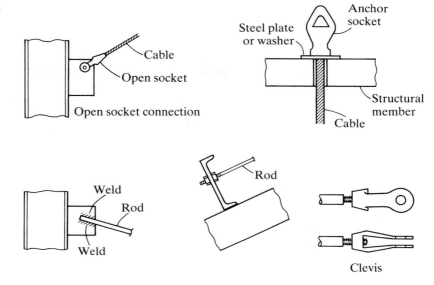

The effective cross-sectional area in the threaded portion of a rod is called the *stress area* and is a function of the unthreaded diameter and the number of threads per inch. The ratio of stress area to nominal area varies but has a lower bound of approximately 0.75. The nominal tensile strength of the threaded rod can therefore be written as

$$P_n = A_s F_u = 0.75 A_b F_u \tag{3.5}$$

where

A_s = stress area
A_b = nominal (unthreaded) area

The AISC Specification, in Chapter J, presents the nominal strength in a somewhat different form:

$$R_n = F_n A_b \qquad \text{(AISC Equation J3-1)}$$

where R_n is the nominal strength and F_n is given in Table J3.2 as $F_{nt} = 0.75 F_u$. This associates the 0.75 factor with the ultimate tensile stress rather than the area, but the result is the same as that given by Equation 3.5.

For LRFD, the resistance factor ϕ is 0.75, so the strength relationship is

$$P_u \le \phi_t P_n \quad \text{or} \quad P_u \le 0.75(0.75 A_b F_u)$$

and the required area is

$$A_b = \frac{P_u}{0.75(0.75 F_u)} \tag{3.6}$$

For ASD, the safety factor Ω is 2.00, leading to the requirement

$$P_a \le \frac{P_n}{2.00} \quad \text{or} \quad P_a \le 0.5 P_n$$

Using P_n from Equation 3.5, we get

$$P_a \leq 0.5(0.75A_bF_u)$$

If we divide both sides by the area A_b, we obtain the allowable stress

$$F_t = 0.5(0.75F_u) = 0.375F_u \tag{3.7}$$

If upset ends are used, the tensile capacity at the major thread diameter must be greater than F_y times the unthreaded body area (AISC Table J3.2, footnote d).

Example 3.14   A threaded rod is to be used as a bracing member that must resist a service tensile load of 2 kips dead load and 6 kips live load. What size rod is required if A36 steel is used?

LRFD Solution   The factored load is

$$P_u = 1.2(2) + 1.6(6) = 12 \text{ kips}$$

From Equation 3.6,

$$\text{Required Area} = A_b = \frac{P_u}{0.75(0.75F_u)} = \frac{12}{0.75(0.75)(58)} = 0.3678 \text{ in.}^2$$

From $A_b = \dfrac{\pi d^2}{4}$,

$$\text{Required } d = \sqrt{\frac{4(0.3678)}{\pi}} = 0.684 \text{ in.}$$

Answer   Use a ¾-inch-diameter threaded rod ($A_b = 0.442$ in.^{2}).

ASD Solution   The required strength is

$$P_a = D + L = 2 + 6 = 8 \text{ kips}$$

From Equation 3.7, the allowable tensile stress is

$$F_t = 0.375F_u = 0.375(58) = 21.75 \text{ ksi}$$

and the required area is

$$A_b = \frac{P_a}{F_t} = \frac{8}{21.75} = 0.3678 \text{ in.}^2$$

Answer   Use a ¾-inch-diameter threaded rod ($A_b = 0.442$ in.^{2}).

To prevent damage during construction, rods should not be too slender. Although there is no specification requirement, a common practice is to use a minimum diameter of ⅝ inch.

FIGURE 3.27

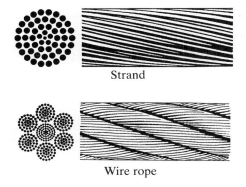

Strand

Wire rope

Flexible cables, in the form of strands or wire rope, are used in applications where high strength is required and rigidity is unimportant. In addition to their use in bridges and cable roof systems, they are also used in hoists and derricks, as guy lines for towers, and as longitudinal bracing in metal building systems. The difference between strand and wire rope is illustrated in Figure 3.27. A strand consists of individual wires wound helically around a central core, and a wire rope is made of several strands laid helically around a core.

Selection of the correct cable for a given loading is usually based on both strength and deformation considerations. In addition to ordinary elastic elongation, an initial stretching is caused by seating or shifting of the individual wires, which results in a permanent stretch. For this reason, cables are often prestretched. Wire rope and strand are made from steels of much higher strength than structural steels and are not covered by the AISC Specification. The breaking strengths of various cables, as well as details of available fixtures for connections, can be obtained from manufacturers' literature.

3.8   TENSION MEMBERS IN ROOF TRUSSES

Many of the tension members that structural engineers design are components of trusses. For this reason, some general discussion of roof trusses is in order. A more comprehensive treatment of the subject is given by Lothars (1972).

When trusses are used in buildings, they usually function as the main supporting elements of roof systems where long spans are required. They are used when the cost and weight of a beam would be prohibitive. (A truss may be thought of as a deep beam with much of the web removed.) Roof trusses are often used in industrial or mill buildings, although construction of this type has largely given way to rigid frames. Typical roof construction with trusses supported by load-bearing walls is illustrated in Figure 3.28. In this type of construction, one end of the connection of the truss to the walls usually can be considered as pinned and the other as roller-supported. Thus the truss can be analyzed as an externally statically determinate structure. The supporting walls can be reinforced concrete, concrete block, brick, or a combination of these materials.

Roof trusses normally are spaced uniformly along the length of the building and are tied together by longitudinal beams called *purlins* and by x-bracing. The primary function of the purlins is to transfer loads to the top chord of the truss, but they can

FIGURE 3.28

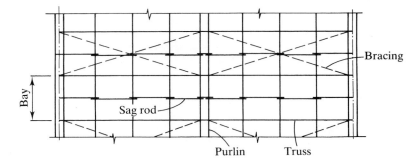

Plan

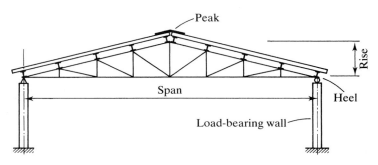

Elevation

also act as part of the bracing system. Bracing is usually provided in the planes of both the top and bottom chords, but it is not required in every bay because lateral forces can be transferred from one braced bay to the other through the purlins.

Ideally, purlins are located at the truss joints so that the truss can be treated as a pin-connected structure loaded only at the joints. Sometimes, however, the roof deck cannot span the distance between joints, and intermediate purlins may be needed. In such cases, top chord members will be subjected to significant bending as well as axial compression and must be designed as beam–columns (Chapter 6).

Sag rods are tension members used to provide lateral support for the purlins. Most of the loads applied to the purlins are vertical, so there will be a component parallel to a sloping roof, which will cause the purlin to bend (sag) in that direction (Figure 3.29).

FIGURE 3.29

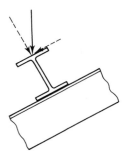

FIGURE 3.30

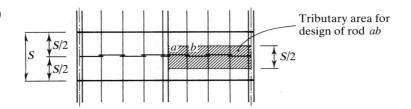

Sag rods can be located at the midpoint, the third points, or at more frequent intervals along the purlins, depending on the amount of support needed. The interval is a function of the truss spacing, the slope of the top chord, the resistance of the purlin to this type of bending (most shapes used for purlins are very weak in this respect), and the amount of support furnished by the roofing. If a metal deck is used, it will usually be rigidly attached to the purlins, and sag rods may not be needed. Sometimes, however, the weight of the purlin itself is enough to cause problems, and sag rods may be needed to provide support during construction before the deck is in place.

If sag rods are used, they are designed to support the component of roof loads parallel to the roof. Each segment between purlins is assumed to support everything below it; thus the top rod is designed for the load on the roof area tributary to the rod, from the heel of the truss to the peak, as shown in Figure 3.30. Although the force will be different in each segment of rod, the usual practice is to use one size throughout. The extra amount of material in question is insignificant, and the use of the same size for each segment eliminates the possibility of a mix-up during construction.

A possible treatment at the peak or ridge is shown in Figure 3.31a. The tie rod between ridge purlins must resist the load from all of the sag rods on either side. The tensile force in this horizontal member has as one of its components the force in the upper sag-rod segment. A free-body diagram of one ridge purlin illustrates this effect, as shown in Figure 3.31b.

FIGURE 3.31

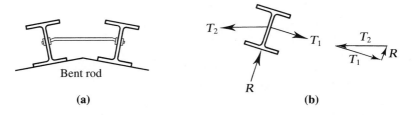

(a)   (b)

Example 3.15   Fink trusses spaced at 20 feet on centers support W6 × 12 purlins, as shown in Figure 3.32a. The purlins are supported at their midpoints by sag rods. Use A36 steel and design the sag rods and the tie rod at the ridge for the following service loads.

Metal deck:	2 psf
Built-up roof:	5 psf
Snow:	18 psf of horizontal projection of the roof surface
Purlin weight:	12 pounds per foot (lb/ft) of length

FIGURE 3.32

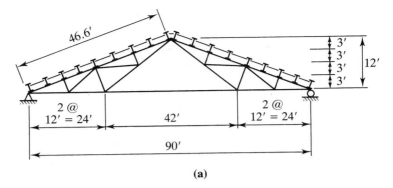

(a)

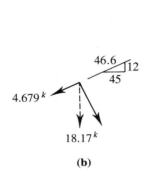

(b)

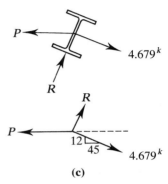

(c)

Solution   Calculate loads.

    Tributary width for each sag rod = $20/2 = 10$ ft

    Tributary area for deck and built-up roof = $10(46.6) = 466$ ft^2

    Dead load (deck and roof) = $(2 + 5)(466) = 3262$ lb

    Total purlin weight = $12(10)(9) = 1080$ lb

    Total dead load = $3262 + 1080 = 4342$ lb

    Tributary area for snow load = $10(45) = 450$ ft^2

    Total snow load = $18(450) = 8100$ lb

LRFD Solution   Check load combinations.

    Combination 2: $1.2D + 0.5S = 1.2(4342) + 0.5(8100) = 9260$ lb

    Combination 3: $1.2D + 1.6S = 1.2(4342) + 1.6(8100) = 18,170$ lb

Combination 3 controls. (By inspection, the remaining combinations will not govern.)
For the component parallel to the roof (Figure 3.32b),

$$T = (18.17)\frac{12}{46.6} = 4.679 \text{ kips}$$

$$\text{Required } A_b = \frac{T}{\phi_t(0.75F_u)} = \frac{4.679}{0.75(0.75)(58)} = 0.1434 \text{ in.}^2$$

Answer   Use a ⅝-inch-diameter threaded rod ($A_b = 0.3068$ in.^{2}).
Tie rod at the ridge (Figure 3.32c):

$$P = (4.679)\frac{46.6}{45} = 4.845 \text{ kips}$$

$$\text{Required } A_b = \frac{4.845}{0.75(0.75)(58)} = 0.1485 \text{ in.}^2$$

Answer   Use a ⅝-inch-diameter threaded rod ($A_b = 0.3068$ in.^{2}).

ASD Solution   By inspection, load combination 3 will control.

$$D + S = 4342 + 8100 = 12,440 \text{ lb}$$

The component parallel to the roof is

$$T = 12.44\left(\frac{12}{46.6}\right) = 3.203 \text{ kips}$$

The allowable tensile stress is $F_t = 0.375F_u = 0.375(58) = 21.75$ ksi.

$$\text{Required } A_b = \frac{T}{F_t} = \frac{3.203}{21.75} = 0.1473 \text{ in.}^2$$

Answer   Use a ⅝-inch-diameter threaded rod ($A_b = 0.3068$ in.^{2}) for the sag rods.
Tie rod at the ridge:

$$P = 3.203\left(\frac{46.6}{45}\right) = 3.317 \text{ kips}$$

$$\text{Required } A_b = \frac{3.317}{21.75} = 0.1525 \text{ in.}^2$$

Answer   Use a ⅝-inch-diameter threaded rod ($A_b = 0.3068$ in.^{2}) for the tie rod at the ridge. ▪

For the usual truss geometry and loading, the bottom chord will be in tension and the top chord will be in compression. Some web members will be in tension and others will be in compression. When wind effects are included and consideration is given to different wind directions, the force in some web members may alternate between tension and compression. In this case, the affected member must be designed to function as both a tension member and a compression member.

In bolted trusses, double-angle sections are frequently used for both chord and web members. This design facilitates the connection of members meeting at a joint by permitting the use of a single gusset plate, as illustrated in Figure 3.33. When structural tee-shapes are used as chord members in welded trusses, the web angles can usually be welded to the stem of the tee. If the force in a web member is small, single

FIGURE 3.33

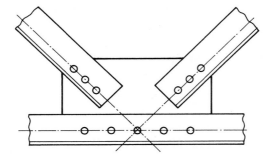

angles can be used, although doing so eliminates the plane of symmetry from the truss and causes the web member to be eccentrically loaded. Chord members are usually fabricated as continuous pieces and spliced if necessary.

The fact that chord members are continuous and joints are bolted or welded would seem to invalidate the usual assumption that the truss is pin-connected. Joint rigidity does introduce some bending moment into the members, but it is usually small and considered to be a secondary effect. The usual practice is to ignore it. Bending caused by loads directly applied to members between the joints, however, must be taken into account. We consider this condition in Chapter 6, "Beam–Columns."

The *working lines* of the members in a properly detailed truss intersect at the *working point* at each joint. For a bolted truss, the bolt lines are the working lines, and in welded trusses the centroidal axes of the welds are the working lines. For truss analysis, member lengths are measured from working point to working point.

Example 3.16   Select a structural tee for the bottom chord of the Warren roof truss shown in Figure 3.34. The trusses are welded and spaced at 20 feet. Assume that the bottom chord connection is made with 9-inch-long longitudinal welds at the flange. Use A992 steel and the following load data (wind is not considered in this example):

Purlins:	M8 × 6.5
Snow:	20 psf of horizontal projection
Metal deck:	2 psf
Roofing:	4 psf
Insulation:	3 psf

FIGURE 3.34

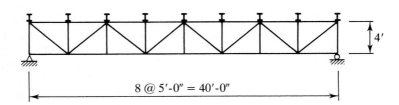

8 @ 5'-0" = 40'-0"

Solution   Calculate loads:

Snow = 20(40)(20) = 16,000 lb

Dead load (exclusive of purlins) =	Deck	2 psf
	Roof	4
	Insulation	3
	Total	9 psf

Total dead load = 9(40)(20) = 7200 lb

Total purlin weight = 6.5(20)(9) = 1170 lb

Estimate the truss weight as 10% of the other loads:

0.10(16,000 + 7200 + 1170) = 2437 lb

Loads at an interior joint are

$$D = \frac{7200}{8} + \frac{2437}{8} + 6.5(20) = 1335 \text{ lb}$$

$$S = \frac{16,000}{8} = 2000 \text{ lb}$$

At an exterior joint, the tributary roof area is half of that at an interior joint. The corresponding loads are

$$D = \frac{7200}{2(8)} + \frac{2437}{2(8)} + 6.5(20) = 732.3 \text{ lb}$$

$$S = \frac{16,000}{2(8)} = 1000 \text{ lb}$$

LRFD Solution   Load combination 3 will control:

$$P_u = 1.2D + 1.6S$$

At an interior joint,

$$P_u = 1.2(1.335) + 1.6(2.0) = 4.802 \text{ kips}$$

At an exterior joint,

$$P_u = 1.2(0.7323) + 1.6(1.0) = 2.479 \text{ kips}$$

The loaded truss is shown in Figure 3.35a.

The bottom chord is designed by determining the force in each member of the bottom chord and selecting a cross section to resist the largest force. In this example, the force in member *IJ* will control. For the free body left of section *a–a* shown in Figure 3.35b,

$$\sum M_E = 19.29(20) - 2.479(20) - 4.802(15 + 10 + 5) - 4F_{IJ} = 0$$

$$F_{IJ} = 48.04 \text{ kips}$$

FIGURE 3.35

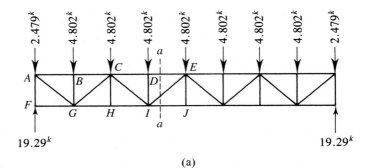

(a)

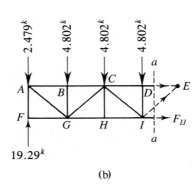

(b)

For the gross section,

$$\text{Required } A_g = \frac{F_{IJ}}{0.90F_y} = \frac{48.04}{0.90(50)} = 1.07 \text{ in.}^2$$

For the net section,

$$\text{Required } A_e = \frac{F_{IJ}}{0.75F_u} = \frac{48.04}{0.75(65)} = 0.985 \text{ in.}^2$$

Try an MT5 ×3.75:

$$A_g = 1.10 \text{ in.}^2 > 1.07 \text{ in.}^2 \qquad \text{(OK)}$$

Compute the shear lag factor U from Equation 3.1.

$$U = 1 - \left(\frac{\overline{x}}{\ell}\right) = 1 - \left(\frac{1.51}{9}\right) = 0.8322$$

$$A_e = A_g U = 1.10(0.8322) = 0.915 \text{ in.}^2 < 0.985 \text{ in.}^2 \quad \text{(N.G.)}$$

Try an MT6 × 5:

$$A_g = 1.46 \text{ in.}^2 > 1.07 \text{ in.}^2 \quad \text{(OK)}$$

$$U = 1 - \left(\frac{\overline{x}}{\ell}\right) = 1 - \left(\frac{1.86}{9}\right) = 0.7933$$

$$A_e = A_g U = 1.46(0.7933) = 1.16 \text{ in.}^2 > 0.985 \text{ in.}^2 \qquad \text{(OK)}$$

If we assume that the bottom chord is braced at the panel points,

$$\frac{L}{r} = \frac{5(12)}{0.594} = 101 < 300 \qquad (\text{OK})$$

Answer Use an MT6×5.

ASD Solution Load combination 3 will control. At an interior joint,

$$P_a = D + S = 1.335 + 2.0 = 3.335 \text{ kips}$$

At an exterior joint,

$$P_a = 0.7323 + 1.0 = 1.732 \text{ kips}$$

The loaded truss is shown in Figure 3.36a.

FIGURE 3.36

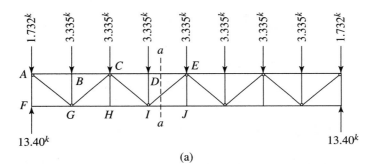

(a)

(b)

Member *IJ* is the bottom chord member with the largest force. For the free body shown in Figure 3.36b,

$$\sum M_E = 13.40(20) - 1.732(20) - 3.335(15 + 10 + 5) - 4F_{IJ} = 0$$
$$F_{IJ} = 33.33 \text{ kips}$$

For the gross section, $F_t = 0.6F_y = 0.6(36) = 21.6 \text{ ksi}$

$$\text{Required } A_g = \frac{F_{IJ}}{F_t} = \frac{33.33}{21.6} = 1.54 \text{ in.}^2$$

For the net section, $F_t = 0.5F_u = 0.5(58) = 29.0$ ksi

$$\text{Required } A_e = \frac{F_U}{F_t} = \frac{33.33}{29.0} = 1.15 \text{ in.}^2$$

Try an MT6 × 5.4:

$$A_g = 1.58 \text{ in.}^2 > 1.54 \text{ in.}^2 \qquad \text{(OK)}$$

$$U = 1 - \frac{\bar{x}}{\ell} = 1 - \frac{1.86}{9} = 0.7933$$

$$A_e = A_g U = 1.58(0.7933) = 1.25 \text{ in.}^2 > 1.15 \text{ in.}^2 \qquad \text{(OK)}$$

Assuming that the bottom chord is braced at the panel points, we get

$$\frac{L}{r} = \frac{5(12)}{0.566} = 106 < 300 \qquad \text{(OK)}$$

Answer   Use an MT6 × 5.4.

3.9   PIN-CONNECTED MEMBERS

When a member is to be pin-connected, a hole is made in both the member and the parts to which it is connected and a pin is placed through the holes. This provides a connection that is as moment-free as can be fabricated. Tension members connected in this manner are subject to several types of failure, which are covered in AISC D5 and D6 and discussed in the following paragraphs.

The eyebar is a special type of pin-connected member in which the end containing the pin hole is enlarged, as shown in Figure 3.37. The design strength is based on yielding of the gross section. Detailed rules for proportioning eyebars are given in AISC D6 and are not repeated here. These requirements are based on experience and test programs for forged eyebars, but they are conservative when applied to eyebars thermally cut from plates (the present fabrication method). Eyebars were widely used in the past as single tension members in bridge trusses or were linked in chainlike fashion in suspension bridges. They are rarely used today.

Pin-connected members should be designed for the following limit states (see Figure 3.38).

FIGURE 3.37

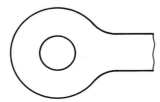

FIGURE 3.38

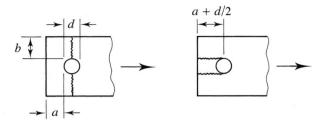

(a) Fracture of net section   **(b) Longitudinal shear**

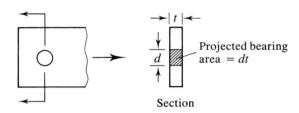

(c) Bearing

1. **Tension** on the net effective area (Figure 3.38a):

$$\phi_t = 0.75, \ \Omega_t = 2.00, \qquad P_n = 2t b_{\text{eff}} F_u \qquad \text{(AISC Equation D5-1)}$$

2. **Shear** on the effective area (Figure 3.38b):

$$\phi_{sf} = 0.75, \ \Omega_{sf} = 2.00, \qquad P_n = 0.6 F_u A_{sf} \qquad \text{(AISC Equation D5-2)}$$

3. **Bearing.** This requirement is given in Chapter J ("Connections, Joints, and Fasteners"), Section J7 (Figure 3.38c):

$$\phi = 0.75, \ \Omega = 2.00, \qquad P_n = 1.8 F_y A_{pb} \qquad \text{(AISC Equation J7-1)}$$

4. **Tension** on the gross section:

$$\phi_t = 0.90, \ \Omega_t = 1.67, \qquad P_n = F_y A_g \qquad \text{(AISC Equation D2-1)}$$

where

t = thickness of connected part

b_{eff} = $2t + 0.63 \leq b$

b = distance from edge of pin hole to edge of member, perpendicular to direction of force

A_{sf} = $2t(a + d/2)$

a = distance from edge of pin hole to edge of member, parallel to direction of force

d = pin diameter

A_{pb} = projected bearing area = dt

Additional requirements for the relative proportions of the pin and the member are covered in AISC D5.2

Problems

Tensile Strength

3.2-1   A PL $\frac{3}{8} \times 7$ tension member is connected with three 1-inch-diameter bolts, as shown in Figure P3.2-1. The steel is A36. Assume that $A_e = A_n$ and compute

a. the design strength for LRFD

b. the allowable strength for ASD

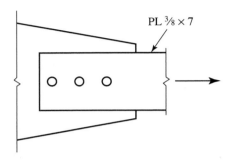

PL $\frac{3}{8} \times 7$

FIGURE P3.2-1

3.2-2   A PL $\frac{3}{8} \times 6$ tension member is welded to a gusset plate as shown in Figure P3.2-2. The steel has a yield stress $F_y = 50$ ksi and an ultimate tensile stress $F_u = 65$ ksi. Assume that $A_e = A_g$ and compute

a. the design strength for LRFD

b. the allowable strength for ASD

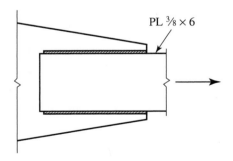

PL $\frac{3}{8} \times 6$

FIGURE P3.2-2

3.2-3   A C8 × 11.5 is connected to a gusset plate with $\frac{7}{8}$-inch-diameter bolts as shown in Figure P3.2-3. The steel is A572 Grade 50. If the member is subjected to dead load and live load only, what is the total service load capacity if the live-to-dead load ratio is 3? Assume That $A_e = 0.85A_n$.

a. Use LRFD.

b. Use ASD.

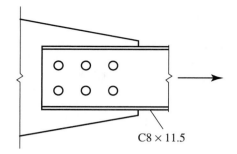

FIGURE P3.2-3

3.2-4 A PL $\frac{1}{2} \times 8$ tension member is connected with six 1-inch-diameter bolts, as shown in Figure P3.2-4. The steel is ASTM A242. Assume that $A_e = A_n$ and compute

a. the design strength for LRFD

b. the allowable strength for ASD

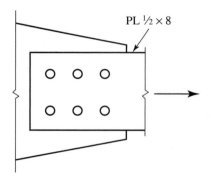

FIGURE P3.2-4

3.2-5 The tension member shown in Figure P3.2-5 must resist a service dead load of 25 kips and a service live load of 45 kips. Does the member have enough strength? The steel is A588 and the bolts are $1\frac{1}{8}$ inches in diameter. Assume that $A_e = A_n$.

a. Use LRFD.

b. Use ASD.

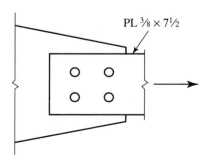

FIGURE P3.2-5

3.2-6   A double-angle tension number, 2L 3 × 2 × ¼ LLBB, of A36 steel is subjected to a dead load of 12 kips and a live load of 36 kips. It is connected to a gusset plate with one line of ¾-inch-diameter bolts through the long legs. Does this member have enough strength? Assume that $A_e = 0.85A_n$.

a. Use LRFD.

b. Use ASD.

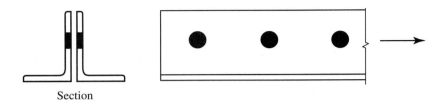

Section

FIGURE P3.2-6

Effective area

3.3-1   Determine the effective area A_e for each case shown in Figure P3.3-1.

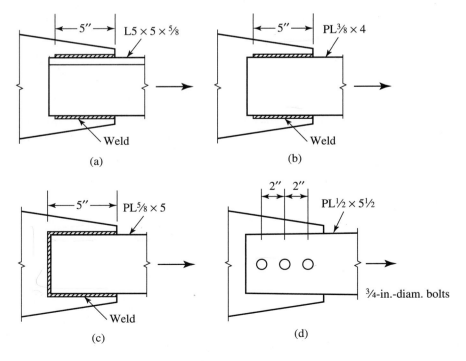

FIGURE P3.3-1

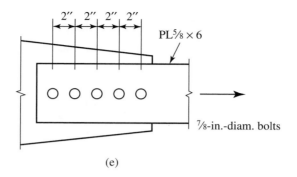

(e)

FIGURE P3.3-1 (*continued*)

3.3-2   A single-angle tension member is connected to a gusset plate, as shown in Figure P3.3-2. The yield stress is $F_y = 50$ ksi and the ultimate tensile stress is $F_u = 70$ ksi. The bolts are ⅞ inch in diameter.

a. Determine the nominal strength based on the effective area. Use Equation 3.1 for U.

b. Repeat part (a), but use the alternate value of U from AISC Table D3.1.

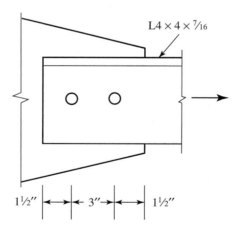

FIGURE P3.3-2

3.3-3   An L4 × 3 × ⅜ tension member is welded to a gusset plate, as shown in Figure P3.3-3. A36 steel is used. Determine the nominal strength based on the effective area. Use Equation 3.1 for U.

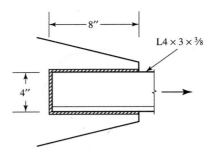

FIGURE P3.3-3

3.3-4 An L5 × 5 × ½ tension member of A242 steel is connected to a gusset plate with six ¾-inch-diameter bolts as shown in Figure P3.3-4. If the member is subject to dead load and live load only, what is the maximum total service load that can be applied if the ratio of live load to dead load is 2.0? Use the alternative value of U from AISC Table D3.1.

a. Use LRFD.

b. Use ASD.

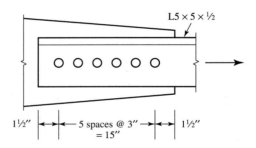

FIGURE P3.3-4

3.3-5 An L6 × 4 × ⅝ tension member of A36 steel is connected to a gusset plate with 1-inch-diameter bolts, as shown in Figure P3.3-5. It is subjected to the following service loads: dead load = 50 kips, live load = 100 kips, and wind load = 45 kips. Use Equation 3.1 for U and determine whether this member is adequate.

a. Use LRFD.

b. Use ASD.

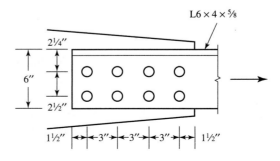

FIGURE P3.3-5

3.3-6 A PL¼ × 5 is used as a tension member and is connected by a pair of longitudinal welds along its edges. The welds are each 7 inches long. A36 steel is used.

a. What is the design strength for LRFD?

b. What is the allowable strength for ASD?

3.3-7 A W12 × 35 of A992 steel is connected through its flanges with ⅞-inch-diameter bolts, as shown in Figure P3.3-7. Use the alternative value of U from AISC Table D3.1 and compute

a. the design tensile strength

b. the allowable tensile strength

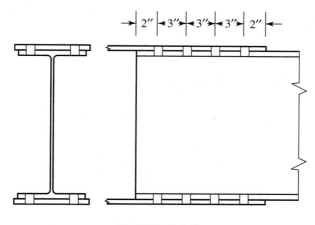

FIGURE P3.3-7

3.3-8 A WT6 × 17.5 is welded to a plate as shown in Figure P3.3-8. $F_y = 50$ ksi and $F_u = 70$ ksi. Determine whether the member can resist the following service loads: $D = 75$ kips, $L_r = 40$ kips, $S = 50$ kips, and $W = 70$ kips.

a. Use LRFD.

b. Use ASD.

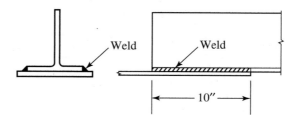

FIGURE P3.3-8

Staggered Fasteners

3.4-1 The tension member shown in Figure P3.4-1 is a ½ × 10 plate of A36 steel. The connection is with ⅞-inch-diameter bolts. Compute the nominal strength based on the net section.

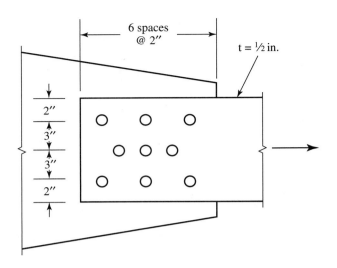

FIGURE P3.4-1

3.4-2 A tension member is composed of two $\frac{1}{2} \times 10$ plates. They are connected to a gusset plate with the gusset plate between the two tension member plates, as shown in Figure P3.4-2. A36 steel and $\frac{3}{4}$-inch-diameter bolts are used. Determine the nominal strength based on the net section.

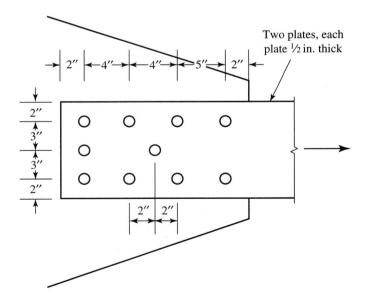

FIGURE P3.4-2

3.4-3 The tension member shown in Figure P3.4-3 is a PL$\frac{3}{8} \times 8$. The bolts are $\frac{1}{2}$ inch in diameter and A36 steel is used.

a. Compute the design strength.
b. Compute the allowable strength.

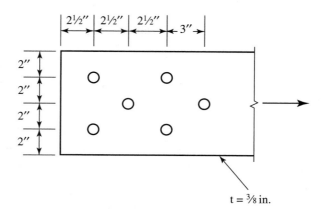

FIGURE P3.4-3

3.4-4 A C9 × 20 tension member is connected with $1\frac{1}{8}$-inch-diameter bolts, as shown in Figure P3.4-4. $F_y = 50$ ksi and $F_u = 70$ ksi. The member is subjected to the following service loads: dead load = 36 kips and live load = 110 kips. Determine whether the member has enough strength.

a. Use LRFD.
b. Use ASD.

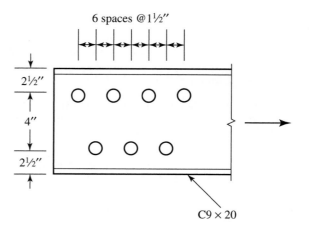

FIGURE P3.4-4

3.4-5 A double-angle shape, 2L7 × 4 × $\frac{3}{8}$, is used as a tension member. The two angles are connected to a gusset plate with $\frac{7}{8}$-inch-diameter bolts through the 7-inch legs, as shown in Figure P3.4-5. A572 Grade 50 steel is used.

a. Compute the design strength.

b. Compute the allowable strength.

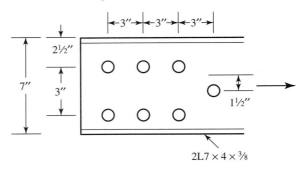

FIGURE P3.4-5

3.4-6   An L4 × 4 × $^7/_{16}$ tension member is connected with $^3/_4$-inch-diameter bolts, as shown in Figure P3.4-6. Both legs of the angle are connected. If A36 steel is used,

a. what is the design strength?

b. what is the allowable strength?

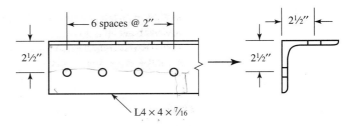

FIGURE P3.4-6

Block Shear

3.5-1   Compute the nominal block shear strength of the tension member shown in Figure P3.5-1. ASTM A572 Grade 50 steel is used. The bolts are $^7/_8$ inch in diameter.

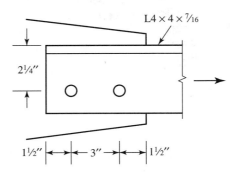

FIGURE P3.5-1

3.5-2 Determine the nominal block shear strength of the tension member shown in Figure P3.5-2. The bolts are 1 inch in diameter, and A36 steel is used.

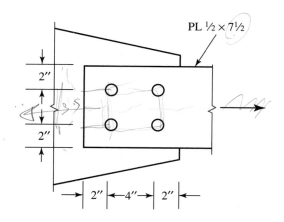

FIGURE P3.5-2

3.5-3 In the connection shown in Figure P3.5-3, the bolts are ¾–inch in diameter, and A36 steel is used for all components. Consider both the tension member and the gusset plate and compute the following:

a. the design block shear strength of the connection

b. the allowable block shear strength of the connection

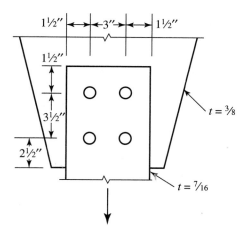

FIGURE P3.5-3

3.5-4 In the connection shown in Figure P3.5-4, ASTM A572 Grade 50 steel is used for the tension member, A36 steel is used for the gusset plate, and the holes are for ¾-inch bolts.

a. What is the maximum factored load that can be applied if LRFD is used? Consider all limit states.

b. What is the maximum total service load that can be applied if ASD is used? Consider all limit states.

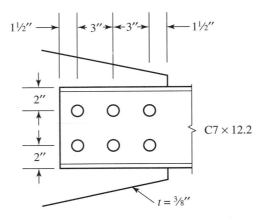

FIGURE P3.5-4

Design of Tension Members

3.6-1   Select a single-angle tension member of A36 steel to resist a dead load of 28 kips and a live load of 84 kips. The length of the member is 18 feet, and it will be connected with a single line of 1-inch-diameter bolts, as shown in Figure P3.6-1. There will be four or more bolts in this line.

a. Use LRFD.

b. Use ASD.

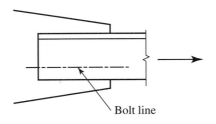

Bolt line

FIGURE P3.6-1

3.6-2   Use A36 steel and select the lightest American Standard Channel for a tension member that will safely support a service dead load of 100 kips and a service live load of 50 kips. The member is 20 feet long, and it is connected through the web with two lines of 1-inch-diameter bolts. The length of the connection is 6 inches.

a. Use LRFD.

b. Use ASD.

3.6-3 Select a double-angle tension member to resist a service dead load of 30 kips and a service live load of 90 kips. The member will be connected with two lines of $\frac{7}{8}$-inch-diameter bolts placed at the usual gage distances (see Figure 3.24). There will be more than three bolts in each line. The member is 25 feet long and will be connected to a $\frac{3}{8}$-inch-thick gusset plate. Use A572 Grade 50 steel.

   a. Use LRFD.

   b. Use ASD.

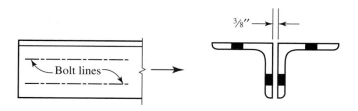

FIGURE P3.6-3

3.6-4 Select an American Standard Channel shape for the following tensile loads: dead load = 54 kips, live load = 80 kips, and wind load = 75 kips. The connection will be with longitudinal welds. Use an estimated shear lag factor of $U = 0.85$ (In a practical design, once the member was selected and the connection designed, the value of U would be computed with Equation 3.1. The member design could then be revised if necessary.) The length is 17.5 ft. Use $F_y = 50$ ksi and $F_u = 65$ ksi.

   a. Use LRFD.

   b. Use ASD.

3.6-5 Use load and resistance factor design and select an American Standard Channel shape to resist a *factored* tensile load of 180 kips. The length is 15 ft, and there will be two lines of $\frac{7}{8}$-in. diameter bolts in the web, as shown in Figure P3.6-5. Estimate the shear lag factor U to be 0.85. (In a practical design, once the member and bolt layout are selected, the value of U could be computed with Equation 3.1. The member design could then be revised if necessary.) Use A36 steel.

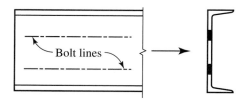

FIGURE P3.6-5

3.6-6 Use load and resistance factor design and select a W shape with a nominal depth of 10 inches (a W10) to resist a dead load of 175 kips and a live load of 175 kips.

The connection will be through the flanges with two lines of 1¼-inch-diameter bolts in each flange, as shown in Figure P3.6-6. Each line contains more than two bolts. The length of the member is 30 feet. Use A242 steel.

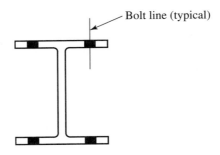

FIGURE P3.6-6

Threaded Rods and Cables

3.7-1 Select a threaded rod to resist a service dead load of 45 kips and a service live load of 5 kips. Use A36 steel.

a. Use LRFD.

b. Use ASD.

3.7-2 A W14 × 48 is supported by two tension rods *AB* and *CD*, as shown in Figure P3.7-2. The 20-kip load is a service live load. Use load and resistance factor design and select threaded rods of A36 steel for the following load cases.

a. The 20-kip load cannot move from the location shown.

b. The 20-kip load can be located anywhere between the two rods.

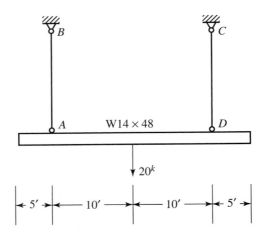

FIGURE P3.7-2

3.7-3   Same as problem 3.7-2, but use allowable *stress* design.

3.7-4   As shown in Figure P3.7-4, members *AC* and *BD* are used to brace the pin-connected structure against a horizontal wind load of 10 kips. Both of these members are assumed to be tension members and not resist any compression. For the load direction shown, member *AC* will resist the load in tension, and member *BD* will be unloaded. Select threaded rods of A36 steel for these members. Use load and resistance factor design.

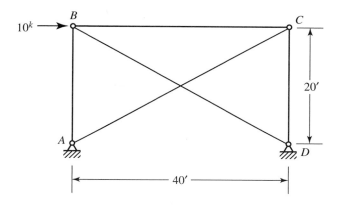

FIGURE P3.7-4

3.7-5   What size A36 threaded rod is required for member *AB*, as shown in Figure P3.7-5? The load is a service live load. (Neglect the weight of member *CB*.)

a. Use LRFD.

b. Use ASD.

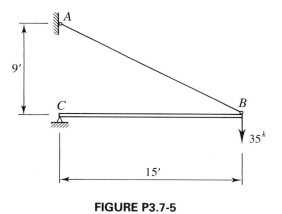

FIGURE P3.7-5

3.7-6   A pipe is supported at 10-foot intervals by a bent, threaded rod, as shown in Figure P3.7-6. If a 10-inch-diameter standard weight steel pipe full of water is used, what size A36 steel rod is required?

a. Use LRFD.

b. Use ASD.

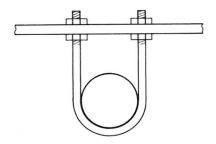

FIGURE P3.7-6

Tension Members in Roof Trusses

3.8-1 Use A36 steel and select a structural tee for the top chord of the welded roof truss shown in Figure P3.8-1. All connections are made with longitudinal plus transverse welds. Assume a connection length of 11 inches. The spacing of trusses in the roof system is 12 feet 6 inches. Design for the following loads.

> Snow: 20 psf of horizontal projection
>
> Roofing: 12 psf
>
> MC8 × 8.5 purlins
>
> Truss weight: 1000 lb (estimated)

a. Use LRFD.

b. Use ASD.

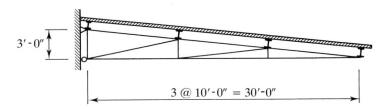

3′-0″

3 @ 10′-0″ = 30′-0″

FIGURE P3.8-1

3.8-2 Use LRFD and select single-angle shapes for the web tension members of the truss loaded as shown in Figure P3.8-2. The loads are *factored loads.* Assume that a single line of at least four ¾-inch-diameter bolts will be used for each connection. Use A572 Grade 50 steel.

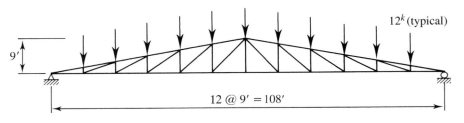

12^k (typical)

9′

12 @ 9′ = 108′

FIGURE P3.8-2

3.8-3 Compute the factored joint loads for the truss of Problem 3.8-2 for the following conditions.

> Trusses spaced at 15 ft
>
> Weight of roofing = 12 psf
>
> Snow load = 18 psf of horizontal projection
>
> $W10 \times 33$ purlins located only at the joints
>
> Total estimated truss weight = 5000 lb

3.8-4 Design the tension members of the roof truss shown in Figure P3.8-4. Use double-angle shapes throughout and assume ⅜-inch-thick gusset plates and welded connections. Assume a shear lag factor of $U = 0.85$. The trusses are spaced at 25 feet. Use A572 Grade 50 steel and design for the following loads.

> Metal deck: 4 psf of roof surface
>
> Build-up roof: 12 psf of roof surface
>
> Purlins: 6 psf of roof surface (estimated)
>
> Snow: 18 psf of horizontal projection
>
> Truss weight: 5 psf of horizontal projection (estimated)

a. Use LRFD.

b. Use ASD.

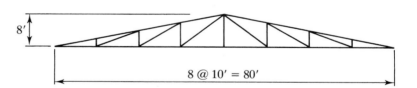

FIGURE P3.8-4

3.8-5 Use A36 steel and design sag rods for the truss of Problem 3.8-4. Assume that, once attached, the metal deck will provide lateral support for the purlins; therefore the sag rods need to be designed for the purlin weight only.

a. Use LRFD.

b. Use ASD.

4 Compression Members

4.1 DEFINITION

Compression members are structural elements that are subjected only to axial compressive forces; that is, the loads are applied along a longitudinal axis through the centroid of the member cross section, and the stress can be taken as $f = P/A$, where f is considered to be uniform over the entire cross section. This ideal state is never achieved in reality, however, because some eccentricity of the load is inevitable. Bending will result, but it can usually be regarded as secondary and can be neglected if the theoretical loading condition is closely approximated. Bending cannot be neglected if there is a *computed* bending moment. We consider situations of this type in Chapter 6, "Beam–Columns."

The most common type of compression member occurring in buildings and bridges is the *column,* a vertical member whose primary function is to support vertical loads. In many instances these members are also called upon to resist bending, and in these cases the member is a *beam–column*. Compression members are also used in trusses and as components of bracing systems. Smaller compression members not classified as columns are sometimes referred to as *struts*.

4.2 COLUMN THEORY

Consider the long, slender compression member shown in Figure 4.1a. If the axial load P is slowly applied, it will ultimately become large enough to cause the member to become unstable and assume the shape indicated by the dashed line. The member is said to have buckled, and the corresponding load is called the *critical buckling load*. If the member is stockier, as shown in Figure 4.1b, a larger load will be required to bring the member to the point of instability. For extremely stocky members, failure may occur by compressive yielding rather than buckling. Prior to failure, the compressive stress P/A will be uniform over the cross section at any point along the length, whether the failure is by yielding or by buckling. The load at which buckling occurs is a function of slenderness, and for very slender members this load could be quite small.

If the member is so slender (we give a precise definition of slenderness shortly) that the stress just before buckling is below the proportional limit — that is, the member is still elastic — the critical buckling load is given by

$$P_{cr} = \frac{\pi^2 EI}{L^2} \tag{4.1}$$

FIGURE 4.1

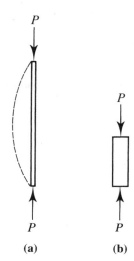

(a)   (b)

FIGURE 4.2

where E is the modulus of elasticity of the material, I is the moment of inertia of the cross-sectional area with respect to the minor principal axis, and L is the length of the member between points of support. For Equation 4.1 to be valid, the member must be elastic, and its ends must be free to rotate but not translate laterally. This end condition is satisfied by hinges or pins, as shown in Figure 4.2. This remarkable relationship was first formulated by Swiss mathematician Leonhard Euler and published in 1759. The critical load is sometimes referred to as the *Euler load* or the *Euler buckling load*. The validity of Equation 4.1 has been demonstrated convincingly by numerous tests. Its derivation is given here to illustrate the importance of the end conditions.

For convenience, in the following derivation, the member will be oriented with its longitudinal axis along the x-axis of the coordinate system given in Figure 4.3. The roller support is to be interpreted as restraining the member from translating either up or down. An axial compressive load is applied and gradually increased. If a temporary transverse load is applied so as to deflect the member into the shape indicated by the dashed line, the member will return to its original position when this temporary load is removed if the axial load is less than the critical buckling load. The critical buckling load, P_{cr}, is defined as the load that is just large enough to maintain the deflected shape when the temporary transverse load is removed.

FIGURE 4.3

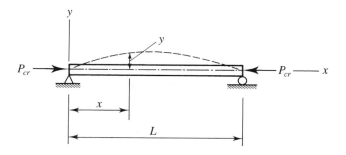

The differential equation giving the deflected shape of an elastic member subjected to bending is

$$\frac{d^2y}{dx^2} = -\frac{M}{EI} \tag{4.2}$$

where x locates a point along the longitudinal axis of the member, y is the deflection of the axis at that point, and M is the bending moment at the point. E and I were previously defined, and here the moment of inertia I is with respect to the axis of bending (buckling). This equation was derived by Jacob Bernoulli and independently by Euler, who specialized it for the column buckling problem (Timoshenko, 1953). If we begin at the point of buckling, then from Figure 4.3 the bending moment is $P_{cr}y$. Equation 4.2 can then be written as

$$y'' + \frac{P_{cr}}{EI}\, y = 0$$

where the prime denotes differentiation with respect to x. This is a second-order, linear, ordinary differential equation with constant coefficients and has the solution

$$y = A\cos(cx) + B\sin(cx)$$

where

$$c = \sqrt{\frac{P_{cr}}{EI}}$$

and A and B are constants. These constants are evaluated by applying the following boundary conditions:

At $x = 0$, $y = 0$: $0 = A\cos(0) + B\sin(0)$ $A = 0$

At $x = L$, $y = 0$: $0 = B\sin(cL)$

This last condition requires that $\sin(cL)$ be zero if B is not to be zero (the trivial solution, corresponding to $P = 0$). For $\sin(cL) = 0$,

$$cL = 0,\ \pi,\ 2\pi,\ 3\pi,\ \ldots = n\pi, \qquad n = 0, 1, 2, 3, \ldots$$

From

$$c = \sqrt{\frac{P_{cr}}{EI}}$$

we obtain

$$cL = \left(\sqrt{\frac{P_{cr}}{EI}}\right)L = n\pi, \quad \frac{P_{cr}}{EI}L^2 = n^2\pi^2 \quad \text{and} \quad P_{cr} = \frac{n^2\pi^2 EI}{L^2}$$

The various values of n correspond to different buckling modes; $n = 1$ represents the first mode, $n = 2$ the second, and so on. A value of zero gives the trivial case of no load. These buckling modes are illustrated in Figure 4.4. Values of n larger than 1 are

FIGURE 4.4

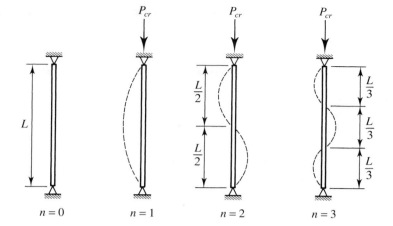

not possible unless the compression member is physically restrained from deflecting at the points where the reversal of curvature would occur.

The solution to the differential equation is therefore

$$y = B \sin\left(\frac{n\pi x}{L}\right)$$

and the coefficient B is indeterminate. This result is a consequence of approximations made in formulating the differential equation; a linear representation of a nonlinear phenomenon was used.

For the usual case of a compression member with no supports between its ends, $n = 1$ and the Euler equation is written as

$$P_{cr} = \frac{\pi^2 EI}{L^2} \tag{4.3}$$

It is convenient to rewrite Equation 4.3 as

$$P_{cr} = \frac{\pi^2 EI}{L^2} = \frac{\pi^2 EAr^2}{L^2} = \frac{\pi^2 EA}{(L/r)^2}$$

where A is the cross-sectional area and r is the radius of gyration with respect to the axis of buckling. The ratio L/r is the slenderness ratio and is the measure of a member's slenderness, with large values corresponding to slender members.

If the critical load is divided by the cross-sectional area, the critical buckling stress is obtained:

$$F_{cr} = \frac{P_{cr}}{A} = \frac{\pi^2 E}{(L/r)^2} \tag{4.4}$$

At this compressive stress, buckling will occur about the axis corresponding to r. Buckling will take place as soon as the load reaches the value given by Equation 4.3, and the column will become unstable about the principal axis corresponding to the largest slenderness ratio. This axis usually is the axis with the smaller moment of inertia (we examine exceptions to this condition later). Thus the minimum moment of inertia and radius of gyration of the cross section should ordinarily be used in Equations 4.3 and 4.4.

Example 4.1 A W12 × 50 is used as a column to support an axial compressive load of 145 kips. The length is 20 feet, and the ends are pinned. Without regard to load or resistance factors, investigate this member for stability. (The grade of steel need not be known: The critical buckling load is a function of the modulus of elasticity, not the yield stress or ultimate tensile strength.)

Solution For a W12 × 50,

$$\text{Minimum } r = r_y = 1.96 \text{ in.}$$

$$\text{Maximum } \frac{L}{r} = \frac{20(12)}{1.96} = 122.4$$

$$P_{cr} = \frac{\pi^2 EA}{(L/r)^2} = \frac{\pi^2 (29,000)(14.6)}{(122.4)^2} = 278.9 \text{ kips}$$

Answer Because the applied load of 145 kips is less than P_{cr}, the column remains stable and has an overall factor of safety against buckling of $278.9/145 = 1.92$. ∎

Early researchers soon found that Euler's equation did not give reliable results for stocky, or less slender, compression members. The reason is that the small slenderness ratio for members of this type causes a large buckling stress (from Equation 4.4). If the stress at which buckling occurs is greater than the proportional limit of the material, the relation between stress and strain is not linear, and the modulus of elasticity E can no longer be used. (In Example 4.1, the stress at buckling is $P_{cr}/A = 278.9/14.6 = 19.10$ ksi, which is well below the proportional limit for any grade of structural steel.) This difficulty was initially resolved by Friedrich Engesser, who proposed in 1889 the use of a variable tangent modulus, E_t, in Equation 4.3. For a material with a stress–strain curve like the one shown in Figure 4.5, E is not a constant for stresses greater than the proportional limit F_{pl}. The tangent modulus E_t is defined as the slope of the tangent to the stress–strain curve for values of f between F_{pl} and F_y. If the compressive stress at buckling, P_{cr}/A, is in this region, it can be shown that

$$P_{cr} = \frac{\pi^2 E_t I}{L^2} \tag{4.5}$$

Equation 4.5 is identical to the Euler equation, except that E_t is substituted for E.

The stress–strain curve shown in Figure 4.5 is different from those shown earlier for ductile steel (in Figures 1.3 and 1.4) because it has a pronounced region of nonlinearity. This curve is typical of a compression test of a short length of W-shape called a *stub column*, rather than the result of a tensile test. The nonlinearity is primarily because of the presence of residual stresses in the W-shape. When a hot-rolled shape cools after

FIGURE 4.5

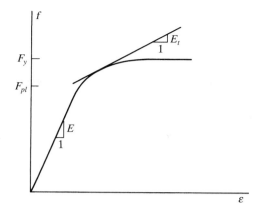

rolling, all elements of the cross section do not cool at the same rate. The tips of the flanges, for example, cool faster than the junction of the flange and the web. This uneven cooling induces stresses that remain permanently. Other factors, such as welding and cold-bending to create curvature in a beam, can contribute to the residual stress, but the cooling process is its chief source.

Note that E_t is smaller than E and for the same L/r corresponds to a smaller critical load, P_{cr}. Because of the variability of E_t, computation of P_{cr} in the inelastic range by the use of Equation 4.5 is difficult. In general, a trial-and-error approach must be used, and a compressive stress–strain curve such as the one shown in Figure 4.5 must be used to determine E_t for trial values of P_{cr}. For this reason, most design specifications, including the AISC Specification, contain empirical formulas for inelastic columns.

Engesser's tangent modulus theory had its detractors, who pointed out several inconsistencies. Engesser was convinced by their arguments, and in 1895 he refined his theory to incorporate a reduced modulus, which has a value between E and E_t. Test results, however, always agreed more closely with the tangent modulus theory. Shanley (1947) resolved the apparent inconsistencies in the original theory, and today the tangent modulus formula, Equation 4.5, is accepted as the correct one for inelastic buckling. Although the load predicted by this equation is actually a lower bound on the true value of the critical load, the difference is slight (Bleich, 1952).

For any material, the critical buckling stress can be plotted as a function of slenderness, as shown in Figure 4.6. The tangent modulus curve is tangent to the Euler curve at the point corresponding to the proportional limit of the material. The composite curve, called a *column strength curve,* completely describes the strength of any column of a given material. Other than F_y, E, and E_t, which are properties of the material, the strength is a function only of the slenderness ratio.

Effective Length

Both the Euler and tangent modulus equations are based on the following assumptions:

1. The column is perfectly straight, with no initial crookedness.
2. The load is axial, with no eccentricity.
3. The column is pinned at both ends.

FIGURE 4.6

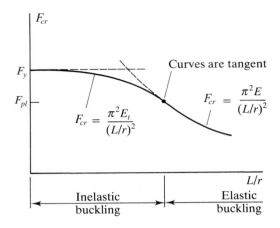

The first two conditions mean that there is no bending moment in the member before buckling. As mentioned previously, some accidental moment will be present, but in most cases it can be ignored. The requirement for pinned ends, however, is a serious limitation, and provisions must be made for other support conditions. The pinned-end condition requires that the member be restrained from lateral translation, but not rotation, at the ends. Constructing a frictionless pin connection is virtually impossible, so even this support condition can only be closely approximated at best. Obviously, all columns must be free to deform axially.

Other end conditions can be accounted for in the derivation of Equation 4.3. In general, the bending moment will be a function of x, resulting in a nonhomogeneous differential equation. The boundary conditions will be different from those in the original derivation, but the overall procedure will be the same. The form of the resulting equation for P_{cr} will also be the same. For example, consider a compression member pinned

FIGURE 4.7

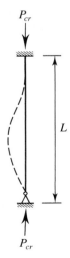

at one end and fixed against rotation and translation at the other, as shown in Figure 4.7. The Euler equation for this case, derived in the same manner as Equation 4.3, is

$$P_{cr} = \frac{2.05\pi^2 EI}{L^2}$$

or

$$P_{cr} = \frac{2.05\pi^2 EA}{(L/r)^2} = \frac{\pi^2 EA}{(0.70L/r)^2}$$

Thus this compression member has the same load capacity as a column that is pinned at both ends and is only 70% as long as the given column. Similar expressions can be found for columns with other end conditions.

The column buckling problem can also be formulated in terms of a fourth-order differential equation instead of Equation 4.2. This proves to be convenient when dealing with boundary conditions other than pinned ends.

For convenience, the equations for critical buckling load will be written as

$$P_{cr} = \frac{\pi^2 EA}{(KL/r)^2} \quad \text{or} \quad P_{cr} = \frac{\pi^2 E_t A}{(KL/r)^2} \tag{4.6a/4.6b}$$

where KL is the *effective length,* and K is the *effective length factor.* The effective length factor for the fixed-pinned compression member is 0.70. For the most favorable condition of both ends fixed against rotation and translation, $K = 0.5$. Values of K for these and other cases can be determined with the aid of Table C-C2.2 in the Commentary to the AISC Specification. The three conditions mentioned thus far are included, as well as some for which end translation is possible. Two values of K are given: a theoretical value and a recommended design value to be used when the ideal end condition is approximated. Hence, unless a "fixed" end is perfectly fixed, the more conservative design values are to be used. Only under the most extraordinary circumstances would the use of the theoretical values be justified. Note, however, that the theoretical and recommended design values are the same for conditions (d) and (f) in Commentary Table C-C2.2. The reason is that any deviation from a perfectly frictionless hinge or pin introduces rotational restraint and tends to reduce K. Therefore, use of the theoretical values in these two cases is conservative.

The use of the effective length KL in place of the actual length L in no way alters any of the relationships discussed so far. The column strength curve shown in Figure 4.6 is unchanged except for renaming the abscissa KL/r. The critical buckling stress corresponding to a given length, actual or effective, remains the same.

4.3 AISC REQUIREMENTS

The basic requirements for compression members are covered in Chapter E of the AISC Specification. The nominal compressive strength is

$$P_n = F_{cr}A_g \qquad \text{(AISC Equation E3-1)}$$

For LRFD,

$$P_u \le \phi_c P_n$$

where

P_u = sum of the factored loads
ϕ_c = resistance factor for compression = 0.90
$\phi_c P_n$ = design compressive strength

For ASD,

$$P_a \le \frac{P_n}{\Omega_c}$$

where

P_a = sum of the service loads
Ω_c = safety factor for compression = 1.67
P_n/Ω_c = allowable compressive strength

If an allowable stress formulation is used,

$$f_a \le F_a$$

where

f_a = computed axial compressive stress = P_a/A_g
F_a = allowable axial compressive stress

$$= \frac{F_{cr}}{\Omega_c} = \frac{F_{cr}}{1.67} = 0.6 F_{cr} \tag{4.7}$$

In order to present the AISC expressions for the critical stress F_{cr}, we first define the Euler load as

$$P_e = \frac{\pi^2 EA}{(KL/r)^2}$$

This is the critical buckling load according to the Euler equation. The Euler stress is

$$F_e = \frac{P_e}{A} = \frac{\pi^2 E}{(KL/r)^2} \qquad \text{(AISC Equation E3-4)}$$

With a slight modification, this expression will be used for the critical stress in the elastic range. To obtain the critical stress for elastic columns, the Euler stress is reduced as follows to account for the effects of initial crookedness:

$$F_{cr} = 0.877 F_e \tag{4.8}$$

For inelastic columns, the tangent modulus equation, Equation 4.6b, is replaced by the exponential equation

$$F_{cr} = \left(0.658^{\frac{F_y}{F_e}} \right) F_y \tag{4.9}$$

With Equation 4.9, a direct solution for inelastic columns can be obtained, avoiding the trial-and-error approach inherent in the use of the tangent modulus equation. At the boundary between inelastic and elastic columns, Equations 4.8 and 4.9 give the same value of F_{cr}. This occurs when KL/r is approximately

$$4.71\sqrt{\frac{E}{F_y}}$$

To summarize,

$$\text{When } \frac{KL}{r} \leq 4.71\sqrt{\frac{E}{F_y}}, \quad F_{cr} = (0.658^{F_y/F_e})F_y \tag{4.10}$$

$$\text{When } \frac{KL}{r} > 4.71\sqrt{\frac{E}{F_y}}, \quad F_{cr} = 0.877F_e \tag{4.11}$$

The AISC Specification provides for separating inelastic and elastic behavior based on either the value of KL/r (as in equations 4.10 and 4.11) or the value of F_e. The limiting value of F_e can be derived as follows. From AISC Equation E3-4,

$$\frac{KL}{r} = \sqrt{\frac{\pi^2 E}{F_e}}$$

For $\quad \dfrac{KL}{r} \leq 4.71\sqrt{\dfrac{E}{F_y}}$,

$$\sqrt{\frac{\pi^2 E}{F_e}} \leq 4.71\sqrt{\frac{E}{F_y}}$$
$$F_e \geq 0.44F_y$$

The complete AISC Specification for compressive strength is as follows:

$$\text{When } \frac{KL}{r} \leq 4.71\sqrt{\frac{E}{F_y}} \quad \text{or} \quad F_e \geq 0.44F_y,$$
$$F_{cr} = (0.658^{F_y/F_e})F_y$$

(AISC Equation E3-2)

$$\text{When } \frac{KL}{r} > 4.71\sqrt{\frac{E}{F_y}} \quad \text{or} \quad F_e \geq 0.44F_y, a$$
$$F_{cr} = 0.877F_e$$

(AISC Equation E3-3)

FIGURE 4.8

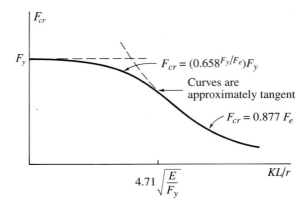

In this book, we will usually use the limit on KL/r, as expressed in Equations 4.10 and 4.11.

These requirements are represented graphically in Figure 4.8.

AISC Equations E3-2 and E3-3 are a condensed version of five equations that cover five ranges of KL/r (Galambos, 1988). These equations are based on experimental and theoretical studies that account for the effects of residual stresses and an initial out-of-straightness of $L/1500$, where L is the member length. A complete derivation of these equations is given by Tide (2001).

Although AISC does not require an upper limit on the slenderness ratio KL/r, an upper limit of 200 is recommended (see user note in AISC E2). This is a practical upper limit, because compression members that are any more slender will have little strength and will not be economical.

Example 4.2   A W14 × 74 of A992 steel has a length of 20 feet and pinned ends. Compute the design compressive strength for LRFD and the allowable compressive strength for ASD.

Solution   Slenderness ratio:

$$\text{Maximum } \frac{KL}{r} = \frac{KL}{r_y} = \frac{1.0(20 \times 12)}{2.48} = 96.77 < 200 \quad \text{(OK)}$$

$$4.71\sqrt{\frac{E}{F_y}} = 4.71\sqrt{\frac{29,000}{50}} = 113$$

Since $96.77 < 113$, use AISC Equation E3-2.

$$F_e = \frac{\pi^2 E}{(KL/r)^2} = \frac{\pi^2(29,000)}{(96.77)^2} = 30.56 \text{ ksi}$$

$$F_{cr} = 0.658^{(F_y/F_e)}F_y = 0.658^{(50/30.56)}(50) = 25.21 \text{ ksi}$$

The nominal strength is

$$P_n = F_{cr}A_g = 25.21(21.8) = 549.6 \text{ kips}$$

LRFD Solution   The design strength is

$$\phi_c P_n = 0.90(549.6) = 495 \text{ kips}$$

ASD Solution   From Equation 4.7, the allowable stress is

$$F_a = 0.6F_{cr} = 0.6(25.21) = 15.13 \text{ ksi}$$

The allowable strength is

$$F_a A_g = 15.13(21.8) = 330 \text{ kips}$$

Answer   Design compressive strength = 495 kips. Allowable compressive strength = 330 kips.

In Example 4.2, $r_y < r_x$, and there is excess strength in the x-direction. Square structural tubes (HSS) are efficient shapes for compression members because $r_y = r_x$ and the strength is the same for both axes. Hollow circular shapes are sometimes used as compression members for the same reason.

The mode of failure considered so far is referred to as *flexural* buckling, as the member is subjected to flexure, or bending, when it becomes unstable. For some cross-sectional configurations, the member will fail by twisting (torsional buckling) or by a combination of twisting and bending (flexural-torsional buckling). We consider these infrequent cases in Section 4.8.

4.4   LOCAL STABILITY

The strength corresponding to any buckling mode cannot be developed, however, if the elements of the cross section are so thin that *local* buckling occurs. This type of instability is a localized buckling or wrinkling at an isolated location. If it occurs, the cross section is no longer fully effective, and the member has failed. I- and H-shaped cross sections with thin flanges or webs are susceptible to this phenomenon, and their use should be avoided whenever possible. Otherwise, the compressive strength given by AISC Equations E3-2 and E3-3 must be reduced. The measure of this susceptibility is the width–thickness ratio of each cross-sectional element. Two types of elements must be considered: unstiffened elements, which are unsupported along one edge parallel to the direction of load, and stiffened elements, which are supported along both edges.

Limiting values of width–thickness ratios are given in AISC B4, "Classification of Sections for Local Buckling," where cross-sectional shapes are classified as *compact, noncompact,* or *slender,* according to the values of the ratios. For uniformly compressed elements, as in an axially loaded compression member, the strength must be reduced if the shape has any slender elements. The width–thickness ratio is given the generic name of λ. Depending on the particular cross-sectional element, λ for I- and H-shapes is either the ratio b/t or h/t_w, both of which are defined presently. If λ is greater than the specified limit, denoted λ_r, the shape is slender, and the potential for local buckling must be accounted for. (We postpone a discussion of the compact and noncompact categories until Chapter 5, "Beams.") For I- and H-shapes, the projecting flange is considered to be an unstiffened element, and its width can be taken as half the full nominal width. Using AISC notation gives

$$\lambda = \frac{b}{t} = \frac{b_f/2}{t_f} = \frac{b_f}{2t_f}$$

where b_f and t_f are the width and thickness of the flange. The upper limit is

$$\lambda_r = 0.56\sqrt{\frac{E}{F_y}}$$

The webs of I- and H-shapes are stiffened elements, and the stiffened width is the distance between the roots of the flanges. The width–thickness parameter is

$$\lambda = \frac{h}{t_w}$$

FIGURE 4.9

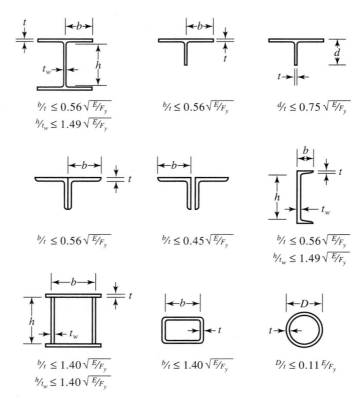

where h is the distance between the roots of the flanges, and t_w is the web thickness. The upper limit is

$$\lambda_r = 1.49 \frac{E}{\sqrt{F_y}}$$

Stiffened and unstiffened elements of various cross-sectional shapes are illustrated in Figure 4.9. The appropriate compression member limit, λ_r, from AISC B4 is given for each case.

Example 4.3   Investigate the column of Example 4.2 for local stability.

Solution   For a W14×74, $b_f = 10.1$ in., $t_f = 0.785$ in., and

$$\frac{b_f}{2t_f} = \frac{10.1}{2(0.785)} = 6.43$$

$$0.56\sqrt{\frac{E}{F_y}} = 0.56\sqrt{\frac{29,000}{50}} = 13.5 > 6.43 \qquad \text{(OK)}$$

$$\frac{h}{t_w} = \frac{d - 2k_{des}}{t_w} = \frac{14.2 - 2(1.38)}{0.450} = 25.4$$

where k_{des} is the *design* value of k. (Different manufacturers will produce this shape with different values of k. The *design* value is the smallest of these values. The *detailing* value is the largest.)

$$1.49\sqrt{\frac{E}{F_y}} = 1.49\sqrt{\frac{29,000}{50}} = 35.9 > 25.4 \qquad \text{(OK)}$$

Answer   Local instability is not a problem.   ◼

In Example 4.3, the width-thickness ratios $b_f/2t_f$ and h/t_w were computed. This is not necessary, however, because these ratios are tabulated in the dimensions and properties table. In addition, shapes that are slender for compression are indicated with a footnote (footnote c).

It is permissible to use a cross-sectional shape that does not satisfy the width-thickness ratio requirements, but such a member may not be permitted to carry as large a load as one that does satisfy the requirements. In other words, the strength could be reduced because of local buckling. The overall procedure for making this investigation is as follows.

- If the width-thickness ratio λ is greater than λ_r, use the provisions of AISC E7 and compute a reduction factor Q.
- Compute KL/r and F_e as usual.
- If $\dfrac{KL}{r} \leq 4.71 \sqrt{\dfrac{E}{QF_y}}$ or $F_e \geq 0.44QF_y$,

$$F_{cr} = Q\left(0.658^{\frac{QF_y}{F_e}}\right)F_y \qquad \text{(AISC Equation E7-2)}$$

- If $\dfrac{KL}{r} > 4.71 \sqrt{\dfrac{E}{QF_y}}$ or $F_e < 0.44QF_y$,

$$F_{cr} = 0.877F_e \qquad \text{(AISC Equation E7-3)}$$

- The nominal strength is $P_n = F_{cr}A_g$ (AISC Equation E7-1)

The reduction factor Q is the product of two factors—Q_s for unstiffened elements and Q_a for stiffened elements. If the shape has no slender unstiffened elements, $Q_s = 1.0$. If the shape has no slender stiffened elements, $Q_a = 1.0$.

Many of the shapes commonly used as columns are not slender, and the reduction will not be needed. This includes most (but not all) W-shapes. However, a large number of hollow structural shapes (HSS), double angles, and tees have slender elements.

AISC Specification Section E7.1 gives the procedure for calculating Q_s for slender unstiffened elements. The procedure is straightforward, and involves comparing the width-thickness ratio with a limiting value and then computing Q_s from an expression that is a function of the width-thickness ratio, F_y, and E.

The computation of Q_a for slender stiffened elements is given in AISC E7.2 and is slightly more complicated than the procedure for unstiffened elements. The general procedure is as follows.

- Compute an effective area of the cross section. This requires a knowledge of the stress in the effective area, so iteration is required. The Specification allows a simplifying assumption, however, so iteration can be avoided.
- Compute $Q_a = A_{eff}/A$, where A_{eff} is the effective area and A is the actual area.

The details of the computation of Q_s and Q_a will not be given here but will be shown in the following example, which illustrates the procedure for an HSS.

Example 4.4   Determine the axial compressive strength of an HSS $8 \times 4 \times \frac{1}{8}$ with an effective length of 15 feet with respect to each principal axis. Use $F_y = 46$ ksi.

Solution   Compute the overall, or flexural, buckling strength.

$$\text{Maximum } \frac{KL}{r} = \frac{KL}{r_y} = \frac{15 \times 12}{1.71} = 105.3 < 200 \qquad \text{(OK)}$$

$$4.71\sqrt{\frac{E}{F_y}} = 4.71\sqrt{\frac{29,000}{46}} = 118$$

Since $105.3 < 118$, use AISC Equation E3-2.

$$F_e = \frac{\pi^2 E}{(KL/r)^2} = \frac{\pi^2 (29,000)}{(105.3)^2} = 25.81 \text{ ksi}$$

$$F_{cr} = 0.658^{(F_y/F_e)} F_y = 0.658^{(46/25.81)}(46) = 21.82 \text{ ksi}$$

The nominal strength is

$$P_n = F_{cr}A_g = 21.82(2.70) = 58.91 \text{ kips}$$

Check width-thickness ratios:

From the dimensions and properties table in the *Manual,* the width-thickness ratio for the larger overall dimension is

$$\frac{h}{t} = 66.0$$

The ratio for the smaller dimension is

$$\frac{b}{t} = 31.5$$

From AISC Table B4.1, case 12 (and Figure 4.9 in this book), the upper limit for nonslender elements is

$$1.40\sqrt{\frac{E}{F_y}} = 1.40\sqrt{\frac{29,000}{46}} = 35.15$$

Since $h/t > 1.40\sqrt{E/F_y}$, the larger dimension element is slender and the local buckling strength must be computed. (Although the limiting width-thickness ratio is labeled b/t in the table, that is a generic notation, and it applies to h/t as well.)

Because this cross-sectional element is a stiffened element, $Q_s = 1.0$, and Q_a must be computed from AISC Section E7.2. The shape is a rectangular section of uniform thickness, so AISC E7.2(b) applies, provided that

$$\frac{b}{t} \geq 1.40\sqrt{\frac{E}{f}},$$

where

$$f = \frac{P_n}{A_{eff}}$$

and A_{eff} is the reduced effective area. The Specification user note for square and rectangular sections permits a value of $f = F_y$ to be used in lieu of determining f by iteration. From AISC Equation E7-18, the effective width of the slender element is

$$b_e = 1.92t\sqrt{\frac{E}{f}}\left[1 - \frac{0.38}{b/t}\sqrt{\frac{E}{f}}\right] \le b \qquad \text{(AISC Equation E7-18)}$$

For the 8-inch side, using $f = F_y$ and the *design* thickness from the dimensions and properties table,

$$b_e = 1.92(0.116)\sqrt{\frac{29,000}{46}}\left[1 - \frac{0.38}{(66.0)}\sqrt{\frac{29,000}{46}}\right] = 4.784 \text{ in.}$$

From AISC B4.2(d) and the discussion in Part 1 of the *Manual,* the unreduced length of the 8-inch side between the corner radii can be taken as

$$b = 8 - 2(1.5t) = 8 - 2(1.5)(0.116) = 7.652 \text{ in.}$$

where the corner radius is taken as 1.5 times the design thickness.

The total loss in area is therefore

$$2(b - b_e)t = 2(7.652 - 4.784)(0.116) = 0.6654 \text{ in.}^2$$

and the reduced area is

$$A_{eff} = 2.70 - 0.6654 = 2.035 \text{ in.}^2$$

The reduction factor is

$$Q_a = \frac{A_{eff}}{A} = \frac{2.035}{2.70} = 0.7537$$

$$Q = Q_s Q_a = 1.0(0.7537) = 0.7537$$

Compute the local buckling strength.

$$4.71\sqrt{\frac{E}{QF_y}} = 4.71\sqrt{\frac{29,000}{0.7537(46)}} = 136.2$$

$$\frac{KL}{r} = 105.3 < 136.2 \qquad \therefore \text{ use AISC Equation E7-2.}$$

$$F_{cr} = Q\left(0.658^{\frac{QF_y}{F_e}}\right)F_y = 0.7537\left(0.658^{\frac{0.7537(46)}{25.81}}\right)46 = 19.76 \text{ ksi}$$

$$P_n = F_{cr}A_g = 19.76(2.70) = 53.35 \text{ kips}$$

Since this is less than the flexural buckling strength of 58.91 kips, local buckling controls.

LRFD Solution   Design strength $= \phi_c P_n = 0.90(53.35) = 48.0$ kips

ASD Solution   Allowable strength $= \dfrac{P_n}{\Omega} = \dfrac{53.35}{1.67} = 32.0$ kips

(Allowable stress $= 0.6F_{cr} = 0.6(19.76) = 11.9$ ksi)

Alternative Solution with f determined by Iteration   As an initial trial value use

$f = F_{cr} = 19.76$ ksi (the value obtained above after using an initial value of $f = F_y$)

$$b_e = 1.92(0.116)\sqrt{\frac{29,000}{19.76}}\left[1 - \frac{0.38}{(66.0)}\sqrt{\frac{29,000}{19.76}}\right] = 6.65 \text{ in.}$$

The total loss in area is

$$2(b - b_e)t = 2(7.652 - 6.65)(0.116) = 0.2325 \text{ in.}^2$$

and the reduced area is

$$A_{eff} = 2.70 - 0.2325 = 2.468 \text{ in.}^2$$

The reduction factor is

$$Q_a = \frac{A_{eff}}{A} = \frac{2.468}{2.70} = 0.9141$$

$$Q = Q_s Q_a = 1.0(0.9141) = 0.9141$$

Compute the local buckling strength.

$$4.71\sqrt{\frac{E}{QF_y}} = 4.71\sqrt{\frac{29,000}{0.9141(46)}} = 123.7$$

$$\frac{KL}{r} = 105.3 < 123.7 \qquad \therefore \text{ use AISC Equation E7-2.}$$

$$F_{cr} = Q\left(0.658^{\frac{QF_y}{F_e}}\right)F_y$$

$$= 0.9141\left(0.658^{\frac{0.9141(46)}{25.81}}\right)46 = 21.26 \text{ ksi} \neq 19.76 \text{ ksi (the assumed value)}$$

Try $f = 21.26$ ksi:

$$b_e = 1.92(0.116)\sqrt{\frac{29{,}000}{21.26}}\left[1 - \frac{0.38}{(66.0)}\sqrt{\frac{29{,}000}{21.26}}\right] = 6.477 \text{ in.}$$

The total loss in area is

$$2(b - b_e)t = 2(7.652 - 6.477)(0.116) = 0.2726 \text{ in.}^2$$

and the reduced area is

$$A_{eff} = 2.70 - 0.2726 = 2.427 \text{ in.}^2$$

The reduction factor is

$$Q_a = \frac{A_{eff}}{A} = \frac{2.427}{2.70} = 0.8989$$

$$Q = Q_s Q_a = 1.0(0.8989) = 0.8989$$

Compute the local buckling strength.

$$4.71\sqrt{\frac{E}{QF_y}} = 4.71\sqrt{\frac{29{,}000}{0.8989(46)}} = 124.7$$

$$\frac{KL}{r} = 105.3 < 124.7 \qquad \therefore \text{ use AISC Equation E7-2.}$$

$$F_{cr} = Q\left(0.658^{\frac{QF_y}{F_e}}\right)F_y = 0.8989\left(0.658^{\frac{0.8989(46)}{25.81}}\right)46$$

$$= 21.15 \text{ ksi} \neq 21.26 \text{ ksi}$$

Try $f = 21.15$ ksi:

$$b_e = 1.92(0.116)\sqrt{\frac{29{,}000}{21.15}}\left[1 - \frac{0.38}{(66.0)}\sqrt{\frac{29{,}000}{21.15}}\right] = 6.489 \text{ in.}$$

The total loss in area is

$$2(b - b_e)t = 2(7.652 - 6.489)(0.116) = 0.2698 \text{ in.}^2$$

and the reduced area is

$$A_{eff} = 2.70 - 0.2698 = 2.430 \text{ in.}^2$$

The reduction factor is

$$Q_a = \frac{A_{eff}}{A} = \frac{2.430}{2.70} = 0.9000$$

$$Q = Q_s Q_a = 1.0(0.9000) = 0.9000$$

Compute the local buckling strength.

$$4.71\sqrt{\frac{E}{QF_y}} = 4.71\sqrt{\frac{29,000}{0.9000(46)}} = 124.7$$

$$\frac{KL}{r} = 105.3 < 124.7 \qquad \therefore \text{use AISC Equation E7-2.}$$

$$F_{cr} = Q\left(0.658^{\frac{QF_y}{F_e}}\right)F_y$$

$$= 0.9000\left(0.658^{\frac{0.9000(46)}{25.81}}\right)46 = 21.16 \text{ ksi} \approx 21.15 \text{ ksi (convergence)}$$

Recall that AISC Equation E7-18 for b_e applies when $b/t \geq 1.40\sqrt{E/f}$. In the present case,

$$1.40\sqrt{\frac{E}{f}} = 1.40\sqrt{\frac{29,000}{21.16}} = 51.8$$

Since $66 > 51.8$, AISC Equation E7-18 does apply.

$$P_n = F_{cr}A_g = 21.16(2.70) = 57.13 \text{ kips} \qquad \therefore \text{local buckling controls.}$$

LRFD Solution   Design strength $= \phi_c P_n = 0.90(57.13) = 51.4$ kips

ASD Solution   Allowable strength $\dfrac{P_n}{\Omega} = \dfrac{57.13}{1.67} = 34.2$ kips

(Allowable stress $= 0.6F_{cr} = 0.6(21.16) = 12.7$ ksi)

4.5   TABLES FOR COMPRESSION MEMBERS

The *Manual* contains many useful tables for analysis and design. For compression members whose strength is governed by flexural buckling (that is, not local buckling), Table 4-22 in Part 4 of the *Manual,* "Design of Compression Members," can be used. This table gives values of $\phi_c F_{cr}$ (for LRFD) and F_{cr}/Ω_c (for ASD) as a function of KL/r for various values of F_y. This table stops at the recommended upper limit of $KL/r = 200$. The available strength tables, however, are the most useful. These tables, which we will refer to as the "column load tables," give the available strengths of selected shapes, both $\phi_c P_n$ for LRFD and P_n/Ω_c for ASD, as a function of the effective length KL. These tables include values of KL up to those corresponding to $KL/r = 200$.

The use of the tables is illustrated in the following example.

Example 4.5   Compute the available strength of the compression member of Example 4.2 with the aid of (a) Table 4-22 and (b) the column load tables.

LRFD Solution
a. From Example 4.2, $KL/r = 96.77$ and $F_y = 50$ ksi. Values of $\phi_c F_{cr}$ in Table 4-22 are given only for integer values of KL/r; for decimal values, KL/r may be rounded *up* or linear interpolation may be used. For uniformity, we use interpolation in this book for all tables unless otherwise indicated. For $KL/r = 96.77$ and $F_y = 50$ ksi,

$$\phi_c F_{cr} = 22.67 \text{ ksi}$$
$$\phi_c P_n = \phi_c F_{cr} A_g = 22.67(21.8) = 494 \text{ kips}$$

b. The column load tables in Part 4 of the *Manual* give the available strength for selected W-, HP-, single-angle, WT-, HSS, pipe, double-angle, and composite shapes. (We cover composite construction in Chapter 9.) The tabular values for the symmetrical shapes (W, HP, HSS and pipe) were calculated by using the minimum radius of gyration for each shape. From Example 4.2, $K = 1.0$, so

$$KL = 1.0(20) = 20 \text{ ft}$$

For a W14 × 74, $F_y = 50$ ksi and $KL = 20$ ft,

$$\phi_c P_n = 494 \text{ kips}$$

ASD Solution
a. From Example 4.2, $KL/r = 96.77$ and $F_y = 50$ ksi. By interpolation, for $KL/r = 96.77$ and $F_y = 50$ ksi,

$$F_{cr}/\Omega_c = 15.07 \text{ ksi}$$

Note that this is the allowable stress, $F_a = 0.6F_{cr}$. Therefore, the allowable strength is

$$\frac{P_n}{\Omega_c} = F_a A_g = 15.07(21.8) = 329 \text{ kips}$$

b. From Example 4.2, $K = 1.0$, so

$$KL = 1.0(20) = 20 \text{ ft}$$

From the column load tables, for a W14 × 74 with $F_y = 50$ ksi and $KL = 20$ ft,

$$\frac{P_n}{\Omega_c} = 329 \text{ kips}$$

The values from Table 4-22 are based on flexural buckling and AISC Equations E3-2 and E3-3. Thus, local stability is assumed, and width-thickness ratio limits must not be exceeded. Although some shapes in the column load tables exceed those limits (and they are identified with a "c" footnote), the tabulated strength has been computed according to the requirements of AISC Section E7, "Members with Slender Elements," and no further reduction is needed.

From a practical standpoint, if a compression member to be analyzed can be found in the column load tables, then these tables should be used. Otherwise, Table 4-22 can be used for the flexural buckling strength. If the member has slender elements, the local buckling strength must be computed using the provisions of AISC E7.

4.6 DESIGN

The selection of an economical rolled shape to resist a given compressive load is simple with the aid of the column load tables. Enter the table with the effective length and move horizontally until you find the desired available strength (or something slightly larger). In some cases, you must continue the search to be certain that you have found the lightest shape. Usually the category of shape (W, WT, etc.) will have been decided upon in advance. Often the overall nominal dimensions will also be known because of architectural or other requirements. As pointed out earlier, all tabulated values correspond to a slenderness ratio of 200 or less. The tabulated unsymmetrical shapes — the structural tees and the single and double angles — require special consideration and are covered in Section 4.8.

Example 4.6 A compression member is subjected to service loads of 165 kips dead load and 535 kips live load. The member is 26 feet long and pinned at each end. Use A992 steel and select a W14 shape.

LRFD Solution Calculate the factored load:

$$P_u = 1.2D + 1.6L = 1.2(165) + 1.6(535) = 1054 \text{ kips}$$
$$\therefore \text{ Required design strength } \phi_c P_n = 1054 \text{ kips.}$$

From the column load tables for $KL = 1.0(26) = 26$ ft, a W14 × 145 has a design strength of 1230 kips.

Answer Use a W14 × 145.

ASD Solution Calculate the total applied load:

$$P_a = D + L = 165 + 535 = 700 \text{ kips}$$
$$\therefore \text{ Required allowable strength } \frac{P_n}{\Omega_c} = 700 \text{ kips}$$

From the column load tables for $KL = 1.0(26) = 26$ ft, a W14 × 132 has an allowable strength of 702 kips.

Answer Use a W14 × 132.

Example 4.7 Select the lightest W-shape that can resist a service dead load of 62.5 kips and a service live load of 125 kips. The effective length is 24 feet. Use ASTM A992 steel.

Solution The appropriate strategy here is to find the lightest shape for each nominal depth in the column load tables and then choose the lightest overall.

LRFD Solution The factored load is

$$P_u = 1.2D + 1.6L = 1.2(62.5) + 1.6(125) = 275 \text{ kips}$$

From the column load tables, the choices are as follows:

W8: There are no W8s with $\phi_c P_n \geq 275$ kips.
W10: W10×54, $\phi_c P_n = 282$ kips
W12: W12×58, $\phi_c P_n = 293$ kips
W14: W14×61, $\phi_c P_n = 293$ kips

Note that the strength is not proportional to the weight (which is a function of the cross-sectional area).

Answer Use a W10×54

ASD Solution The total applied load is

$$P_a = D + L = 62.5 + 125 = 188 \text{ kips}$$

From the column load tables, the choices are as follows:

W8: There are no W8s with $P_n/\Omega_c \geq 188$ kips.

W10: W10×54, $\dfrac{P_n}{\Omega_c} = 188$ kips

W12: W12×58, $\dfrac{P_n}{\Omega_c} = 195$ kips

W14: W14×61, $\dfrac{P_n}{\Omega_c} = 195$ kips

Note that the strength is not proportional to the weight (which is a function of the cross-sectional area).

Answer Use a W10 × 54.

For shapes not in the column load tables, a trial-and-error approach must be used. The general procedure is to assume a shape and then compute its strength. If the strength is too small (unsafe) or too large (uneconomical), another trial must be made. A systematic approach to making the trial selection is as follows:

1. Assume a value for the critical buckling stress F_{cr}. Examination of AISC Equations E3-2 and E3-3 shows that the theoretically maximum value of F_{cr} is the yield stress F_y.
2. Determine the required area. For LRFD,

$$\phi_c F_{cr} A_g \geq P_u$$

$$A_g \geq \frac{P_u}{\phi_c F_{cr}}$$

For ASD,

$$0.6F_{cr} \geq \frac{P_a}{A_g}$$

$$A_g \geq \frac{P_a}{0.6F_{cr}}$$

3. Select a shape that satisfies the area requirement.
4. Compute F_{cr} and the strength for the trial shape.
5. Revise if necessary. If the available strength is very close to the required value, the next tabulated size can be tried. Otherwise, repeat the entire procedure, using the value of F_{cr} found for the current trial shape as a value for Step 1.
6. Check local stability (check width–thickness ratios). Revise if necessary.

Example 4.8   Select a W18 shape of A992 steel that can resist a service dead load of 100 kips and a service live load of 300 kips. The effective length KL is 26 feet.

LRFD Solution $P_u = 1.2D + 1.6L = 1.2(100) + 1.6(300) = 600$ kips
Try $F_{cr} = 33$ ksi (an arbitrary choice of two-thirds F_y):

$$\text{Required } A_g = \frac{P_u}{\phi_c F_{cr}} = \frac{600}{0.90(33)} = 20.2 \text{ in.}^2$$

Try a W18 × 71:

$$A_g = 20.8 \text{ in.}^2 > 20.2 \text{ in.}^2 \quad \text{(OK)}$$

$$\frac{KL}{r_{min}} = \frac{26 \times 12}{1.70} = 183.5 < 200 \quad \text{(OK)}$$

$$F_e = \frac{\pi^2 E}{(KL/r)^2} = \frac{\pi^2(29,000)}{(183.5)^2} = 8.5 \text{ ksi}$$

$$4.71\sqrt{\frac{E}{F_y}} = 4.71\sqrt{\frac{29,000}{50}} = 113$$

Since $\dfrac{KL}{r} > 4.71\sqrt{\dfrac{E}{F_y}}$, AISC Equation E3-3 applies.

$$F_{cr} = 0.877F_e = 0.877(8.5) = 7.455 \text{ ksi}$$
$$\phi_c P_n = \phi_c F_{cr} A_g = 0.90(7.455)(20.8) = 140 \text{ kips} < 600 \text{ kips} \qquad \text{(N.G.)}$$

Because the initial estimate of F_{cr} was so far off, assume a value about halfway between 33 and 7.455 ksi. Try $F_{cr} = 20$ ksi.

$$\text{Required } A_g = \frac{P_u}{\phi_c F_{cr}} = \frac{600}{0.90(20)} = 33.3 \text{ in.}^2$$

Try a W18 × 119:

$$A_g = 35.1 \text{ in.}^2 > 33.3 \text{ in.}^2 \quad \text{(OK)}$$

$$\frac{KL}{r_{\min}} = \frac{26 \times 12}{2.69} = 116.0 < 200 \qquad \text{(OK)}$$

$$F_e = \frac{\pi^2 E}{(KL/r)^2} = \frac{\pi^2(29,000)}{(116.0)^2} = 21.27 \text{ ksi}$$

Since $\dfrac{KL}{r} > 4.71\sqrt{\dfrac{E}{F_y}} = 113$, AISC Equation E3-3 applies.

$$F_{cr} = 0.877F_e = 0.877(21.27) = 18.65 \text{ ksi}$$
$$\phi_c P_n = \phi_c F_{cr} A_g = 0.90(18.65)(35.1) = 589 \text{ kips} < 600 \text{ kips} \qquad \text{(N.G.)}$$

This is very close, so try the next larger size.

Try a W18 × 130:

$$A_g = 38.2 \text{ in.}^2$$

$$\frac{KL}{r_{\min}} = \frac{26 \times 12}{2.70} = 115.6 < 200 \qquad \text{(OK)}$$

$$F_e = \frac{\pi^2 E}{(KL/r)^2} = \frac{\pi^2(29,000)}{(115.6)^2} = 21.42 \text{ ksi}$$

Since $\dfrac{KL}{r} > 4.71\sqrt{\dfrac{E}{F_y}} = 113$, AISC Equation E3-3 applies.

$$F_{cr} = 0.877F_e = 0.877(21.42) = 18.79 \text{ ksi}$$
$$\phi_c P_n = \phi_c F_{cr} A_g = 0.90(18.79)(38.2) = 646 \text{ kips} > 600 \text{ kips} \qquad \text{(OK.)}$$

This shape is not slender (there is no footnote in the dimensions and properties table to indicate that it is), so local buckling does not have to be investigated.

Answer   Use a W18 × 130.

ASD Solution   The ASD solution procedure is essentially the same as for LRFD, and the same trial values of F_{cr} will be used here.

$$P_a = D + L = 100 + 300 = 400 \text{ kips}$$

Try $F_{cr} = 33$ ksi (an arbitrary choice of two-thirds F_y):

$$\text{Required } A_g = \frac{P_a}{0.6F_{cr}} = \frac{400}{0.6(33)} = 20.2 \text{ in.}^2$$

Try a W18 × 71:

$$A_g = 20.8 \text{ in.}^2 > 20.2 \text{ in.}^2 \qquad \text{(OK)}$$

$$\frac{KL}{r_{min}} = \frac{26 \times 12}{1.70} = 183.5 < 200 \qquad \text{(OK)}$$

$$F_e = \frac{\pi^2 E}{(KL/r)^2} = \frac{\pi^2 (29,000)}{(183.5)^2} = 8.5 \text{ ksi}$$

$$4.71\sqrt{\frac{E}{F_y}} = 4.71\sqrt{\frac{29,000}{50}} = 113$$

Since $\dfrac{KL}{r} > 4.71\sqrt{\dfrac{E}{F_y}}$, AISC Equation E3-3 applies.

$$F_{cr} = 0.877F_e = 0.877(8.5) = 7.455 \text{ ksi}$$

$$\frac{P_n}{\Omega_c} = 0.6F_{cr}A_g = 0.6(7.455)(20.8) = 93.0 \text{ kips} < 400 \text{ kips} \qquad \text{(N.G.)}$$

Because the initial estimate of F_{cr} was so far off, assume a value about halfway between 33 and 7.455 ksi. Try $F_{cr} = 20$ ksi.

$$\text{Required } A_g = \frac{P_a}{0.6F_{cr}} = \frac{400}{0.6(20)} = 33.3 \text{ in.}^2$$

Try a W18 × 119:

$$A_g = 35.1 \text{ in.}^2 > 33.3 \text{ in.}^2 \qquad \text{(OK)}$$

$$\frac{KL}{r_{min}} = \frac{26 \times 12}{2.69} = 116.0 < 200 \qquad \text{(OK)}$$

$$F_e = \frac{\pi^2 E}{(KL/r)^2} = \frac{\pi^2 (29,000)}{(116.0)^2} = 21.27 \text{ ksi}$$

Since $\dfrac{KL}{r} > 4.71\sqrt{\dfrac{E}{F_y}} = 113$, AISC Equation E3-3 applies.

$$F_{cr} = 0.877 F_e = 0.877(21.27) = 18.65 \text{ ksi}$$

$$0.6 F_{cr} A_g = 0.6(18.65)(35.1) = 393 \text{ kips} < 400 \text{ kips} \qquad \text{(N.G.)}$$

This is very close, so try the next larger size.

Try a W18 × 130:

$$A_g = 38.2 \text{ in.}^2$$

$$\frac{KL}{r_{min}} = \frac{26 \times 12}{2.70} = 115.6 < 200 \qquad \text{(OK)}$$

$$F_e = \frac{\pi^2 E}{(KL/r)^2} = \frac{\pi^2 (29,000)}{(115.6^2)} = 21.42 \text{ ksi}$$

Since $\dfrac{KL}{r} > 4.71\sqrt{\dfrac{E}{F_y}} = 113$, AISC Equation E3-3 applies.

$$F_{cr} = 0.877 F_e = 0.877(21.42) = 18.79 \text{ ksi}$$

$$0.6 F_{cr} A_g = 0.6(18.79)(38.2) = 431 \text{ kips} < 400 \text{ kips} \qquad \text{(OK)}$$

This shape is not slender (there is no footnote in the dimensions and properties table to indicate that it is), so local buckling does not have to be investigated.

Answer   Use a W18 × 130.

4.7   MORE ON EFFECTIVE LENGTH

We introduced the concept of effective length in Section 4.2, "Column Theory." All compression members are treated as pin-ended regardless of the actual end conditions but with an effective length KL that may differ from the actual length. With this modification, the load capacity of compression members is a function of only the slenderness ratio and modulus of elasticity. For a given material, the load capacity is a function of the slenderness ratio only.

If a compression member is supported differently with respect to each of its principal axes, the effective length will be different for the two directions. In Figure 4.10, a W-shape is used as a column and is braced by horizontal members in two perpendicular directions at the top. These members prevent translation of the column in all directions, but the connections, the details of which are not shown, permit small rotations to take place. Under these conditions, the member can be treated as pin-connected at the top. For the same reasons, the connection to the support at the bottom may also be treated as a pin connection. Generally speaking, a rigid, or fixed, condition is very difficult to achieve, and unless some special provisions are made,

FIGURE 4.10

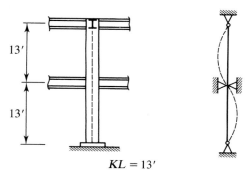

$KL = 13'$

(a) Minor Axis Buckling

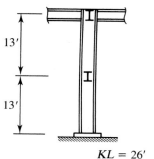

$KL = 26'$

(b) Major Axis Buckling

ordinary connections will usually closely approximate a hinge or pin connection. At midheight, the column is braced, but only in one direction.

Again, the connection prevents translation, but no restraint against rotation is furnished. This brace prevents translation perpendicular to the weak axis of the cross section but provides no restraint perpendicular to the strong axis. As shown schematically in Figure 4.10, if the member were to buckle about the major axis, the effective length would be 26 feet, whereas buckling about the minor axis would have to be in the second buckling mode, corresponding to an effective length of 13 feet. Because its strength decreases with increasing KL/r, a column will buckle in the direction corresponding to the largest slenderness ratio, so $K_x L/r_x$ must be compared with $K_y L/r_y$. In Figure 4.10, the ratio $26(12)/r_x$ must be compared with $13(12)/r_y$ (where r_x and r_y are in inches), and the larger ratio would be used for the determination of the axial compressive strength.

Example 4.9   A W12 × 58, 24 feet long, is pinned at both ends and braced in the weak direction at the third points, as shown in Figure 4.11. A992 steel is used. Determine the available compressive strength.

Solution
$$\frac{K_x L}{r_x} = \frac{24(12)}{5.28} = 54.55$$

$$\frac{K_y L}{r_y} = \frac{8(12)}{2.51} = 38.25$$

$K_x L/r_x$, the larger value, controls.

LRFD Solution   From Table 4-22 with $KL/r = 54.55$,

$\phi_c F_{cr} = 36.24$ ksi

$\phi_c P_n = \phi_c F_{cr} A_g = 36.24(17.0) = 616$ kips

Answer   Design strength = 616 kips.

FIGURE 4.11

24'

x-direction

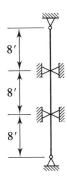

8'

8'

8'

y-direction

ASD Solution   From Table 4-22 with $KL/r = 54.55$,

$$\frac{F_{cr}}{\Omega_c} = 24.09 \text{ ksi}$$

$$\frac{P_n}{\Omega_c} = \frac{F_{cr}}{\Omega_c} A_g = 24.09(17.0) = 410 \text{ kips}$$

Answer   Allowable strength $= 410$ kips.

■

    The available strengths given in the column load tables are based on the effective length with respect to the y-axis. A procedure for using the tables with K_xL, however, can be developed by examining how the tabular values were obtained. Starting with a value of KL, the strength was obtained by a procedure similar to the following.

- KL was divided by r_y to obtain KL/r_y.
- F_{cr} was computed.
- The available strengths, $\phi_c P_n$ for LRFD and P_n/Ω_c for ASD, were computed.

Thus the tabulated strengths are based on the values of KL being equal to K_yL. If the capacity with respect to x-axis buckling is desired, the table can be entered with

$$KL = \frac{K_xL}{r_x/r_y}$$

and the tabulated load will be based on

$$\frac{KL}{r_y} = \frac{K_xL/(r_x/r_y)}{r_y} = \frac{K_xL}{r_x}$$

The ratio r_x/r_y is given in the column load tables for each shape listed.

Example 4.10   The compression member shown in Figure 4.12 is pinned at both ends and supported in the weak direction at midheight. A service load of 400 kips, with equal parts of dead and live load, must be supported. Use $F_y = 50$ ksi and select the lightest W-shape.

LRFD Solution   Factored load $= P_u = 1.2(200) + 1.6(200) = 560$ kips

Assume that the weak direction controls and enter the column load tables with $KL = 9$ feet. Beginning with the smallest shapes, the first one found that will work is a W8 × 58 with a design strength of 634 kips.

FIGURE 4.12

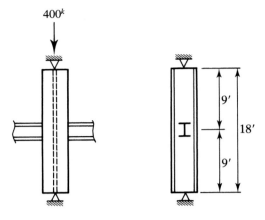

Check the strong axis:

$$\frac{K_x L}{r_x / r_y} = \frac{18}{1.74} = 10.34 \text{ ft} > 9 \text{ ft}$$

$\therefore K_x L$ controls for this shape.

Enter the tables with $KL = 10.34$ feet. A W8 × 58 has an interpolated strength of

$$\phi_c P_n = 596 \text{ kips} > 560 \text{ kips} \quad \text{(OK)}$$

Next, investigate the W10 shapes. Try a W10 × 49 with a design strength of 569 kips.

Check the strong axis:

$$\frac{K_x L}{r_x / r_y} = \frac{18}{1.71} = 10.53 \text{ ft} > 9 \text{ ft}$$

$\therefore K_x L$ controls for this shape.

Enter the tables with $KL = 10.53$ feet. A W10 × 54 is the lightest W10, with an interpolated design strength of 596 kips.

Continue the search and investigate a W12 × 53 ($\phi_c P_n = 610$ kips for $KL = 9$ ft):

$$\frac{K_x L}{r_x / r_y} = \frac{18}{2.11} = 8.53 \text{ ft} < 9 \text{ ft}$$

$\therefore K_y L$ controls for this shape, and $\phi_c P_n = 610$ kips.

Determine the lightest W14. The lightest one with a possibility of working is a W14 × 61. It is heavier than the lightest one found so far, so it will not be considered.

Answer   Use a W12 × 53.

ASD Solution   The required load capacity is $P = 400$ kips. Assume that the weak direction controls and enter the column load tables with $KL = 9$ feet. Beginning with the smallest shapes, the first one found that will work is a W8 × 58 with an allowable strength of 422 kips.

Check the strong axis:

$$\frac{K_x L}{r_x / r_y} = \frac{18}{1.74} = 10.34 \text{ ft} > 9 \text{ ft}$$

$\therefore K_x L$ controls for this shape.

Enter the tables with $KL = 10.34$ feet. A W8 × 58 has an interpolated strength of

$$\frac{P_n}{\Omega_c} = 396 \text{ kips} < 400 \text{ kips} \qquad \text{(N.G.)}$$

The next lightest W8 that will work is a W8 × 67.

$$\frac{K_x L}{r_x / r_y} = \frac{18}{1.75} = 10.29 \text{ ft} > 9 \text{ ft}$$

The interpolated allowable strength is

$$\frac{P_n}{\Omega_c} = 460 \text{ kips} > 400 \text{ kips} \qquad \text{(OK)}$$

Next, investigate the W10 shapes. Try a W10 × 60.

$$\frac{K_x L}{r_x / r_y} = \frac{18}{1.71} = 10.53 \text{ ft} > 9 \text{ ft}$$

The interpolated strength is

$$\frac{P_n}{\Omega_c} = 443 \text{ kips} > 400 \text{ kips} \qquad \text{(OK)}$$

Check the W12 shapes. Try a W12 × 53 ($P_n / \Omega_c = 406$ kips for $KL = 9$ ft):

$$\frac{K_x L}{r_x / r_y} = \frac{18}{2.11} = 8.53 \text{ ft} < 9 \text{ ft}$$

$\therefore K_y L$ controls for this shape, and $P_n / \Omega_c = 406$ kips.

Find the lightest W14. The lightest one with a possibility of working is a W14 × 61. Since it is heavier than the lightest one found so far, it will not be considered.

Answer   Use a W12 × 53.

Whenever possible, the designer should provide extra support for the weak direction of a column. Otherwise, the member is inefficient: It has an excess of strength in one direction. When K_xL and K_yL are different, K_yL will control unless r_x/r_y is smaller than K_xL/K_yL. When the two ratios are equal, the column has equal strength in both directions. For most of the W-shapes in the column load tables, r_x/r_y ranges between 1.6 and 1.8, but it is as high as 3.1 for some shapes.

Example 4.11

The column shown in Figure 4.13 is subjected to a service dead load of 140 kips and a service live load of 420 kips. Use A992 steel and select a W-shape.

Solution

$K_xL = 20$ ft and maximum $K_yL = 8$ ft. The effective length K_xL will control whenever

$$\frac{K_xL}{r_x/r_y} > K_yL$$

or

$$r_x/r_y < \frac{K_xL}{K_yL}$$

In this example,

$$\frac{K_xL}{K_yL} = \frac{20}{8} = 2.5$$

so K_xL will control if $r_x/r_y < 2.5$. Since this is true for almost every shape in the column load tables, K_xL probably controls in this example.

Assume $r_x/r_y = 1.7$:

$$\frac{K_xL}{r_x/r_y} = \frac{20}{1.7} = 11.76 > K_yL$$

FIGURE 4.13

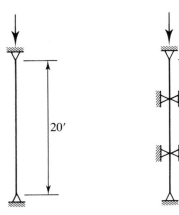

20'

6'

8'

6'

Support in
strong direction

Support in
weak direction

LRFD Solution $P_u = 1.2D + 1.6L = 1.2(140) + 1.6(420) = 840$ kips

Enter the column load tables with $KL = 12$ feet. There are no W8 shapes with enough load capacity.

Try a $W10 \times 88$ ($\phi_c P_n = 936$ kips):

$$\text{Actual } \frac{K_x L}{r_x / r_y} = \frac{20}{1.73} = 11.56 \text{ ft} < 12 \text{ ft}$$

$\therefore \phi_c P_n >$ required 840 kips.

(By interpolation, $\phi_c P_n = 951$ kips.)

Check a $W12 \times 79$:

$$\frac{K_x L}{r_x / r_y} = \frac{20}{1.75} = 11.43 \text{ ft.}$$

$\phi_c P_n = 900$ kips > 840 kips   (OK)

Investigate W14 shapes. For $r_x / r_y = 2.44$ (the approximate ratio for all likely possibilities),

$$\frac{K_x L}{r_x / r_y} = \frac{20}{2.44} = 8.197 \text{ ft} > K_y L = 8 \text{ ft}$$

For $KL = 9$ ft, a $W14 \times 74$, with a capacity of 853 kips, is the lightest W14-shape. Since 9 feet is a conservative approximation of the actual effective length, this shape is satisfactory.

Answer   Use a $W14 \times 74$ (lightest of the three possibilities).

ASD Solution $P_a = D + L = 140 + 420 = 560$ kips

Enter the column load tables with $KL = 12$ feet. There are no W8 shapes with enough load capacity. Investigate a $W10 \times 88$ (for $KL = 12$ ft, 12ft, $P_n / \Omega_c = 623$ kips):

$$\text{Actual } \frac{K_x L}{r_x / r_y} = \frac{20}{1.73} = 11.56 \text{ ft} < 12 \text{ ft}$$

$\therefore \dfrac{P_n}{\Omega_c} >$ required 560 kips

(By interpolation, $P_n / \Omega_c = 633$ kips.)

Check a W12 × 79:

$$\frac{K_xL}{r_x/r_y} = \frac{20}{1.75} = 11.43 \text{ ft} < 12 \text{ ft}$$

$$\frac{P_n}{\Omega_c} = 599 \text{ kips} > 560 \text{ kips} \qquad \text{(OK)}$$

Investigate W14 shapes. Try a W14 × 74:

$$\frac{K_xL}{r_x/r_y} = \frac{20}{2.44} = 8.20 > K_yL = 8 \text{ ft}$$

For $KL = 8.20$ ft,

$$\frac{P_n}{\Omega_c} = 581 \text{ kips} > 560 \text{ kips} \qquad \text{(OK)}$$

Answer   Use a W14 × 74 (lightest of the three possibilities).

For isolated columns that are not part of a continuous frame, Table C-C2.2 in the Commentary to the Specification will usually suffice. Consider, however, the rigid frame in Figure 4.14. The columns in this frame are not independent members but part of a continuous structure. Except for those in the lower story, the columns are restrained at both ends by their connection to beams and other columns. This frame is also unbraced, meaning that horizontal displacements of the frame are possible and all columns are subject to sidesway. If Table C-C2.2 is used for this frame, the lower-story columns are best approximated by condition (f), and a value of $K = 2$ might be used. For a column such as *AB,* a value of $K = 1.2$, corresponding to condition (c), could be selected. A more rational procedure, however, will account for the degree of restraint provided by connecting members.

The rotational restraint provided by the beams, or girders, at the end of a column is a function of the rotational stiffnesses of the members intersecting at the joint. The rotational stiffness of a member is proportional to EI/L, where I is the moment of inertia of the cross section with respect to the axis of bending. Gaylord, Gaylord, and Stallmeyer (1992) show that the effective length factor K depends on the ratio of column stiffness to girder stiffness at each end of the member, which can be expressed as

$$G = \frac{\sum E_c I_c/L_c}{\sum E_g I_g/L_g} = \frac{\sum I_c/L_c}{\sum I_g/L_g} \tag{4.12}$$

where

$\sum E_c I_c/L_c$ = sum of the stiffnesses of all columns at the end of the column under consideration

FIGURE 4.14

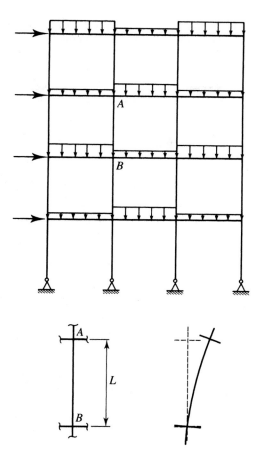

$\Sigma E_g I_g / L_g$ = sum of the stiffnesses of all girders at the end of the column under consideration

$E_c = E_g = E$, the modulus of elasticity of structural steel.

If a very slender column is connected to girders having large cross sections, the girders will effectively prevent rotation of the column. The ends of the column are approximately fixed, and K is relatively small. This condition corresponds to small values of G given by Equation 4.12. However, the ends of stiff columns connected to flexible beams can more freely rotate and approach the pinned condition, giving relatively large values of G and K.

The relationship between G and K has been quantified in the Jackson–Mooreland Alignment Charts (Johnston, 1976), which are reproduced in Figures C-C2.3 and C-C2.4 in the Commentary. To obtain a value of K from one of these nomograms, first calculate the value of G at each end of the column, letting one value be G_A and the other be G_B. Connect G_A and G_B with a straight line, and read the value of K on the middle scale. The effective length factor obtained in this manner is with respect to the axis of bending, which is the axis perpendicular to the plane of the frame. A separate analysis must be made for buckling about the other axis. Normally the beam-to-column connections in this direction will not transmit moment; sidesway is prevented by bracing; and K can be taken as 1.0.

Example 4.12   The rigid frame shown in Figure 4.15 is unbraced. Each member is oriented so that its web is in the plane of the frame. Determine the effective length factor K_x for columns AB and BC.

Solution   Column *AB:*

For joint *A*,

$$G = \frac{\sum I_c/L_c}{\sum I_g/L_g} = \frac{833/12 + 1070/12}{1350/20 + 1830/18} = \frac{158.6}{169.2} = 0.94$$

For joint *B*,

$$G = \frac{\sum I_c/L_c}{\sum I_g/L_g} = \frac{1070/12 + 1070/15}{169.2} = \frac{160.5}{169.2} = 0.95$$

Answer   From the alignment chart for sidesway uninhibited (AISC Figure C-C2.4), with $G_A = 0.94$ and $G_B = 0.95$, $K_x = 1.3$ for column *AB*.

For column *BC:*

For joint *B*, as before,

$G = 0.95$

FIGURE 4.15

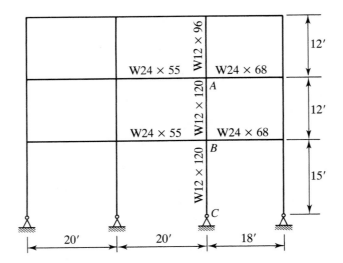

For joint *C*, a pin connection, the situation is analogous to that of a very stiff column attached to infinitely flexible girders — that is, girders of zero stiffness. The ratio of column stiffness to girder stiffness would therefore be infinite for a perfectly frictionless hinge. This end condition can only be approximated in practice, so the discussion accompanying the alignment chart recommends that *G* be taken as 10.0.

Answer From the alignment chart with $G_A = 0.95$ and $G_B = 10.0$, $K_x = 1.85$ for column *BC*.

As pointed out in Example 4.12, for a pinned support, *G* should be taken as 10.0; for a fixed support, *G* should be taken as 1.0. The latter support condition corresponds to an infinitely stiff girder and a flexible column, corresponding to a theoretical value of $G = 0$. The discussion accompanying the alignment chart in the Commentary recommends a value of $G = 1.0$ because true fixity will rarely be achieved.

Unbraced frames are able to support lateral loads because of their moment-resisting joints. Often the frame is augmented by a bracing system of some sort; such frames are called *braced frames*. The additional resistance to lateral loads can take the form of diagonal bracing or rigid shear walls, as illustrated in Figure 4.16.

FIGURE 4.16

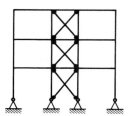 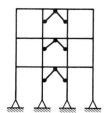

(a) Diagonal Bracing

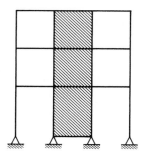

(b) Shear Walls
(masonry, reinforced concrete,
or steel plate)

In either case, the tendency for columns to sway is blocked within a given panel, or bay, for the full height of the frame. This support system forms a cantilever structure that is resistant to horizontal displacements and also provides horizontal support for the other bays. Depending on the size of the structure, more than one bay may require bracing.

A frame must resist not only the tendency to sway under the action of lateral loads but also the tendency to buckle, or become unstable, under the action of vertical loads. Bracing to stabilize a structure against vertical loading is called *stability bracing*. Appendix 6 of the AISC Specification, "Stability Bracing for Columns and Beams," covers this type of bracing. Two categories are covered: *relative* and *nodal*. With relative bracing, a brace point is restrained relative to adjacent brace points. A relative brace is connected not only to the member to be braced but also to other members, as with diagonal bracing. With relative bracing, both the brace and other members contribute to stabilizing the member to be braced. Nodal bracing provides isolated support at specific locations on the member and is not relative to other brace points or other members. The provisions of AISC Appendix 6 give equations for the required strength and stiffness (resistance to deformation) of stability bracing. The provisions for columns are from the *Guide to Stability Design Criteria* (Galambos, 1998). The required strength and stiffness for stability can be added directly to the requirements for bracing to resist lateral loading. Stability bracing is discussed further in Chapter 5, "Beams," and Chapter 6, "Beam–Columns."

Columns that are members of braced rigid frames are prevented from sidesway and have some degree of rotational restraint at their ends. Thus they are in a category that lies somewhere between cases (a) and (d) in Table C-C2.2 of the Commentary, and K is between 0.5 and 1.0. A value of 1.0 is therefore always conservative for members of braced frames and is the value prescribed by AISC C1.3a unless an analysis is made. Such an analysis can be made with the alignment chart for braced frames. Use of this nomogram would result in an effective length factor somewhat less than 1.0, and some savings could be realized.[*]

As with any design aid, the alignment charts should be used only under the conditions for which they were derived. These conditions are discussed in Section C2 of the Commentary to the Specification and are not enumerated here. Most of the conditions will usually be approximately satisfied; if they are not, the deviation will be on the conservative side. One condition that usually is not satisfied is the requirement that all behavior be elastic. If the slenderness ratio KL/r is less than $4.71\sqrt{E/F_y}$, the column will buckle inelastically, and the effective length factor obtained from the alignment chart will be overly conservative. A large number of columns are in this category. A convenient procedure for determining K for inelastic columns allows the alignment charts to be used (Yura, 1971; Disque, 1973). To demonstrate the procedure,

---

[*]If a frame is braced against sidesway, the beam-to-column connections need not be moment-resisting, and the bracing system could be designed to resist all sidesway tendency. If the connections are not moment-resisting, however, there will be no continuity between columns and girders, and the alignment chart cannot be used. For this type of braced frame, K_x should be taken as 1.0.

we begin with the critical buckling load for an inelastic column given by Equation 4.6b. Dividing it by the cross-sectional area gives the buckling stress:

$$F_{cr} = \frac{\pi^2 E_t}{(KL/r)^2}$$

The rotational stiffness of a column in this state would be proportional to $E_t I_c / L_c$, and the appropriate value of G for use in the alignment chart is

$$G_{\text{inelastic}} = \frac{\sum E_t I_c / L_c}{\sum E I_g / L_g} = \frac{E_t}{E} G_{\text{elastic}}$$

Because E_t is less than E, $G_{\text{inelastic}}$ is less than G_{elastic}, and the effective length factor K will be reduced, resulting in a more economical design. To evaluate E_t/E, called the *stiffness reduction factor* (denoted by τ_a), consider the following relationship for a column with pinned ends:

$$\frac{F_{cr(\text{inelastic})}}{F_{cr(\text{elastic})}} = \frac{\pi^2 E_t / (L/r)^2}{\pi^2 E / (L/r)^2} = \frac{E_t}{E} \tag{4.13}$$

AISC uses an approximation for the inelastic portion of the column strength curve, so Equation 4.13 is an approximation when AISC Equations E3-2 and E3-3 are used for F_{cr}.

We can approximate F_{cr} by the compressive strength:

$$F_{cr} = \frac{P_u}{\phi_c A_g} \quad \text{for LRFD}$$

$$= \frac{\Omega_c P_a}{A_g} \quad \text{for ASD}$$

Then in the elastic range, $F_{cr(\text{inelastic})}$ is approximately

$$\frac{P_u}{\phi_c A_g} = 0.658^{(F_y/F_e)} F_y \quad \text{for LRFD}$$

and

$$\frac{\Omega_c P_a}{A_g} = 0.658^{(F_y/F_e)} F_y \quad \text{for ASD}$$

We can solve for F_e, then compute

$$F_{cr(\text{elastic})} = 0.877 F_e$$

The stiffness reduction factor τ_a can then be computed.

Example 4.13   Compute the stiffness reduction factor τ_a for an axial compressive stress of 25 ksi and $F_y = 50$ ksi.

LRFD Solution

$$\frac{P_u}{A_g} = 25 \text{ ksi}$$

$$F_{cr(\text{inelastic})} = \frac{P_u}{\phi_c A_g} = \frac{25}{0.90} = 27.78 \text{ ksi} = 0.658^{(F_y/F_e)} F_y$$

or

$$27.78 = 0.658^{(50/F_e)}(50), \qquad F_e = 35.61 \text{ ksi}$$

$$F_{cr(\text{inelastic})} = 0.877 F_e = 0.877(35.61) = 31.23 \text{ ksi}$$

The stiffness reduction factor is therefore

Answer

$$\tau_a = \frac{F_{cr(\text{inelastic})}}{F_{cr(\text{elastic})}} = \frac{27.78}{31.23} = 0.890$$

ASD Solution

$$F_{cr(\text{inelastic})} = \frac{\Omega_c P_a}{A_g} = 1.67(25) = 41.75 \text{ ksi}$$

$$41.75 = 0.658^{(50/F_e)}(50), \qquad F_e = 116.1 \text{ ksi}$$

$$F_{cr(\text{elastic})} = 0.877 F_e = 0.877(116.1) = 101.8 \text{ ksi}$$

Answer

$$\tau_a = \frac{F_{cr(\text{inelastic})}}{F_{cr(\text{elastic})}} = \frac{41.75}{101.8} = 0.410 \text{ ksi}$$

Values of the stiffness reduction factor τ_a as a function of P_u/A_g and P_a/A_g are given in Table 4-21 in Part 4 of the *Manual*.

Example 4.14   A rigid unbraced frame is shown in Figure 4.17. All members are oriented so that bending is about the strong axis. Lateral support is provided at each joint by simply connected bracing in the direction perpendicular to the frame. Determine the effective length factors with respect to each axis for member *AB*. The service dead load is 35.5 kips, and the service live load is 142 kips. A992 steel is used.

Solution   Compute elastic *G* factors:

For joint *A*,

$$\frac{\Sigma(I_c/L_c)}{\Sigma(I_g/L_g)} = \frac{171/12}{88.6/20 + 88.6/18} = \frac{14.25}{9.352} = 1.52$$

FIGURE 4.17

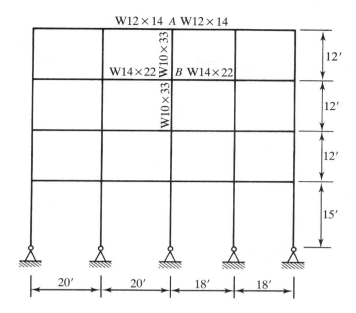

For joint B,

$$\frac{\Sigma(I_c/L_c)}{\Sigma(I_g/L_g)} = \frac{2(171/12)}{199/20 + 199/18} = \frac{28.5}{21.01} = 1.36$$

From the alignment chart for unbraced frames, $K_x = 1.45$, based on elastic behavior. Determine whether the column behavior is elastic or inelastic.

$$\frac{K_x L}{r_x} = \frac{1.45(12 \times 12)}{4.19} = 49.83$$

$$4.71\sqrt{\frac{E}{F_y}} = 4.71\sqrt{\frac{29,000}{50}} = 113$$

Since

$$\frac{K_x L}{r_x} < 4.71\sqrt{\frac{E}{F_y}}$$

behavior is inelastic, and the inelastic K factor can be used.

LRFD Solution   The factored load is

$$P_u = 1.2D + 1.6L = 1.2(35.5) + 1.6(142) = 269.8 \text{ kips}$$

Enter Table 4-21 in Part 4 of the *Manual* with

$$\frac{P_u}{A_g} = \frac{269.8}{9.71} = 27.79 \text{ ksi}$$

and obtain the stiffness reduction factor $\tau_a = 0.8105$ by interpolation.

For joint A,

$$G_{\text{inelastic}} = \tau_a \times G_{\text{elastic}} = 0.8105(1.52) = 1.23$$

For joint B,

$$G_{\text{inelastic}} = 0.8105(1.36) = 1.10$$

Answer   From the alignment chart, $K_x = 1.35$. Because of the support conditions normal to the frame, K_y can be taken as 1.0.

ASD Solution   The applied load is

$$P_a = D + L = 35.5 + 142 = 177.5 \text{ kips}$$

Enter Table 4-21 in Part 4 of the *Manual* with

$$\frac{P_a}{A_g} = \frac{177.5}{9.71} = 18.28 \text{ ksi}$$

and obtain the stiffness reduction factor $\tau_a = 0.8198$ by interpolation.
For joint A,

$$G_{\text{inelastic}} = \tau_a \times G_{\text{elastic}} = 0.8198(1.52) = 1.25$$

For joint B,

$$G_{\text{inelastic}} = 0.8198(1.36) = 1.12$$

Answer   From the alignment chart, $K_x = 1.35$. Because of the support conditions normal to the frame, K_y can be taken as 1.0.   ■

If the end of a column is fixed ($G = 1.0$) or pinned ($G = 10.0$), the value of G at that end should *not* be multiplied by the stiffness reduction factor.

4.8   TORSIONAL AND FLEXURAL-TORSIONAL BUCKLING

When an axially loaded compression member becomes unstable overall (that is, not locally unstable), it can buckle in one of three ways, as shown in Figure 4.18).

1. **Flexural buckling.** We have considered this type of buckling up to now. It is a deflection caused by bending, or flexure, about the axis corresponding to the largest slenderness ratio (Figure 4.18a). This is usually the minor principal axis — the one with the smallest radius of gyration. Compression members with any type of cross-sectional configuration can fail in this way.
2. **Torsional buckling.** This type of failure is caused by twisting about the longitudinal axis of the member. It can occur only with doubly symmetrical cross

FIGURE 4.18

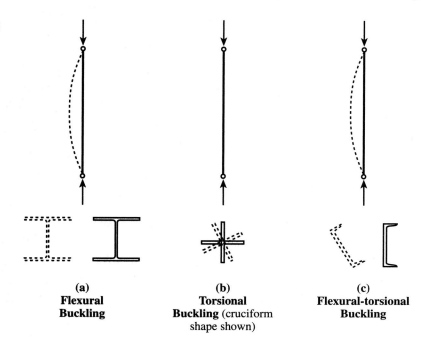

(a)
Flexural
Buckling

(b)
Torsional
Buckling (cruciform
shape shown)

(c)
Flexural-torsional
Buckling

sections with very slender cross-sectional elements (Figure 4.18b). Standard hot-rolled shapes are not susceptible to torsional buckling, but members built up from thin plate elements may be and should be investigated. The cruciform shape shown is particularly vulnerable to this type of buckling. This shape can be fabricated from plates as shown in the figure, or built up from four angles placed back to back.

3. **Flexural-torsional buckling.** This type of failure is caused by a combination of flexural buckling and torsional buckling. The member bends and twists simultaneously (Figure 4.18c). This type of failure can occur only with unsymmetrical cross sections, both those with one axis of symmetry — such as channels, structural tees, double-angle shapes, and equal-leg single angles — and those with no axis of symmetry, such as unequal-leg single angles.

The AISC Specification requires an analysis of torsional or flexural-torsional buckling when appropriate. Section E4(a) of the Specification covers double-angle and tee-shaped members, and Section E4(b) provides a more general approach that can be used for any shape. We discuss the general approach first. It is based on first determining a value of F_e, which is analogous to the Euler buckling stress. This stress can then be used with the flexural buckling equations, AISC Equations E3-2 and E3-3. The stress F_e can be defined as the elastic buckling stress corresponding to the controlling mode of failure, whether flexural, torsional, or flexural-torsional.

The equations for F_e given in AISC E4(b) are based on well-established theory given in *Theory of Elastic Stability* (Timoshenko and Gere, 1961). Except for some changes in notation, they are the same equations as those given in that work, with no simplifications. For doubly symmetrical shapes (torsional buckling),

$$F_e = \left[\frac{\pi^2 E C_w}{(K_z L)^2} + GJ \right] \frac{1}{I_x + I_y} \qquad \text{(AISC Equation E4-4)}$$

For singly symmetrical shapes (flexural-torsional buckling),

$$F_e = \frac{F_{ey} + F_{ez}}{2H} \left(1 - \sqrt{1 - \frac{4 F_{ey} F_{ez} H}{(F_{ey} + F_{ez})^2}} \right) \qquad \text{(AISC Equation E4-5)}$$

where y is the axis of symmetry.

For shapes with *no* axis of symmetry (flexural-torsional buckling),

$$(F_e - F_{ex})(F_e - F_{ey})(F_e - F_{ez}) - F_e^2 (F_e - F_{ey}) \left(\frac{x_0}{\bar{r}_0} \right)^2$$

$$- F_e^2 (F_e - F_{ex}) \left(\frac{y_0}{\bar{r}_0} \right)^2 = 0 \qquad \text{(AISC Equation E4-6)}$$

This last equation is a cubic; F_e is the smallest root. Fortunately, there will be little need for solving this equation, because completely unsymmetrical shapes are rarely used as compression members.

In the above equations, the z-axis is the longitudinal axis. The previously undefined terms in these three equations are defined as

C_w = warping constant (in.^{6})
K_z = effective length factor for *torsional* buckling, which is based on the amount of end restraint against twisting about the longitudinal axis
G = shear modulus (ksi) = 11,200 ksi for structural steel
J = torsional constant (equal to the polar moment of inertia only for circular cross sections) (in.^{4})

$$F_{ex} = \frac{\pi^2 E}{(K_x L / r_x)^2} \qquad \text{(AISC Equation E4-9)}$$

$$F_{ey} = \frac{\pi^2 E}{(K_y L / r_y)^2} \qquad \text{(AISC Equation E4-10)}$$

where y is the axis of symmetry for singly symmetrical shapes.

$$F_{ez} = \left(\frac{\pi^2 E C_w}{(K_z L)^2} + GJ \right) \frac{1}{A_g \bar{r}_0^2} \qquad \text{(AISC Equation E4-11)}$$

$$H = 1 - \frac{x_0^2 + y_0^2}{\bar{r}_0^2} \qquad \text{(AISC Equation E4-8)}$$

TABLE 4.1

Shape	Constants
W, M, S, HP, WT, MT, ST	J, C_w (In addition, the Manual Companion CD gives values of $\bar{r}_0$, and H for WT, MT, and ST shapes)
C	J, C_w, $\bar{r}_0$, H
MC, Angles	J, C_w $\bar{r}_0$, (In addition, the Manual Companion CD gives values of H for MC and angle shapes.)
Double Angles	$\bar{r}_0$, H (J and C_w are double the values given for single angles.)

where z is the longitudinal axis and x_0, y_0 are the coordinates of the shear center of the cross section with respect to the centroid (in inches). The shear center is the point on the cross section through which a transverse load on a beam must pass if the member is to bend without twisting.

$$\bar{r}_0^2 = x_0^2 + y_0^2 + \frac{I_x + I_y}{A_g} \qquad \text{(AISC Equation E4-7)}$$

Values of the constants used in the equations for F_e can be found in the dimensions and properties tables in Part 1 of the *Manual*. Table 4.1 shows which constants are given for various types of shapes. Table 4.1 shows that the *Manual* does not give the constants $\bar{r}_0$ and H for tees, although they are given on the Companion CD. They are easily computed, however, if x_0 and y_0 are known. Since x_0 and y_0 are the coordinates of the shear center with respect to the centroid of the cross section, the location of the shear center must be known. For a tee shape, it is located at the intersection of the centerlines of the flange and the stem. Example 4.15 illustrates the computation of $\bar{r}_0$ and H.

As previously pointed out, the need for a torsional buckling analysis of a doubly symmetrical shape will be rare. Similarly, shapes with no axis of symmetry are rarely used for compression members, and flexural-torsional buckling analysis of these types of members will seldom, if ever, need to be done. For these reasons, we limit further consideration to flexural-torsional buckling of shapes with one axis of symmetry. Furthermore, the most commonly used of these shapes, the double angle, is a built-up shape, and we postpone consideration of it until Section 4.9.

For singly symmetrical shapes, the flexural-torsional buckling stress F_e is found from AISC Equation E4-5. In this equation, y is defined as the axis of symmetry (regardless of the orientation of the member), and flexural-torsional buckling will take place only about this axis (flexural buckling about this axis will not occur). The x-axis is subject only to flexural buckling. Therefore, for singly symmetrical shapes, there are two possibilities for the strength: either flexural-torsional buckling about the y-axis (the axis of symmetry) or flexural buckling about the x-axis (Timoshenko and Gere, 1961 and Zahn and Iwankiw, 1989). To determine which one controls, compute the strength corresponding to each axis and use the smaller value.

Example 4.15 Compute the compressive strength of a WT12 × 81 of A992 steel. The effective length with respect to the *x*-axis is 25 feet 6 inches, the effective length with respect to the *y*-axis is 20 feet, and the effective length with respect to the *z*-axis is 20 feet. Use the general approach of AISC E4(b).

Solution Compute the flexural buckling strength for the *x*-axis:

$$\frac{K_x L}{r_x} = \frac{25.5 \times 12}{3.50} = 87.43$$

$$F_e = \frac{\pi^2 E}{(KL/r)^2} = \frac{\pi^2 (29,000)}{(87.43)^2} = 37.44 \text{ ksi}$$

$$4.71 \sqrt{\frac{E}{F_y}} = 4.71 \sqrt{\frac{29,000}{50}} = 113$$

Since $\dfrac{KL}{r} < 4.71 \sqrt{\dfrac{E}{F_y}}$, AISC Equation E3-2 applies.

$$F_{cr} = 0.658^{(F_y/F_e)} F_y = 0.658^{(50/37.44)}(50) = 28.59 \text{ ksi}$$

The nominal strength is

$$P_n = F_{cr} A_g = 28.59(23.9) = 683.3 \text{ kips}$$

Compute the flexural-torsional buckling strength about the *y*-axis (the axis of symmetry):

$$\frac{K_y L}{r_y} = \frac{20 \times 12}{3.05} = 78.69$$

$$F_{ey} = \frac{\pi^2 E}{(KL/r)^2} = \frac{\pi^2 (29,000)}{(78.69)^2} = 46.22 \text{ ksi}$$

Because the shear center of a tee is located at the intersection of the centerlines of the flange and the stem,

$$x_0 = 0$$

$$y_0 = \bar{y} - \frac{t_f}{2} = 2.70 - \frac{1.22}{2} = 2.090 \text{ in.}$$

$$\bar{r}_0^2 = x_0^2 + y_0^2 + \frac{I_x + I_y}{A_g} = 0 + (2.090)^2 + \frac{293 + 221}{23.9} = 25.87 \text{ in.}^2$$

$$F_{ez} = \left[\frac{\pi^2 E C_w}{(K_z L)^2} + GJ \right] \frac{1}{A_g \bar{r}_0^2}$$

$$= \left[\frac{\pi^2 (29,000)(43.8)}{(20 \times 12)^2} + 11,200(9.22) \right] \frac{1}{23.9(25.87)} = 167.4 \text{ ksi}$$

$$F_{ey} + F_{ez} = 46.22 + 167.4 = 213.6 \text{ ksi}$$

$$H = 1 - \frac{x_0^2 + y_0^2}{\bar{r}_0^2} = 1 - \frac{0 + (2.090)^2}{25.87} = 0.8312$$

(Note that, for tees, the values of $\bar{r}_0$ and H can be found on the manual companion CD.)

$$F_e = \left(\frac{F_{ey} + F_{ez}}{2H} \right) \left[1 - \sqrt{1 - \frac{4 F_{ey} F_{ez} H}{(F_{ey} + F_{ez})^2}} \right]$$

$$= \frac{213.6}{2(0.8312)} \left[1 - \sqrt{1 - \frac{4(46.22)(167.4)(0.8312)}{(213.6)^2}} \right] = 43.63 \text{ ksi}$$

To determine which compressive strength equation to use, compare this value of F_e with

$$0.44 F_y = 0.44(50) = 22.0 \text{ ksi}$$

Since 43.63 ksi > 22.0 ksi, use AISC Equation E3-2.

$$F_{cr} = 0.658^{(F_y/F_e)} F_y = 0.658^{(50/37.44)} (50) = 28.59 \text{ ksi}$$

The nominal strength is

$$P_n = F_{cr} A_g = 30.95(23.9) = 739.7 \text{ kips}$$

The flexural buckling strength controls, and the nominal strength is 683.3 kips.

Answer   For LRFD, the design strength is $\phi_c P_n = 0.90(683.3) = 615$ kips.

For ASD, the allowable stress is $F_a = 0.6 F_{cr} = 0.6(28.59) = 17.15$ ksi and the allowable strength is $F_a A_g = 17.15(23.9) = 410$ kips.

■

The procedure for flexural-torsional buckling analysis of double angles and tees given in AISC Section E4(a) is a modification of the procedure given in AISC E4(b). There is also some notational change: F_e becomes F_{cr}, F_{ey} becomes F_{cry}, and F_{ez} becomes F_{crz}.

To obtain F_{crz}, we can drop the first term of AISC Equation E4-11 to get

$$F_{crz} = \frac{GJ}{A_g \bar{r}_0^2} \qquad \text{(AISC Equation E4-3)}$$

This approximation is acceptable because for double angles and tees, the first term is negligible compared to the second term.

The flexural buckling stress F_{cry} is computed with the usual equations of AISC E3, using KL/r corresponding to the y-axis (the axis of symmetry).

The nominal strength can then be computed as

$$P_n = F_{cr} A_g \qquad \text{(AISC Equation E3-1)}$$

where

$$F_{cr} = \left(\frac{F_{cry} + F_{crz}}{2H} \right) \left[1 - \sqrt{1 - \frac{4 F_{cry} F_{crz} H}{(F_{cry} + F_{crz})^2}} \right]$$ (AISC Equation E4-2)

All other terms from Section E4(b) remain unchanged. This procedure, to be used with double angles and tees only, is more accurate than the procedure given in E4(b).

Example 4.16   Compute the strength of the shape in Example 4.15 by using the equations of AISC E4(a).

Solution   From Example 4.15, the nominal flexural buckling strength for the x-axis is 683.3 kips (with $F_{cr} = 28.59$ ksi). The following values were also computed in Example 4.15:

$K_y L / r_y = 78.69$

$\bar{r}_0^2 = 25.87$ in.^{2}

$H = 0.8312$

Compute F_{cry} using AISC E3. From AISC Equation E3-4,

$$F_e = \frac{\pi^2 E}{(KL/r)^2} = \frac{\pi^2 E}{(K_y L / r_y)^2} = \frac{\pi^2 (29,000)}{(78.69)^2} = 46.22 \text{ ksi}$$

Since $K_y L / r_y < 4.71 \sqrt{\dfrac{E}{F_y}} = 113$

$$F_{cry} = 0.658^{(F_y/F_e)} F_y = 0.658^{(50/46.22)} (50) = 31.79 \text{ ksi}$$

$$F_{crz} = \frac{GJ}{A_g \bar{r}_0^2} = \frac{11,200(9.22)}{23.9(25.87)} = 167.0 \text{ ksi}$$

$$F_{cry} + F_{crz} = 31.79 + 167.0 = 198.8 \text{ ksi}$$

$$F_{cr} = \left(\frac{F_{cry} + F_{crz}}{2H} \right) \left[1 - \sqrt{1 - \frac{4 F_{cry} F_{crz} H}{(F_{cry} + F_{crz})^2}} \right]$$

$$= \frac{198.8}{2(0.8312)} \left[1 - \sqrt{1 - \frac{4(31.79)(167.0)(0.8312)}{(198.8)^2}} \right] = 30.63 \text{ ksi}$$

$$P_n = F_{cr} A_g = 30.63(23.9) = 732.1 \text{ kips}$$

The flexural buckling strength controls, and the nominal strength is 683.3 kips.

Answer   For LRFD, the design strength is $\phi_c P_n = 0.90(683.3) = 615$ kips.

For ASD, the allowable stress is $F_a = 0.6F_{cr} = 0.6(28.59) = 17.15$ ksi, and the allowable strength is $F_a A_g = 17.15(23.9) = 410$ kips.

◼

The flexural-torsional buckling results of Examples 4.15 and 4.16 show that the error in using the general approach of Section E4(b) for this shape is on the unconservative side. The procedure used in Example 4.16, which is based on AISC Specification E4(a), should always be used for double angles and tees. In practice, however, the strength of most double angles and tees can be found in the column load tables. These tables give two sets of values of the available strength, one based on flexural buckling about the x-axis and one based on flexural-torsional buckling about the y axis. The flexural-torsional buckling strengths are based on the recommended procedure of AISC E4(a).

Available compressive strength tables are also provided for single-angle members. The values of strength in these tables are not based on flexural-torsional buckling theory, but on the provisions of AISC E5.

When using the column load tables for unsymmetrical shapes, there is no need to account for slender compression elements, because that has already been done. If an analysis is being done for a member not in the column load tables, then any element slenderness must be accounted for.

4.9   BUILT-UP MEMBERS

If the cross-sectional properties of a built-up compression member are known, its analysis is the same as for any other compression member, provided the component parts of the cross section are properly connected. AISC E6 contains many details related to this connection, with separate requirements for members composed of two or more rolled shapes and for members composed of plates or a combination of plates and shapes. Before considering the connection problem, we will review the computation of cross-sectional properties of built-up shapes.

The design strength of a built-up compression member is a function of the slenderness ratio KL/r. Hence the principal axes and the corresponding radii of gyration about these axes must be determined. For homogeneous cross sections, the principal axes coincide with the centroidal axes. The procedure is illustrated in Example 4.17. The components of the cross section are assumed to be properly connected.

Example 4.17   The column shown in Figure 4.19 is fabricated by welding a ⅜-inch by 4-inch cover plate to the flange of a W18 × 65. Steel with $F_y = 50$ ksi is used for both components. The effective length is 15 feet with respect to both axes. Assume that the components are connected in such a way that the member is fully effective and compute the strength based on flexural buckling.

FIGURE 4.19

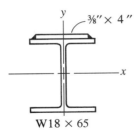

W18 × 65

TABLE 4.2

Component	A	y	Ay
Plate	1.500	0.1875	0.2813
W	19.10	9.575	182.9
Σ	20.60		183.2

Solution   With the addition of the cover plate, the shape is slightly unsymmetrical, but the flexural-torsional effects will be negligible.

The vertical axis of symmetry is one of the principal axes, and its location need not be computed. The horizontal principal axis will be found by application of the *principle of moments:* The sum of moments of component areas about any axis (in this example, a horizontal axis along the top of the plate will be used) must equal the moment of the total area. We use Table 4.2 to keep track of the computations.

$$\bar{y} = \frac{\Sigma Ay}{\Sigma A} = \frac{183.2}{20.60} = 8.893 \text{ in.}$$

With the location of the horizontal centroidal axis known, the moment of inertia with respect to this axis can be found by using the *parallel-axis theorem:*

$$I = \bar{I} + Ad^2$$

where
$\bar{I}$ = moment of inertia about the centroidal axis of a component area
A = area of the component
I = moment of inertia about an axis parallel to the centroidal axis of the component area
d = perpendicular distance between the two axes

The contributions from each component area are computed and summed to obtain the moment of inertia of the composite area. These computations are shown in Table 4.3, which is an expanded version of Table 4.2. The moment of inertia about the *x*-axis is

$$I_x = 1193 \text{ in.}^4$$

TABLE 4.3

Component	A	y	Ay	$\bar{I}$	d	$\bar{I} + Ad^2$
Plate	1.500	0.1875	0.2813	0.01758	8.706	113.7
W	19.10	9.575	182.9	1070	0.6820	1079
Σ	20.60		183.2			1193

For the vertical axis,

$$I_y = \frac{1}{12}\left(\frac{3}{8}\right)(4)^3 + 54.8 = 56.80 \text{ in.}^4$$

Since $I_y < I_x$, the y-axis controls.

$$r_{min} = r_y = \sqrt{\frac{I_y}{A}} = \sqrt{\frac{56.80}{20.60}} = 1.661 \text{ in.}$$

$$\frac{KL}{r_{min}} = \frac{15 \times 12}{1.661} = 108.4$$

$$F_e = \frac{\pi^2 E}{(KL/r)^2} = \frac{\pi^2(29,000)}{(108.4)^2} = 24.36 \text{ ksi}$$

$$4.71\sqrt{\frac{E}{F_y}} = 4.71\sqrt{\frac{29,000}{50}} = 113$$

Since $\dfrac{KL}{r} < 4.71\sqrt{\dfrac{E}{F_y}}$, use AISC Equation E3-2.

$$F_{cr} = 0.658^{(F_y/F_e)} F_y = 0.658^{(50/24.36)}(50) = 21.18 \text{ ksi}$$

The nominal strength is

$$P_n = F_{cr}A_g = 21.18(20.60) = 436.3 \text{ kips}$$

LRFD Solution   The design strength is

$$\phi_c P_n = 0.90(436.3) = 393 \text{ kips}$$

ASD Solution   From Equation 4.7, the allowable stress is

$$F_a = 0.6F_{cr} = 0.6(21.18) = 12.71 \text{ ksi}$$

The allowable strength is

$$F_a A_g = 12.71(20.60) = 262 \text{ kips}$$

Answer   Design compressive strength = 393 kips. Allowable compressive strength = 262 kips.

FIGURE 4.20

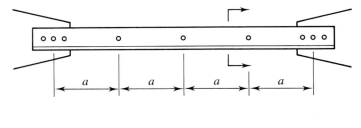

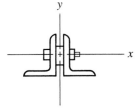

Connection Requirements for Built-Up Members Composed of Rolled Shapes

The most common built-up shape is one that is composed of rolled shapes, namely, the double-angle shape. This type of member will be used to illustrate the requirements for this category of built-up members. Figure 4.20 shows a truss compression member connected to gusset plates at each end. To maintain the back-to-back separation of the angles along the length, fillers (spacers) of the same thickness as the gusset plate are placed between the angles at equal intervals. The intervals must be small enough that the member functions as a unit. If the member buckles about the x-axis (flexural buckling), the connectors are not subjected to any calculated load, and the connection problem is simply one of maintaining the relative positions of the two components. To ensure that the built-up member acts as a unit, AISC E6.2 requires that the slenderness of an individual component be no greater than three-fourths of the slenderness of the built-up member; that is,

$$\frac{Ka}{r_i} \leq \frac{3}{4} \frac{KL}{r} \tag{4.14}$$

where

$$
\begin{aligned}
a &= \text{spacing of the connectors} \\
r_i &= \text{smallest radius of gyration of the component} \\
Ka/r_i &= \text{effective slenderness ratio of the component} \\
KL/r &= \text{maximum slenderness ratio of the built-up member}
\end{aligned}
$$

If the member buckles about the axis of symmetry — that is, if it is subjected to flexural-torsional buckling about the y-axis — the connectors are subjected to shearing forces. This condition can be visualized by considering two planks used as a beam, as shown in Figure 4.21. If the planks are unconnected, they will slip along the surface of contact when loaded and will function as two separate beams. When connected by bolts (or any other fasteners, such as nails), the two planks will behave as a

FIGURE 4.21

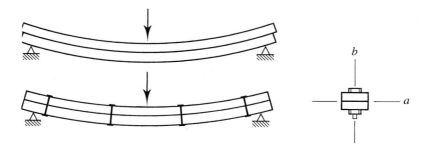

unit, and the resistance to slip will be provided by shear in the bolts. This behavior takes place in the double-angle shape when bending about its y-axis. If the plank beam is oriented so that bending takes place about its other axis (the b-axis), then both planks bend in exactly the same manner, and there is no slippage and hence no shear. This behavior is analogous to bending about the x-axis of the double-angle shape. When the fasteners are subjected to shear, a modified slenderness ratio larger than the actual value may be required.

AISC E6 considers two categories of connectors: (1) snug-tight bolts and (2) welds or fully-tensioned bolts. We cover these connection methods in detail in Chapter 7, "Simple Connections." The column load tables for double angles are based on the use of welds or fully tightened bolts. For this category,

$$\left(\frac{KL}{r}\right)_m = \sqrt{\left(\frac{KL}{r}\right)_0^2 + 0.82\frac{\alpha^2}{(1+\alpha^2)}\left(\frac{a}{r_{ib}}\right)^2} \qquad \text{(AISC Equation E6-2)}$$

where
$(KL/r)_m$ = modified slenderness ratio
$(KL/r)_0$ = original unmodified slenderness ratio
r_{ib} = radius of gyration of component about axis parallel to member axis of buckling
α = separation ratio = $h/2r_{ib}$
h = distance between component centroids (perpendicular to member axis of buckling)

When the connectors are snug-tight bolts,

$$\left(\frac{KL}{r}\right)_m = \sqrt{\left(\frac{KL}{r}\right)_0^2 + \left(\frac{a}{r_i}\right)^2} \qquad \text{(AISC Equation E6-1)}$$

The column load tables for double-angle shapes show the number of intermediate connectors required for the given y-axis flexural-torsional buckling strength. The number of connectors needed for the x-axis flexural buckling strength must be determined from the requirement of Equation 4.14 that the slenderness of one angle between connectors must not exceed three-fourths of the overall slenderness of the double-angle shape.

Example 4.18

Compute the available strength of the compression member shown in Figure 4.22. Two angles, $5 \times 3 \times \frac{1}{2}$, are oriented with the long legs back-to-back (2L5 × 3 × ½ LLBB) and separated by $\frac{3}{8}$ inch. The effective length KL is 16 feet, and there are three fully tightened intermediate connectors. A36 steel is used.

FIGURE 4.22

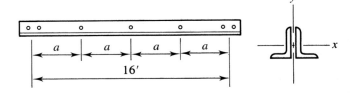

Solution

Compute the flexural buckling strength for the x-axis:

$$\frac{K_x L}{r_x} = \frac{16(12)}{1.58} = 121.5$$

$$F_e = \frac{\pi^2 E}{(KL/r)^2} = \frac{\pi^2 (29,000)}{(121.5)^2} = 19.39 \text{ ksi}$$

$$4.71 \sqrt{\frac{E}{F_y}} = 4.71 \sqrt{\frac{29,000}{36}} = 134$$

Since $\dfrac{KL}{r} < 4.71 \sqrt{\dfrac{E}{F_y}}$, use AISC Equation E3-2.

$$F_{cr} = 0.658^{(F_y/F_e)} F_y = 0.658^{(36/19.39)}(36) = 16.55 \text{ ksi}$$

The nominal strength is

$$P_n = F_{cr} A_g = 16.55(2 \times 3.75) = 124.1 \text{ kips}$$

To determine the flexural-torsional buckling strength for the y-axis, use the modified slenderness ratio, which is based on the spacing of the connectors. The unmodified slenderness ratio is

$$\left(\frac{KL}{r}\right)_0 = \frac{KL}{r_y} = \frac{16(12)}{1.24} = 154.8$$

The spacing of the connectors is

$$a = \frac{16(12)}{4 \text{ spaces}} = 48 \text{ in.}$$

Then, from Equation 4.14,

$$\frac{Ka}{r_i} = \frac{Ka}{r_z} = \frac{48}{0.642} = 74.77 < 0.75(154.8) = 116.1 \quad \text{(OK)}$$

$$r_{ib} = r_y = 0.824 \text{ in.}$$

$$h = 2(0.746) + \frac{3}{8} = 1.867 \text{ in.}$$

$$\alpha = \frac{h}{2r_{ib}} = \frac{1.867}{2 \times 0.824} = 1.133$$

From AISC Equation E6-2, the modified slenderness ratio is

$$\left(\frac{KL}{r}\right)_m = \sqrt{\left(\frac{KL}{r}\right)_o^2 + 0.82 \frac{\alpha^2}{(1 + \alpha^2)} \left(\frac{a}{r_{ib}}\right)^2}$$

$$= \sqrt{(154.8)^2 + 0.82 \frac{(1.133)^2}{\left[1 + (1.133)^2\right]} \left(\frac{48}{0.824}\right)^2} = 159.8$$

This value should be used in place of KL/r_y for the computation of F_{cry}:

$$F_e = \frac{\pi^2 E}{(KL/r)^2} = \frac{\pi^2 (29,000)}{(159.8)^2} = 11.21 \text{ ksi}$$

Since $\dfrac{KL}{r} > 4.71\sqrt{\dfrac{E}{F_y}} = 134$,

$$F_{cry} = 0.877 F_e = 0.877(11.21) = 9.831 \text{ ksi}$$

From AISC Equation E4-3,

$$F_{crz} = \frac{GJ}{A_g \bar{r}_o^2} = \frac{11,200(2 \times 0.322)}{7.50(2.51)^2} = 152.6 \text{ ksi}$$

$$F_{cry} + F_{crz} = 9.831 + 152.6 = 162.4 \text{ ksi}$$

$$F_{cr} = \left(\frac{F_{cry} + F_{crz}}{2H}\right)\left[1 - \sqrt{1 - \frac{4 F_{cry} F_{crz} H}{(F_{cry} + F_{crz})^2}}\right]$$

$$= \frac{162.4}{2(0.646)}\left[1 - \sqrt{1 - \frac{4(9.832)(152.6)(0.646)}{(162.4)^2}}\right] = 9.606 \text{ ksi}$$

The nominal strength is

$$P_n = F_{cr}A_g = 9.606(7.50) = 72.05 \text{ kips}$$

Therefore the flexural-torsional buckling strength controls.

LRFD Solution   The design strength is

$$\phi_c P_n = 0.90(72.05) = 64.9 \text{ kips}$$

ASD Solution   From Equation 4.7, the allowable stress is

$$F_a = 0.6F_{cr} = 0.6(9.606) = 5.764 \text{ ksi}$$

The allowable strength is

$$F_a A_g = 5.764(7.50) = 43.2 \text{ kips}$$

Answer   Design compressive strength = 64.9 kips. Allowable compressive strength = 43.2 kips.

Example 4.19   Design a 14-foot-long compression member to resist a service dead load of 12 kips and a service live load of 23 kips. Use a double-angle shape with the short legs back-to-back, separated by $\frac{3}{8}$-inch. The member will be braced at midlength against buckling about the x-axis (the axis parallel to the long legs). Specify the number of intermediate connectors needed (the midlength brace will provide one such connector). Use A36 steel.

LRFD Solution   The factored load is

$$P_u = 1.2D + 1.6L = 1.2(12) + 1.6(23) = 51.2 \text{ kips}$$

From the column load tables, select 2L $3\frac{1}{2} \times 3 \times \frac{1}{4}$ SLBB, weighing 10.8 lb/ft. The capacity of this shape is 53.2 kips, based on buckling about the y-axis with an effective length of 14 feet. (The strength corresponding to flexural buckling about the x-axis is 63.1 kips, based on an effective length of $\frac{14}{2} = 7$ feet.) Note that this shape is a slender-element cross section, but this is taken into account in the tabular values.

Bending about the y-axis subjects the fasteners to shear, so a sufficient number of fasteners must be provided to account for this action. The table reveals that three intermediate connectors are required. (This number also satisfies Equation 4.14.)

Answer   Use 2L $3\frac{1}{2} \times 3 \times \frac{1}{4}$ SLBB with three intermediate connectors within the 14-foot length.

ASD Solution   The total load is

$$P_a = D + L = 12 + 23 = 35 \text{ kips}$$

From the column load tables, select 2L $3\frac{1}{4} \times 3 \times \frac{1}{4}$ SLBB, weighing 10.8 lb/ft. The capacity is 35.4 kips, based on buckling about the y axis, with an effective length of

14 feet. (The strength corresponding to flexural buckling about the x axis is 42.0 kips, based on an effective length of $^{14}/_2 = 7$ feet.) Note that this shape is a slender-element section, but this is taken into account in the tabular values.

Bending about the y axis subjects the fasteners to shear, so a sufficient number of fasteners must be provided to account for this action. The table reveals that three intermediate connectors are required. (This number also satisfies Equation 4.14.)

Answer   Use 2L $3\frac{1}{2} \times 3 \times \frac{1}{4}$ SLBB with three intermediate connectors within the 14-foot length.   ■

Connection Requirements for Built-Up Members Composed of Plates or Both Plates and Shapes

When a built-up member consists of two or more rolled shapes separated by a substantial distance, plates must be used to connect the shapes. AISC E6 contains many details regarding the connection requirements and the proportioning of the plates. Additional connection requirements are given for other built-up compression members composed of plates or plates and shapes.

Problems

AISC Requirements

4.3-1   Use AISC Equation E3-2 or E3-3 and determine the nominal axial compressive strength for the following cases:

    a. $L = 10$ ft
    b. $L = 30$ ft

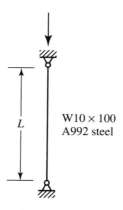

W10 × 100
A992 steel

FIGURE P4.3-1

4.3-2   Compute the nominal axial compressive strength of the member shown in Figure P4.3-2. Use AISC Equation E3-2 or E3-3.

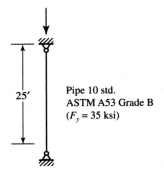

25′

Pipe 10 std.
ASTM A53 Grade B
($F_y = 35$ ksi)

FIGURE P4.3-2

4.3-3   Compute the nominal compressive strength of the member shown in Figure P4.3-3. Use AISC Equation E3-2 or E3-3.

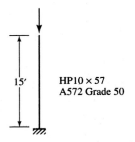

15′

HP10 × 57
A572 Grade 50

FIGURE P4.3-3

4.3-4   Determine the available strength of the compression member shown in Figure P4.3-4, in each of the following ways:

a. Use AISC Equation E3-2 or E3-3. Compute both the design strength for LRFD and the allowable strength for ASD.
b. Use Table 4-22 from Part 4 of the *Manual*. Compute both the design strength for LRFD and the allowable strength for ASD.

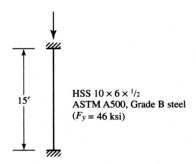

15′

HSS 10 × 6 × ¹/₂
ASTM A500, Grade B steel
($F_y = 46$ ksi)

FIGURE P4.3-4

4.3-5 Determine the available axial compressive strength by each of the following methods:

   a. Use AISC Equation E3-2 or E3-3. Compute both the design strength for LRFD and the allowable strength for ASD.

   b. Use Table 4-22 from Part 4 of the *Manual*. Compute both the design strength for LRFD and the allowable strength for ASD.

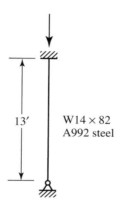

13′   W14 × 82
A992 steel

FIGURE P4.3-5

4.3-6 A W18 × 119 is used as a compression member with one end fixed and the other end pinned. The length is 12 feet. What is the available compressive strength if A992 steel is used?

   a. Use AISC Equation E3-2 or E3-3. Compute both the design strength for LRFD and the allowable strength for ASD.

   b. Use Table 4-22 from Part 4 of the *Manual*. Compute both the design strength for LRFD and the allowable strength for ASD.

4.3-7 An HSS $10 \times 8 \times \frac{3}{16}$ is used as a compression member with one end pinned and the other end fixed against rotation but free to translate. The length is 12 feet. Compute the nominal compressive strength for A500 Grade B steel ($F_y = 46$ ksi). *Note that this is a slender-element compression member, and the equations of AISC Section E7 must be used.*

4.3-8 A W21 × 101 is used as compression member with one end fixed and the other end free. The length is 10 feet. What is the nominal compressive strength if $F_y = 50$ ksi? *Note that this is a slender-element compression member, and the equations of AISC Section E7 must be used.*

4.3-9 Determine the maximum axial compressive service load that can be supported if the live load is twice as large as the dead load. Use AISC Equation E3-2 or E3-3.

   a. Use LRFD.

   a. Use ASD.

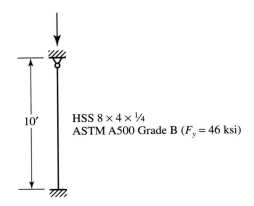

FIGURE P4.3-9

4.3-10   Determine whether the compression member shown in Figure P4.3-10 is adequate to support the given service loads.

a. Use LRFD.
b. Use ASD.

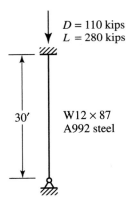

FIGURE P4.3-10

Design

4.6-1   a. Select a W14 of A992 steel. Use the column load tables.

    1. Use LRFD.
    2. Use ASD.

b. Select a W16 of A992 steel. Use the trial-and-error approach covered in Section 4.6.

    1. Use LRFD.
    2. Use ASD.

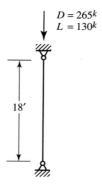

$D = 265^k$
$L = 130^k$

18′

FIGURE P4.6-1

4.6-2 A 20-foot long column is pinned at the bottom and fixed against rotation but free to translate at the top. It must support a service dead load of 110 kips and a service live load of 110 kips.

a. Select a W12 of A992 steel. Use the column load tables.

  1. Use LRFD.
  2. Use ASD.

b. Select a W18 of A992 steel. Use the trial-and-error approach covered in Section 4.6.

  1. Use LRFD.
  2. Use ASD.

4.6-3 Select a rectangular (not square) HSS ($F_y = 46$ ksi).

a. Use LRFD.
b. Use ASD.

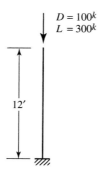

$D = 100^k$
$L = 300^k$

12′

FIGURE P4.6-3

4.6-4 Select a steel pipe of A53 Grade B steel ($F_y = 35$ ksi). Specify whether your selection is Standard, Extra-Strong, or Double-Extra Strong.

a. Use LRFD.
b. Use ASD.

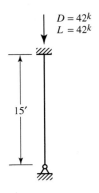

FIGURE P4.6-4

4.6-5 Select an HP-shape for the conditions of Problem 4.6-3. Use $F_y = 50$ ksi.

    a. Use LRFD.

    b. Use ASD.

4.6-6 Select a rectangular (not square) HSS for the conditions of Problem 4.6-4.

    a. Use LRFD.

    b. Use ASD.

4.6-7 For the conditions shown in Figure P4.6-7, use LRFD and

    a. select a W12 of A992 steel.

    b. select a steel pipe.

    c. select a square HSS.

    d. select a rectangular HSS.

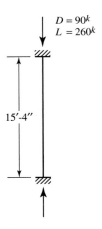

FIGURE P4.6-7

4.6-8 Same as Problem 4.6-7, but use ASD.

4.6-9   For the conditions shown in Figure P4.6-7, select an A992 W-shape with a nominal depth of 21 inches.

a. Use LRFD.
b. Use ASD.

Effective Length

4.7-1   A W16 × 100 with $F_y = 60$ ksi is used as a compression member. The length is 13 feet. Compute the nominal strength for $K_x = 2.1$ and $K_y = 1.0$.

4.7-2   An HSS 10 × 6 × 5/16 with $F_y = 46$ ksi is used as a column. The length is 15 feet. Both ends are pinned, and there is support against weak axis buckling at a point 6 feet from the top. Determine

a. the design strength for LRFD.
b. the allowable *stress* for ASD.

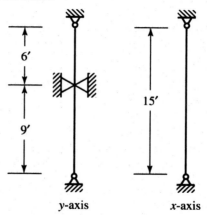

FIGURE P4.7-2

4.7-3   A W12 × 79 of A572 Grade 60 steel is used as a compression member. It is 28 feet long, pinned at each end, and has additional support in the weak direction at a point 12 feet from the top. Can this member resist a service dead load of 180 kips and a service live load of 320 kips?

a. Use LRFD.
b. Use ASD.

4.7-4   Use A992 steel and select a W14 shape for an axially loaded column to meet the following specifications: The length is 22 feet, both ends are pinned, and there is bracing in the weak direction at a point 10 feet from the top. The service dead load is 142 kips, and the service live load is 356 kips.

a. Use LRFD.
b. Use ASD.

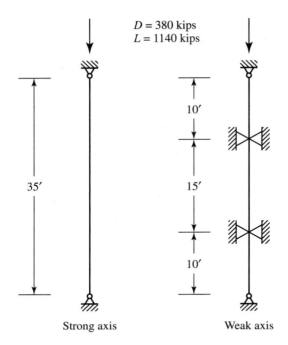

$D = 380$ kips
$L = 1140$ kips

10'

35'

15'

10'

Strong axis

Weak axis

FIGURE P4.7-5

4.7-5 Use A992 steel and select a W shape.

a. Use LRFD.
b. Use ASD.

4.7-6 Select a rectangular (not square) HSS for use as a 15-foot-long compression member that must resist a service dead load of 35 kips and a service live load of 80 kips. The member will be pinned at each end, with additional support in the weak direction at midheight. Use A500 Grade B steel ($F_y = 46$ ksi).

a. Use LRFD.
b. Use ASD.

4.7-7 Select the best rectangular (not square) HSS for a column to support a service dead load of 33 kips and a service live load of 82 kips. The member is 27 feet long and is pinned at the ends. It is supported in the weak direction at a point 12 feet from the top. Use $F_y = 46$ ksi.

a. Use LRFD.
b. Use ASD.

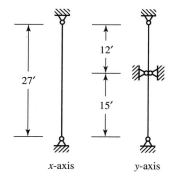

FIGURE P4.7-7

4.7-8 The frame shown in Figure P4.7-8 is unbraced, and bending is about the x-axis of the members. All beams are W18 × 35, and all columns are W10 × 54.

    a. Determine the effective length factor K_x for column AB. Do not consider the stiffness reduction factor.

    b. Determine the effective length factor K_x for column BC. Do not consider the stiffness reduction factor.

    c. If $F_y = 50$ ksi, is the stiffness reduction factor applicable to these columns?

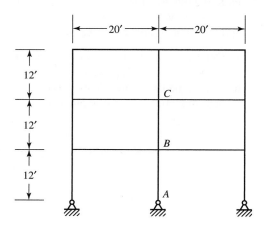

FIGURE P4.7-8

4.7-9 The given frame is unbraced, and bending is about the x axis of each member. The axial dead load supported by column AB is 204 kips, and the axial live load is 408 kips. $F_y = 50$ ksi. Determine K_x for member AB. Use the stiffness reduction factor if possible.

    a. Use LRFD.

    b. Use ASD.

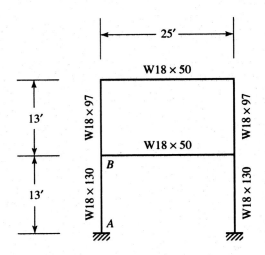

FIGURE P4.7-9

4.7-10 The rigid frame shown in Figure P4.7-10 is unbraced. The members are oriented so that bending is about the strong axis. Support conditions in the direction perpendicular to the plane of the frame are such that $K_y = 1.0$. The beams are W18 × 50, and the columns are W12 × 72. A992 steel is used. The axial compressive dead load is 50 kips, and the axial compressive live load is 150 kips.

a. Determine the axial compressive design strength of column AB. Use the stiffness reduction factor if applicable.

b. Determine the allowable axial compressive strength of column AB. Use the stiffness reduction factor if applicable.

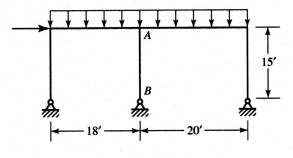

FIGURE P4.7-10

4.7.11 The frame shown in Figure P4.7-11 is unbraced against sidesway. Relative moments of inertia of the members have been assumed for preliminary design purposes. Use the alignment chart and determine K_x for members AB, BC, DE, and EF.

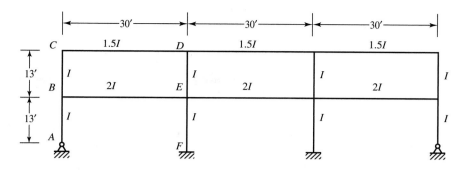

FIGURE P4.7-11

4.7-12 The frame shown in Figure P4.7-12 is unbraced against sidesway. Assume that all columns are W14 × 61 and that all girders are W18 × 76. ASTM A992 steel is used for all members. The members are oriented so that bending is about the x-axis. Assume that $K_y = 1.0$

a. Use the alignment chart to determine K_x for member *GF*. Use the stiffness reduction factor if applicable. For member *GF*, the service dead load is 80 kips and the service live load is 159 kips.
b. Compute the nominal compressive strength of member GF.
c. Estimate K_x from Table C-C2.2 in the Commentary and compare your estimate with the results of part (a).

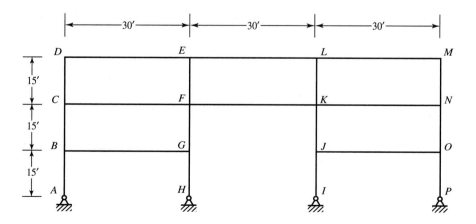

FIGURE P4.7-12

4.7-13 The frame shown in Figure P4.7-13 is unbraced against sidesway. The columns are HSS 6 × 6 × ⅝, and the beams are W12 × 22. ASTM A500 grade B steel ($F_y = 46$ ksi)

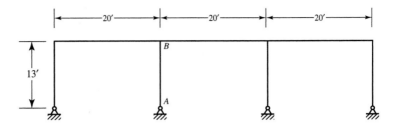

FIGURE P4.7-13

is used for the columns, and $F_y = 50$ ksi for the beams. The beams are oriented so that bending is about the x-axis. Assume that $K_y = 1.0$.

a. Use the alignment chart to determine K_x for column AB. Use the stiffness reduction factor if applicable. For column AB, the service dead load is 17 kips and the service live load is 50 kips.

b. Compute the nominal compressive strength of column AB.

4.7-14   The rigid frame shown in Figure P4.7-14 is unbraced in the plane of the frame. In the direction perpendicular to the frame, the frame is braced at the joints. The connections at these points of bracing are simple (moment-free) connections. Roof girders are W14 × 30, and floor girders are W16 × 36. Member BC is a W10 × 45. Use A992 steel and select a W-shape for AB. Assume that the controlling load combination causes no moment in AB. The service dead load is 25 kips and the service live load is 75 kips.

a. Use LRFD.

b. Use ASD.

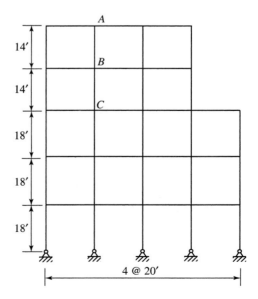

FIGURE P4.7-14

Torsional and Flexural-Torsional Buckling

4.8-1   Compute the nominal compressive strength for a WT10.5 × 91 with an effective length of 18 feet with respect to each axis. Use A992 steel and the procedure of AISC Section E4 (not the column load tables).

4.8-2   Use A572 Grade 50 steel and compute the nominal strength of the column shown in Figure P4.8-2. The member ends are fixed in all directions (*x*, *y*, and *z*).

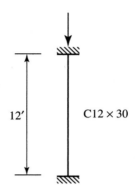

12′   C12 × 30

FIGURE P4.8-2

4.8-3   Select a WT section for the compression member shown in Figure P4.8-3. The load is the total service load, with a live-to-dead load ratio of 2.5:1. Use $F_y = 50$ ksi.

a. Use LRFD.
b. Use ASD.

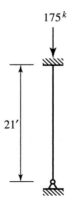

175^k

21′

FIGURE P4.8-3

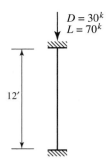

$$D = 30^k$$
$$L = 70^k$$

12′

FIGURE P4.8-4

4.8-4 Select an American Standard Channel for the compression member shown in Figure P4.8-4. Use A36 steel. The member ends are fixed in all directions (x, y, and z).

a. Use LRFD.
b. Use ASD.

Built-Up Members

4.9-1 Verify the value of r_y given in Part 1 of the *Manual* for the double-angle shape 2L5 × 3½ × ½ LLBB. The angles will be connected to a ⅜-inch-thick gusset plate.

4.9-2 Verify the values of y_2, r_x, and r_y given in Part 1 of the *Manual* for the combination shape consisting of a W12 × 26 with a C10 × 15.3 cap channel.

4.9-3 A column is built up from four 6 × 6 × ⅝ angle shapes as shown in Figure P4.9-3. The plates are not continuous but are spaced at intervals along the column length and function to maintain the separation of the angles. They do not contribute to the cross-sectional properties. Compute r_x and r_y.

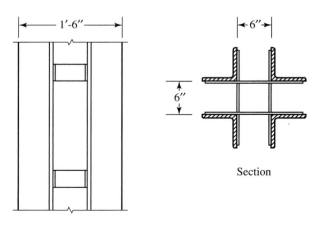

1′-6″   6″   6″

Section

FIGURE P4.9-3

4.9-4 An unsymmetrical compression member consists of $\frac{1}{2} \times 12$ top flange, a $\frac{1}{2} \times 7$ bottom flange, and a $\frac{3}{8} \times 16$ web (the shape is symmetrical about a vertical centroidal axis). Compute the radius of gyration about each of the principal axes.

4.9-5 A column for a multistory building is fabricated from ASTM A588 plates as shown in Figure P4.9-5. Compute the nominal axial compressive strength based on flexural buckling (do not consider torsional buckling). Assume that the components of the cross section are connected in such a way that the section is fully effective.

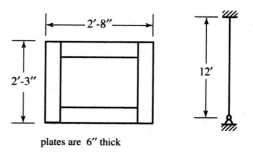

plates are 6″ thick

FIGURE P4.9-5

4.9-6 Compute the nominal compressive strength based on flexural buckling for the built-up shape shown in Figure P4.9-6 (do not consider torsional buckling). Assume that the components of the cross section are connected in such a way that the section is fully effective. ASTM A242 steel is used.

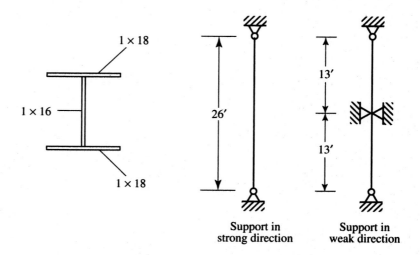

FIGURE P4.9-6

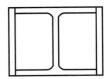

FIGURE P4.9-7

4.9-7 Two plates $\frac{9}{16} \times 10$ are welded to a W10 × 49 to form a built-up shape, as shown in Figure P4.9-7. Assume that the components are connected so that the cross section is fully effective. $F_y = 50$ ksi, and $K_xL = K_yL = 25$ ft.

  a. Compute the nominal axial compressive strength based on flexural buckling (do not consider torsional bucking).
  b. What is the percentage increase in strength from the unreinforced W10 × 49?

4.9-8 A structural tee shape is fabricated by splitting an HP14 × 117, as shown in Figure P4.9-8. Compute the nominal axial compressive strength based on flexural buckling (do not consider flexural-torsional buckling). Account for the area of the fillets at the web-to-flange junction. A572 Grade so steel is used. The effective length is 10 feet with respect to both axes.

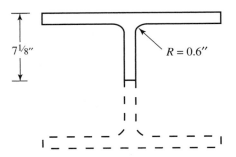

FIGURE P4.9-8

4.9-9 A column cross section is built up from four L5 × 5 × ¾, as shown in Figure P4.9-9. The angles are held in position by lacing bars, whose primary function is to hold the angles in position. The lacing is not considered to contribute to the cross-sectional area, which is why it is shown by dashed lines. AISC Section E6 covers the design of lacing. The effective length is 30 feet with respect to both axes, and A572 Grade 50 steel is used. Investigate flexural buckling only (no torsional buckling) and compute

  a. the design strength for LRFD.
  b. the allowable strength for ASD.

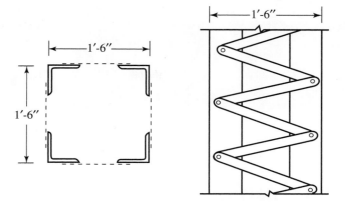

FIGURE P4.9-9

4.9-10   Compute the available strengths (for both LRFD and ASD) for the following double-angle shape: 2L6 × 4 × ⅝, long legs ⅜-inch back-to-back, F_y = 50 ksi. The effective length KL is 18 feet for all axes, and there are two intermediate fully tightened
bolts. Use the procedure of AISC E4 (not the column load tables). Compare the flexural and the flexural-torsional buckling strengths.

4.9-11   For the conditions shown in Figure P4.9-11, select a double-angle section (⅜-inch gusset plate connection). Use A36 steel. Specify the number of intermediate connectors.

   a. Use LRFD.
   b. Use ASD.

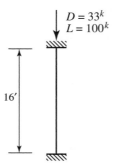

FIGURE P4.9-11

4.9-12   Use LRFD and select a double-angle shape for the top chord of the truss of Problem
3.8-2. Use $K_x = K_y$ =1.0. Assume ⅜-inch gusset plates, and use A36 steel.

5

Beams

5.1 INTRODUCTION

Beams are structural members that support transverse loads and are therefore subjected primarily to flexure, or bending. If a substantial amount of axial load is also present, the member is referred to as a *beam–column* (beam–columns are considered in Chapter 6). Although some degree of axial load will be present in any structural member, in many practical situations this effect is negligible and the member can be treated as a beam. Beams are usually thought of as being oriented horizontally and subjected to vertical loads, but that is not necessarily the case. A structural member is considered to be a beam if it is loaded so as to cause bending.

Commonly used cross-sectional shapes include the W-, S-, and M-shapes. Channel shapes are sometimes used, as are beams built up from plates, in the form of I or box shapes. For reasons to be discussed later, doubly symmetric shapes such as the standard rolled W-, M-, and S-shapes are the most efficient.

Coverage of beams in the AISC Specification is spread over two chapters: Chapter F, "Design of Members for Flexure," and Chapter G, "Design of Members for Shear." Several categories of beams are covered in the Specification; in this book, we cover the most common cases in the present chapter, and we cover a special case, plate girders, in Chapter 10.

Figure 5.1 shows two types of beam cross sections; a hot-rolled doubly-symmetric I-shape and a welded doubly-symmetric built-up I-shape. The hot-rolled I-shape is the one most commonly used for beams. Welded shapes usually fall into the category classified as plate girders.

For flexure (shear will be covered later), the required and available strengths are moments. For load and resistance factor design (LRFD), Equation 2.6 can be written as

$$M_u \le \phi_b M_n \tag{5.1}$$

where

M_u = required moment strength = maximum moment caused by the controlling load combination from ASCE 7

ϕ_b = resistance factor for bending (flexure) = 0.90

M_n = nominal moment strength

The right-hand side of Equation 5.1 is the design strength, sometimes called the *design moment*.

FIGURE 5.1

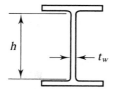 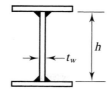

For allowable strength design (ASD), Equation 2.7 can be written as

$$M_a \le \frac{M_n}{\Omega_b}$$

(5.2)

where

M_a = required moment strength = maximum moment corresponding to the controlling load combination from ASCE 7

Ω_b = safety factor for bending = 1.67

Equation 5.2 can also be written as

$$M_a \le \frac{M_n}{1.67} = 0.6M_n$$

Dividing both sides by the elastic section modulus S (which will be reviewed in the next section), we get an equation for *allowable stress design:*

$$\frac{M_a}{S} \le \frac{0.6M_n}{S}$$

or

$$f_b \le F_b$$

where

f_b = maximum computed bending stress

F_b = allowable bending stress

5.2   BENDING STRESS AND THE PLASTIC MOMENT

To be able to determine the nominal moment strength M_n, we must first examine the behavior of beams throughout the full range of loading, from very small loads to the point of collapse. Consider the beam shown in Figure 5.2a, which is oriented so that bending is about the major principal axis (for an I shape, it will be the x–x axis). For a linear elastic material and small deformations, the distribution of bending stress will be as shown in Figure 5.2b, with the stress assumed to be uniform across the width of the beam. (Shear is considered separately in Section 5.8.) From elementary mechanics of materials, the stress at any point can be found from the flexure formula:

$$f_b = \frac{My}{I_x}$$

(5.3)

FIGURE 5.2

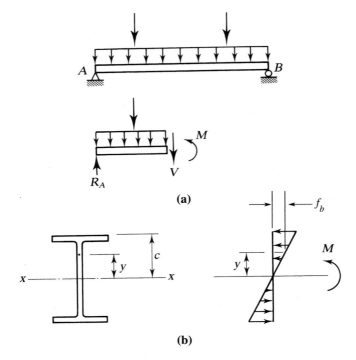

(a)

(b)

where M is the bending moment at the cross section under consideration, y is the perpendicular distance from the neutral plane to the point of interest, and I_x is the moment of inertia of the area of the cross section with respect to the neutral axis. For a homogeneous material, the neutral axis coincides with the centroidal axis. Equation 5.3 is based on the assumption of a linear distribution of strains from top to bottom, which in turn is based on the assumption that cross sections that are plane before bending remain plane after bending. In addition, the beam cross section must have a vertical axis of symmetry, and the loads must be in the longitudinal plane containing this axis. Beams that do not satisfy these criteria are considered in Section 5.15. The maximum stress will occur at the extreme fiber, where y is maximum. Thus there are two maxima: maximum compressive stress in the top fiber and maximum tensile stress in the bottom fiber. If the neutral axis is an axis of symmetry, these two stresses will be equal in magnitude. For maximum stress, Equation 5.3 takes the form:

$$f_{max} = \frac{Mc}{I_x} = \frac{M}{I_x/c} = \frac{M}{S_x} \tag{5.4}$$

where c is the perpendicular distance from the neutral axis to the extreme fiber, and S_x is the elastic section modulus of the cross section. For any cross-sectional shape, the section modulus will be a constant. For an unsymmetrical cross section, S_x will have two values: one for the top extreme fiber and one for the bottom. Values of S_x for standard rolled shapes are tabulated in the dimensions and properties tables in the *Manual*.

FIGURE 5.3

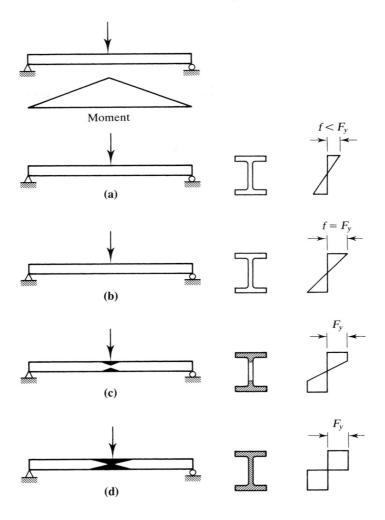

Equations 5.3 and 5.4 are valid as long as the loads are small enough that the material remains within its linear elastic range. For structural steel, this means that the stress f_{max} must not exceed F_y and that the bending moment must not exceed

$$M_y = F_y S_x$$

where M_y is the bending moment that brings the beam to the point of yielding.

In Figure 5.3, a simply supported beam with a concentrated load at midspan is shown at successive stages of loading. Once yielding begins, the distribution of stress on the cross section will no longer be linear, and yielding will progress from the extreme fiber toward the neutral axis. At the same time, the yielded region will extend longitudinally from the center of the beam as the bending moment reaches M_y at more locations. These yielded regions are indicated by the dark areas in Figure 5.3c and d. In Figure 5.3b, yielding has just begun. In Figure 5.3c, the yielding has progressed

FIGURE 5.4

Plastic hinge

into the web, and in Figure 5.3d the entire cross section has yielded. The additional moment required to bring the beam from stage b to stage d is 10 to 20% of the yield moment, M_y, for W-shapes. When stage d has been reached, any further increase in the load will cause collapse, since all elements of the cross section have reached the yield plateau of the stress–strain curve and unrestricted plastic flow will occur. A *plastic hinge* is said to have formed at the center of the beam, and this hinge along with the actual hinges at the ends of the beam constitute an unstable mechanism. During plastic collapse, the mechanism motion will be as shown in Figure 5.4. Structural analysis based on a consideration of collapse mechanisms is called *plastic analysis*. An introduction to plastic analysis and design is presented in the Appendix of this book.

The plastic moment capacity, which is the moment required to form the plastic hinge, can easily be computed from a consideration of the corresponding stress distribution. In Figure 5.5, the compressive and tensile stress resultants are shown, where A_c is the cross-sectional area subjected to compression, and A_t is the area in tension. These are the areas above and below the plastic neutral axis, which is not necessarily the same as the elastic neutral axis. From equilibrium of forces,

$$C = T$$
$$A_c F_y = A_t F_y$$
$$A_c = A_t$$

Thus the plastic neutral axis divides the cross section into two equal areas. For shapes that are symmetrical about the axis of bending, the elastic and plastic neutral axes are the same. The plastic moment, M_p, is the resisting couple formed by the two equal and opposite forces, or

$$M_p = F_y(A_c)a = F_y(A_t)a = F_y\left(\frac{A}{2}\right)a = F_y Z$$

where

A = total cross-sectional area
a = distance between the centroids of the two half-areas
$Z = \left(\dfrac{A}{2}\right)a$ = plastic section modulus

FIGURE 5.5

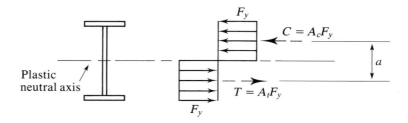

Example 5.1   For the built-up shape shown in Figure 5.6, determine (a) the elastic section modulus S and the yield moment M_y and (b) the plastic section modulus Z and the plastic moment M_p. Bending is about the x-axis, and the steel is A572 Grade 50.

FIGURE 5.6

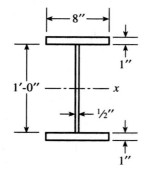

Solution   a. Because of symmetry, the elastic neutral axis (the x-axis) is located at mid-depth of the cross section (the location of the centroid). The moment of inertia of the cross section can be found by using the parallel axis theorem, and the results of the calculations are summarized in Table 5.1.

TABLE 5.1

Component	$\bar{I}$	A	d	$\bar{I} + Ad^2$
Flange	0.6667	8	6.5	338.7
Flange	0.6667	8	6.5	338.7
Web	72	—	—	72.0
Sum				749.4

The elastic section modulus is

$$S = \frac{I}{c} = \frac{749.4}{1 + (12/2)} = \frac{749.4}{7} = 107 \text{ in.}^3$$

and the yield moment is

$$M_y = F_y S = 50(107) = 5350 \text{ in.-kips} = 446 \text{ ft-kips}$$

Answer $S = 107$ in.^{3} and $M_y = 446$ ft-kips.

b. Because this shape is symmetrical about the x-axis, this axis divides the cross section into equal areas and is therefore the plastic neutral axis. The centroid of the top half-area can be found by the principle of moments. Taking moments about the x-axis (the neutral axis of the entire cross section) and tabulating the computations in Table 5.2, we get

$$\bar{y} = \frac{\Sigma Ay}{\Sigma A} = \frac{61}{11} = 5.545 \text{ in.}$$

TABLE 5.2

Component	A	y	Ay
Flange	8	6.5	52
Web	3	3	9
Sum	11		61

FIGURE 5.7

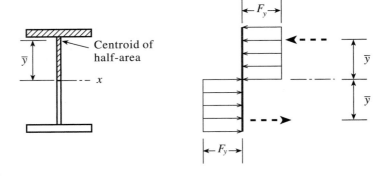

Figure 5.7 shows that the moment arm of the internal resisting couple is

$$a = 2\bar{y} = 2(5.545) = 11.09 \text{ in.}$$

and that the plastic section modulus is

$$\left(\frac{A}{2}\right)a = 11(11.09) = 122 \text{ in.}^3$$

The plastic moment is

$$M_p = F_y Z = 50(122) = 6100 \text{ in.-kips} = 508 \text{ ft-kips}$$

Answer $Z = 122$ in.^{3} and $M_p = 508$ ft-kips.

Example 5.2   Compute the plastic moment, M_p, for a W10 × 60 of A992 steel.

Solution   From the dimensions and properties tables in Part 1 of the *Manual*,

$$A = 17.6 \text{ in.}^2$$

$$\frac{A}{2} = \frac{17.6}{2} = 8.8 \text{ in.}^2$$

The centroid of the half-area can be found in the tables for WT-shapes, which are cut from W-shapes. The relevant shape here is the WT5 × 30, and the distance from the outside face of the flange to the centroid is 0.884 inch, as shown in Figure 5.8.

FIGURE 5.8

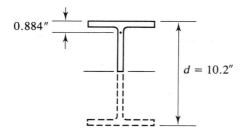

$$a = d - 2(0.884) = 10.2 - 2(0.884) = 8.432 \text{ in.}$$

$$Z = \left(\frac{A}{2}\right)a = 8.8(8.432) = 74.20 \text{ in.}^3$$

This result compares favorably with the value of 74.6 given in the dimensions and properties tables (the difference results from rounding of the tabular values).

Answer $M_p = F_y Z = 50(74.20) = 3710$ in.-kips $= 309$ ft-kips. ∎

5.3 STABILITY

If a beam can be counted on to remain stable up to the fully plastic condition, the nominal moment strength can be taken as the plastic moment capacity; that is,

$$M_n = M_p$$

Otherwise, M_n will be less than M_p.

As with a compression member, instability can be in an overall sense or it can be local. Overall buckling is illustrated in Figure 5.9a. When a beam bends, the compression region (above the neutral axis) is analogous to a column, and in a manner similar to a column, it will buckle if the member is slender enough. Unlike a column, however, the compression portion of the cross section is restrained by the tension portion, and the outward deflection (flexural buckling) is accompanied by twisting (torsion). This form of instability is called *lateral-torsional buckling* (LTB). Lateral-torsional buckling can be prevented by bracing the beam against twisting at sufficiently close intervals. This can be accomplished with either of two types of stability bracing: lateral bracing, illustrated schematically in Figure 5.9b, and torsional bracing, represented in Figure 5.9c. Lateral bracing, which prevents lateral translation, should be applied as close to the compression flange as possible. Torsional bracing prevents twist directly; it can be either nodal or continuous, and it can take the form of either cross frames or diaphragms. The nodal and relative categories were defined in Chapter 4, "Compression Members." Appendix 6 of the AISC Specification gives the strength and stiffness requirements for beam bracing. These provisions are based on the work of Yura (2001). As we shall see, the moment strength depends in part on the unbraced length, which is the distance between points of bracing.

FIGURE 5.9

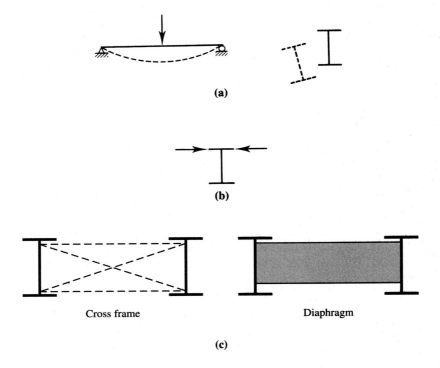

Whether the beam can sustain a moment large enough to bring it to the fully plastic condition also depends on whether the cross-sectional integrity is maintained. This integrity will be lost if one of the compression elements of the cross section buckles. This type of buckling can be either compression flange buckling, called *flange local buckling* (FLB), or buckling of the compression part of the web, called *web local buckling* (WLB). As discussed in Chapter 4, "Compression Members," whether either type of local buckling occurs will depend on the width–thickness ratios of the compression elements of the cross section.

Figure 5.10 further illustrates the effects of local and lateral-torsional buckling. Five separate beams are represented on this graph of load versus central deflection.

FIGURE 5.10

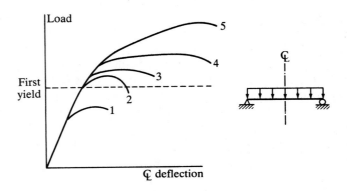

Curve 1 is the load-deflection curve of a beam that becomes unstable (in any way) and loses its load-carrying capacity before first yield (see Figure 5.3b) is attained. Curves 2 and 3 correspond to beams that can be loaded past first yield but not far enough for the formation of a plastic hinge and the resulting plastic collapse. If plastic collapse can be reached, the load-deflection curve will have the appearance of either curve 4 or curve 5. Curve 4 is for the case of uniform moment over the full length of the beam, and curve 5 is for a beam with a variable bending moment (moment gradient). Safe designs can be achieved with beams corresponding to any of these curves, but curves 1 and 2 represent inefficient use of material.

5.4 CLASSIFICATION OF SHAPES

AISC classifies cross-sectional shapes as compact, noncompact, or slender, depending on the values of the width–thickness ratios. For I shapes, the ratio for the projecting flange (an *unstiffened* element) is $b_f/2t_f$, and the ratio for the web (a *stiffened* element) is h/t_w. The classification of shapes is found in Section B4 of the Specification, "Local Buckling," in Table B4.1. It can be summarized as follows. Let

λ = width–thickness ratio

λ_p = upper limit for compact category

λ_r = upper limit for noncompact category

Then

if $\lambda \leq \lambda_p$ and the flange is continuously connected to the web, the shape is compact;

if $\lambda_p < \lambda \leq \lambda_r$, the shape is noncompact; and

if $\lambda > \lambda_r$, the shape is slender.

The category is based on the worst width–thickness ratio of the cross section. For example, if the web is compact and the flange is noncompact, the shape is classified as noncompact. Table 5.3 has been extracted from AISC Table B4.1 and is specialized for hot-rolled I-shaped cross sections.

Table 5.3 also applies to channels, except that λ for the flange is b_f/t_f.

TABLE 5.3
Width–Thickness
Parameters*

Element	λ	λ_p	λ_r
Flange	$\dfrac{b_f}{2t_f}$	$0.38\sqrt{\dfrac{E}{F_y}}$	$1.0\sqrt{\dfrac{E}{F_y}}$
Web	$\dfrac{h}{t_w}$	$3.76\sqrt{\dfrac{E}{F_y}}$	$5.70\sqrt{\dfrac{E}{F_y}}$

*For hot-rolled I shapes in flexure.

5.5   BENDING STRENGTH OF COMPACT SHAPES

A beam can fail by reaching M_p and becoming fully plastic, or it can fail by

1. lateral-torsional buckling (LTB), either elastically or inelastically;
2. flange local buckling (FLB), elastically or inelastically; or
3. web local buckling (WLB), elastically or inelastically.

If the maximum bending stress is less than the proportional limit when buckling occurs, the failure is said to be *elastic*. Otherwise, it is *inelastic*. (See the related discussion in Section 4.2, "Column Theory.")

For convenience, we first categorize beams as compact, noncompact, or slender, and then determine the moment resistance based on the degree of lateral support. The discussion in this section applies to two types of beams: (1) hot-rolled I shapes bent about the strong axis and loaded in the plane of the weak axis, and (2) channels bent about the strong axis and either loaded through the shear center or restrained against twisting. (The shear center is the point on the cross section through which a transverse load must pass if the beam is to bend without twisting.) Emphasis will be on I shapes. C-shapes are different only in that the width–thickness ratio of the flange is b_f/t_f rather than $b_f/2t_f$.

We begin with *compact shapes*, defined as those whose webs are continuously connected to the flanges and that satisfy the following width–thickness ratio requirements for the flange and the web:

$$\frac{b_f}{2t_f} \leq 0.38\sqrt{\frac{E}{F_y}} \quad \text{and} \quad \frac{h}{t_w} \leq 3.76\sqrt{\frac{E}{F_y}}$$

The web criterion is met by all standard I and C shapes listed in the *Manual* for $F_y \leq 65$ ksi; therefore, in most cases only the flange ratio need be checked (note that built-up welded I shapes can have noncompact or slender webs). Most shapes will also satisfy the flange requirement and will therefore be classified as compact. The noncompact shapes are identified in the dimensions and properties table with a footnote (footnote f). Note that compression members have different criteria than flexural members, so a shape could be compact for flexure but slender for compression. As discussed in Chapter 4, shapes with slender compression elements are identified with a footnote (footnote c). If the beam is compact and has continuous lateral support, or if the unbraced length is very short, the nominal moment strength, M_n, is the full plastic moment capacity of the shape, M_p. For members with inadequate lateral support, the moment resistance is limited by the lateral-torsional buckling strength, either inelastic or elastic.

The first category, laterally supported compact beams, is quite common and is the simplest case. AISC F2.1 gives the nominal strength as

$$M_n = M_p \qquad \qquad \text{(AISC Equation F2-1)}$$

where

$$M_p = F_y Z_x$$

Example 5.3 The beam shown in Figure 5.11 is a W16 × 31 of A992 steel. It supports a reinforced concrete floor slab that provides continuous lateral support of the compression flange. The service dead load is 450 lb/ft. This load is superimposed on the beam; it does not include the weight of the beam itself. The service live load is 550 lb/ft. Does this beam have adequate moment strength?

FIGURE 5.11

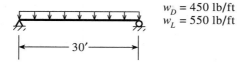

$w_D = 450$ lb/ft
$w_L = 550$ lb/ft

30'

Solution First, determine the nominal flexural strength. Check for compactness.

$$\frac{b_f}{2t_f} = 6.28 \qquad \text{(from Part 1 of the } \textit{Manual}\text{)}$$

$$0.38\sqrt{\frac{E}{F_y}} = 0.38\sqrt{\frac{29,000}{50}} = 9.15 > 6.28 \quad \therefore \text{ the flange is compact.}$$

$$\frac{h}{t_w} < 3.76\sqrt{\frac{E}{F_y}} \quad \therefore \text{ the web is compact.}$$

(The web is compact for all shapes in the *Manual* for $F_y \leq 65$ ksi.)

This shape can also be identified as compact because there is no footnote in the dimensions and properties tables to indicate otherwise. Because the beam is compact and laterally supported, the nominal flexural strength is

$$M_n = M_p = F_y Z_x = 50(54.0) = 2700 \text{ in.-kips} = 225.0 \text{ ft kips.}$$

Compute the maximum bending moment. The total service dead load, including the weight of the beam, is

$$w_D = 450 + 31 = 481 \text{ lb/ft}$$

For a simply supported, uniformly loaded beam, the maximum bending moment occurs at midspan and is equal to

$$M_{max} = \frac{1}{8}wL^2$$

where w is the load in units of force per unit length, and L is the span length. Then

$$M_D = \frac{1}{8}w_D L^2 = \frac{1}{8}(0.481)(30)^2 = 54.11 \text{ ft-kips}$$

$$M_L = \frac{1}{8}(0.550)(30)^2 = 61.88 \text{ ft-kips}$$

LRFD Solution   The dead load is less than 8 times the live load, so load combination 2 controls:

$$M_u = 1.2M_D + 1.6M_L = 1.2(54.11) + 1.6(61.88) = 164 \text{ ft-kips.}$$

Alternatively, the loads can be factored at the outset:

$$w_u = 1.2w_D + 1.6w_L = 1.2(0.481) + 1.6(0.550) = 1.457 \text{ kips/ft}$$

$$M_u = \frac{1}{8}w_u L^2 = \frac{1}{8}(1.457)(30)^2 = 164 \text{ ft-kips}$$

The design strength is

$$\phi_b M_n = 0.90(225.0) = 203 \text{ ft-kips} > 164 \text{ ft-kips} \qquad \text{(OK)}$$

Answer   The design moment is greater than the factored-load moment, so the W16 × 31 is satisfactory.

ASD Solution   ASD load combination 2 controls.

$$M_a = M_D + M_L = 54.11 + 61.88 = 116.0 \text{ ft-kips}$$

Alternatively, the loads can be added before the moment is computed:

$$w_a = w_D + w_L = 0.481 + 0.550 = 1.031 \text{ kips/ft}$$

$$M_a = \frac{1}{8}w_a L^2 = \frac{1}{8}(1.031)(30)^2 = 116.0 \text{ ft-kips}$$

The allowable moment is

$$\frac{M_n}{\Omega_b} = \frac{M_n}{1.67} = 0.6M_n = 0.6(225.0) = 135 \text{ ft-kips} > 116 \text{ ft-kips} \qquad \text{(OK)}$$

Allowable stress solution: The applied stress is

$$f_b = \frac{M_a}{S_x} = \frac{116.0(12)}{47.2} = 29.5 \text{ ksi}$$

The allowable stress is

$$F_b = \frac{0.6M_n}{S_x} = \frac{0.6(225.0)(12)}{47.2} = 34.3 \text{ ksi}$$

Since $f_b < F_b$, the beam has enough strength.

Answer   The W16 × 31 is satisfactory.   ∎

The allowable stress solution can be simplified if a slight approximation is made. The allowable stress can be written as

$$F_b = \frac{0.6M_n}{S_x} = \frac{0.6F_y Z_x}{S_x}$$

If an average value of $Z_x/S_x = 1.1$ is used (this is conservative),

$$F_b = 0.6F_y(1.1) = 0.66F_y$$

If this value is used in Example 5.3,

$$F_b = 0.66(50) = 33.0 \text{ ksi}$$

which is conservative by about 4%. Thus, for compact, laterally-supported beams, the allowable stress can be conservatively taken as $0.66F_y$. (This value of allowable stress has been used in AISC allowable stress design specifications since 1963.)

We can formulate an allowable stress approach that requires no approximation if we use the plastic section modulus instead of the elastic section modulus. From

$$\frac{M_n}{\Omega_b} \geq M_a$$

and

$$\frac{M_n}{\Omega_b} = \frac{F_y Z_y}{1.67} = 0.6 F_y Z_x$$

The required plastic section modulus is

$$Z_x \geq \frac{M_a}{0.6F_y}$$

Thus, if the bending stress is based on the plastic section modulus Z_x,

$$f_b = \frac{M_a}{Z_x} \quad \text{and} \quad F_b = 0.6F_y$$

This approach is useful when designing compact, laterally-supported beams.

The moment strength of compact shapes is a function of the unbraced length, L_b, defined as the distance between points of lateral support, or bracing. In this book, we indicate points of lateral support with an "×," as shown in Figure 5.12. The relationship between the nominal strength, M_n, and the unbraced length is shown in Figure 5.13. If the unbraced length is no greater than L_p, to be defined presently, the beam is considered to have full lateral support, and $M_n = M_p$. If L_b is greater than L_p but less than or equal to the parameter L_r, the strength is based on inelastic LTB. If L_b is greater than L_r, the strength is based on elastic LTB.

FIGURE 5.12

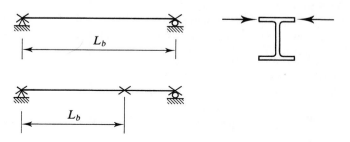

FIGURE 5.13

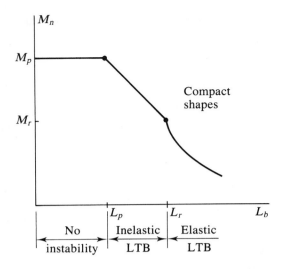

The equation for the theoretical elastic lateral-torsional buckling strength can be found in *Theory of Elastic Stability* (Timoshenko and Gere, 1961). With some notational changes, the nominal moment strength is

$$M_n = F_{cr}S_x$$

where F_{cr} is the elastic buckling stress and is given by

$$F_{cr} = \frac{\pi}{L_bS_x}\sqrt{EI_yGJ + \left(\frac{\pi E}{L_b}\right)^2 I_yC_w}, \text{ ksi} \qquad (5.5)$$

where

L_b = unbraced length (in.)
I_y = moment of inertia about the weak axis of the cross section (in.^{4})
G = shear modulus of structural steel = 11,200 ksi
J = torsional constant (in.^{4})
C_w = warping constant (in.^{6})

(The constants G, J, and C_w were defined in Chapter 4 in the discussion of torsional and lateral-torsional buckling of columns.)

Equation 5.5 is valid as long as the bending moment within the unbraced length is uniform (nonuniform moment is accounted for with a factor C_b, which is explained later). The AISC Specification gives a different, but equivalent, form for the elastic buckling stress F_{cr}. AISC gives the nominal moment strength as

$$M_n = F_{cr}S_x \le M_p \qquad \text{(AISC Equation F2-3)}$$

where

$$F_{cr} = \frac{C_b\pi^2 E}{(L_b/r_{ts})^2}\sqrt{1 + 0.078\frac{Jc}{S_xh_o}\left(\frac{L_b}{r_{ts}}\right)^2} \qquad \text{(AISC Equation F2-4)}$$

and

C_b = factor to account for nonuniform bending within the unbraced length L_b. This factor will be covered following Example 5.4.

$$r_{ts}^2 = \frac{\sqrt{I_y C_w}}{S_x} \qquad \text{(AISC Equation F2-7)}$$

$$c = 1.0 \text{ for doubly-symmetric I-shapes} \qquad \text{(AISC Equation F2-8a)}$$

$$= \frac{h_o}{2}\sqrt{\frac{I_y}{C_w}} \quad \text{for channels} \qquad \text{(AISC Equation F2-8b)}$$

h_o = distance between flange centroids = $d - t_f$

If the moment when lateral-torsional buckling occurs is greater than the moment corresponding to first yield, the strength is based on inelastic behavior. The moment corresponding to first yield is

$$M_r = 0.7F_y S_x$$

where the yield stress has been reduced by 30% to account for the effect of residual stress. As shown in Figure 5.13, the boundary between elastic and inelastic behavior will be for an unbraced length of L_r, which is the value of L_b obtained from AISC Equation F2-4 when F_{cr} is set equal to $0.7F_y$ with $C_b = 1.0$. The following equation results:

$$L_r = 1.95 r_{ts}\frac{E}{0.7F_y}\sqrt{\frac{Jc}{S_x h_o}}\sqrt{1 + \sqrt{1 + 6.76\left(\frac{0.7F_y S_x h_o}{EJc}\right)^2}} \qquad \text{(AISC Equation F2-6)}$$

As with columns, inelastic buckling of beams is more complicated than elastic buckling, and empirical formulas are often used. The following equation is used by AISC:

$$M_n = C_b\left[M_p - (M_p - 0.7F_y S_x)\left(\frac{L_b - L_p}{L_r - L_p}\right)\right] \leq M_p \qquad \text{(AISC Equation F2-2)}$$

where the $0.7F_y S_x$ term is the yield moment adjusted for residual stress, and

$$L_p = 1.76 r_y \sqrt{\frac{E}{F_y}} \qquad \text{(AISC Equation F2-5)}$$

Summary of Nominal Flexural Strength

The nominal bending strength for compact I and C-shaped sections can be summarized as follows:

For $L_b \leq L_p$,

$$M_n = M_p \qquad \text{(AISC Equation F2-1)}$$

For $L_p < L_b \le L_r$,

$$M_n = C_b \left[M_p - (M_p - 0.7F_y S_x) \left(\frac{L_b - L_p}{L_r - L_p} \right) \right] \le M_p \qquad \text{(AISC Equation F2-2)}$$

For $L_b > L_r$,

$$M_n = F_{cr} S_x \le M_p \qquad \text{(AISC Equation F2-3)}$$

where

$$F_{cr} = \frac{C_b \pi^2 E}{(L_b/r_{ts})^2} \sqrt{1 + 0.078 \frac{Jc}{S_x h_o} \left(\frac{L_b}{r_{ts}} \right)^2} \qquad \text{(AISC Equation F2-4)}$$

Example 5.4 Determine the flexural strength of a W14 × 68 of A242 steel subject to

    a. Continuous lateral support.
    b. An unbraced length of 20 ft with $C_b = 1.0$.
    c. An unbraced length of 30 ft with $C_b = 1.0$.

Solution To determine the yield stress of a W14 × 68 of A242 steel, we refer to Table 2-3 in Part 2 of the *Manual*. The yield stress is a function of the flange thickness, which for this shape is 0.720 inch. This corresponds to footnote 1, so a W14 × 68 is available in A242 steel with a yield stress F_y of 50 ksi. Next, determine whether this shape is compact, noncompact, or slender:

$$\frac{b_f}{2t_f} = 6.97 \quad \text{(from Part 1 of the Manual)}$$

$$0.38 \sqrt{\frac{E}{F_y}} = 0.38 \sqrt{\frac{29,000}{50}} = 9.15 > 6.97 \qquad \therefore \text{ the flange is compact}$$

The web is compact for all shapes in the *Manual* for $F_y \le 65$ ksi; therefore, a W14 × 68 is compact for $F_y = 50$ ksi. (This determination could also be made by observing that there is no footnote in the dimensions and properties tables to indicate that the shape is not compact.)

    a. Because the beam is compact and laterally supported, the nominal flexural strength is

$$M_n = M_p = F_y Z_x = 50(115) = 5750 \text{ in.-kips} = 479.2 \text{ ft kips}$$

LRFD Solution The design strength is

$$\phi_b M_n = 0.90(479.2) = 431 \text{ ft-kips}$$

ASD Solution   The allowable moment strength is

$$\frac{M_n}{\Omega_b} = \frac{M_n}{1.67} = 0.6M_n = 0.6(479.2) = 288 \text{ ft-kips}$$

b. $L_b = 20$ ft and $C_b = 1.0$. First, determine L_p and L_r:

$$L_p = 1.76r_y\sqrt{\frac{E}{F_y}} = 1.76(2.46)\sqrt{\frac{29,000}{50}} = 104.3 \text{ in.} = 8.692 \text{ ft}$$

The following terms will be needed in the computation of L_r:

$$r_{ts}^2 = \frac{\sqrt{I_yC_w}}{S_x} = \frac{\sqrt{121(5380)}}{103} = 7.833 \text{ in.}^2$$

$$r_{ts} = \sqrt{7.833} = 2.799 \text{ in.}$$

(r_{ts} can also be found in the dimensions and properties tables. For a W14 × 68, it is given as 2.80 in.)

$$h_o = d - t_f = 14.0 - 0.720 = 13.28 \text{ in.}$$

(h_o can also be found in the dimensions and properties tables. For a W14 × 68, it is given as 13.3 in.)

For a doubly-symmetric I-shape, $c = 1.0$. From AISC Equation F2-6,

$$L_r = 1.95r_{ts}\frac{E}{0.7F_y}\sqrt{\frac{Jc}{S_xh_o}}\sqrt{1+\sqrt{1+6.76\left(\frac{0.7F_yS_xh_o}{EJc}\right)^2}}$$

$$= 1.95(2.799)\frac{29,000}{0.7(50)}\sqrt{\frac{3.01(1.0)}{103(13.28)}}\sqrt{1+\sqrt{1+6.76\left(\frac{0.7(50)(103)(13.28)}{29,000(3.01)(1.0)}\right)^2}}$$

$$= 351.3 \text{ in.} = 29.28 \text{ ft}$$

Since $L_p < L_b < L_r$,

$$M_n = C_b\left[M_p - (M_p - 0.7F_yS_x)\left(\frac{L_b - L_p}{L_r - L_p}\right)\right] \le M_p$$

$$= 1.0\left[5,750 - (5,750 - 0.7 \times 50 \times 103)\left(\frac{20 - 8.692}{29.28 - 8.692}\right)\right]$$

$$= 4572 \text{ in.-kips} = 381.0 \text{ ft-kips} < M_p = 479.2 \text{ ft.-kips}$$

LRFD Solution   The design strength is

$$\phi_bM_n = 0.90(381.0) = 343 \text{ ft-kips}$$

ASD Solution   The allowable moment strength is

$$\frac{M_n}{\Omega_b} = 0.6M_n = 0.6(381.0) = 229 \text{ ft-kips}$$

c. $L_b = 30$ ft   and $C_b = 1.0$

$L_b > L_r = 29.28$ ft, so elastic lateral-torsional buckling controls.

From AISC Equation F2-4,

$$F_{cr} = \frac{C_b \pi^2 E}{(L_b/r_{ts})^2} \sqrt{1 + 0.078 \frac{Jc}{S_x h_o}\left(\frac{L_b}{r_{ts}}\right)^2}$$

$$= \frac{1.0\pi^2 (29,000)}{\left(\dfrac{30 \times 12}{2.799}\right)^2} \sqrt{1 + 0.078 \frac{3.01(1.0)}{103(13.28)}\left(\frac{30 \times 12}{2.799}\right)^2} = 33.90 \text{ ksi}$$

From AISC Equation F2-3,

$$M_n = F_{cr}S_x = 33.90(103) = 3492 \text{ in.-kips} = 291.0 \text{ ft-kips} < M_p = 479.2 \text{ ft-kips}$$

LRFD Solution $\phi_b M_n = 0.90(291.0) = 262$ ft-kips

ASD Solution $M_n/\Omega_b = 0.6M_n = 0.6(291.0) = 175$ ft-kips   ■

If the moment within the unbraced length L_b is uniform (constant), there is no *moment gradient* and $C_b = 1.0$. If there is a moment gradient, the value of C_b is given by

$$C_b = \frac{12.5M_{max}}{2.5M_{max} + 3M_A + 4M_B + 3M_C} R_m \leq 3.0 \qquad \text{(AISC Equation F1-1)}$$

where
M_{max} = absolute value of the maximum moment within the unbraced length (including the end points)
M_A = absolute value of the moment at the quarter point of the unbraced length
M_B = absolute value of the moment at the midpoint of the unbraced length
M_C = absolute value of the moment at the three-quarter point of the unbraced length
R_m = 1.0 for doubly-symmetric cross sections (such as W shapes) and singly symmetric shapes (such as channels) subject to single-curvature bending

$$= 0.5 + 2\left(\frac{I_{yc}}{I_y}\right)^2 \text{ for singly-symmetric shapes subject to reverse-curvature}$$

bending

I_{yc} = moment of inertia of the compression flange about the y axis. For doubly-symmetric shapes, $I_{yc} \approx I_y/2$. For reverse-curvature bending of singly-symmetric I-shaped sections, I_{yc} is the moment of inertia of the smaller flange.

All of the beams found in this textbook will be either doubly- or singly-symmetric. Most will be simply supported, so they will be subject to single curvature. Thus, R_m will usually be equal to 1.0.

When the bending moment is uniform, the value of C_b is

$$C_b = \frac{12.5M}{2.5M + 3M + 4M + 3M} = 1.0$$

where R_m has been taken as 1.0.

Example 5.5   Determine C_b for a uniformly loaded, simply supported W-shape with lateral support at its ends only.

Solution   Because of symmetry, the maximum moment is at midspan, so

$$M_{\max} = M_B = \frac{1}{8}wL^2$$

Also because of symmetry, the moment at the quarter point equals the moment at the three-quarter point. From Figure 5.14,

$$M_A = M_C = \frac{wL}{2}\left(\frac{L}{4}\right) - \frac{wL}{4}\left(\frac{L}{8}\right) = \frac{wL^2}{8} - \frac{wL^2}{32} = \frac{3}{32}wL^2$$

FIGURE 5.14

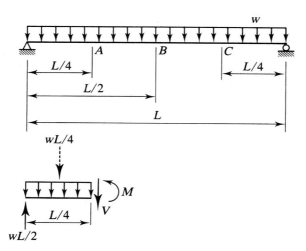

Since this is a W shape (doubly-symmetric) and is simply supported with gravity loading (single-curvature bending), $R_m = 1.0$. From AISC Equation (F1-1),

$$C_b = \frac{12.5M_{max}}{2.5M_{max} + 3M_A + 4M_B + 3M_C}$$

$$= \frac{12.5\left(\frac{1}{8}\right)}{2.5\left(\frac{1}{8}\right) + 3\left(\frac{3}{32}\right) + 4\left(\frac{1}{8}\right) + 3\left(\frac{3}{32}\right)} = 1.14$$

Answer $C_b = 1.14$.

In future examples, unless the unusual case of reverse-curvature bending is present, the equation for C_b will be shown as above—that is, without R_m explicitly shown.

Figure 5.15 shows the value of C_b for several common cases of loading and lateral support. Values of C_b for other cases can be found in Part 3 of the *Manual*, "Design of Flexural Members."

FIGURE 5.15

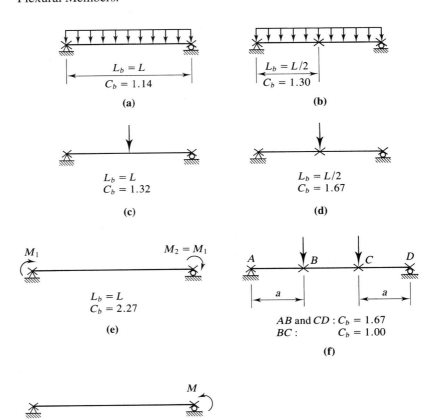

FIGURE 5.16

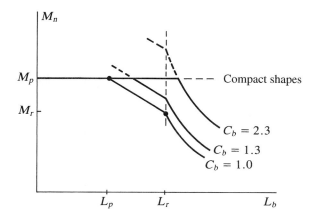

For unbraced cantilever beams, AISC specifies a value of C_b of 1.0. A value of 1.0 is always conservative, regardless of beam configuration or loading, but in some cases it may be excessively conservative.

The effect of C_b on the nominal strength is illustrated in Figure 5.16. Although the strength is directly proportional to C_b, this graph clearly shows the importance of observing the upper limit of M_p, regardless of which equation is used for M_n.

Part 3 of the *Steel Construction Manual,* "Design of Flexural Members," contains several useful tables and charts for the analysis and design of beams. For example, Table 3-2, "W Shapes, Selection by Z_x," (hereafter referred to as the "Z_x table"), lists shapes commonly used as beams, arranged in order of available flexural strength—both $\phi_b M_{px}$ and M_{px}/Ω_b. Other useful constants that are tabulated include L_p and L_r (which is particularly tedious to compute). These two constants can also be found in several other tables in Part 3 of the *Manual.* We cover additional design aids in other sections of this chapter.

5.6   BENDING STRENGTH OF NONCOMPACT SHAPES

As previously noted, most standard W-, M-, S-, and C-shapes are compact. A few are noncompact because of the flange width–thickness ratio, but none are slender.

In general, a noncompact beam may fail by lateral-torsional buckling, flange local buckling, or web local buckling. Any of these types of failure can be in either the elastic range or the inelastic range. The strength corresponding to each of these three limit states must be computed, and the smallest value will control.

From AISC F3, for flange local buckling, if $\lambda_p < \lambda \leq \lambda_r$, the flange is noncompact, buckling will be inelastic, and

$$M_n = M_p - (M_p - 0.7F_y S_x)\left(\frac{\lambda - \lambda_p}{\lambda_r - \lambda_p}\right)$$ (AISC Equation F3-1)

where

$$\lambda = \frac{b_f}{2t_f}$$

$$\lambda_p = 0.38\sqrt{\frac{E}{F_y}}$$

$$\lambda_r = 1.0\sqrt{\frac{E}{F_y}}$$

The webs of all hot-rolled shapes in the *Manual* are compact, so the noncompact shapes are subject only to the limit states of lateral-torsional buckling and flange local buckling. Built-up welded shapes, however, can have noncompact or slender webs as well as noncompact or slender flanges. These cases are covered in AISC Sections F4 and F5. Built-up shapes, including plate girders, are covered in Chapter 10 of this textbook.

Example 5.6   A simply supported beam with a span length of 45 feet is laterally supported at its ends and is subjected to the following service loads:

> Dead load = 400 lb/ft (including the weight of the beam)
> Live load = 1000 lb/ft

If $F_y = 50$ ksi, is a W14 × 90 adequate?

Solution   Determine whether the shape is compact, noncompact, or slender:

$$\lambda = \frac{b_f}{2t_f} = 10.2$$

$$\lambda_p = 0.38\sqrt{\frac{E}{F_y}} = 0.38\sqrt{\frac{29,000}{50}} = 9.15$$

$$\lambda_r = 1.0\sqrt{\frac{E}{F_y}} = 1.0\sqrt{\frac{29,000}{50}} = 24.1$$

Since $\lambda_p < \lambda < \lambda_r$, this shape is noncompact. Check the capacity based on the limit state of flange local buckling:

$$M_p = F_y Z_x = 50(157) = 7850 \text{ in.-kips}$$

$$M_n = M_p - (M_p - 0.7F_y S_x)\left(\frac{\lambda - \lambda_p}{\lambda_r - \lambda_p}\right)$$

$$= 7850 - (7850 - 0.7 \times 50 \times 143)\left(\frac{10.2 - 9.15}{24.1 - 9.15}\right) = 7650 \text{ in.-kips} = 637.5 \text{ ft-kips}$$

Check the capacity based on the limit state of lateral-torsional buckling. From the Z_x table,

$$L_p = 15.2 \text{ ft} \quad \text{and} \quad L_r = 42.6 \text{ ft}$$
$$L_b = 45 \text{ ft} > L_r \qquad \therefore \text{ failure is by } \textit{elastic} \text{ LTB}$$

From Part 1 of the *Manual*,

$$I_y = 362 \text{ in.}^4$$
$$r_{ts} = 4.11 \text{ in.}$$
$$h_o = 13.3 \text{ in.}$$
$$J = 4.06 \text{ in.}^4$$
$$C_w = 16{,}000 \text{ in.}^6$$

For a uniformly loaded, simply supported beam with lateral support at the ends,

$$C_b = 1.14 \qquad \text{(Fig. 5.15a)}$$

For a doubly-symmetric I-shape, $c = 1.0$. AISC Equation F2-4 gives

$$F_{cr} = \frac{C_b \pi^2 E}{(L_b/r_{ts})^2} \sqrt{1 + 0.078 \frac{Jc}{S_x h_o} \left(\frac{L_b}{r_{ts}}\right)^2}$$

$$= \frac{1.14 \pi^2 (29{,}000)}{\left(\dfrac{45 \times 12}{4.11}\right)^2} \sqrt{1 + 0.078 \frac{4.06(1.0)}{143(13.3)} \left(\frac{45 \times 12}{4.11}\right)^2} = 37.20 \text{ ksi}$$

From AISC Equation F2-3,

$$M_n = F_{cr} S_x = 37.20(143) = 5320 \text{ in.-kips} < M_p = 7850 \text{ in.-kips}$$

This is smaller than the nominal strength based on flange local buckling, so lateral-torsional buckling controls.

LRFD Solution   The design strength is

$$\phi_b M_n = 0.90(5320) = 4788 \text{ in.-kips} = 399 \text{ ft-kips}$$

The factored load and moment are

$$w_u = 1.2 w_D + 1.6 w_L = 1.2(0.400) + 1.6(1.000) = 2.080 \text{ kips/ft}$$
$$M_u = \frac{1}{8} w_u L^2 = \frac{1}{8}(2.080)(45)^2 = 527 \text{ ft-kips} > 399 \text{ ft-kips} \quad \text{(N.G.)}$$

Answer   Since $M_u > \phi_b M_n$, the beam does not have adequate moment strength.

ASD Solution   The allowable stress is

$$F_b = 0.6 F_{cr} = 0.6(37.20) = 22.3 \text{ ksi}$$

The applied bending moment is

$$M_a = \frac{1}{8} w_a L^2 = \frac{1}{8}(0.400 + 1.000)(45)^2 = 354.4 \text{ ft-kips}$$

and the applied stress is

$$f_b = \frac{M_a}{S_x} = \frac{354.4(12)}{143} = 29.7 \text{ ksi} > 22.3 \text{ ksi} \quad \text{(N.G.)}$$

Answer   Since $f_b > F_b$, the beam does not have adequate moment strength.   ■

The identification of noncompact shapes is facilitated by the Z_x table in that non-compact shapes are identified by an "f" footnote (this same identification is used in the dimensions and properties tables). Noncompact shapes are also treated differently in the Z_x table in the following way. The tabulated value of L_p is the value of unbraced length at which the nominal strength based on inelastic lateral-torsional buckling equals the nominal strength based on flange local buckling, that is, the maximum un-braced length for which the nominal strength can be taken as the strength based on flange local buckling. (Recall that L_p for compact shapes is the maximum unbraced length for which the nominal strength can be taken as the plastic moment.) For the shape in Example 5.6, equate the nominal strength based on FLB to the strength based on inelastic LTB (AISC Equation F2-2), with $C_b = 1.0$:

$$M_n = M_p - (M_p - 0.7F_y S_x)\left(\frac{L_b - L_p}{L_r - L_p}\right) \tag{5.6}$$

The value of L_r was given in Example 5.6 and is unchanged. The value of L_p, however, must be computed from AISC Equation F2-5:

$$L_p = 1.76r_y\sqrt{\frac{E}{F_y}} = 1.76(3.70)\sqrt{\frac{29,000}{50}} = 156.8 \text{ in.} = 13.07 \text{ ft}$$

Returning to Equation 5.6, we obtain

$$7650 = 7850 - (7850 - 0.7 \times 50 \times 143)\left(\frac{L_b - 13.07}{42.6 - 13.07}\right)$$

$$L_b = 15.2 \text{ ft}$$

This is the value tabulated as L_p for a W14 × 90 with $F_y = 50$ ksi. Note that

$$L_p = 1.76r_y\sqrt{\frac{E}{F_y}}$$

could still be used for noncompact shapes. If doing so resulted in the equation for in-elastic LTB being used when L_b was not really large enough, the strength based on FLB would control anyway.

In addition to the different meaning of L_p for noncompact shapes in the Z_x table, the available strength values, $\phi_b M_{px}$ and M_p/Ω_b, are based on flange local buckling rather than the plastic moment.

5.7 SUMMARY OF MOMENT STRENGTH

The procedure for computation of nominal moment strength for I-shaped sections bent about the x axis will now be summarized. All terms in the following equations have been previously defined, and AISC equation numbers will not be shown. This summary is for compact and noncompact shapes (noncompact flanges) only (no slender shapes).

1. Determine whether the shape is compact.
2. If the shape is compact, check for lateral-torsional buckling as follows:

   Using $L_p = 1.76 r_y \sqrt{\dfrac{E}{F_y}}$,

   If $L_b \leq L_p$, there is no LTB, and $M_n = M_p$
   If $L_p < L_b \leq L_r$, there is inelastic LTB, and

   $$M_n = C_b \left[M_p - (M_p - 0.7 F_y S_x) \left(\frac{L_b - L_p}{L_r - L_p} \right) \right] \leq M_p$$

   If $L_b > L_r$, there is elastic LTB, and

   $$M_n = F_{cr} S_x \leq M_p$$

   where

   $$F_{cr} = \frac{C_b \pi^2 E}{(L_b/r_{ts})^2} \sqrt{1 + 0.078 \frac{Jc}{S_x h_o} \left(\frac{L_b}{r_{ts}} \right)^2}$$

3. If the shape is noncompact because of the flange, the nominal strength will be the smaller of the strengths corresponding to flange local buckling and lateral-torsional buckling.

   a. Flange local buckling:
      If $\lambda \leq \lambda_p$, there is no FLB
      If $\lambda_p < \lambda \leq \lambda_r$, the flange is noncompact, and

      $$M_n = M_p - (M_p - 0.7 F_y S_x) \left(\frac{\lambda - \lambda_p}{\lambda_r - \lambda_p} \right)$$

   b. Lateral-torsional buckling:
      Using $L_p = 1.76 r_y \sqrt{\dfrac{E}{F_y}}$,
      If $L_b \leq L_p$, there is no LTB

If $L_p < L_b \le L_r$, there is inelastic LTB, and

$$M_n = C_b \left[M_p - (M_p - 0.7F_y S_x) \left(\frac{L_b - L_p}{L_r - L_p} \right) \right] \le M_p$$

If $L_b > L_r$, there is elastic LTB, and

$$M_n = F_{cr} S_x \le M_p$$

where

$$F_{cr} = \frac{C_b \pi^2 E}{(L_b/r_{ts})^2} \sqrt{1 + 0.078 \frac{Jc}{S_x h_o} \left(\frac{L_b}{r_{ts}} \right)^2}$$

5.8 SHEAR STRENGTH

Beam shear strength is covered in Chapter G of the AISC Specification, "Design of Members for Shear." Both hot-rolled shapes and welded built-up shapes are covered. We discuss hot-rolled shapes in the present chapter of this book and built-up shapes in Chapter 10, "Plate Girders." The AISC provisions for hot-rolled shapes are covered in Section G2.1.

Before covering the AISC provisions for shear strength, we will first review some basic concepts from mechanics of materials. Consider the simple beam of Figure 5.17. At a distance x from the left end and at the neutral axis of the cross section, the state of stress is as shown in Figure 5.17d. Because this element is located at the neutral axis, it is not subjected to flexural stress. From elementary mechanics of materials, the shearing stress is

$$f_v = \frac{VQ}{Ib} \tag{5.7}$$

where
  f_v = vertical and horizontal shearing stress at the point of interest
  V = vertical shear force at the section under consideration
  Q = first moment, about the neutral axis, of the area of the cross section
    between the point of interest and the top or bottom of the cross section
  I = moment of inertia about the neutral axis
  b = width of the cross section at the point of interest

Equation 5.7 is based on the assumption that the stress is constant across the width b, and it is therefore accurate only for small values of b. For a rectangular cross section of depth d and width b, the error for $d/b = 2$ is approximately 3%. For $d/b = 1$, the error is 12% and for $d/b = \frac{1}{4}$, it is 100% (Higdon, Ohlsen, and Stiles, 1960). For this reason, Equation 5.7 cannot be applied to the flange of a W-shape in the same manner as for the web.

FIGURE 5.17

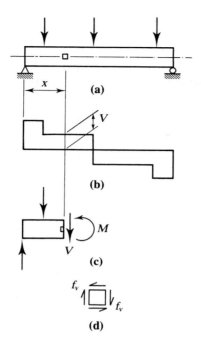

(a)

(b)

(c)

(d)

Figure 5.18 shows the shearing stress distribution for a W-shape. Superimposed on the actual distribution is the average stress in the web, V/A_w, which does not differ much from the maximum web stress. Clearly, the web will completely yield long before the flanges begin to yield. Because of this, yielding of the web represents one of the shear limit states. Taking the shear yield stress as 60% of the tensile yield stress, we can write the equation for the stress in the web at failure as

$$f_v = \frac{V_n}{A_w} = 0.6F_y$$

where A_w = area of the web. The nominal strength corresponding to this limit state is therefore

$$V_n = 0.6F_yA_w \tag{5.8}$$

FIGURE 5.18

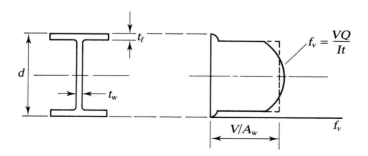

and will be the nominal strength in shear provided that there is no shear buckling of the web. Whether that occurs will depend on h/t_w, the width–thickness ratio of the web. If this ratio is too large — that is, if the web is too slender — the web can buckle in shear, either inelastically or elastically.

AISC Specification Requirements for Shear

For LRFD, the relationship between required and available strength is

$$V_u \le \phi_v V_n$$

where
  V_u = maximum shear based on the controlling combination of factored loads
  ϕ_v = resistance factor for shear

For ASD, the relationship is

$$V_a \le \frac{V_n}{\Omega_v}$$

where
  V_a = maximum shear based on the controlling combination of service loads
  Ω_v = safety factor for shear

As we will see, the values of the resistance factor and safety factor will depend on the web width-thickness ratio.

Section G2.1 of the AISC Specification covers both beams with stiffened webs and beams with unstiffened webs. In most cases, hot-rolled beams will not have stiffeners, and we will defer treatment of stiffened webs until Chapter 10. The basic strength equation is

$$V_n = 0.6F_y A_w C_v \qquad \text{(AISC Equation G2-1)}$$

where
  A_w = area of the web $\approx dt_w$
  d = overall depth of the beam
  C_v = ratio of critical web stress to shear yield stress

The value of C_v depends on whether the limit state is web yielding, web inelastic buckling, or web elastic buckling.

For the special case of hot-rolled I shapes with

$$\frac{h}{t_w} \le 2.24\sqrt{\frac{E}{F_y}}$$

The limit state is shear yielding, and

  $C_v = 1.0$ (AISC Equation G2-2)
  $\phi_v = 1.00$
  $\Omega_v = 1.50$

Most W shapes with $F_y \le 50$ ksi fall into this category (see User Note in AISC G2.1[a]).

For all other doubly and singly symmetric shapes, except for round HSS (see AISC G6),

$$\phi_v = 0.90$$
$$\Omega_v = 1.67$$

and C_v is determined as follows:

For $\dfrac{h}{t_w} \leq 1.10\sqrt{\dfrac{k_v E}{F_y}}$, there is no web instability, and

$$C_v = 1.0 \qquad\qquad\qquad\qquad \text{(AISC Equation G2-3)}$$

(This corresponds to Equation 5.8 for shear yielding.)

For $1.10\sqrt{\dfrac{K_v E}{F_y}} < \dfrac{h}{t_w} \leq 1.37\sqrt{\dfrac{K_v E}{F_y}}$, inelastic web buckling can occur, and

$$C_v = \frac{1.10\sqrt{\dfrac{k_v E}{F_y}}}{h/t_w} \qquad\qquad\qquad \text{(AISC Equation G2-4)}$$

For $\dfrac{h}{t_w} > 1.37\sqrt{\dfrac{k_v E}{F_y}}$, the limit state is elastic web buckling:

$$C_v = \frac{1.51 E k_v}{(h/t_w)^2 F_y} \qquad\qquad\qquad \text{(AISC Equation G2-5)}$$

where
$$k_v = 5$$

This value of k_v is for unstiffened webs with $h/t_w < 260$. Although section G2.1 of the Specification does not give $h/t_w = 260$ as an upper limit, no value of k_v is given when $h/t_w \geq 260$. In addition, AISC F13.2, "Proportioning Limits for I-Shaped Members," states that h/t_w in unstiffened girders shall not exceed 260.

AISC Equation G2-5 is based on elastic stability theory, and AISC Equation G2-4 is an empirical equation for the inelastic region, providing a transition between the limit states of web yielding and elastic web buckling.

The relationship between shear strength and the web width-thickness ratio is analogous to that between flexural strength and the width–thickness ratio (for FLB) and between flexural strength and the unbraced length (for LTB). This relationship is illustrated in Figure 5.19.

Allowable Stress Formulation

The allowable strength relation

$$V_a \leq \frac{V_n}{\Omega_v}$$

FIGURE 5.19

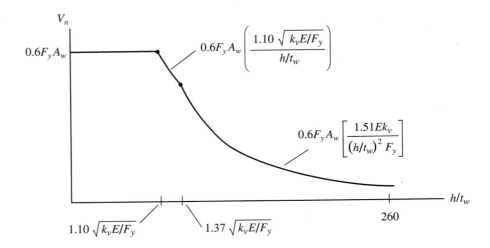

can also be written in terms of stress as

$$f_v \le F_v$$

where

$$f_v = \frac{V_a}{A_w} = \text{applied shear stress}$$

$$F_v = \frac{V_n/\Omega_v}{A_w} = \frac{0.6F_yA_wC_v/\Omega_v}{A_w} = \text{allowable shear stress}$$

For the most common case of hot-rolled I-shapes with $h/t_w \le 2.24\sqrt{E/F_y}$,

$$F_v = \frac{0.6F_yA_w(1.0)/1.50}{A_w} = 0.4F_y$$

Shear is rarely a problem in rolled steel beams; the usual practice is to design a beam for flexure and then to check it for shear.

Example 5.7   Check the beam in Example 5.6 for shear.

Solution   From the dimensions and properties tables in Part 1 of the *Manual*, the web width–thickness ratio of a W14 × 90 is

$$\frac{h}{t_w} = 25.9$$

and the web area is $A_w = dt_w = 14.0(0.440) = 6.160$ in.^{2}

$$2.24\sqrt{\frac{E}{F_y}} = 2.24\sqrt{\frac{29,000}{50}} = 54.0$$

Since

$$\frac{h}{t_w} < 2.24\sqrt{\frac{E}{F_y}}$$

the strength is governed by shear yielding of the web and $C_v = 1.0$. (As pointed out in the Specification User Note, this will be the case for most W shapes with $F_y \leq 50$ ksi.) The nominal shear strength is

$$V_n = 0.6F_y A_w C_v = 0.6(50)(6.160)(1.0) = 184.8 \text{ kips}$$

LRFD Solution   Determine the resistance factor ϕ_v.

Since $\dfrac{h}{t_w} < 2.24\sqrt{\dfrac{E}{F_y}}$,

$$\phi_v = 1.00$$

and the design shear strength is

$$\phi_v V_n = 1.00(184.8) = 185 \text{ kips}$$

From Example 5.6, $w_u = 2.080$ kips/ft and $L = 45$ ft. For a simply supported, uniformly loaded beam, the maximum shear occurs at the support and is equal to the reaction.

$$V_u = \frac{w_u L}{2} = \frac{2.080(45)}{2} = 46.8 \text{ kips} < 185 \text{ kips} \quad \text{(OK)}$$

ASD Solution   Determine the safety factor Ω_v.

Since $\dfrac{h}{t_w} < 2.24\sqrt{\dfrac{E}{F_y}}$,

$$\Omega_v = 1.50$$

and the allowable shear strength is

$$\frac{V_n}{\Omega_v} = \frac{184.8}{1.50} = 123 \text{ kips}$$

From Example 5.6, the total service load is

$$w_a = w_D + w_L = 0.400 + 1.000 = 1.4 \text{ kips/ft}$$

The maximum shear is

$$V_a = \frac{w_a L}{2} = \frac{1.4(45)}{2} = 31.5 \text{ kips} < 123 \text{ kips} \quad \text{(OK)}$$

Alternately, a solution in terms of stress can be done. Since shear yielding controls ($C_v = 1.0$) and $\Omega_v = 1.50$, the allowable shear stress is

$$F_v = 0.4F_y = 0.4(50) = 20 \text{ ksi}$$

The required shear strength (stress) is

$$f_a = \frac{V_a}{A_w} = \frac{31.5}{6.160} = 5.11 \text{ ksi} < 20 \text{ ksi} \quad \text{(OK)}$$

Answer The required shear strength is less than the available shear strength, so the beam is satisfactory. ∎

Values of $\phi_v V_n$ and V_n/Ω_v are given in several tables in Part 3 of the *Manual*, including the Z_x table, so computation of shear strength is unnecessary for hot-rolled shapes.

Block Shear

Block shear, which was considered earlier in conjunction with tension member connections, can occur in certain types of beam connections. To facilitate the connection of beams to other beams so that the top flanges are at the same elevation, a short length of the top flange of one of the beams may be cut away, or *coped*. If a coped beam is connected with bolts as in Figure 5.20, segment *ABC* will tend to tear out.

FIGURE 5.20

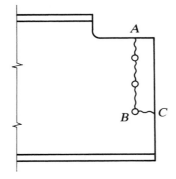

The applied load in this case will be the vertical beam reaction, so shear will occur along line AB and there will be tension along BC. Thus the block shear strength will be a limiting value of the reaction.

We covered the computation of block shear strength in Chapter 3, but we will review it here. Failure is assumed to occur by rupture (fracture) on the shear area (subject to an upper limit) and rupture on the tension area. AISC J4.3, "Block Shear Strength," gives the following equation for block shear strength:

$$R_n = 0.6F_uA_{nv} + U_{bs}F_uA_{nt} \le 0.6F_yA_{gv} + U_{bs}F_uA_{nt} \qquad \text{(AISC Equation J4-5)}$$

where

A_{gv} = gross area in shear (in Figure 5.20, length AB times the web thickness)
A_{nv} = net area along the shear surface or surfaces
A_{nt} = net area along the tension surface (in Figure 5.20, along BC)
U_{bs} = 1.0 when the tensile stress is uniform (for most coped beams)
    = 0.5 when the tension stress is not uniform (coped beams with two lines
       of bolts or with nonstandard distance from bolts to end of beam)

For LRFD, $\phi = 0.75$. For ASD, $\Omega = 2.00$.

Example 5.8   Determine the maximum reaction, based on block shear, that can be resisted by the beam shown in Figure 5.21. Treat the bolt end distance of $1\frac{1}{4}$ inch as standard.

FIGURE 5.21

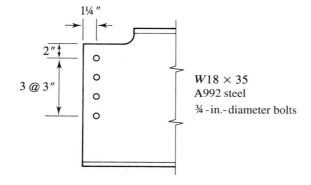

$W18 \times 35$
A992 steel
¾-in.-diameter bolts

Solution   The effective hole diameter is $\dfrac{3}{4} + \dfrac{1}{8} = \dfrac{7}{8}$ in.   The shear areas are

$$A_{gv} = t_w(2+3+3+3) = 0.300(11) = 3.300 \text{ in.}^2$$

$$A_{nv} = 0.300\left[11 - 3.5\left(\frac{7}{8}\right)\right] = 2.381 \text{ in.}^2$$

The net tension area is

$$A_{nt} = 0.300\left[1.25 - \frac{1}{2}\left(\frac{7}{8}\right)\right] = 0.2438 \text{ in.}^2$$

Since the block shear will occur in a coped beam with a standard bolt end distance, $U_{bs} = 1.0$. From AISC Equation J4-5,

$$R_n = 0.6F_uA_{nv} + U_{bs}F_uA_{nt} = 0.6(65)(2.381) + 1.0(65)(0.2438) = 108.7 \text{ kips}$$

with an upper limit of

$$0.6F_yA_{gv} + U_{bs}F_uA_{nt} = 0.6(65)(3.300) + 1.0(65)(0.2438) = 144.5 \text{ kips}$$

The nominal block shear strength is therefore 108.7 kips.

LRFD Solution   The maximum factored load reaction is the design strength: $\phi R_n = 0.75(108.7) = 81.5 \text{ kips}$

ASD Solution   The maximum service load reaction is the allowable strength: $\dfrac{R_n}{\Omega} = \dfrac{108.7}{2.00} = 54.4 \text{ kips}$

■

5.9   DEFLECTION

In addition to being safe, a structure must be *serviceable*. A serviceable structure is one that performs satisfactorily, not causing any discomfort or perceptions of unsafety for the occupants or users of the structure. For a beam, being serviceable usually means that the deformations, primarily the vertical sag, or deflection, must be limited. Excessive deflection is usually an indication of a very flexible beam, which can lead to problems with vibrations. The deflection itself can cause problems if elements attached to the beam can be damaged by small distortions. In addition, users of the structure may view large deflections negatively and wrongly assume that the structure is unsafe.

For the common case of a simply supported, uniformly loaded beam such as that in Figure 5.22, the maximum vertical deflection is

$$\Delta = \frac{5}{384}\frac{wL^4}{EI}$$

Deflection formulas for a variety of beams and loading conditions can be found in Part 3, "Design of Flexural Members," of the *Manual*. For more unusual situations, standard analytical methods such as the method of virtual work may be used. Deflection is a serviceability limit state, not one of strength, so deflections should always be computed with *service* loads.

FIGURE 5.22

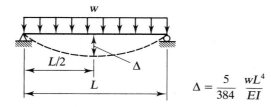

$$\Delta = \frac{5}{384} \frac{wL^4}{EI}$$

TABLE 5.4
Deflection Limits

Type of member	Max. live load defl.	Max. dead + live load defl.	Max. snow or wind load defl.
Roof beam:			
Supporting plaster ceiling	$L/360$	$L/240$	$L/360$
Supporting nonplaster ceiling	$L/240$	$L/180$	$L/240$
Not supporting a ceiling	$L/180$	$L/120$	$L/180$
Floor beam	$L/360$	$L/240$	—

The appropriate limit for the maximum deflection depends on the function of the beam and the likelihood of damage resulting from the deflection. The AISC Specification furnishes little guidance other than a statement in Chapter L, "Design for Serviceability," that deflections should not be excessive. Appropriate limits for deflection can usually be found from the governing building code, expressed as a fraction of the span length L, such as $L/360$. Sometimes a numerical limit, such as 1 inch, is appropriate. The limits given in the International Building Code (ICC, 2003) are typical. Table 5.4 shows some of the deflection limits given by that code.

The limits shown in Table 5.4 for deflection due to dead load plus live load do not apply to steel beams, because the dead load deflection is usually compensated for by some means, such as *cambering*. Camber is a curvature in the opposite direction of the dead load deflection curve and can be accomplished by bending the beam, with or without heat. When the dead load is applied to the cambered beam, the curvature is removed, and the beam becomes level. Therefore, only the live load deflection is of concern in the completed structure. Dead load deflection can also be accounted for by pouring a variable depth slab with a level top surface, the variable depth being a consequence of the deflection of the beam (this is referred to as *ponding* of the concrete). Detailed coverage of control of dead load deflection is given in an AISC seminar series (AISC, 1997a) and several papers (Ruddy, 1986; Ricker, 1989; and Larson and Huzzard, 1990).

Example 5.9

Compute the dead load and live load deflections for the beam shown in Figure 5.23. If the maximum permissible *live* load deflection is $L/360$, is the beam satisfactory?

Solution

It is more convenient to express the deflection in inches than in feet, so units of inches are used in the deflection formula. The dead load deflection is

$$\Delta_D = \frac{5}{384} \frac{w_D L^4}{EI} = \frac{5}{384} \frac{(0.500/12)(30 \times 12)^4}{29,000(510)} = 0.616 \text{ in.}$$

FIGURE 5.23

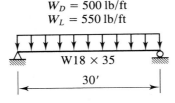

$W_D = 500$ lb/ft
$W_L = 550$ lb/ft

W18 × 35

30'

The live load deflection is

$$\Delta_L = \frac{5}{384} \frac{w_L L^4}{EI} = \frac{5}{384} \frac{(0.550/12)(30 \times 12)^4}{29,000(510)} = 0.678 \text{ in.}$$

The maximum permissible live load deflection is

$$\frac{L}{360} = \frac{30(12)}{360} = 1.0 \text{ in.} > 0.678 \text{ in.} \quad \text{(OK)}$$

Answer   The beam satisfies the deflection criterion.

Ponding is one deflection problem that does affect the safety of a structure. It is a potential hazard for flat roof systems that can trap rainwater. If drains become clogged during a storm, the weight of the water will cause the roof to deflect, thus providing a reservoir for still more water. If this process proceeds unabated, collapse can occur. The AISC specification requires that the roof system have sufficient stiffness to prevent ponding, and it prescribes limits on stiffness parameters in Appendix 2, "Design for Ponding."

5.10   DESIGN

Beam design entails the selection of a cross-sectional shape that will have enough strength and that will meet serviceability requirements. As far as strength is concerned, flexure is almost always more critical than shear, so the usual practice is to design for flexure and then check shear. The design process can be outlined as follows.

1. Compute the required moment strength (i.e., the factored load moment M_u for LRFD or the unfactored moment M_a for ASD). The weight of the beam is part of the dead load but is unknown at this point. A value may be assumed and verified after a shape is selected, or the weight may be ignored initially and checked after a shape has been selected. Because the beam weight is usually a small part of the total load, if it is ignored at the beginning of a design problem, the selected shape will usually be satisfactory when the moment is recomputed.

2. Select a shape that satisfies this strength requirement. This can be done in one of two ways.
   a. Assume a shape, compute the available strength, and compare it with the required strength. Revise if necessary. The trial shape can be easily selected in only a limited number of situations (as in Example 5.10).
   b. Use the beam design charts in Part 3 of the *Manual*. This method is preferred, and we explain it following Example 5.10.
3. Check the shear strength.
4. Check the deflection.

Example 5.10   Select a standard hot-rolled shape of A992 steel for the beam shown in Figure 5.24. The beam has continuous lateral support and must support a uniform service live load of 4.5 kips/ft. The maximum permissible live load deflection is $L/240$.

FIGURE 5.24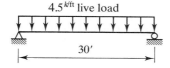

4.5 $^{k/ft}$ live load

30′

LRFD Solution   Ignore the beam weight initially then check for its effect after a selection is made.

$$w_u = 1.2w_D + 1.6w_L = 1.2(0) + 1.6(4.5) = 7.2 \text{ kips/ft}$$

Required moment strength $M_u = \dfrac{1}{8}w_u L^2 = \dfrac{1}{8}(7.2)(30)^2 = 810.0$ ft-kips

$$= \text{required } \phi_b M_n$$

Assume that the shape will be compact. For a compact shape with full lateral support,

$$M_n = M_p = F_y Z_x$$

From $\phi_b M_n \geq M_u$,

$$\phi_b F_y Z_x \geq M_u$$

$$Z_x \geq \frac{M_u}{\phi_b F_y} = \frac{810.0(12)}{0.90(50)} = 216 \text{ in.}^3$$

The Z_x table lists hot-rolled shapes normally used as beams in order of decreasing plastic section modulus. Furthermore, they are grouped so that the shape at the top of each group (in bold type) is the lightest one that has enough section modulus to satisfy a required section modulus that falls within the group. In this example, the shape that comes closest to meeting the section modulus requirement is a W21 × 93, with

$Z_x = 221$ in.^{3}, but the lightest one is a W24 × 84, with $Z_x = 224$ in.^{3}. Because section modulus is not directly proportional to area, it is possible to have more section modulus with less area, and hence less weight.

Try a W24 × 84. This shape is compact, as assumed (noncompact shapes are marked as such in the table); therefore $M_n = M_p$, as assumed.

Account for the beam weight.

$$w_u = 1.2w_D + 1.6w_L = 1.2(0.084) + 1.6(4.5) = 7.301 \text{ kips/ft}$$

$$\text{Required moment strength} = M_u = \frac{1}{8}w_uL^2 = \frac{1}{8}(7.301)(30)^2 = 821.4 \text{ ft-kips}$$

The required section modulus is

$$Z_x = \frac{M_u}{\phi_b F_y} = \frac{821.4(12)}{0.90(50)} = 219 \text{ in.}^3 < 224 \text{ in.}^3 \quad \text{(OK)}$$

In lieu of basing the search on the required section modulus, the design strength $\phi_b M_p$ could be used, because it is directly proportional to Z_x and is also tabulated. Next, check the shear:

$$V_u = \frac{w_u L}{2} = \frac{7.301(30)}{2} = 110 \text{ kips}$$

From the Z_x table,

$$\phi_v V_n = 340 \text{ kips} > 110 \text{ kips} \quad \text{(OK)}$$

Finally, check the deflection. The maximum permissible live load deflection is $L/240 = (30 \times 12)/240 = 1.5$ inch.

$$\Delta_L = \frac{5}{384} \frac{w_L L^4}{EI_x} = \frac{5}{384} \frac{(4.5/12)(30 \times 12)^4}{29,000(2370)} = 1.19 \text{ in.} < 1.5 \text{ in.} \quad \text{(OK)}$$

Answer Use a W24 × 84.

ASD Solution Ignore the beam weight initially, then check for its effect after a selection is made.

$$w_a = w_D + w_L = 0 + 4.5 = 4.5 \text{ kips/ft}$$

$$\text{Required moment strength} = M_a = \frac{1}{8}w_aL^2 = \frac{1}{8}(4.5)(30)^2 = 506.3 \text{ ft-kips}$$

$$= \text{required } \frac{M_n}{\Omega_b}$$

Assume that the shape will be compact. For a compact shape with full lateral support,

$$M_n = M_p = F_y Z_x$$

From $\dfrac{M_p}{\Omega_b} \geq M_a$,

$$\frac{F_y Z_x}{\Omega_b} \geq M_a$$

$$Z_x \geq \frac{\Omega_b M_a}{F_y} = \frac{1.67(506.3 \times 12)}{50} = 203 \text{ in.}^3$$

The Z_x table lists hot-rolled shapes normally used as beams in order of decreasing plastic section modulus. They are arranged in groups, with the lightest shape in each group at the top of that group. For the current case, the shape with a section modulus closest to 203 in.^{3} is a W18 × 97, but the lightest shape with sufficient section modulus is a W24 × 84, with $Z_x = 224$ in.^{3}

Try a W24 × 84. This shape is compact, as assumed (if it were noncompact, there would be a footnote in the Z_x table). Therefore, $M_n = M_p$ as assumed. Account for the beam weight:

$$w_a = w_D + w_L = 0.084 + 4.5 = 4.584 \text{ kips/ft}$$

$$M_a = \frac{1}{8} w_a L^2 = \frac{1}{8}(4.584)(30)^2 = 515.7 \text{ ft-kips}$$

The required plastic section modulus is

$$Z_x = \frac{\Omega_b M_a}{F_y} = \frac{1.67(515.7 \times 12)}{50} = 207 \text{ in.}^3 < 224 \text{ in.}^3 \quad \text{(OK)}$$

Instead of searching for the required section modulus, the search could be based on the required value of M_p/Ω_b, which is also tabulated. Because M_p/Ω_b is proportional to Z_x, the results will be the same.

Another approach is to use the allowable stress for compact laterally supported shapes. From Section 5.5 of this book, with the flexural stress based on the plastic section modulus,

$$F_b = 0.6F_y = 0.6(50) = 30.0 \text{ ksi}$$

and the required section modulus (before the beam weight is included) is

$$Z_x = \frac{M_a}{F_b} = \frac{506.3 \times 12}{30} = 203 \text{ in.}^3$$

Next, check the shear. The required shear strength is

$$V_a = \frac{w_a L}{2} = \frac{4.584(30)}{2} = 68.8 \text{ kips}$$

From the Z_x table, the available shear strength is

$$\frac{V_n}{\Omega_v} = 227 \text{ kips} > 68.8 \text{ kips} \quad \text{(OK)}$$

Check deflection. The maximum permissible live load deflection is

$$\frac{L}{240} = \frac{30 \times 12}{240} = 1.5 \text{ in.}$$

$$\Delta_L = \frac{5}{384}\frac{w_L L^4}{EI_x} = \frac{5}{384}\frac{(4.5/12)(30 \times 12)^4}{29{,}000(2370)} = 1.19 \text{ in.} < 1.5 \text{ in.} \quad \text{(OK)}$$

Answer   Use a W24 × 84. ∎

In Example 5.10, it was first assumed that a compact shape would be used, and then the assumption was verified. However, if the search is made based on available strength ($\phi_b M_p$ or M_p/Ω_b) rather than section modulus, it is irrelevant whether the shape is compact or noncompact. This is because for noncompact shapes, the tabulated values of $\phi_b M_p$ and M_p/Ω_b are based on flange local buckling and not the plastic moment (see Section 5.6). This means that for laterally supported beams, the Z_x table can be used for design without regard to whether the shape is compact or noncompact.

Beam Design Charts

Many graphs, charts, and tables are available for the practicing engineer, and these aids can greatly simplify the design process. For the sake of efficiency, they are widely used in design offices, but you should approach their use with caution and not allow basic principles to become obscured. It is not our purpose to describe in this book all available design aids in detail, but some are worthy of note, particularly the curves of moment strength versus unbraced length given in Part 3 of the *Manual*.

These curves will be described with reference to Figure 5.25, which shows a graph of nominal moment strength as a function of unbraced length L_b for a particular compact shape. Such a graph can be constructed for any cross-sectional shape and specific values of F_y and C_b by using the appropriate equations for moment strength.

The design charts in the *Manual* comprise a family of graphs similar to the one shown in Figure 5.25. Two sets of curves are available, one for W shapes with $F_y = 50$ ksi and one for C and MC shapes with $F_y = 36$ ksi. Each graph gives the flexural strength of a standard hot-rolled shape. Instead of giving the nominal strength M_n, however, both the allowable moment strength M_n/Ω_b and the design moment strength $\phi_b M_n$ are given. Two scales are shown on the vertical axis—one for M_n/Ω_b and one for $\phi_b M_n$. All curves were generated with $C_b = 1.0$. For other values of C_b, simply multiply the moment from the chart by C_b. However, the strength can never exceed the value represented by the horizontal line at the left side of the graph. For a compact shape, this represents the strength corresponding to yielding (reaching the plastic

FIGURE 5.25

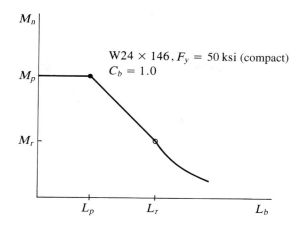

moment M_p). If the curve is for a noncompact shape, the horizontal line represents the flange local buckling strength.

Use of the charts is illustrated in Figure 5.26, where two such curves are shown. Any point on this graph, such as the intersection of the two dashed lines, represents an available moment strength and an unbraced length. If the moment is a required moment capacity, then any curve above the point corresponds to a beam with a larger moment capacity. Any curve to the right is for a beam with exactly the required moment capacity, although for a larger unbraced length. In a design problem, therefore, if the charts are entered with a given unbraced length and a required strength, curves above and to the right of the point correspond to acceptable beams. If a dashed portion of a curve is encountered, then a curve for a lighter shape lies above or to the right of the dashed curve. Points on the curves corresponding to L_p are indicated by a solid circle; L_r is represented by an open circle.

In the LRFD solution of Example 5.10, the required design strength was 810.0 ft-kips, and there was continuous lateral support. For continuous lateral support, L_b can be taken as zero. From the charts, the first solid curve above the 810.0 ft-kip

FIGURE 5.26

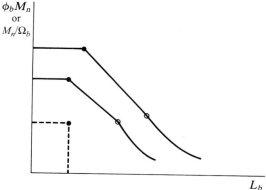

mark is for a W24 × 84, the same as selected in Example 5.10. Although $L_b = 0$ is not on this particular chart, the smallest value of L_b shown is less than L_p for all shapes on that page.

The beam curve shown in Figure 5.25 is for a compact shape, so the value of M_n for sufficiently small values of L_b is M_p. As discussed in Section 5.6, if the shape is noncompact, the maximum value of M_n will be based on flange local buckling. The maximum unbraced length for which this condition is true will be different from the value of L_p obtained with AISC Equation F2-5. The moment strength of noncompact shapes is illustrated graphically in Figure 5.27, where the maximum nominal strength is denoted M_p', and the maximum unbraced length for which this strength is valid is denoted L_p'.

Although the charts for compact and noncompact shapes are similar in appearance, M_p and L_p are used for compact shapes, whereas M_p' and L_p' are used for noncompact shapes. (This notation is not used in the charts or in any of the other design aids in the *Manual*.) Whether a shape is compact or noncompact is irrelevant to the *use* of the charts.

FIGURE 5.27

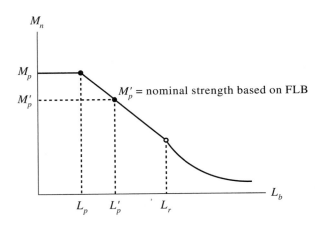

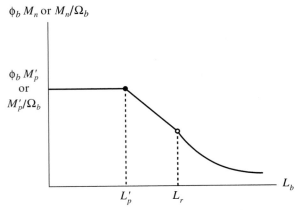

Form of the strength curve in the charts

Example 5.11   The beam shown in Figure 5.28 must support two concentrated *live* loads of 20 kips each at the quarter points. The maximum live load deflection must not exceed $L/240$. Lateral support is provided at the ends of the beam. Use A992 steel and select a W-shape.

FIGURE 5.28

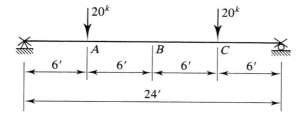

Solution   If the weight of the beam is neglected, the central half of the beam is subjected to a uniform moment, and

$$M_A = M_B = M_C = M_{\max}, \quad \therefore \ C_b = 1.0$$

Even if the weight is included, it will be negligible compared to the concentrated loads, and C_b can still be taken as 1.0, permitting the charts to be used without modification.

LRFD Solution   Temporarily ignoring the beam weight, the factored-load moment is

$$M_u = 6(1.6 \times 20) = 192 \text{ ft-kips}$$

From the charts, with $L_b = 24$ ft, **try W12 × 53**:

$$\phi_b M_n = 209 \text{ ft-kips} > 192 \text{ ft-kips} \quad \text{(OK)}$$

Now, we account for the beam weight:

$$M_u = 192 + \frac{1}{8}(1.2 \times 0.053)(24)^2 = 197 \text{ ft-kips} < 209 \text{ ft-kips} \quad \text{(OK)}$$

The shear is

$$V_u = 1.6(20) + \frac{1.2(0.053)(24)}{2} = 32.8 \text{ ft-kips}$$

From the Z_x table (or the uniform load table),

$$\phi_v V_n = 125 \text{ kips} > 32.8 \text{ kips} \quad \text{(OK)}$$

The maximum permissible live load deflection is

$$\frac{L}{240} = \frac{24(12)}{240} = 1.20 \text{ in.}$$

From Table 3-23, "Shears, Moments, and Deflections," in Part 3 of the *Manual,* the maximum deflection (at midspan) for two equal and symmetrically placed loads is

$$\Delta = \frac{Pa}{24EI}(3L^2 - 4a^2)$$

where

P = magnitude of concentrated load

a = distance from support to load

L = span length

$$\Delta = \frac{20(6 \times 12)}{24EI}[3(24 \times 12)^2 - 4(6 \times 12)^2] = \frac{13.69 \times 10^6}{EI}$$

$$= \frac{13.69 \times 10^6}{29,000(425)} = 1.11 \text{ in.} < 1.20 \text{ in.} \quad \text{(OK)}$$

Answer   Use a W12 × 53.

ASD Solution   The required flexural strength (not including the beam weight) is

$$M_a = 6(20) = 120 \text{ ft-kips}$$

From the charts, with L_b = 24 ft, **try W12 × 53**.

$$\frac{M_n}{\Omega_b} = 139 \text{ ft-kips} > 120 \text{ ft-kips} \quad \text{(OK)}$$

Account for the beam weight:

$$M_a = 6(20) + \frac{1}{8}(0.053)(24)^2 = 124 \text{ ft-kips} < 139 \text{ ft-kips} \quad \text{(OK)}$$

The required shear strength is

$$V_a = 20 + \frac{0.053(24)}{2} = 20.6 \text{ kips}$$

From the Z_x table (or the uniform load table),

$$\frac{V_n}{\Omega_v} = 83.2 \text{ kips} > 20.6 \text{ kips} \quad \text{(OK)}$$

Since deflections are computed with service loads, the deflection check is the same for both LRFD and ASD. From the LRFD solution,

$$\Delta = 1.11 \text{ in.} < 1.20 \text{ in.} \quad \text{(OK)}$$

Answer   Use a W12 × 53.

Although the charts are based on $C_b = 1.0$, they can easily be used for design when C_b is not 1.0; simply divide the required strength by C_b before entering the charts. We illustrate this technique in Example 5.12.

Example 5.12   Use A992 steel and select a rolled shape for the beam in Figure 5.29. The concentrated load is a service live load, and the uniform load is 30% dead load and 70% live load. Lateral bracing is provided at the ends and at midspan. There is no restriction on deflection.

FIGURE 5.29

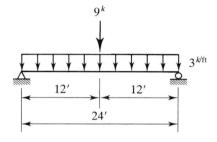

Solution   Neglect the beam weight and check it later.

$$w_D = 0.30(3) = 0.9 \text{ kips/ft}$$
$$w_L = 0.70(3) = 2.1 \text{ kips/ft}$$

LRFD Solution

$$w_u = 1.2(0.9) + 1.6(2.1) = 4.44 \text{ kips/ft}$$
$$P_u = 1.6(9) = 14.4 \text{ kips}$$

The factored loads and reactions are shown in Figure 5.30. Next, determine the moments required for the computation of C_b. The bending moment at a distance x from the left end is

$$M = 60.48x - 4.44x\left(\frac{x}{2}\right) = 60.48x - 2.22x^2 \quad \text{(for } x \leq 12 \text{ ft)}$$

FIGURE 5.30

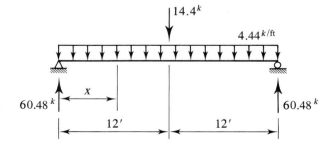

For $x = 3$ ft, $M_A = 60.48(3) - 2.22(3)^2 = 161.5$ ft-kips

For $x = 6$ ft, $M_B = 60.48(6) - 2.22(6)^2 = 283.0$ ft-kips

For $x = 9$ ft, $M_C = 60.48(9) - 2.22(9)^2 = 364.5$ ft-kips

For $x = 12$ ft, $M_{max} = M_u = 60.48(12) - 2.22(12)^2 = 406.1$ ft-kips

$$C_b = \frac{12.5 M_{max}}{2.5 M_{max} + 3 M_A + 4 M_B + 3 M_C}$$

$$= \frac{12.5(406.1)}{2.5(406.1) + 3(161.5) + 4(283.0) + 3(364.5)} = 1.36$$

Enter the charts with an unbraced length $L_b = 12$ ft and a bending moment of

$$\frac{M_u}{C_b} = \frac{406.1}{1.36} = 299 \text{ ft-kips}$$

Try W21 × 48:

$$\phi_b M_n = 311 \text{ ft-kips} \qquad \text{(for } C_b = 1\text{)}$$

Since $C_b = 1.36$, the actual design strength is $1.36(311) = 423$ ft-kips. But the design strength cannot exceed $\phi_b M_p$, which is only 398 ft-kips (obtained from the chart), so the actual design strength must be taken as

$$\phi_b M_n = \phi_b M_p = 398 \text{ ft-kips} < M_u = 406.1 \text{ ft-kips} \qquad \text{(N.G.)}$$

For the next trial shape, move up in the charts to the next solid curve and **try W18 × 55**. For $L_b = 12$ ft, the design strength from the chart is 335 ft-kips for $C_b = 1$. The strength for $C_b = 1.36$ is

$$\phi_b M_n = 1.36(335) = 456 \text{ ft-kips} > \phi_b M_p = 420 \text{ ft-kips}$$

$$\therefore \quad \phi_b M_n = \phi_b M_p = 420 \text{ ft-kips} > M_u = 406.1 \text{ ft-kips} \qquad \text{(OK)}$$

Check the beam weight.

$$M_u = 406.1 + \frac{1}{8}(1.2 \times 0.055)(24)^2 = 411 \text{ ft-kips} < 420 \text{ ft-kips} \qquad \text{(OK)}$$

The maximum shear is

$$V_u = 60.48 + \frac{1.2(0.055)}{2}(24) = 61.3 \text{ kips}$$

From the Z_x tables,

$$\phi_v V_n = 212 \text{ kips} > 61.3 \text{ kips} \qquad \text{(OK)}$$

Answer   Use a W18 × 55.

ASD Solution   The applied loads are

$$w_a = 3 \text{ kips/ft} \quad \text{and} \quad P_a = 9 \text{ kips}$$

The left-end reaction is

$$\frac{w_a L + P_a}{2} = \frac{3(24) + 9}{2} = 40.5 \text{ kips}$$

and the bending moment at a distance x from the left end is

$$M = 40.5x - 3x\left(\frac{x}{2}\right) = 40.5x - 1.5x^2 \quad (\text{for } x \le 12 \text{ ft})$$

Compute the moments required for the computation of C_b:

For $x = 3$ ft, $M_A = 40.5(3) - 1.5(3)^2 = 108.0$ ft-kips
For $x = 6$ ft, $M_B = 40.5(6) - 1.5(6)^2 = 189.0$ ft-kips
For $x = 9$ ft, $M_C = 40.5(9) - 1.5(9)^2 = 243.0$ ft-kips
For $x = 12$ ft, $M_{\max} = 40.5(12) - 1.5(12)^2 = 270.0$ ft-kip

$$C_b = \frac{12.5 M_{\max}}{2.5 M_{\max} + 3 M_A + 4 M_B + 3 M_C}$$

$$= \frac{12.5(270)}{2.5(270) + 3(108) + 4(189) + 3(243)} = 1.36$$

Enter the charts with an unbraced length $L_b = 12$ ft and a bending moment of

$$\frac{M_a}{C_b} = \frac{270}{1.36} = 199 \text{ ft-kips}$$

Try W21 × 48. For $C_b = 1$,

$$M_n/\Omega_b = 207 \text{ ft-kips}$$

For $C_b = 1.36$, the actual allowable strength is $1.36(207) = 282$ ft-kips, but the strength cannot exceed M_p/Ω_n, which is only 265 ft-kips (this can be obtained from the chart), so the actual allowable strength must be taken as

$$\frac{M_n}{\Omega_b} = \frac{M_p}{\Omega_b} = 265 \text{ ft-kips} < M_a = 270 \text{ ft-kips} \quad (\text{N.G.})$$

Move up in the charts to the next solid curve and **try W18 × 55.** For $L_b = 12$ ft, the allowable strength for $C_b = 1$ is 223 ft-kips. The strength for $C_b = 1.36$ is

$$\frac{M_n}{\Omega_b} = 1.36(223) = 303 \text{ ft-kips} > \frac{M_p}{\Omega_b} = 280 \text{ ft-kips}$$

$$\therefore \frac{M_n}{\Omega_b} = \frac{M_p}{\Omega_b} = 280 \text{ ft-kips} > M_a = 270 \text{ ft-kips} \quad (\text{OK})$$

Account for the beam weight.

$$M_a = 270 + \frac{1}{8}(0.055)(24)^2 = 274 \text{ ft-kips} \ < 280 \text{ ft-kips} \quad \text{(OK)}$$

The maximum shear is

$$V_a = \frac{9 + 3.055(24)}{2} = 41.2 \text{ kips}$$

From the Z_x table (or the uniform load table)

$$\frac{V_n}{\Omega_v} = 141 \text{ kips} \ > 41.2 \text{ kips} \quad \text{(OK)}$$

Answer   Use a W18 × 55. ◾

In Example 5.12, the value of C_b is the same (to three significant figures) for both the factored and the unfactored moments. The two computed values will always be nearly the same, and for this reason, it makes no practical difference which moments are used.

If deflection requirements control the design of a beam, a minimum required moment of inertia is computed, and the lightest shape having this value is sought. This task is greatly simplified by the moment of inertia selection tables in Part 3 of the *Manual*. We illustrate the use of these tables in Example 5.13, following a discussion of the design procedure for a beam in a typical floor or roof system.

5.11   FLOOR AND ROOF FRAMING SYSTEMS

When a distributed load acts on an area such as a floor in a building, certain portions of that load are supported by various components of the floor system. The actual distribution is difficult to determine, but it can be approximated quite easily. The basic idea is that of *tributary areas*. In the same way that tributaries flow into a river and contribute to the volume of water in it, the loads on certain areas of a structural surface "flow" into a structural component. The concept of tributary areas was first discussed in Section 3.8 in the coverage of tension members in roof trusses.

Figure 5.31 shows a typical floor framing plan for a multistory building. Part (a) of the figure shows one of the rigid frames comprising the building. Part (b) shows what would be seen if a horizontal section were cut through the building above one of the floors and the lower portion viewed from above. The gridwork thus exposed consists of the column cross sections (in this case, wide-flange structural steel shapes), girders connecting the columns in the east-west direction, and intermediate floor beams such as *EF* spanning between the girders. Girders are defined as beams that support other beams, although sometimes the term is applied to large beams in general.

FIGURE 5.31

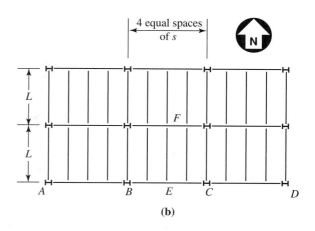

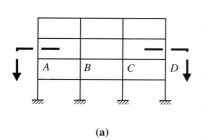

(a)

(b)

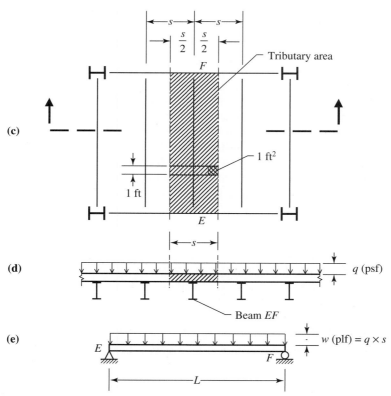

The floor beams, which fill in the panels defined by the columns, are sometimes called *filler beams*. The columns and girders along any of the east-west lines make up an individual frame. The frames are connected by the beams in the north-south direction, completing the framework for the building. There may also be secondary components, such as bracing for stability, that are not shown in Figure 5.31.

Figure 5.31(c) shows a typical bay of the floor framing system. When columns are placed in a rectangular grid, the region between four columns is called a *bay*. The bay size, such as 30 ft × 40 ft, is a measure of the geometry of a building. Figure 5.31(d) is a cross section of this bay, showing the floor beams as wide-flange steel shapes supporting a reinforced concrete floor slab.

The overall objective of a structure is to transmit loads to the foundation. As far as floor loads are concerned, this transmission of loads is accomplished as follows:

1. Floor loads, both live and dead, are supported by the floor slab.
2. The weight of the slab, along with the loads it supports, is supported by the floor beams.
3. The floor beams transmit their loads, including their own weight, to the girders.
4. The girders and their loads are supported by the columns.
5. The column loads are supported by the columns of the story below. The column loads accumulate from the top story to the foundation.

(The route taken by the loads from one part of the structure to another is sometimes called the *load path*.) This is a fairly accurate representation of what happens, but it is not exact. For example, part of the slab and its load will be supported directly by the girders, but most of it will be carried by the floor beams.

Figure 5.31(c) shows a shaded area around floor beam *EF*. This is the tributary area for this member, and it consists of half the floor between beam *EF* and the adjacent beam on each side. Thus, the total width of floor being supported is equal to the beam spacing *s* if the spacing is uniform. If the load on the floor is uniformly distributed, we can express the uniform load on beam *EF* as a force per unit length (for example, pounds per linear foot [plf]) by multiplying the floor load in force per unit area (for example, pounds per square foot [psf]) by the tributary width *s*. Figure 5.31(e) shows the final beam model (for the usual floor framing connections, the beams can be treated as simply supported).

For convenience, the weight of a reinforced concrete floor slab is usually expressed in pounds per square foot of floor surface. This way, the slab weight can be combined with other loads similarly expressed. If the floor consists of a metal deck and concrete fill, the combined weight can usually be obtained from the deck manufacturer's literature. If the floor is a slab of uniform thickness, the weight can be calculated as follows. Normal-weight concrete weighs approximately 145 pounds per cubic foot. If 5 pcf is added to account for the reinforcing steel, the total weight is 150 pcf. The volume of slab contained in one square foot of floor is 1 ft² × the slab thickness *t*. For a thickness expressed in inches, the slab weight is therefore $(t/12)(150)$ psf. For lightweight concrete, a unit weight of 115 pounds per cubic foot can be used in lieu of more specific data.

Example 5.13   Part of a floor framing system is shown in Figure 5.32. A 4-inch-thick reinforced concrete floor slab of normal-weight concrete is supported by floor beams spaced at 7 feet. The floor beams are supported by girders, which in turn are supported by the columns. In addition to the weight of the structure, loads consist of a uniform live load of 80 psf and moveable partitions, to be accounted for by using a uniformly distributed load of 20 pounds per square foot of floor surface. The maximum live load deflection must not exceed 1/360 of the span length. Use A992 steel and

FIGURE 5.32

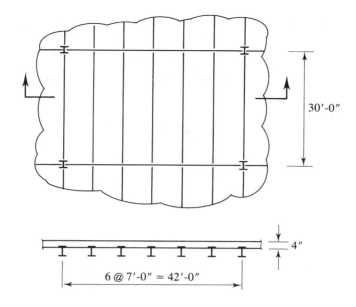

design the floor beams. Assume that the slab provides continuous lateral support of the floor beams.

Solution   The slab weight is

$$w_{slab} = \frac{t}{12}(150) = \frac{4}{12}(150) = 50 \text{ psf}$$

Assume that each beam supports a 7-ft width (tributary width) of floor.

> Slab: $50(7) = 350 \text{ lb/ft}$
> Partitions: $20(7) = 140 \text{ lb/ft}$
> Live load: $80(7) = 560 \text{ lb/ft}$

The beam weight will be accounted for once a trial selection has been made.

Since the partitions are moveable, they will be treated as live load. This is consistent with the provisions of the International Building Code (ICC, 2003). The dead and live loads are, therefore,

> $w_D = 0.350 \text{ lb/ft}$ (excluding the beam weight)
> $w_L = 0.560 + 0.140 = 0.700 \text{ lb/ft}$

LRFD Solution   The total factored load is

$$w_u = 1.2w_D + 1.6w_L = 1.2(0.350) + 1.6(0.700) = 1.540 \text{ kips/ft}$$

The typical floor-beam connection will provide virtually no moment restraint, and the beams can be treated as simply supported. Hence

$$M_u = \frac{1}{8}w_uL^2 = \frac{1}{8}(1.540)(30)^2 = 173 \text{ ft-kips}$$

Since the beams have continuous lateral support, the Z_x table can be used.

Try a W14 × 30:

$$\phi_b M_n = 177 \text{ ft-kips} > 173 \text{ ft-kips} \qquad (\text{OK})$$

Check the beam weight.

$$M_u = 173 + \frac{1}{8}(1.2 \times 0.030)(30)^2 = 177 \text{ ft-kips} \qquad (\text{OK})$$

The maximum shear is

$$V_u \approx \frac{1.540(30)}{2} = 23.1 \text{ kips}$$

From the Z_x table,

$$\phi_v V_n = 112 \text{ kips} > 23.1 \text{ kips} \qquad (\text{OK})$$

The maximum permissible deflection is

$$\frac{L}{360} = \frac{30(12)}{360} = 1.0 \text{ in.}$$

$$\Delta_L = \frac{5}{384} \frac{w_L L^4}{EI} = \frac{5}{384} \frac{(0.700/12)(30 \times 12)^4}{29{,}000(291)} = 1.51 \text{ in.} > 1.0 \text{ in.} \qquad (\text{N.G.})$$

Solving the deflection equation for the required moment of inertia yields

$$I_{\text{required}} = \frac{5 w_L L^4}{384 E \Delta_{\text{required}}} = \frac{5(0.700/12)(30 \times 12)^4}{384(29{,}000)(1.0)} = 440 \text{ in.}^4$$

Part 3 of the *Manual* contains selection tables for both I_x and I_y. These tables are organized in the same way as the Z_x table, so selection of the lightest shape with sufficient moment of inertia is simple. From the I_x table, **try a W18 × 35:**

$$I_x = 510 \text{ in.}^4 > 440 \text{ in.}^4 \qquad (\text{OK})$$
$$\phi_b M_n = 249 \text{ ft-kips} > 177 \text{ ft-kips} \qquad (\text{OK})$$
$$\phi_v V_n = 159 \text{ kips} > 23.1 \text{ kips} \qquad (\text{OK})$$

Answer Use a W18 × 35.

ASD Solution Account for the beam weight after a selection has been made.

$$w_a = w_D + w_L = 0.350 + 0.700 = 1.05 \text{ kips/ft}$$

If we treat the beam connection as a simple support, the required moment strength is

$$M_a = \frac{1}{8} w_a L^2 = \frac{1}{8}(1.05)(30)^2 = 118 \text{ ft-kips} = \text{required } \frac{M_n}{\Omega_b}$$

For a beam with full lateral support, the Z_x table can be used.

Try a W16 × 31:

$$\frac{M_n}{\Omega_b} = 135 \text{ ft-kips} > 118 \text{ ft-kips} \quad \text{(OK)}$$

(A W14 × 30 has an allowable moment strength of exactly 118 ft-kips, but the beam weight has not yet been accounted for.)

Account for the beam weight:

$$M_a = 118 + \frac{1}{8}(0.031)(30)^2 = 122 \text{ ft-kips} < 135 \text{ ft-kips} \quad \text{(OK)}$$

The required shear strength is

$$V_a = \frac{w_a L}{2} = \frac{(1.05 + 0.031)(30)}{2} = 16.2 \text{ kips}$$

From the Z_z table, the available shear strength is

$$\frac{V_n}{\Omega_v} = 87.3 \text{ kips} > 16.2 \text{ kips} \quad \text{(OK)}$$

Check deflection. The maximum permissible live load deflection is

$$\frac{L}{360} = \frac{30 \times 12}{360} = 1.0 \text{ in.}$$

$$\Delta_L = \frac{5}{384} \frac{w_L L^4}{EI_x} = \frac{5}{384} \frac{(0.700/12)(30 \times 12)^4}{29,000(375)} = 1.17 \text{ in.} > 1.0 \text{ in.} \quad \text{(N.G.)}$$

Solve the deflection equation for the required moment of inertia:

$$I_{required} = \frac{5 w_L L^4}{384 E \Delta_{required}} = \frac{5(0.700/12)(30 \times 12)^4}{384(29,000)(1.0)} = 440 \text{ in.}^4$$

Part 3 of the *Manual* contains selection tables for both I_x and I_y. From the I_x table, try a W18 × 35:

$$I_x = 510 \text{ in.}^4 > 440 \text{ in.}^2 \quad \text{(OK)}$$

$$\frac{M_n}{\Omega_b} = 166 \text{ ft-kips} > 122 \text{ ft-kips} \quad \text{(OK)}$$

$$\frac{V_n}{\Omega_v} = 106 \text{ kips} > 16.2 \text{ kips} \quad \text{(OK)}$$

Answer   Use a W18 × 35.

Note that in Example 5.13, the design was controlled by serviceability rather than strength. This is not unusual, but the recommended sequence in beam design is still to select a shape for moment and then check shear and deflection. Although there is no limit on the dead load deflection in this example, this deflection may be needed if the beam is to be cambered.

$$\Delta_D = \frac{5}{384} \frac{w_{\text{slab+beam}} L^4}{EI} = \frac{5}{384} \frac{[(0.350 + 0.035)/12](30 \times 12)^4}{29,000(510)} = 0.474 \text{ in.}$$

5.12 HOLES IN BEAMS

If beam connections are made with bolts, holes will be punched or drilled in the beam web or flanges. In addition, relatively large holes are sometimes cut in beam webs to provide space for utilities such as electrical conduits and ventilation ducts. Ideally, holes should be placed in the web only at sections of low shear, and holes should be made in the flanges at points of low bending moment. That will not always be possible, so the effect of the holes must be accounted for.

For relatively small holes such as those for bolts, the effect will be small, particularly for flexure, for two reasons. First, the reduction in the cross section is usually small. Second, adjacent cross sections are not reduced, and the change in cross section is actually more of a minor discontinuity than a "weak link."

Holes in a beam flange are of concern for the tension flange only, since bolts in the compression flange will transmit the load through the bolts. This is the same rationale that is used for compression members, where the net area is not considered. The AISC Specification requires that bolt holes in beam flanges be accounted for when the nominal tensile rupture strength (fracture strength) of the flange is less than the nominal tensile yield strength—that is, when

$$F_u A_{fn} < F_y A_{fg} \tag{5.9}$$

where

A_{fn} = net tension flange area
A_{fg} = gross tension flange area

If $F_y / F_u > 0.8$, the Specification requires that the right hand side of Equation 5.9 be increased by 10%. Equation 5.9 can be written more generally as follows:

$$F_u A_{fn} < Y_t F_y A_{fg} \tag{5.10}$$

where

$Y_t = 1.0$ for $F_y / F_u \le 0.8$
    $= 1.1$ for $F_y / F_u > 0.8$

Note that, for A992 steel, the preferred steel for W shapes, the *maximum* value of F_y / F_u is 0.85. This means that unless more information is available, use $Y_t = 1.1$. If the condition of Equation 5.10 exists—that is, if

$$F_u A_{fn} < Y_t F_y A_{fg}$$

then AISC F13.1 requires that the nominal flexural strength be limited by the condition of flexural rupture. This limit state corresponds to a flexural stress of

$$f_b = \frac{M_n}{S_x(A_{fn}/A_{fg})} = F_u \tag{5.11}$$

where $S_x(A_{fn}/A_{fg})$ can be considered to be a "net" elastic section modulus. The relationship of Equation 5.11 corresponds to a nominal flexural strength of

$$M_n = \frac{F_u A_{fn}}{A_{fg}} S_x$$

The AISC requirement for holes in beam flanges can be summarized as follows: If

$$F_u A_{fn} < Y_t F_y A_{fg}$$

The nominal flexural strength cannot exceed

$$M_n = \frac{F_u A_{fn}}{A_{fg}} S_x \qquad \text{(AISC Equation F13-1)}$$

where
$$\begin{aligned} Y_t &= 1.0 \text{ for } F_y/F_u \le 0.8 \\ &= 1.1 \text{ for } F_y/F_u > 0.8 \end{aligned}$$

The constant Y_t should be taken as 1.1 for A992 steel or if the maximum value of F_y/F_u is not known.

Example 5.14   The shape shown in Figure 5.33 is a W18 × 71 with holes in each flange for 1-inch-diameter bolts. The steel is A992. Compute the nominal flexural strength for an unbraced length of 10 feet. Use $C_b = 1.0$.

FIGURE 5.33

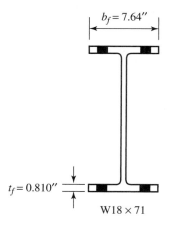

$b_f = 7.64''$

$t_f = 0.810''$

W18 × 71

Solution   To determine the nominal flexural strength M_n, all applicable limit states must be checked. From the Z_x table, a W18 × 71 is seen to be a compact shape (no footnote to indicate otherwise). Also from the Z_x table, $L_p = 6.00$ ft and $L_r = 19.6$ ft Therefore for an unbraced length $L_b = 10$ ft,

$$L_p < L_b < L_r$$

and the beam is subject to inelastic lateral-torsional buckling. The nominal strength for this limit state is given by

$$M_n = C_b \left[M_p - (M_p - 0.7F_y S_x) \left(\frac{L_b - L_p}{L_r - L_p} \right) \right] \le M_p \qquad \text{(AISC Equation F2-2)}$$

where

$$M_p = F_y Z_x = 50(146) = 7300 \text{ in.-kips}$$

$$M_n = 1.0 \left[7300 - (7300 - 0.7 \times 50 \times 127) \left(\frac{10-6}{19.6-6} \right) \right] = 6460 \text{ in.-kips}$$

Check to see if the flange holes need to be accounted for. The gross area of one flange is

$$A_{fg} = t_f b_f = 0.810(7.64) = 6.188 \text{ in.}^2$$

The effective hole diameter is

$$d_h = 1 + \frac{1}{8} = 1\frac{1}{8} \text{ in.}$$

and the net flange area is

$$A_{fn} = A_{fg} - t_f \Sigma d_h = 6.188 - 0.810(2 \times 1.125) = 4.366 \text{ in.}^2$$

$$F_u A_{fn} = 65(4.366) = 283.8 \text{ kips}$$

Determine Y_t. For A992 steel, the maximum F_y/F_u ratio is 0.85. Since this is greater than 0.8, use $Y_t = 1.1$.

$$Y_t F_y A_{fg} = 1.1(50)(6.188) = 340.3 \text{ kips}$$

Since $F_u A_{fn} < Y_t F_y A_{fg}$, the holes must be accounted for. From AISC Equation F13-1,

$$M_n = \frac{F_u A_{fn}}{A_{fg}} S_x = \frac{283.8}{6.188}(127) = 5825 \text{ in.-kips}$$

This value is less than the LTB value of 6460 in.-kips, so it controls.

Answer $M_n = 5825$ in.-kips $= 485$ ft-kips.

Beams with large holes in their webs will require special treatment and are beyond the scope of this book. *Design of Steel and Composite Beams with Web Openings* is a useful guide to this topic (Darwin, 1990).

5.13   OPEN-WEB STEEL JOISTS

Open-web steel joists are prefabricated trusses of the type shown in Figure 5.34. Many of the smaller ones use a continuous circular bar to form the web members and are commonly called *bar* joists. They are used in floor and roof systems in a wide variety of structures. For a given span, an open-web joist will be lighter in weight than a rolled shape, and the absence of a solid web allows for the easy passage of duct work and electrical conduits. Depending on the span length, open-web joists may be more economical than rolled shapes, although there are no general guidelines for making this determination.

Open-web joists are available in standard depths and load capacities from various manufacturers. Some open-web joists are designed to function as floor or roof joists, and others are designed to function as girders, supporting the concentrated reactions from joists. The AISC Specification does not cover open-web steel joists; a separate organization, the Steel Joist Institute (SJI), exists for this purpose. All aspects of steel joist usage, including their design and manufacture, are addressed in the publication *Standard Specifications, Load Tables, and Weight Tables for Steel Joists and Joist Girders* (SJI, 2005).

An open-web steel joist can be selected with the aid of the standard load tables (SJI, 2005). These tables give load capacities in pounds per foot of length for various standard joists. Tables are available for both LRFD and ASD, in either U.S. Customary units or metric units. One of the LRFD tables is reproduced in Figure 5.35. For each combination of span and joist, a pair of load values is given. The top number is the total load capacity in pounds per foot. The bottom number is the live load per foot that will produce a deflection of 1/360 of the span length. For span lengths in the shaded areas, special bridging (interconnection of joists) is required. The ASD tables use the same format, but the loads are unfactored. The first number in the designation is the nominal depth in inches. The table also gives the approximate weight in pounds per foot of length. Steel fabricators who furnish open-web steel joists must certify that a particular joist of a given designation, such as a 10K1 of span length 20 feet, will have a safe load capacity of at least the value given in the table. Different manufacturers' 10K1 joists may have

FIGURE 5.34

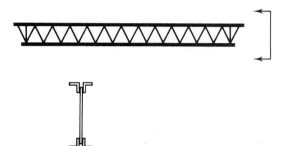

FIGURE 5.35

LRFD

STANDARD LOAD TABLE FOR OPEN WEB STEEL JOISTS, K-SERIES
Based on a 50 ksi Maximum Yield Strength - Loads Shown in Pounds per Linear Foot (plf)

Joist Designation	8K1	10K1	12K1	12K3	12K5	14K1	14K3	14K4	14K6	16K2	16K3	16K4	16K5	16K6	16K7	16K9
Depth (in.)	8	8	8	8	8	14	14	14	14	16	16	16	16	16	16	16
Approx. Wt (lbs./ft.)	5.1	5.0	5.0	5.7	7.1	5.2	6.0	6.7	7.7	5.5	6.3	7.0	7.5	8.1	8.6	10.0
Span (ft.)																
8	825/550															
9	825/550															
10	825/480	825/550														
11	798/377	825/542														
12	666/288	825/455	825/550	825/550	825/550											
13	565/225	718/363	825/510	825/510	825/510											
14	486/179	618/289	750/425	825/463	825/463	825/550	825/550	825/550	825/550							
15	421/145	537/234	651/344	814/428	825/434	766/475	825/507	825/507	825/507							
16	369/119	469/192	570/282	714/351	825/396	672/390	825/467	825/467	825/467	825/550	825/550	825/550	825/550	825/550	825/550	825/550
17		415/159	504/234	630/291	825/366	592/324	742/404	825/443	825/443	768/488	825/526	825/526	825/526	825/526	825/526	825/526
18		369/134	448/197	561/245	760/317	528/272	661/339	795/397	825/408	684/409	762/456	825/490	825/490	825/490	825/490	825/490
19		331/113	402/167	502/207	681/269	472/230	592/287	712/336	825/383	612/347	682/386	820/452	825/455	825/455	825/455	825/455
20		298/97	361/142	453/177	613/230	426/197	534/246	642/287	787/347	552/297	615/330	739/386	825/426	825/426	825/426	825/426
21			327/123	409/153	555/198	385/170	483/212	582/248	712/299	499/255	556/285	670/333	754/405	822/406	822/406	822/406
22			298/106	373/132	505/172	351/147	439/184	529/215	648/259	454/222	505/247	609/289	687/323	747/351	825/385	825/385
23			271/93	340/116	462/150	321/128	402/160	483/188	592/226	415/194	462/216	556/252	627/282	682/307	760/339	825/363
24			249/81	312/101	423/132	294/113	367/141	442/165	543/199	381/170	424/189	510/221	576/248	627/269	697/298	825/346
25						270/100	339/124	408/145	501/175	351/150	390/167	469/195	529/219	576/238	642/263	771/311
26						249/88	313/110	376/129	462/156	324/133	360/148	433/173	489/194	532/211	592/233	711/276
27						231/79	289/98	349/115	427/139	300/119	334/132	402/155	453/173	493/188	549/208	658/246
28						214/70	270/88	324/103	397/124	279/106	310/118	373/138	421/155	459/168	510/186	612/220
29										259/95	289/106	348/124	391/139	427/151	475/167	570/198
30										241/86	270/96	324/112	366/126	399/137	444/151	532/178
31										226/78	252/87	304/101	342/114	373/124	415/137	498/161
32										213/71	237/79	285/92	321/103	349/112	388/124	466/147

Source: *Standard Specifications, Load Tables, and Weight Tables for Steel Joists and Joist Girders.* Myrtle Beach, S.C.: Steel Joist Institute, 2005. Reprinted with permission.

different member cross sections, but they all must have a nominal depth of 10 inches and, for a span length of 20 feet, a factored load capacity of at least 361 pounds per foot.

The open-web steel joists that are designed to function as floor or roof joists (in contrast to girders) are available as open-web steel joists (K-series, both standard and KCS), longspan steel joists (LH-series), and deep longspan steel joists (DLH-series). Standard load tables are given for each of these categories. The higher you move up the series, the greater the available span lengths and load-carrying capacities become. At the lower end, an 8K1 is available with a span length of 8 feet and a factored load capacity of 825 pounds per foot, whereas a 72DLH19 can support a load of 745 pounds per foot on a span of 144 feet.

With the exception of the KCS joists, all open-web steel joists are designed as simply supported trusses with uniformly distributed loads on the top chord. This loading subjects the top chord to bending as well as axial compression, so the top chord is designed as a beam–column (see Chapter 6). To ensure stability of the top chord, the floor or roof deck must be attached in such a way that continuous lateral support is provided.

KCS joists are designed to support both concentrated loads and distributed loads (including nonuniform distributions). To select a KCS joist, the engineer must compute a maximum moment and shear in the joist and enter the KCS tables with these values. (The KCS joists are designed to resist a uniform moment and a constant shear.) If concentrated loads must be supported by an LH or a DLH joist, a special analysis should be requested from the manufacturer.

Both top and bottom chord members of K-series joists must be made of steel with a yield stress of 50 ksi, and the web members may have a yield stress of either 36 ksi or 50 ksi. All members of LH- and DLH-series joists can be made with steel of any yield stress between 36 ksi and 50 ksi inclusive. The load capacity of K-series joists must be verified by the manufacturer by testing. No testing program is required for LH- or DLH-series joists.

Joist girders are designed to support open-web steel joists. For a given span, the engineer determines the number of joist spaces, then from the joist girder weight tables selects a depth of girder. The joist girder is designated by specifying its depth, the number of joist spaces, the load at each loaded top-chord panel point of the joist girder, and a letter to indicate whether the load is factored ("F") or unfactored ("K"). For example, using LRFD and U.S. Customary units, a 52G9N10.5F is 52 inches deep, provides for 9 equal joist spaces on the top chord, and will support 10.5 kips of factored load at each joist location. The joist girder weight tables give the weight in pounds per linear foot for the specified joist girder for a specific span length.

Example 5.15   Use the load table given in Figure 5.35 to select an open-web steel joist for the following floor system and loads:

> Joist spacing = 3 ft 0 in.
>
> Span length = 20 ft 0 in.

The loads are

> 3-in. floor slab
>
> Other dead load: 20 psf
>
> Live load: 50 psf

The live load deflection must not exceed $L/360$.

Solution For the dead loads of

$$\text{Slab:} \quad 150\left(\frac{3}{12}\right) = 37.5 \text{ psf}$$

Other dead load: $= 20$ psf

Joist weight: $= \underline{3}$ psf (est.)

Total: $= 60.5$ psf

$$w_D = 60.5(3) = 181.5 \text{ lb/ft}$$

For the live load of 50 psf,

$$w_L = 50(3) = 150 \text{ lb/ft}$$

The factored load is

$$w_u = 1.2w_D + 1.6w_L = 1.2\,(181.5) + 1.6(150) = 458 \text{ lb/ft}$$

Figure 5.35 indicates that the following joists satisfy the load requirement: a 12K5, weighing approximately 7.1 lb/ft; a 14K3, weighing approximately 6.0 lb/ft; and a 16K2, weighing approximately 5.5 lb/ft. No restriction was placed on the depth, so we choose the lightest joist, a 16K2.

To limit the live load deflection to $L/360$, the live load must not exceed

297 lb/ft > 150 lb/ft   (OK)

Answer Use a 16K2.

◼

The standard load tables also include a K-series economy table, which facilitates the selection of the lightest joist for a given load.

5.14 BEAM BEARING PLATES AND COLUMN BASE PLATES

The design procedure for column base plates is similar to that for beam bearing plates, and for that reason we consider them together. In addition, the determination of the thickness of a column base plate requires consideration of flexure, so it logically belongs in this chapter rather than in Chapter 4. In both cases, the function of the plate is to distribute a concentrated load to the supporting material.

Two types of beam bearing plates are considered: one that transmits the beam reaction to a support such as a concrete wall and one that transmits a load to the top flange of a beam. Consider first the beam support shown in Figure 5.36. Although many beams are connected to columns or other beams, the type of support shown here is occasionally used, particularly at bridge abutments. The design of the bearing plate consists of three steps.

1. Determine dimension N so that web yielding and web crippling are prevented.
2. Determine dimension B so that the area $B \times N$ is sufficient to prevent the supporting material (usually concrete) from being crushed in bearing.
3. Determine the thickness t so that the plate has sufficient bending strength.

FIGURE 5.36

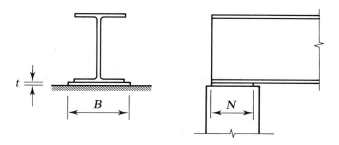

Web yielding, web crippling, and concrete bearing strength are addressed by AISC in Chapter J, "Design of Connections."

Web Yielding

Web yielding is the compressive crushing of a beam web caused by the application of a compressive force to the flange directly above or below the web. This force could be an end reaction from a support of the type shown in Figure 5.36, or it could be a load delivered to the top flange by a column or another beam. Yielding occurs when the compressive stress on a horizontal section through the web reaches the yield point. When the load is transmitted through a plate, web yielding is assumed to take place on the nearest section of width t_w. In a rolled shape, this section will be at the toe of the fillet, a distance k from the outside face of the flange (this dimension is tabulated in the dimensions and properties tables in the *Manual*). If the load is assumed to distribute itself at a slope of 1 : 2.5, as shown in Figure 5.37, the area at the support subject to yielding is $(2.5k + N)t_w$. Multiplying this area by the yield stress gives the nominal strength for web yielding at the support:

$$R_n = (2.5k + N)F_y t_w \qquad \text{(AISC Equation J10-3)}$$

The bearing length N at the support should not be less than k.

At the interior load, the length of the section subject to yielding is

$$2(2.5k) + N = 5k + N$$

FIGURE 5.37

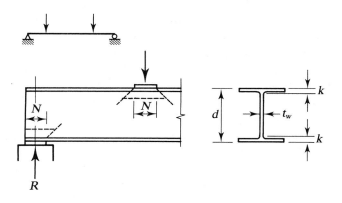

and the nominal strength is

$$R_n = (5k + N)F_y t_w \qquad \text{(AISC Equation J10-2)}$$

For LRFD, the design strength is ϕR_n, where $\phi = 1.0$.
For ASD, the allowable strength is R_n/Ω, where $\Omega = 1.50$.

Web Crippling

Web crippling is buckling of the web caused by the compressive force delivered through the flange. For an interior load, the nominal strength for web crippling is

$$R_n = 0.80t_w^2 \left[1 + 3\left(\frac{N}{d}\right)\left(\frac{t_w}{t_f}\right)^{1.5} \right] \sqrt{\frac{EF_y t_f}{t_w}} \qquad \text{(AISC Equation J10-4)}$$

For a load at or near the support (no greater than half the beam depth from the end), the nominal strength is

$$R_n = 0.40t_w^2 \left[1 + 3\left(\frac{N}{d}\right)\left(\frac{t_w}{t_f}\right)^{1.5} \right] \sqrt{\frac{EF_y t_f}{t_w}} \qquad \text{for } \frac{N}{d} \le 0.2 \quad \text{(AISC Equation J10-5a)}$$

or

$$R_n = 0.40t_w^2 \left[1 + \left(\frac{4N}{d} - 0.2\right)\left(\frac{t_w}{t_f}\right)^{1.5} \right] \sqrt{\frac{EF_y t_f}{t_w}} \qquad \text{for } \frac{N}{d} > 0.2$$

$$\text{(AISC Equation J10-5b)}$$

The resistance factor for this limit state is $\phi = 0.75$. The safety factor is $\Omega = 2.00$.

Concrete Bearing Strength

The material used for a beam support can be concrete, brick, or some other material, but it usually will be concrete. This material must resist the bearing load applied by the steel plate. The nominal bearing strength specified in AISC J8 is the same as that given in the American Concrete Institute's Building Code (ACI, 2005) and may be used if no other building code requirements are in effect. If the plate covers the full area of the support, the nominal strength is

$$P_p = 0.85f_c'A_1 \qquad \text{(AISC Equation J8-1)}$$

If the plate does not cover the full area of the support,

$$P_p = 0.85f_c'A_1\sqrt{\frac{A_2}{A_1}} \le 1.7f_c'A_1 \qquad \text{(AISC Equation J8-2)}$$

FIGURE 5.38

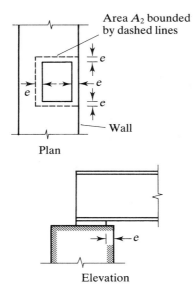

Plan

Elevation

where

$f_c' = $ 28-day compressive strength of the concrete
$A_1 = $ bearing area
$A_2 = $ full area of the support

If area A_2 is not concentric with A_1, then A_2 should be taken as the largest concentric area that is geometrically similar to A_1, as illustrated in Figure 5.38.

For LRFD, the design bearing strength is $\phi_c P_p$, where $\phi_c = 0.60$. For ASD, the allowable bearing strength is P_p / Ω_c, where $\Omega_c = 2.50$.

Plate Thickness

Once the length and width of the plate have been determined, the average bearing pressure is treated as a uniform load on the bottom of the plate, which is assumed to be supported at the top over a central width of $2k$ and length N, as shown in Figure 5.39. The plate is then considered to bend about an axis parallel to the beam span. Thus the plate is treated as a cantilever of span length $n = (B - 2k)/2$ and a width of N. For convenience, a 1-inch width is considered, with a uniform load in pounds per linear inch numerically equal to the bearing pressure in pounds per square inch.

From Figure 5.39, the maximum bending moment in the plate is

$$M = \frac{R}{BN} \times n \times \frac{n}{2} = \frac{Rn^2}{2BN}$$

where R is the beam reaction and R/BN is the average bearing pressure between the plate and the concrete. For a rectangular cross section bent about the minor axis, the nominal moment strength M_n is equal to the plastic moment capacity M_p. As illustrated

FIGURE 5.39

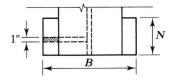

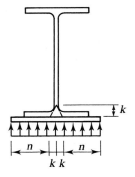

in Figure 5.40, for a rectangular cross section of unit width and depth t, the plastic moment is

$$M_p = F_y\left(1 \times \frac{t}{2}\right)\left(\frac{t}{2}\right) = F_y\frac{t^2}{4}$$

For LRFD, Since the design strength must at least equal the factored-load moment,

$$\phi_b M_p \geq M_u$$

$$0.90F_y\frac{t^2}{4} \geq \frac{R_u n^2}{2BN}$$

$$t \geq \sqrt{\frac{2R_u n^2}{0.90BNF_y}} \tag{5.12}$$

or

$$t \geq \sqrt{\frac{2.22R_u n^2}{BNF_y}} \tag{5.13}$$

FIGURE 5.40

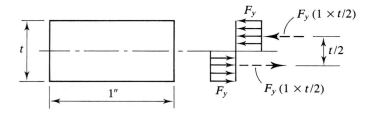

where R_u is the factored-load beam reaction.

For ASD, the allowable flexural strength must at least equal the applied moment:

$$\frac{M_p}{\Omega_b} \geq M_a$$

$$\frac{F_y t^2 / 4}{1.67} \geq \frac{R_a n^2}{2BN}$$

$$t \geq \sqrt{\frac{3.34 R_a n^2}{BNF_y}}$$

(5.14)

where R_a is the service-load beam reaction.

Example 5.16   Design a bearing plate to distribute the reaction of a W21×68 with a span length of 15 feet 10 inches center-to-center of supports. The total service load, including the beam weight, is 9 kips/ft, with equal parts dead and live load. The beam is to be supported on reinforced concrete walls with $f'_c = 3500$ psi. For the beam, $F_y = 50$ ksi, and $F_y = 36$ ksi for the plate.

LRFD Solution   The factored load is

$$w_u = 1.2w_D + 1.6w_L = 1.2(4.5) + 1.6(4.5) = 12.60 \text{ kips/ft}$$

and the reaction is

$$R_u = \frac{w_u L}{2} = \frac{12.60(15.83)}{2} = 99.73 \text{ kips}$$

Determine the length of bearing N required to prevent web yielding. From AISC Equation J10-3, the nominal strength for this limit state is

$$R_n = (2.5k + N)F_y t_w$$

For $\phi R_n \geq R_u$,

$$1.0[2.5(1.19) + N](50)(0.430) \geq 99.73$$

resulting in the requirement

$$N \geq 1.66 \text{ in.}$$

(Note that two values of k are given in the dimensions and properties tables: a decimal value, called the *design* dimension, and a fractional value, called the *detailing* dimension. We always use the design dimension in calculations.)

Use AISC Equation J10-5 to determine the value of N required to prevent web crippling. Assume $N/d > 0.2$ and try the second form of the equation, J10-5(b). For $\phi R_n \geq R_u$,

$$\phi(0.40)t_w^2 \left[1 + \left(\frac{4N}{d} - 0.2 \right) \left(\frac{t_w}{t_f} \right)^{1.5} \right] \sqrt{\frac{EF_y t_f}{t_w}} \geq R_u$$

$$0.75(0.40)(0.430)^2 \left[1 + \left(\frac{4N}{21.1} - 0.2 \right) \left(\frac{0.430}{0.685} \right)^{1.5} \right] \sqrt{\frac{29,000(50)(0.685)}{0.430}} \geq 99.73$$

This results in the requirement

$N \geq 3.00$ in.

Check the assumption:

$$\frac{N}{d} = \frac{3.00}{21.1} = 0.14 < 0.2 \quad \text{(N.G.)}$$

For $N/d \leq 0.2$,

$$\phi(0.40)t_w^2 \left[1 + 3 \left(\frac{N}{d} \right) \left(\frac{t_w}{t_f} \right)^{1.5} \right] \sqrt{\frac{EF_y t_f}{t_w}} \geq R_u \qquad \text{[AISC Equation J10-5(a)}$$

$$0.75(0.40)(0.430)^2 \left[1 + 3 \left(\frac{N}{21.1} \right) \left(\frac{0.430}{0.685} \right)^{1.5} \right] \sqrt{\frac{29,000(50)(0.685)}{0.430}} \geq 99.73$$

resulting in the requirement

$N \geq 2.59$ in.

and

$$\frac{N}{d} = \frac{2.59}{21.1} = 0.12 < 0.2 \quad \text{(OK)}$$

Try $N = 6$ in. Determine dimension B from a consideration of bearing strength. If we conservatively assume that the full area of the support is used, the required plate area A_1 can be found as follows:

$\phi_c P_p \geq R_u$

From AISC Equation J8-1, $P_p = 0.85f_c' A_1$. Then

$$\phi_c(0.85f_c' A_1) \geq R_u$$
$$0.60(0.85)(3.5)A_1 \geq 99.73$$
$$A_1 \geq 55.87 \text{ in.}^2$$

The minimum value of dimension B is

$$B = \frac{A_1}{N} = \frac{55.87}{6} = 9.31 \text{ in.}$$

The flange width of a W21 × 68 is 8.27 inches, making the plate slightly wider than the flange, which is desirable. Rounding up to the nearest inch, **try B = 10 in.** Compute the required plate thickness:

$$n = \frac{B - 2k}{2} = \frac{10 - 2(1.19)}{2} = 3.810 \text{ in.}$$

From Equation 5.13,

$$t = \sqrt{\frac{2.22 R_u n^2}{B N F_y}} = \sqrt{\frac{2.22(99.73)(3.810)^2}{10(6)(36)}} = 1.22 \text{ in.}$$

Answer   Use a PL $1\frac{1}{4}$ × 6 × 10.

ASD Solution

$$w_a = w_D + w_L = 9 \text{ kips/ft}$$

$$R_a = \frac{w_a L}{2} = \frac{9(15.83)}{2} = 71.24 \text{ kips}$$

Determine the length of bearing N required to prevent web yielding. From AISC Equation J10-3, the nominal strength is

$$R_n = (2.5k + N)F_y t_w$$

For $R_n/\Omega \geq R_a$,

$$\frac{[2.5(1.19) + N](50)(0.430)}{1.50} \geq 71.24$$

$$N \geq 1.20 \text{ in.}$$

Determine the value of N required to prevent web crippling. Assume $N/d \leq 0.2$ and use AISC Equation J10-5a:

$$\frac{R_n}{\Omega} = \frac{1}{\Omega}(0.40)t_w^2 \left[1 + 3\left(\frac{N}{d}\right)\left(\frac{t_w}{t_f}\right)^{1.5} \right] \sqrt{\frac{E F_y t_f}{t_w}} \geq R_a$$

$$\frac{1}{2.00}(0.40)(0.430)^2 \left[1 + 3\left(\frac{N}{21.1}\right)\left(\frac{0.430}{0.685}\right)^{1.5} \right] \sqrt{\frac{29,000(50)(0.685)}{0.430}} \geq 71.24$$

$$N \geq 3.78 \text{ in.}$$

$$\frac{N}{d} = \frac{3.78}{21.1} = 0.179 < 0.2 \quad \text{(OK)}$$

Try $N = 6$ in. Conservatively assume that the full area of the support is used and determine B from a consideration of bearing strength. Using AISC Equation J8-1, we obtain

$$\frac{P_p}{\Omega_c} = \frac{0.85 f_c' A_1}{\Omega_c} \geq R_a$$

$$\frac{0.85(3.5)A_1}{2.50} \geq 71.24$$

$$A_1 \geq 59.87 \text{ in.}^2$$

The minimum value of dimension B is

$$B = \frac{A_1}{N} = \frac{59.87}{6} = 9.98 \text{ in.}$$

Try $B = 10$ in.:

$$n = \frac{B - 2k}{2} = \frac{10 - 2(1.19)}{2} = 3.810 \text{ in.}$$

From Equation 5.14,

$$t \geq \sqrt{\frac{3.34 R_a n^2}{BNF_y}} = \sqrt{\frac{3.34(71.24)(3.810)^2}{10(6)(36)}} = 1.27 \text{ in.}$$

Answer   Use a PL $1\frac{1}{2} \times 6 \times 10$

If the beam is not laterally braced at the load point (in such a way as to prevent relative lateral displacement between the loaded compression flange and the tension flange), the Specification requires that *web sidesway buckling* be investigated (AISC J10.4). When loads are applied to both flanges, *web compression buckling* must be checked (AISC J10.5).

Column Base Plates

As with beam bearing plates, the design of column base plates requires consideration of bearing pressure on the supporting material and bending of the plate. A major difference is that bending in beam bearing plates is in one direction, whereas column base plates are subjected to two-way bending. Moreover, web crippling and web yielding are not factors in column base plate design.

The background and development of the plate thickness equation is presented here in LRFD terms. After some simple modifications, the corresponding ASD equation will be given.

Column base plates can be categorized as large or small, where small plates are those whose dimensions are approximately the same as the column dimensions. Furthermore, small plates behave differently when lightly loaded than when they are more heavily loaded.

FIGURE 5.41

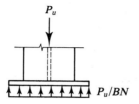

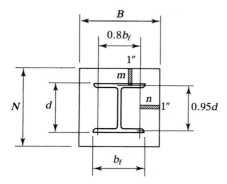

The thickness of *large* plates is determined from consideration of bending of the portions of the plate that extend beyond the column outline. Bending is assumed to take place about axes at middepth of the plate near the edges of the column flanges. Two of the axes are parallel to the web and $0.80b_f$ apart, and two axes are parallel to the flanges and $0.95d$ apart. Of the two 1-inch cantilever strips, labeled m and n in Figure 5.41, the larger is used in place of n in Equation 5.12 to compute the plate thickness, or

$$t \geq \sqrt{\frac{2P_u \ell^2}{0.90BNF_y}}$$

or

$$t \geq \ell \sqrt{\frac{2P_u}{0.90BNF_y}} \tag{5.15}$$

where ℓ is the larger of m and n. This approach is referred to as the *cantilever method*.

Lightly loaded *small* base plates can be designed by using the Murray–Stockwell method (Murray, 1983). In this approach, the portion of the column load that falls within the confines of the column cross section — that is, over an area $b_f d$ — is assumed to be uniformly distributed over the H-shaped area shown in Figure 5.42. Thus the bearing pressure is concentrated near the column outline. The plate thickness is determined from a flexural analysis of a cantilever strip of unit width and of length c. This approach results in the equation

$$t \geq c \sqrt{\frac{2P_o}{0.90A_H F_y}} \tag{5.16}$$

FIGURE 5.42

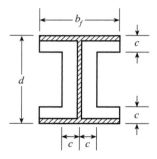

where

$$P_o = \frac{P_u}{BN} \times b_f d$$

        = load within the area $b_f d$

        = load on H-shaped area

A_H = H-shaped area

   $c =$ dimension needed to give a stress of $\dfrac{P_o}{A_H}$ equal to

the design bearing stress of the supporting material

Note that Equation 5.16 has the same form as Equation 5.15 but with the stress P_u/BN replaced by P_o/A_H.

For more heavily loaded base plates (the boundary between lightly loaded and heavily loaded plates is not well defined), Thornton (1990a) proposed an analysis based on two-way bending of the portion of the plate between the web and the flanges. As shown in Figure 5.43, this plate segment is assumed to be fixed at the web, simply supported at the flanges, and free at the other edge. The required thickness is

$$t \geq n' \sqrt{\frac{2P_u}{0.90BNF_y}}$$

FIGURE 5.43

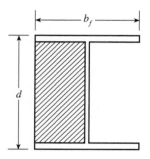

(a) Area of Plate Considered

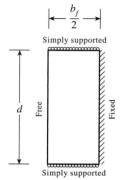

(b) Approximated Size and Edge Conditions

where

$$n' = \frac{1}{4} \sqrt{db_f} \tag{5.17}$$

These three approaches were combined by Thornton (1990b), and a summary of the resulting unified procedure follows. The required plate thickness is

$$t \geq \ell \sqrt{\frac{2P_u}{0.90 BNF_y}} \tag{5.18}$$

where

$$\ell = \max(m, n, \lambda n')$$

$$m = \frac{N - 0.95d}{2}$$

$$n = \frac{B - 0.8b_f}{2}$$

$$\lambda = \frac{2\sqrt{X}}{1 + \sqrt{1 - X}} \leq 1$$

$$X = \left(\frac{4db_f}{(d + b_f)^2} \right) \frac{P_u}{\phi_c P_p}$$

$$n' = \frac{1}{4} \sqrt{db_f}$$

$$\phi_c = 0.60$$

P_p = nominal bearing strength from AISC Equation J8-1 or J8-2

There is no need to determine whether the plate is large or small, lightly loaded, or heavily loaded. As a simplification, λ can always be conservatively taken as 1.0 (Thornton, 1990b).

This procedure is the same as that given in Part 14 of the *Manual*, "Design of Beam Bearing Plates, Column Base Plates, Anchor Rods, and Column Splices."

For ASD, we rewrite Equation 5.18 by substituting P_a for P_u and $1/\Omega$ for ϕ:

$$t \geq \ell \sqrt{\frac{2P_a}{(1/\Omega) BNF_y}}$$

or

$$t \geq \ell \sqrt{\frac{2P_a}{BNF_y / 1.67}} \tag{5.19}$$

In the equation for X, we again substitute P_a for P_u and $1/\Omega$ for ϕ:

$$X = \left(\frac{4db_f}{(d + b_f)^2} \right) \frac{P_a}{P_p / \Omega_c} \tag{5.20}$$

where $\Omega_c = 2.50$.

The equations for ℓ, m, n, λ, and n' are the same as for LRFD.

Example 5.17   A W10×49 is used as a column and is supported by a concrete pier as shown in Figure 5.44. The top surface of the pier is 18 inches by 18 inches. Design an A36 base plate for a column dead load of 98 kips and a live load of 145 kips. The concrete strength is $f'_c = 3000$ psi.

FIGURE 5.44

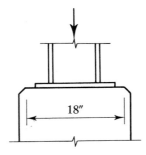

18″

LRFD Solution   The factored load is

$$P_u = 1.2D + 1.6L = 1.2(98) + 1.6(145) = 349.6 \text{ kips}$$

Compute the required bearing area. From AISC Equation J8-2,

$$P_p = 0.85 f'_c A_1 \sqrt{\frac{A_2}{A_1}} \le 1.7 f'_c A_1$$

For $\phi_u P_p \ge P_u$,

$$0.60\left[0.85(3)A_1 \sqrt{\frac{18(18)}{A_1}}\right] \ge 349.6$$

$$A_1 \ge 161 \text{ in.}^2$$

Check upper limit:

$$\phi_c 1.7 f'_c A_1 = 0.60(1.7)(3)(161) = 493 \text{ kips} > 349.6 \text{ kips} \quad \text{(OK)}$$

Also, the plate must be at least as large as the column, so

$$b_f d = 10.0(10.0) = 100 \text{ in.}^2 < 161 \text{ in.}^2 \quad \text{(OK)}$$

For $B = N = 13$ in., A_1 provided $= 13(13) = 169$ in.^{2}

The dimensions of the cantilever strips are

$$m = \frac{N - 0.95d}{2} = \frac{13 - 0.95(10)}{2} = 1.75 \text{ in.}$$

$$n = \frac{B - 0.8b_f}{2} = \frac{13 - 0.8(10)}{2} = 2.5 \text{ in.}$$

From Equation 5.17,

$$n' = \frac{1}{4}\sqrt{db_f} = \frac{1}{4}\sqrt{10.0(10.0)} = 2.5 \text{ in.}$$

As a conservative simplification, let $\lambda = 1.0$, giving

$$\ell = \max(m, n, \lambda n') = \max(1.75, 2.5, 2.5) = 2.5 \text{ in.}$$

From Equation 5.18, the required plate thickness is

$$t = \ell \sqrt{\frac{2P_u}{0.90BNF_y}} = 2.5\sqrt{\frac{2(349.6)}{0.90(13)(13)(36)}} = 0.893 \text{ in.}$$

Answer   Use a PL $1 \times 13 \times 13$.

ASD Solution   The applied load is $P_a = D + L = 98 + 145 = 243$ kips.

Compute the required bearing area. Using AISC Equation J8-2,

$$\frac{P_p}{\Omega_c} \geq P_a$$

$$\frac{1}{2.50}\left[0.85(3)A_1\sqrt{\frac{18(18)}{A_1}}\right] \geq 243$$

$$A_1 \geq 175 \text{ in.}^2$$

The upper limit is

$$\frac{1}{\Omega_c}(1.7f_c'A_1) = \frac{1}{2.50}[(1.7)(3)(175)] = 357 \text{ kips} > 243 \text{ kips} \quad \text{(OK)}$$

The plate must be as large as the column:

$$b_f d = 10.0(10.0) = 100 \text{ in.}^2 < 175 \text{ in.}^2 \quad \text{(OK)}$$

Try $B = N = 13\frac{1}{2}$ in., with A_1 provided $= 13.5(13.5) = 182$ in.^{2} > 175 in.^{2}

$$m = \frac{N - 0.95d}{2} = \frac{13.5 - 0.95(10.0)}{2} = 2.0 \text{ in.}$$

$$n = \frac{B - 0.8b_f}{2} = \frac{13.5 - 0.8(10)}{2} = 2.750 \text{ in.}$$

$$n' = \frac{1}{4}\sqrt{db_f} = \frac{1}{4}\sqrt{10.0(10.0)} = 2.5 \text{ in.}$$

Conservatively, let $\lambda = 1.0$, resulting in

$$\ell = \max(m, n, \lambda n') = \max(2.0, 2.750, 2.5) = 2.750 \text{ in.}$$

From Equation 5.20,

$$t \ge \ell \sqrt{\frac{2P_a}{BNF_y/1.67}} = 2.750 \sqrt{\frac{2(243)}{13.5(13.5)(36)/1.67}} = 0.967 \text{ in.}$$

Answer   Use a PL $1 \times 13\frac{1}{2} \times 13\frac{1}{2}$.

5.15   BIAXIAL BENDING

Biaxial bending occurs when a beam is subjected to a loading condition that produces bending about both the major (strong) axis and the minor (weak) axis. Such a case is illustrated in Figure 5.45, where a single concentrated load acts normal to the longitudinal axis of the beam but is inclined with respect to each of the principal axes of the cross section. Although this loading is more general than those previously considered, it is still a special case: The load passes through the shear center of the cross section. The shear center is that point through which the loads must act if there is to be no twisting, or torsion, of the beam. The location of the shear center can be determined from

FIGURE 5.45

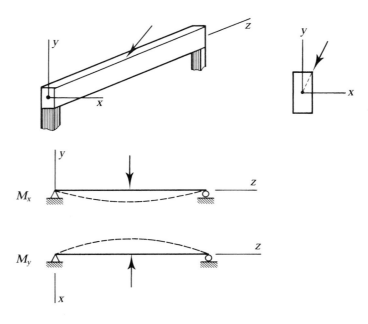

FIGURE 5.46

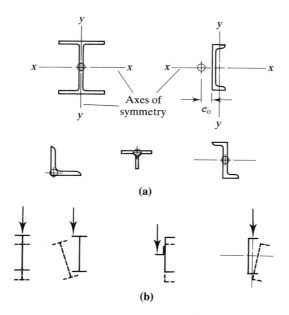

Axes of symmetry

(a)

(b)

elementary mechanics of materials by equating the internal resisting torsional moment, derived from the shear flow on the cross section, to the external torque.

The location of the shear center for several common cross sections is shown in Figure 5.46a, where the shear center is indicated by a circle. The value of e_o, which locates the shear center for channel shapes, is tabulated in the *Manual*. The shear center is always located on an axis of symmetry; thus the shear center will be at the centroid of a cross section with two axes of symmetry. Figure 5.46b shows the deflected position of two different beams when loads are applied through the shear center and when they are not.

Case I: Loads Applied Through the Shear Center

If loads act through the shear center, the problem is one of simple bending in two perpendicular directions. As illustrated in Figure 5.47, the load can be resolved into rectangular components in the x- and y-directions, each producing bending about a different axis.

FIGURE 5.47

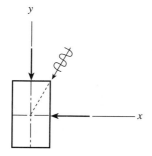

To deal with this combined loading, we temporarily look ahead to Chapter H of the Specification, "Design of Members for Combined Forces and Torsion" (also the subject of Chapter 6 of this textbook, "Beam–Columns"). The Specification deals with combined loading primarily through the use of interaction formulas, which account for the relative importance of each load effect in relation to the strength corresponding to that effect. For example, if there is bending about one axis only, we can write for that axis

required moment strength ≤ available moment strength

or

$$\frac{\text{required moment strength}}{\text{available moment strength}} \leq 1.0$$

If there is bending about both the x and y axes, the interaction approach requires that the sum of ratios for the two effects be less than 1.0; that is,

$$\frac{\text{required } x \text{ - axis moment strength}}{\text{available } x \text{ - axis moment strength}} + \frac{\text{required } y \text{ - axis moment strength}}{\text{available } y \text{ - axis moment strength}} \leq 1.0$$

(5.21)

In effect, this approach allows the designer to allocate to one direction what has not been "used up" in the other direction. AISC Section H1 incorporates a comparable ratio for axial loads and gives two interaction formulas, one for small axial loads and one for large axial loads (we discuss the reason for this in Chapter 6). With biaxial bending and no axial load, the formula for small axial load is the appropriate one, and it reduces to the form of Equation 5.21. (The exact notation and form of the AISC Equations are covered in Chapter 6.)

For LRFD, Equation 5.21 becomes

$$\frac{M_{ux}}{\phi_b M_{nx}} + \frac{M_{uy}}{\phi_b M_{ny}} \leq 1.0$$

(5.22)

where

M_{ux} = factored-load moment about the x axis
M_{nx} = nominal moment strength for x-axis bending
M_{uy} = factored-load moment about the y axis
M_{ny} = nominal moment strength for the y axis

For ASD,

$$\frac{M_{ax}}{M_{nx}/\Omega_b} + \frac{M_{ay}}{M_{ny}/\Omega_b} \leq 1.0$$

(5.23)

where

M_{ax} = service load moment about the x axis
M_{ay} = service load moment about the y axis

To this point, the strength of I-shaped cross sections bent about the weak axis has not been considered. Doing so is relatively simple. Any shape bent about its weak axis cannot buckle in the other direction, so lateral-torsional buckling is not a limit state. If the shape is compact, then

$$M_{ny} = M_{py} = F_y Z_y \leq 1.6 F_y S_y \qquad \text{(AISC Equation F6-1)}$$

where

M_{ny} = nominal moment strength about the y axis
M_{py} = plastic moment strength about the y axis
$F_y S_y$ = yield moment for the y axis

(The y subscripts in M_{ny} and M_{py} are not in the Specification; they have been added here.) The limit of 1.6 $F_y S_y$ is to prevent excessive working load deformation and is satisfied when

$$\frac{Z_y}{S_y} \leq 1.6$$

If the shape is noncompact because of the flange width-thickness ratio, the strength will be given by

$$M_{ny} = M_{py} - (M_{py} - 0.7 F_y S_y)\left(\frac{\lambda - \lambda_p}{\lambda_r - \lambda_p}\right) \qquad \text{(AISC Equation F6-2)}$$

This is the same as AISC Equation F3-1 for flange local buckling, except for the axis of bending.

Example 5.18

A W21 × 68 is used as a simply supported beam with a span length of 12 feet. Lateral support of the compression flange is provided only at the ends. Loads act through the shear center, producing moments about the x and y axes. The service load moments about the x axis are M_{Dx} = 48 ft-kips and M_{Lx} = 144 ft-kips. Service load moments about the y axis are M_{Dy} = 6 ft-kips and M_{Ly} = 18 ft-kips. If A992 steel is used, does this beam satisfy the provisions of the AISC Specification? Assume that all moments are uniform over the length of the beam.

Solution

First, compute the nominal flexural strength for x-axis bending. The following data for a W21 × 68 are obtained from the Z_x table. The shape is compact (no footnote to indicate otherwise) and

$$L_p = 6.36 \text{ ft}, \ L_r = 18.7 \text{ ft}$$

The unbraced length L_b = 12 ft, so $L_p < L_b < L_r$, and the controlling limit state is inelastic lateral-torsional buckling. Then

$$M_{nx} = C_b\left[M_{px} - (M_{px} - 0.7 F_y S_x)\left(\frac{L_b - L_p}{L_r - L_p}\right)\right] \leq M_{px}$$

$$M_{px} = F_y Z_x = 50(160) = 8000 \text{ in.-kips}$$

Because the bending moment is uniform, $C_b = 1.0$.

$$M_{nx} = 1.0 \left[8000 - (8000 - 0.7 \times 50 \times 140) \left(\frac{12 - 6.36}{18.7 - 6.36} \right) \right]$$

$$= 6583 \text{ in.-kips} = 548.6 \text{ ft-kips}$$

For the y axis, since the shape is compact, there is no flange local buckling and

$$M_{ny} = M_{py} = F_y Z_y = 50(24.4) = 1220 \text{ in.-kips} = 101.7 \text{ ft-kips}$$

Check the upper limit:

$$\frac{Z_y}{S_y} = \frac{24.4}{15.7} = 1.55 < 1.6 \qquad \therefore M_{ny} = M_{py} = 101.7 \text{ in.-kips}$$

LRFD Solution   For x-axis bending,

$$M_{ux} = 1.2M_{Dx} + 1.6M_{Lx} = 1.2(48) + 1.6(144) = 288.0 \text{ ft-kips}$$

For y-axis bending,

$$M_{uy} = 1.2M_{Dy} + 1.6M_{Ly} = 1.2(6) + 1.6(18) = 36.0 \text{ ft-kips}$$

Check interaction equation 5.22:

$$\frac{M_{ux}}{\phi_b M_{nx}} + \frac{M_{uy}}{\phi_b M_{ny}} = \frac{288.0}{0.90(548.6)} + \frac{36.0}{0.90(101.7)} = 0.977 < 1.0 \qquad \text{(OK)}$$

(Note that $\phi_b M_{nx}$ can be obtained from the beam design charts.)

Answer   The W21 × 68 is satisfactory.

ASD Solution   For x-axis bending.

$$M_{ax} = M_{Dx} + M_{Lx} = 48 + 144 = 192 \text{ ft-kips}$$

For y-axis bending,

$$M_{ay} = M_{Dy} + M_{Ly} = 6 + 18 = 24 \text{ ft-kips}$$

Check interaction equation 5.23:

$$\frac{M_{ax}}{M_{nx}/\Omega_b} + \frac{M_{ay}}{M_{ny}/\Omega_b} = \frac{192}{548.6/1.67} + \frac{24}{101.7/1.67} = 0.979 < 1.0 \quad \text{(OK)}$$

(Note that M_{nx}/Ω_b can be obtained from the beam design charts.)

Answer   The W21 × 68 is satisfactory.

Case II: Loads Not Applied Through the Shear Center

When loads are not applied through the shear center of a cross section, the result is flexure plus torsion. If possible, the structure or connection geometry should be modified to remove the eccentricity. The problem of torsion in rolled shapes is a complex one, and we resort to approximate, although conservative, methods for dealing with it. A more detailed treatment of this topic, complete with design aids, can be found in *Torsional Analysis of Structural Steel Members* (AISC, 1997b). A typical loading condition that gives rise to torsion is shown in Figure 5.48a. The resultant load is applied to the center of the top flange, but its line of action does not pass through the shear center of the section. As far as equilibrium is concerned, the force can be moved to the shear center provided that a couple is added. The equivalent system thus obtained will consist of the given force acting through the shear center plus a twisting moment, as shown. In Figure 5.48b, there is only one component of load to contend with, but the concept is the same.

Figure 5.49 illustrates a simplified way of treating these two cases. In Figure 5.49a, the top flange is assumed to provide the total resistance to the horizontal component of the load. In Figure 5.49b, the twisting moment Pe is resisted by a couple consisting of equal forces acting at each flange. As an approximation, each flange can be considered to resist each of these forces independently. Consequently, the problem is reduced to a case of bending of two shapes, each one loaded through its shear center. In each of the two situations depicted in Figure 5.49, only about half the cross section is considered to be effective with respect to its *y*-axis; therefore, when considering the strength of a single flange, use half the tabulated value of Z_y for the shape.

FIGURE 5.48

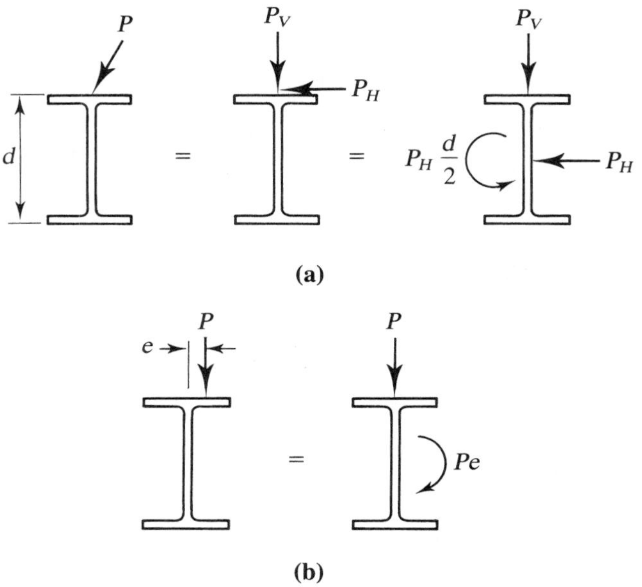

(a)

(b)

FIGURE 5.49

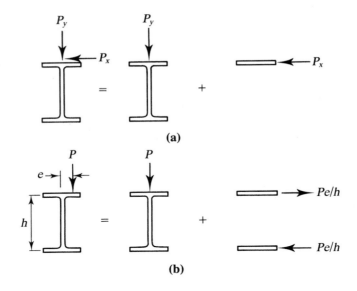

(a)

(b)

Design of Roof Purlins

Roof purlins that are part of a sloping roof system can be subjected to biaxial bending of the type just described. For the roof purlin shown in Figure 5.50, the load is vertical, but the axes of bending are inclined. This condition corresponds to the loading of Figure 5.49a. The component of load normal to the roof will cause bending about the x-axis, and the parallel component bends the beam about its y-axis. If the purlins are simply supported at the trusses (or rigid frame rafters), the maximum bending moment about each axis is $wL^2/8$, where w is the appropriate component of load. If sag rods are used, they will provide lateral support with respect to x-axis bending and will act as transverse supports for y-axis bending, requiring that the purlin be treated as a continuous beam. For uniform sag-rod spacings, the formulas for continuous beams in Part 3 of the *Manual* can be used.

FIGURE 5.50

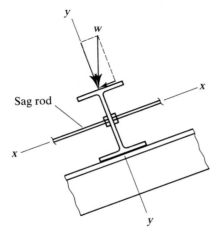

Example 5.19

A roof system consists of trusses of the type shown in Figure 5.51 spaced 15 feet apart. Purlins are to be placed at the joints and at the midpoint of each top-chord member. Sag rods will be located at the center of each purlin. The total gravity load, including an estimated purlin weight, is 42 psf of roof surface, with a live-load–to–dead-load ratio of 1.0. Assuming that this is the critical loading condition, use A36 steel and select a channel shape for the purlins.

FIGURE 5.51

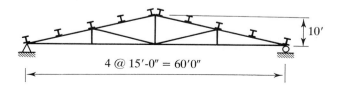

4 @ 15'-0" = 60'0"

10'

LRFD Solution

For the given loading condition, dead load plus a roof live load with no wind or snow, load combination 3 will control:

$$w_u = 1.2w_D + 1.6L_r = 1.2(21) + 1.6(21) = 58.80 \text{ psf}$$

The width of roof surface tributary to each purlin is

$$\frac{15}{2}\frac{\sqrt{10}}{3} = 7.906 \text{ ft}$$

Then

$$\text{Purlin load} = 58.80(7.906) = 464.9 \text{ lb/ft}$$

$$\text{Normal component} = \frac{3}{\sqrt{10}}(464.9) = 441.0 \text{ lb/ft}$$

$$\text{Parallel component} = \frac{1}{\sqrt{10}}(464.9) = 147.0 \text{ lb/ft}$$

and

$$M_{ux} = \frac{1}{8}(0.4410)(15)^2 = 12.40 \text{ ft-kips}$$

With sag rods placed at the midpoint of each purlin, the purlins are two-span continuous beams with respect to weak axis bending. From Table 3-22c," Continuous Beams," the maximum moment in a two-span continuous beam with equal spans is at the interior support and is given by

$$M = 0.125w\ell^2$$

where

w = uniform load intensity
ℓ = span length

The maximum moment about the y axis is therefore

$$M_{uy} = 0.125(0.1470)(15/2)^2 = 1.034 \text{ ft-kips}$$

To select a trial shape, use the beam design charts and choose a shape with a relatively large margin of strength with respect to major axis bending. For an unbraced length of $^{15}\!\!/_2 = 7.5$ ft, **try a C10 × 15.3**.

For $C_b = 1.0$, $\phi_b M_{nx} = 33.0$ ft-kips. From Figure 5.15b, C_b is 1.30 for the load and lateral support conditions of this beam. Therefore,

$$\phi_b M_{nx} = 1.30(33.0) = 42.90 \text{ ft-kips}$$

From the uniform load table for C shapes,

$$\phi_b M_{px} = 43.0 \text{ ft-kips} > 42.90 \text{ ft-kips}$$
$$\therefore \text{ use } \phi_b M_{nx} = 42.9 \text{ ft-kips}$$

This shape is compact (no footnote in the uniform load tables), so

$$\phi_b M_{ny} = \phi_b M_{py} = \phi_b F_y Z_y = 0.90(36)(2.34) = 75.82 \text{ in.-kips} = 6.318 \text{ ft-kips}$$

But

$$\frac{Z_y}{S_y} = \frac{2.34}{1.15} = 2.03 > 1.6$$

$$\therefore \phi_b M_{ny} = \phi_b(1.6 F_y S_y) = 0.90(1.6)(36)(1.15)$$
$$= 59.62 \text{ in.-kips} = 4.968 \text{ ft-kips}$$

Because the load is applied to the top flange, use only half this capacity to account for the torsional effects. From Equation 5.22,

$$\frac{M_{ux}}{\phi_b M_{nx}} + \frac{M_{uy}}{\phi_b M_{ny}} = \frac{12.40}{42.9} + \frac{1.034}{4.968/2} = 0.705 < 1.0 \quad \text{(OK)}$$

The shear is

$$V_u = \frac{0.4410(15)}{2} = 3.31 \text{ kips}$$

From the uniform load tables,

$$\phi_v V_n = 46.7 \text{ kips} > 3.31 \text{ kips} \qquad \text{(OK)}$$

Answer   Use a C10 × 15.3.

ASD Solution   For dead load plus a roof live load, load combination 3 will control:

$$q_a = q_D + q_{Lr} = 42 \text{ psf}$$

The width of roof surface tributary to each purlin is

$$\frac{15}{2}\frac{\sqrt{10}}{3} = 7.906 \text{ ft}$$

Then

$$\text{Purlin load} = 42(7.906) = 332.1 \text{ lb/ft}$$

$$\text{Normal component} = \frac{3}{\sqrt{10}}(332.1) = 315.1 \text{ lb/ft}$$

$$\text{Parallel component} = \frac{1}{\sqrt{10}}(332.1) = 105.0 \text{ lb/ft}$$

and

$$M_{ax} = \frac{1}{8}wL^2 = \frac{1}{8}(0.3151)(15)^2 = 8.862 \text{ ft-kips}$$

With sag rods placed at the midpoint of each purlin, the purlins are two-span continuous beams with respect to weak axis bending. From Table 3-22c, "Continuous Beams," the maximum moment in a two-span continuous beam with equal spans is at the interior support and is given by

$$M = 0.125w\ell^2$$

where
$$w = \text{uniform load intensity}$$
$$\ell = \text{span length}$$

The maximum moment about the y axis is therefore

$$M_{ay} = 0.125(0.1050)(15/2)^2 = 0.7383 \text{ ft-kips}$$

To select a trial shape, use the beam design charts and choose a shape with a relatively large margin of strength with respect to major axis bending. For an unbraced length of $15/2 = 7.5$ ft, try a **C10 × 15.3**.

For $C_b = 1.0$, $M_{nx}/\Omega_b = 22.0$ ft-kips. From Figure 5.15b, $C_b = 1.30$ for the load and lateral support conditions of this beam. Therefore,

$$M_{nx}/\Omega_b = 1.30(22.0) = 28.60 \text{ ft-kips}$$

From the uniform load tables,

$$M_{px}/\Omega_b = 28.6 \text{ ft-kips}$$
$$\therefore \text{ use } M_{nx}/\Omega_b = 28.6 \text{ ft-kips}$$

This shape is compact (no footnote in the uniform load tables), so

$$M_{ny}/\Omega_b = M_{py}/\Omega_b = F_y Z_y /\Omega_b = 36(2.34)/1.67 = 50.44 \text{ in.-kips} = 4.203 \text{ ft-kips}$$

But

$$\frac{Z_y}{S_y} = \frac{2.34}{1.15} = 2.03 > 1.6$$

$$\therefore M_{ny}/\Omega_b = 1.6F_y S_y/\Omega_b = 1.6(36)(1.15)/1.67 = 39.66 \text{ in.-kips} = 3.300 \text{ ft-kips}$$

Because the load is applied to the top flange, use only half of this capacity to account for the torsional effects. From Equation 5.23,

$$\frac{M_{ax}}{M_{nx}/\Omega_b} + \frac{M_{ay}}{M_{ny}/\Omega_b} = \frac{8.862}{28.6} + \frac{0.7383}{3.300/2} = 0.757 < 1.0 \qquad \text{(OK)}$$

The maximum shear is

$$V_a = \frac{0.3151(15)}{2} = 2.36 \text{ kips}$$

From the uniform load tables,

$$\frac{V_n}{\Omega_v} = 31.0 \text{ kips} > 2.36 \text{ kips} \qquad \text{(OK)}$$

Answer   Use a C10 × 15.3.

5.16   BENDING STRENGTH OF VARIOUS SHAPES

W-, S-, M-, and C-shapes are the most commonly used hot-rolled shapes for beams, and their bending strength has been covered in the preceding sections. Other shapes are sometimes used as flexural members, however, and this section provides a summary of some of the relevant AISC provisions. All equations are from Chapter F of the Specification. Nominal strength is given for compact and noncompact hot-rolled shapes, but not for slender shapes or shapes built up from plate elements. No numerical examples are given in this section, but Example 6.11 includes the computation of the flexural strength of a structural tee-shape.

I.   Square and Rectangular HSS and Box-Shaped Members:

    a. Width–thickness parameters (see Figure 5.52):
       i. Flange:

$$\lambda = \frac{b}{t} \qquad \lambda_p = 1.12\sqrt{\frac{E}{F_y}} \qquad \lambda_r = 1.40\sqrt{\frac{E}{F_y}}$$

FIGURE 5.52

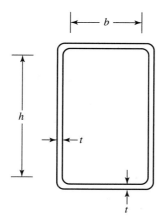

ii. Web:

$$\lambda = \frac{h}{t} \qquad \lambda_p = 2.42\sqrt{\frac{E}{F_y}} \qquad \lambda_r = 5.70\sqrt{\frac{E}{F_y}}$$

If the actual dimensions b and h are not known, they may be estimated as the total width or depth minus three times the thickness. The *design* thickness, which is 0.93 times the *nominal* thickness, should be used. (b/t and h/t ratios for HSS are given in the *Manual* in Part 1, "Dimensions and Properties.")

b. Bending about the major axis (loaded in the plane of symmetry):

i. Compact shapes: For compact shapes, the strength will be based on the limit state of yielding.

$$M_n = M_p = F_y Z \qquad \text{(AISC Equation F7-1)}$$

(Because of the high torsional resistance of closed cross-sectional shapes, lateral-torsional buckling of HSS need not be considered, even for rectangular shapes bent about the strong axis.)

ii. Noncompact shapes: The nominal strength M_n will be the smaller value computed from the limit states of flange local buckling (FLB) and web local buckling (WLB). For FLB,

$$M_n = M_p - (M_p - F_y S)\left(3.57\frac{b}{t}\sqrt{\frac{F_y}{E}} - 4.0\right) \leq M_p$$

(AISC Equation F7-2)

For WLB,

$$M_n = M_p - (M_p - F_y S_x)\left(0.305\frac{h}{t_w}\sqrt{\frac{F_y}{E}} - 0.738\right) \leq M_p$$

(AISC Equation F7-5)

c. Bending about the minor axis: The provisions are the same as for bending about the major axis.

II. Round HSS:

a. Width–thickness parameters:

$$\lambda = \frac{D}{t} \qquad \lambda_p = \frac{0.07E}{F_y} \qquad \lambda_r = \frac{0.31E}{F_y}$$

where D is the outer diameter. Note: The value of D/t must be less than $0.45E/F_y$.

b. Nominal bending strength: There is no LTB limit state for circular (or square) shapes. The strength is limited by local buckling.

Compact shapes:

$$M_n = M_p = F_y Z \qquad \text{(AISC Equation F8-1)}$$

Noncompact shapes:

$$M_n = \left(\frac{0.021E}{D/t} + F_y \right) S \qquad \text{(AISC Equation F8-2)}$$

III. Tees and Double Angles Loaded in the Plane of Symmetry:

a. Width–thickness parameters:

i. Tees:

1. Flange:

$$\lambda = \frac{b_f}{2t_f} \qquad \lambda_p = 0.38\sqrt{\frac{E}{F_y}} \qquad \lambda_r = 1.0\sqrt{\frac{E}{F_y}}$$

2. Web: The shape classification (compact, noncompact, or slender) depends only on the flange width–thickness ratio.

ii. Double angles:

Table B4.1 of the Specification does not include λ, λ_p, or λ_r for double-angle shapes in flexure. One option is to use the values given for tees.

b. Compact shapes: The strength will be smaller values for the limit states of yielding and LTB. For yielding,

$$M_n = M_p \qquad \text{(AISC Equation F9-1)}$$

where

$$M_p = F_y Z \le 1.6 M_y \qquad \text{for stems in tension} \qquad \text{(AISC Equation F9-2)}$$
$$M_p = F_y Z \le M_y \qquad \text{for stems in compression} \qquad \text{(AISC Equation F9-3)}$$

where M_y = yield moment = $F_y S$.

For LTB,

$$M_n = M_{cr} = \frac{\pi\sqrt{EI_y GJ}}{L_b}\left[B + \sqrt{1+B^2}\right]$$ (AISC Equation F9-4)

where

$$B = \pm 2.3\left(\frac{d}{L_b}\right)\sqrt{\frac{I_y}{J}}$$ (AISC Equation F9-5)

The positive sign is used for B when the stem is in tension, and the negative sign is used when the stem is in compression *anywhere* along the unbraced length.

c. Noncompact shapes: The strength will be the lower value obtained for flange local buckling and LTB. For FLB,

$$M_n = F_{cr}S_{xc}$$ (AISC Equation F9-6)

where
S_{xc} = elastic section modulus referred to the compression flange

$$F_{cr} = F_y\left[1.19 - 0.50\left(\frac{b_f}{2t_f}\right)\sqrt{\frac{F_y}{E}}\right]$$ (AISC Equation F9-7)

For LTB,

$$M_n = M_{cr} = \frac{\pi\sqrt{EI_y GJ}}{L_b}\left[B + \sqrt{1+B^2}\right]$$ (AISC Equation F9-4)

where

$$B = \pm 2.3\left(\frac{d}{L_b}\right)\sqrt{\frac{I_y}{J}}$$ (AISC Equation F9-5)

The positive sign is used for B when the stem is in tension, and the negative sign is used when the stem is in compression *anywhere* along the unbraced length.

IV. Solid Rectangular Bars: The applicable limit states are yielding and LTB for major axis bending; local buckling is not a limit state for either major or minor axis bending.

a. Bending about the major axis:

For $\dfrac{L_b d}{t^2} \le \dfrac{0.08E}{F_y}$,

$$M_n = M_p \le 1.6M_y$$ (AISC Equation F11-1)

where M_y = yield moment = $F_y S$.

For $\dfrac{0.08E}{F_y} < \dfrac{L_b d}{t^2} \leq \dfrac{1.9E}{F_y}$,

$$M_n = C_b\left[1.52 - 0.274\left(\dfrac{L_b d}{t^2}\right)\dfrac{F_y}{E}\right]M_y \leq M_p \qquad \text{(AISC Equation F11-2)}$$

For $\dfrac{L_b d}{t^2} > \dfrac{1.9E}{F_y}$

$$M_n = F_{cr}S_x \leq M_p \qquad \text{(AISC Equation F11-3)}$$

where

$$F_{cr} = \dfrac{1.9EC_b}{L_b d/t^2} \qquad \text{(AISC Equation F11-4)}$$

t = thickness of bar (dimension parallel to axis of bending)
d = width of bar

    b. Bending about the minor axis:

$$M_n = M_p = F_y Z \leq 1.6 M_y \qquad \text{(AISC Equation F11-1)}$$

where M_y = yield moment = $F_y S$.

V. Solid Circular Bars:

$$M_n = M_p = F_y Z \leq 1.6 M_y \qquad \text{(AISC Equation F11-1)}$$

(For a circle, $Z/S = 1.7 > 1.6$, so the upper limit *always* controls.)

VI. Solid Square Bars:

$$M_n = M_p = F_y Z \leq 1.6 M_y \qquad \text{(AISC Equation F11-1)}$$

(For a square, $Z/S = 1.5 < 1.6$, so the upper limit *never* controls.)

For flexural members not covered in this summary (single angles, slender shapes, unsymmetrical shapes, and shapes built up from plate elements), refer to Chapter F of the AISC Specification. (Shapes built up from plate elements are also covered in Chapter 10 of this book.)

Problems

Bending Stress and the Plastic Moment

5.2-1 A flexural member is fabricated from two flange plates $\frac{1}{2} \times 7\frac{1}{2}$ and a web plate $\frac{3}{8} \times 17$. The yield stress of the steel is 50 ksi.

a. Compute the plastic section modulus Z and the plastic moment M_p with respect to the major principal axis.
b. Compute the elastic section modulus S and the yield moment M_y with respect to the major principal axis.

5.2-2 An unsymmetrical flexural member consists of a $\frac{1}{2} \times 12$ top flange, a $\frac{1}{2} \times 7$ bottom flange, and a $\frac{3}{8} \times 16$ web.

a. Determine the distance $\bar{y}$ from the top of the shape to the horizontal *plastic* neutral axis.
b. If A572 Grade 50 steel is used, what is the plastic moment M_p for the horizontal plastic neutral axis?
c. Compute the plastic section modulus Z with respect to the minor principal axis.

5.2-3 Verify the value of Z_x for a W18 × 50 that is tabulated in the dimensions and properties tables in Part 1 of the *Manual*.

5.2-4 Verify the value of Z_x given in the *Manual* for an S10 × 35.

Classification of Shapes

5.4-1 For W-, M-, and S-shapes with $F_y = 60$ ksi:

a. List the shapes in Part 1 of the *Manual* that are noncompact (when used as flexural members). State whether they are noncompact because of the flange, the web, or both.
b. List the shapes in Part 1 of the *Manual* that are slender. State whether they are slender because of the flange, the web, or both.

5.4-2 Repeat Problem 5.4-1 for steel with $F_y = 65$ ksi.

5.4-3 Determine the smallest value of yield stress F_y for which a W-, M-, or S-shape from Part 1 of the *Manual* will become slender. To which shapes does this value apply? What conclusion can you draw from your answer?

Bending Strength of Compact Shapes

5.5-1 The beam shown in Figure P5.5-1 is a W10 × 77 and has continuous lateral support. The load P is a service live load. If $F_y = 50$ ksi, what is the maximum permissible value of P?

a. Use LRFD.
b. Use ASD.

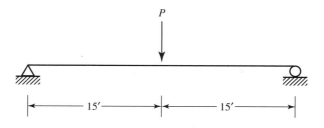

FIGURE P5.5-1

5.5-2   The given beam has continuous lateral support. If the live load is twice the dead load, what is the maximum total *service* load, in kips/ft, that can be supported? A992 steel is used.

   a. Use LRFD.
   b. Use ASD.

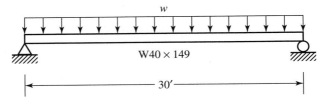

FIGURE P5.5-2

5.5-3   A simply supported beam is subjected to a uniform service dead load of 1.0 kips/ft (including the weight of the beam), a uniform service live load of 2.0 kips/ft, and a concentrated service dead load of 40 kips. The beam is 40 feet long, and the concentrated load is located 15 feet from the left end. The beam has continuous lateral support, and A572 Grade 50 steel is used. Is a W30 × 108 adequate?

   a. Use LRFD.
   b. Use ASD.

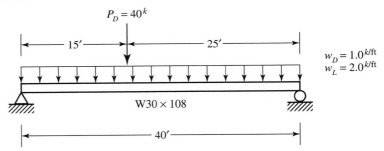

FIGURE P5.5-3

5.5-4   The beam shown in Figure P5.5-4 has continuous lateral support of both flanges. The uniform load is a service load consisting of 50% dead load and 50% live load. The dead load includes the weight of the beam. If A992 steel is used, is a W12×35 adequate?

   a. Use LRFD.
   b. Use ASD.

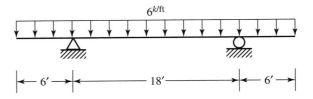

FIGURE P5.5-4

5.5-5   The beam shown in Figure P5.5-5 is a two-span beam with a pin (hinge) in the center of the left span, making the beam statically determinate. There is continuous lateral support. The concentrated loads are service live loads. Determine whether a W18×60 of A992 steel is adequate.

a. Use LRFD.

b. Use ASD.

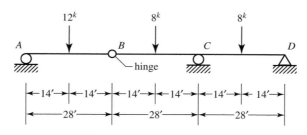

FIGURE P5.5-5

5.5-6   A W12 × 30 of A992 steel has an unbraced length of 10 feet. Using $C_b = 1.0$,

a. Compute L_p and L_r. Use the equations in Chapter F of the AISC Specification. Do not use any of the design aids in the *Manual*.

b. Compute the flexural design strength, $\phi_b M_n$.

c. Compute the allowable flexural strength M_n/Ω_b.

5.5-7   A W18 × 46 is used for a beam with an unbraced length of 10 feet. Using $F_y = 50$ ksi and $C_b = 1$, compute the nominal flexural strength. Use the AISC equations in Chapter F of the Specification. Do not use any of the design aids in the *Manual*.

5.5-8   A W18 × 71 is used as a beam with an unbraced length of 9 feet. Use $F_y = 65$ ksi and $C_b = 1$ and compute the nominal flexural strength. Compute everything with the equations in Chapter F of the AISC Specification.

5.5-9   The beam shown in Figure P5.5-9 is a W36 × 182. It is laterally supported at A and B. The 300 kip load is a service live load. Using the unfactored service loads,

a. Compute C_b. Do not include the beam weight in the loading.

b. Compute C_b. Include the beam weight in the loading.

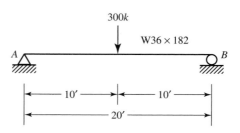

FIGURE P5.5-9

5.5-10   If the beam in Problem 5.5-9 is braced at *A*, *B*, and *C*, compute C_b for the unbraced length *AC* (same as C_b for unbraced length *CB*). Do not include the beam weight in the loading.

a. Use the unfactored service loads.
b. Use factored loads.

5.5-11   The beam shown in Figure P5.5-11 has lateral support at *a*, *b*, *c*, and *d*. Compute C_b for segment *b–c*.

a. Use the unfactored service loads.
b. Use factored loads.

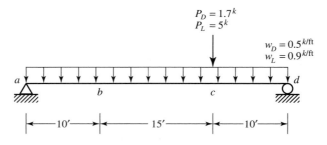

FIGURE P5.5-11

5.5-12   A W21 × 68 of A992 steel is used as a simply supported beam with a span length of 50 feet. The only load in addition to the beam weight is a uniform live load. If lateral support is provided at 10-foot intervals, what is the maximum service live load, in kips/ft, that can be supported?

a. Use LRFD.
b. Use ASD.

5.5-13   The beam shown in Figure P5.5-13 is laterally braced only at the ends. The 30-kip load is a service live load. Use $F_y = 50$ ksi and determine whether a W14 × 38 is adequate.

a. Use LRFD.
b. Use ASD.

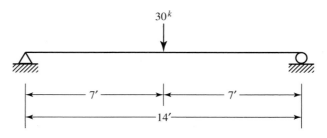

FIGURE P5.5-13

5.5-14 Repeat Problem 5.5-13 for an MC18 × 58 (Assume that the load is applied through the shear center so that there is no torsional loading.) Use $F_y = 36$ ksi.

5.5-15 Determine whether a W24 × 104 of A992 steel is adequate for the beam shown in Figure P5.5-15. The uniform load does not include the weight of the beam. Lateral support is provided at *A*, *B*, and *C*.

   a. Use LRFD.
   b. Use ASD.

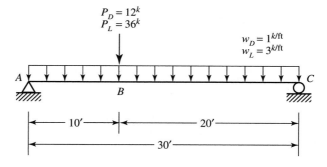

FIGURE P5.5-15

5.5-16 The beam shown in Figure P5.5-16 is laterally braced at *A*, *B*, *C*, and *D*. Is a W14 × 132 adequate for $F_y = 50$ ksi?

   a. Use LRFD.
   b. Use ASD.

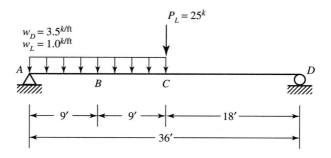

FIGURE P5.5-16

Bending Strength of Noncompact Shapes

5.6-1 A W12 × 65 is used as a simply supported, uniformly loaded beam with a span length of 50 feet and continuous lateral support. The yield stress, F_y, is 50 ksi. If the ratio of live load to dead load is 3, compute the available strength and determine the maximum total service load, in kips/ft, that can be supported.

   a. Use LRFD.
   b. Use ASD.

5.6-2 A W14 × 99 of A992 steel is used as a beam with lateral support at 10-foot intervals. Assume that $C_b = 1.0$ and compute the nominal flexural strength.

5.6-3 A built-up shape consisting of two ¾ × 18 flanges and a ¾ × 52 web is used as a beam. If A572 Grade 50 steel is used, what is the nominal flexural strength based on flange local buckling? For width–thickness ratio limits for welded shapes, refer to Table B4.1 in Chapter B of the AISC Specification, "Design Requirements."

5.6-4 A built-up shape consisting of two ¾ × 16 flanges and a ½ × 40 web is used as a beam with continuous lateral support. If A572 Grade 50 steel is used, what is the nominal flexural strength? For width–thickness ratio limits for welded shapes, refer to Table B4.1 in Chapter B of the AISC Specification, "Design Requirements."

Shear Strength

5.8-1 Compute the nominal shear strength of an M10 × 7.5 of A572 Grade 65 steel.

5.8-2 Compute the nominal shear strength of an M12 × 11.8 of A242 steel.

5.8-3 The beam shown in Figure P5.8-3 is a W16 × 31 of A992 steel and has continuous lateral support. The two concentrated loads are service live loads. Neglect the weight of the beam and determine whether the beam is adequate.

a. Use LRFD.
b. Use ASD.

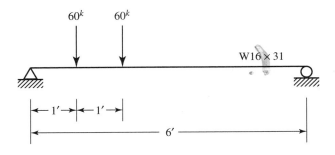

FIGURE P5.8-3

5.8-4 The cantilever beam shown in Figure P5.8-4 is a W16 × 45 of A992 steel. There is no lateral support other than at the fixed end. Use an unbraced length equal to the span length and determine whether the beam is adequate. The uniform load is a service dead load that includes the beam weight, and the concentrated load is a service live load.

a. Use LRFD.
b. Use ASD.

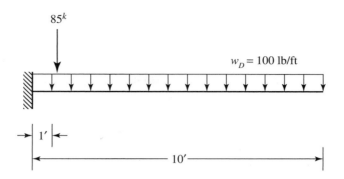

85^k

$w_D = 100$ lb/ft

1′

10′

FIGURE P5.8-4

Design

5.10-1   Use A992 steel and select a W-shape for the following beam:

- Simply supported with a span length of 30 feet
- Laterally braced only at the ends
- Service dead load = 0.75 kips/ft
- The service live load consists of a 34-kip concentrated load at the center of the span

There is no limit on the deflection.

a. Use LRFD.
b. Use ASD.

5.10-2   Use A992 steel and select the most economical W shape for the beam in Figure P5.10-2. The beam weight is not included in the service loads shown. Do not check deflection. Assume continuous lateral support.

a. Use LRFD.
b. Use ASD.

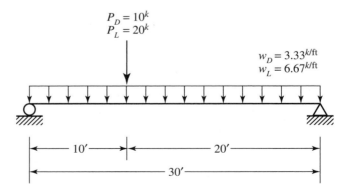

$P_D = 10^k$
$P_L = 20^k$

$w_D = 3.33^{k/ft}$
$w_L = 6.67^{k/ft}$

10′   20′

30′

FIGURE P5.10-2

5.10-3   Same as Problem 5.10-2, except that lateral support is provided only at the ends and at the concentrated load.

5.10-4   The beam shown in Figure P5.10-4 has lateral support only at the ends. The uniform load is a *superimposed* dead load, and the concentrated load is a live load. Use A992 steel and select a W-shape. The *live load* deflection must not exceed $L/360$.

    a. Use LRFD.
    b. Use ASD.

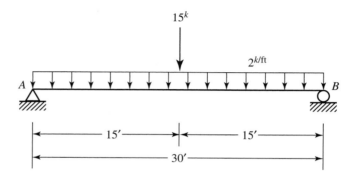

FIGURE P5.10-4

5.10-5   The given beam is laterally supported at the ends and at the ⅓ points (points 1, 2, 3, and 4). The concentrated load is a service live load. Use $F_y = 50$ ksi and select a W-shape. Do not check deflections.

    a. Use LRFD.
    b. Use ASD.

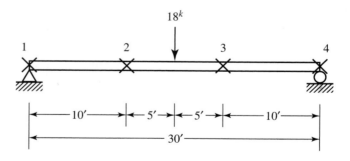

FIGURE P5.10-5

5.10-6   The beam shown in Figure P5.10-6 has lateral support at the ends only. The concentrated loads are live loads. Use A992 steel and select a shape. Do not check deflections. Use $C_b = 1.0$ (this is conservative).

    a. Use LRFD.
    b. Use ASD.

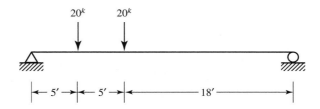

FIGURE P5.10-6

5.10-7 The beam shown in Figure P5.10-7 is part of a roof system. Assume that there is partial lateral support equivalent to bracing at the ends and at midspan. The loading consists of 170 lb/ft dead load (not including the weight of the beam), 100 lb/ft roof live load, 280 lb/ft snow load, and 180 lb/ft wind load acting *upward*. The dead, live, and snow loads are gravity loads and always act downward, whereas the wind load on the roof will always act upward. Use A992 steel and select a shape. The total deflection must not exceed $L/180$.

a. Use LRFD.
b. Use ASD.

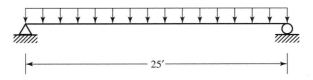

FIGURE P5.10-7

Floor and Roof Framing Systems

5.11-1 Use $F_y = 50$ ksi and select a shape for a typical floor beam AB. Assume that the floor slab provides continuous lateral support. The maximum permissible live load deflection is $L/180$. The service dead loads consist of a 5-inch-thick reinforced-concrete floor slab (normal weight concrete), a partition load of 20 psf, and 10 psf to account for a suspended ceiling and mechanical equipment. The service live load is 60 psf.

a. Use LRFD.
b. Use ASD.

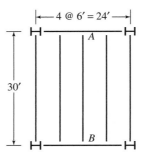

FIGURE P5.11-1

5.11-2   Select a W-shape for the following conditions:

         beam spacing = 5 ft-6 in.
         span length = 30 ft
         slab thickness = $4\frac{1}{2}$ in. (normal-weight concrete)
         partition load = 20 psf
         weight of ceiling = 5 psf
         live load = 150 psf
         F_y = 50 ksi

The maximum live load deflection cannot exceed $L/360$.

a. Use LRFD.
b. Use ASD.

5.11-3   Select a W-shape for the following conditions:

         beam spacing = 12 ft
         span length = 25 ft
         slab and deck combination weight = 43 psf
         partition load = 20 psf
         ceiling weight = 5 psf
         flooring weight = 2 psf
         live load = 160 psf
         F_y = 50 ksi

The maximum live load deflection cannot exceed $L/360$.

a. Use LRFD.
b. Use ASD.

5.11-4   Select a W-shape for the following conditions:

         beam spacing = 10 ft
         span length = 20 ft
         slab and deck weight = 51 psf
         partition load = 20 psf
         miscellaneous dead load = 10 psf
         live load = 80 psf
         F_y = 50 ksi

The maximum live load deflection cannot exceed $L/360$.

a. Use LRFD.
b. Use ASD.

5.11-5   Select an A992 W-shape for beam AB of the floor system shown in Figure P5.11-5. In addition to the weight of the beam, the dead load consists of a 5-inch-thick reinforced concrete slab (normal-weight concrete). The live load is 80 psf, and there is a 20-psf partition load. The total deflection must not exceed $L/240$.

a. Use LRFD.
b. Use ASD.

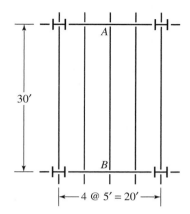

30′

←— 4 @ 5′ = 20′ —→

FIGURE P5.11-5

5.11-6   Design a typical girder for the floor system of Problem 5.11-5. Do not check deflections. Assume that the girder is supporting beams on each side, and assume that the beams weigh 35 lb/ft. Consider the beam reactions to act as point loads on the girder.

   a. Use LRFD.
   b. Use ASD.

5.11-7   Same as Problem 5.11-6, but let all the loads on the girder act as a uniform load (be sure to include the weight of the beams).

Holes in Beams

5.12-1   A W16 × 31 of A992 steel has two holes in each flange for ⅞-inch-diameter bolts.

   a. Assuming continuous lateral support, verify that the holes must be accounted for and determine the nominal flexural strength.
   b. What is the percent reduction in strength?

5.12-2   A W21 × 48 of A992 steel has two holes in each flange for ¾-inch-diameter bolts.

   a. Assuming continuous lateral support, verify that the holes must be accounted for and determine the nominal flexural strength.
   b. What is the percent reduction in strength?

5.12-3   A W18 × 35 of A992 steel has two holes in the tension flange for ¾-inch-diameter bolts.

   a. Assuming continuous lateral support, verify that the holes must be accounted for and determine the reduced nominal flexural strength.
   b. What is the percent reduction in strength?

Open-Web Steel Joists

5.13-1   A floor system consists of open-web steel joists spaced at 3 feet and spanning 20 feet. The live load is 80 psf, and there is a 4-inch-thick normal-weight reinforced concrete floor slab. Other dead load is 5 psf. Assume that the slab provides continuous lateral support. Use Figure 5.35 and select a K-series joist.

5.13-2 Use Figure 5.35 and select an open-web steel joist for the following floor system. The span length is 22 feet and the joist spacing is 4 feet. The loads consist of a 40 psf live load, a partition load of 20 psf, a slab and metal deck system weighing 32 psf, and a ceiling and light fixture weight of 5 psf. Assume that the slab provides continuous lateral support. Is there likely to be a deflection problem? Why or why not?

Beam Bearing Plates and Column Base Plates

5.14-1 A W14 × 61 must support a concentrated service live load of 85 kips applied to the top flange. Assume that the load is at a distance of at least half the beam depth from the support and design a bearing plate. Use $F_y = 50$ ksi for the beam and $F_y = 36$ ksi for the plate.

    a. Use LRFD.

    b. Use ASD.

5.14-2 Design a bearing plate of A36 steel to support a beam reaction consisting of 28 kips dead load and 56 kips live load. Assume that the bearing plate will rest on concrete with a surface area larger than the bearing area by an amount equal to 1 inch of concrete on all sides of the plate. The beam is a W30 × 99 with $F_y = 50$ ksi, and the concrete strength is $f_c' = 3$ ksi.

    a. Use LRFD.

    b. Use ASD.

5.14-3 Design a base plate for a W12 × 87 column supporting a service dead load of 65 kips and a service live load of 195 kips. The support will be a 16-inch × 16-inch concrete pier. Use A36 steel and $f_c' = 3.5$ ksi.

    a. Use LRFD.

    b. Use ASD.

5.14-4 Design a column base plate for a W10 × 33 column supporting a service dead load of 20 kips and a service live load of 50 kips. The column is supported by a 12-inch × 12-inch concrete pier. Use A36 steel and $f_c' = 3$ ksi.

    a. Use LRFD.

    b. Use ASD.

Biaxial Bending

5.15-1 A W18 × 55 is loaded as shown in Figure P5.15-1, with forces at midspan that cause bending about both the strong and weak axes. The loads shown are service loads, consisting of equal parts dead load and live load. Determine whether the AISC Specification is satisfied. The steel is A572 Grade 50, and lateral bracing is provided only at the ends.

    a. Use LRFD.

    b. Use ASD.

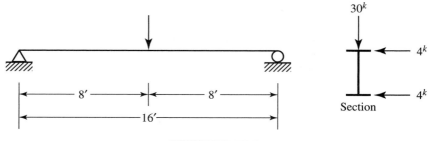

FIGURE P5.15-1

5.15-2 The 24-kip concentrated load shown in Figure P5.15-2 is a service live load. Neglect the weight of the beam and determine whether the beam satisfies the AISC Specification if A992 steel is used. Lateral support is provided at the ends only.

  a. Use LRFD.
  b. Use ASD.

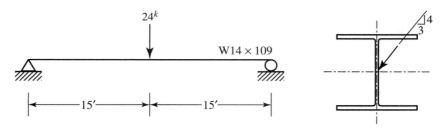

FIGURE P5.15-2

5.15-3 The beam shown in Figure P5.15-3 is a W21×68 of A992 steel and has lateral support only at the ends. Check it for compliance with the AISC Specification.

  a. Use LRFD.
  b. Use ASD.

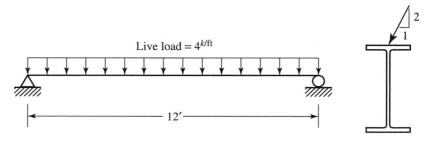

FIGURE P5.15-3

5.15-4   Check the beam shown in Figure P5.15-4 for compliance with the AISC Specification. Lateral support is provided only at the ends, and A992 steel is used. The 15-kip service loads are 30% dead load and 70% live load.

a. Use LRFD.
b. Use ASD.

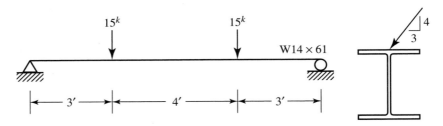

FIGURE P5.15-4

5.15-5   The beam shown in Figure P5.15-5 is simply supported and has lateral support only at its ends. Neglect the beam weight and determine whether it is satisfactory for each of the loading conditions shown. A992 steel is used, and the 1.2 kip/ft is a service live load.

a. Use LRFD.
b. Use ASD.

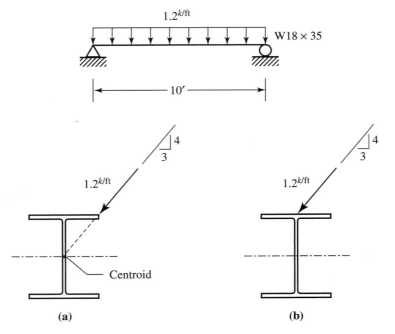

FIGURE P5.15-5

5.15-6   The truss shown in Figure P5.15-6 is part of a roof system supporting a total gravity load of 40 psf of roof surface, half dead load and half snow. Trusses are spaced at 10 feet on centers. Assume that wind load is not a factor and investigate the adequacy of a W6 × 12 of A992 steel for use as a purlin. No sag rods are used. Assume lateral support at the ends only.

   a. Use LRFD.
   b. Use ASD.

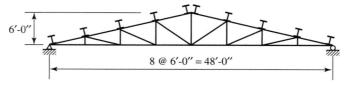

FIGURE P5.15-6

5.15-7   The truss shown in Figure P5.15-7 is one of several roof trusses spaced 18 feet apart. Purlins are located at the joints and halfway between the joints. Sag rods are located midway between the trusses. The weight of the roofing materials is 15 psf, and the snow load is 20 psf of horizontal projection of the roof surface. Use A992 steel and select a W-shape for the purlins.

   a. Use LRFD.
   b. Use ASD.

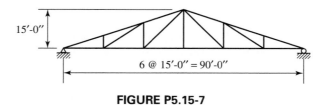

FIGURE P5.15-7

5-15-8   Same as Problem 5.15-7, except that the sag rods are at the third points.

6

Beam–Columns

6.1 DEFINITION

While many structural members can be treated as axially loaded columns or as beams with only flexural loading, most beams and columns are subjected to some degree of both bending and axial load. This is especially true of statically indeterminate structures. Even the roller support of a simple beam can experience friction that restrains the beam longitudinally, inducing axial tension when transverse loads are applied. In this particular case, however, the secondary effects are usually small and can be neglected. Many columns can be treated as pure compression members with negligible error. If the column is a one-story member and can be treated as pinned at both ends, the only bending will result from minor accidental eccentricity of the load.

For many structural members, however, there will be a significant amount of both effects, and such members are called *beam–columns*. Consider the rigid frame in Figure 6.1. For the given loading condition, the horizontal member *AB* must not only support the vertical uniform load but must also assist the vertical members in resisting the concentrated lateral load P_1. Member *CD* is a more critical case, because it must resist the load $P_1 + P_2$ without any assistance from the vertical members. The reason is that the x-bracing, indicated by dashed lines, prevents sidesway in the lower story. For the direction of P_2 shown, member *ED* will be in tension and member *CF* will be slack, provided that the bracing elements have been designed to resist only tension. For this condition to occur, however, member *CD* must transmit the load $P_1 + P_2$ from *C* to *D*.

The vertical members of this frame must also be treated as beam–columns. In the upper story, members *AC* and *BD* will bend under the influence of P_1. In addition, at *A* and *B,* bending moments are transmitted from the horizontal member through the rigid joints. This transmission of moments also takes place at *C* and *D* and is true in any rigid frame, although these moments are usually smaller than those resulting from lateral loads. Most columns in rigid frames are actually beam–columns, and the effects of bending should not be ignored. However, many isolated one-story columns can be realistically treated as axially loaded compression members.

Another example of beam–columns can sometimes be found in roof trusses. Although the top chord is normally treated as an axially loaded compression member, if purlins are placed between the joints, their reactions will cause bending, which must be accounted for. We discuss methods for handling this problem later in this chapter.

FIGURE 6.1

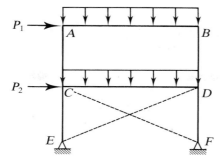

6.2 INTERACTION FORMULAS

The relationship between required and available strengths may be expressed as

$$\frac{\text{required strength}}{\text{available strength}} \leq 1.0 \tag{6.1}$$

For compression members, the strengths are axial forces. For example, for LRFD,

$$\frac{P_u}{\phi_c P_n} \leq 1.0$$

and for ASD,

$$\frac{P_a}{P_n/\Omega_c} \leq 1.0$$

These expressions can be written in the general form

$$\frac{P_r}{P_c} \leq 1.0$$

where
P_r = required axial strength
P_c = available axial strength

If more than one type of resistance is involved, Equation 6.1 can be used to form the basis of an interaction formula. As we discussed in Chapter 5 in conjunction with bi-axial bending, the sum of the load-to-resistance ratios must be limited to unity. For example, if both bending and axial compression are acting, the interaction formula would be

$$\frac{P_r}{P_c} + \frac{M_r}{M_c} \leq 1.0$$

where

M_r = required moment strength

= M_u for LRFD

= M_a for ASD

M_c = available moment strength

= $\phi_b M_n$ for LRFD

= $\dfrac{M_n}{\Omega_b}$ for ASD

For biaxial bending, there will be two moment ratios:

$$\frac{P_r}{P_c} + \left(\frac{M_{rx}}{M_{cx}} + \frac{M_{ry}}{M_{cy}} \right) \le 1.0 \qquad (6.2)$$

where the x and y subscripts refer to bending about the x and y axes.

Equation 6.2 is the basis for the AISC formulas for members subject to bending plus axial compressive load. Two formulas are given in the Specification: one for small axial load and one for large axial load. If the axial load is small, the axial load term is reduced. For large axial load, the bending term is slightly reduced. The AISC requirements are given in Chapter H, "Design of Members for Combined Forces and Torsion," and are summarized as follows:

For $\dfrac{P_r}{P_c} \ge 0.2,$

$$\frac{P_r}{P_c} + \frac{8}{9} \left(\frac{M_{rx}}{M_{cx}} + \frac{M_{ry}}{M_{cy}} \right) \le 1.0 \qquad \text{(AISC Equation H1-1a)}$$

For $\dfrac{P_r}{P_c} < 0.2,$

$$\frac{P_r}{2P_c} + \left(\frac{M_{rx}}{M_{cx}} + \frac{M_{ry}}{M_{cy}} \right) \le 1.0 \qquad \text{(AISC Equation H1-1b)}$$

These requirements may be expressed in either LRFD or ASD form.

LRFD Interaction Equations

For $\dfrac{P_u}{\phi_c P_n} \ge 0.2,$

$$\frac{P_u}{\phi_c P_n} + \frac{8}{9} \left(\frac{M_{ux}}{\phi_b M_{nx}} + \frac{M_{uy}}{\phi_b M_{ny}} \right) \le 1.0 \qquad (6.3)$$

For $\dfrac{P_u}{\phi_c P_n} < 0.2$,

$$\frac{P_u}{2\phi_c P_n} + \left(\frac{M_{ux}}{\phi_b M_{nx}} + \frac{M_{uy}}{\phi_b M_{ny}} \right) \le 1.0 \tag{6.4}$$

ASD Interaction Equations

For $\dfrac{P_a}{P_n / \Omega_c} \ge 0.2$,

$$\frac{P_a}{P_n / \Omega_c} + \frac{8}{9} \left(\frac{M_{ax}}{M_{nx} / \Omega_b} + \frac{M_{ay}}{M_{ny} / \Omega_b} \right) \le 1.0 \tag{6.5}$$

For $\dfrac{P_a}{P_n / \Omega_c} < 0.2$,

$$\frac{P_a}{2 P_n / \Omega_c} + \left(\frac{M_{ax}}{M_{nx} / \Omega_b} + \frac{M_{ay}}{M_{ny} / \Omega_b} \right) \le 1.0 \tag{6.6}$$

Example 6.1 illustrates the application of Equations 6.3–6.6.

Example 6.1   The beam–column shown in Figure 6.2 is pinned at both ends and is subjected to the loads shown. Bending is about the strong axis. Determine whether this member satisfies the appropriate AISC Specification interaction equation.

Solution   As we demonstrate in Section 6.3, the applied moments in AISC Equations H1-1a and b will sometimes be increased by *moment amplification*. The purpose of this example is to illustrate how interaction formulas work.

FIGURE 6.2

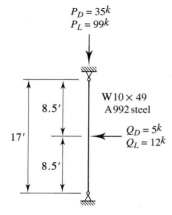

LRFD Solution From the column load tables, the axial compressive design strength of a W10 ×49 with $F_y = 50$ ksi and an effective length of $K_y L = 1.0 \times 17 = 17$ feet is

$$\phi_c P_n = 405 \text{ kips}$$

Since bending is about the strong axis, the design moment, $\phi_b M_n$, for $C_b = 1.0$ can be obtained from the beam design charts in Part 3 of the *Manual*.

For an unbraced length $L_b = 17$ ft,

$$\phi_b M_n = 197 \text{ ft-kips}$$

For the end conditions and loading of this problem, $C_b = 1.32$ (see Figure 5.15c).

For $C_b = 1.32$, the design strength is

$$\phi_b M_n = C_b \times 197 = 1.32(197) = 260 \text{ ft-kips}$$

This moment is larger than $\phi_b M_p = 226.5$ ft-kips (also obtained from the beam design charts), so the design moment must be limited to $\phi_b M_p$. Therefore,

$$\phi_b M_n = 226.5 \text{ ft-kips}$$

Factored loads:

$$P_u = 1.2P_D + 1.6P_L = 1.2(35) + 1.6(99) = 200.4 \text{ kips}$$
$$Q_u = 1.2Q_D + 1.6Q_L = 1.2(5) + 1.6(12) = 25.2 \text{ kips}$$

The maximum bending moment occurs at midheight, so

$$M_u = \frac{25.2(17)}{4} = 107.1 \text{ ft-kips}$$

Determine which interaction equation controls:

$$\frac{P_u}{\phi_c P_n} = \frac{200.4}{405} = 0.4948 > 0.2 \quad \therefore \text{ use Equation 6.3 (AISC Eq. H1-1a)}$$

$$\frac{P_u}{\phi_c P_n} + \frac{8}{9}\left(\frac{M_{ux}}{\phi_b M_{nx}} + \frac{M_{uy}}{\phi_b M_{ny}}\right) = \frac{200.4}{405} + \frac{8}{9}\left(\frac{107.1}{226.5} + 0\right) = 0.915 < 1.0 \quad \text{(OK)}$$

Answer This member satisfies the AISC Specification.

ASD Solution From the column load tables, the allowable compressive strength of a W10 × 49 with $F_y = 50$ ksi and $K_y L = 1.0 \times 17 = 17$ feet is

$$\frac{P_n}{\Omega_c} = 270 \text{ kips}$$

From the design charts in Part 3 of the Manual, for $L_b = 17$ ft and $C_b = 1.0$,

$$\frac{M_n}{\Omega_b} = 131 \text{ ft-kips}$$

From Figure 5.15c, $C_b = 1.32$. For $C_b = 1.32$,

$$\frac{M_n}{\Omega_b} = C_b \times 131 = 1.32(131) = 172.9 \text{ ft-kips}$$

This is larger than $M_p/\Omega_b = 151$ ft-kips, so the allowable moment must be limited to M_p/Ω_b. Therefore,

$$\frac{M_n}{\Omega_b} = 151 \text{ ft-kips}$$

The total axial compressive load is

$$P_a = P_D + P_L = 35 + 99 = 134 \text{ kips}$$

The total transverse load is

$$Q_a = Q_D + Q_L = 5 + 12 = 17 \text{ kips}$$

The maximum bending moment is at midheight

$$M_a = \frac{17(17)}{4} = 72.25 \text{ ft-kips}$$

Determine which interaction equation controls:

$$\frac{P_a}{P_n/\Omega_c} = \frac{134}{270} = 0.4963 > 0.2 \qquad \therefore \text{ use Equation 6.5 (AISC Equation H1-1a)}$$

$$\frac{P_a}{P_n/\Omega_c} + \frac{8}{9}\left(\frac{M_{ax}}{M_{nx}/\Omega_b} + \frac{M_{ay}}{M_{ny}/\Omega_b}\right) = \frac{134}{270} + \frac{8}{9}\left(\frac{72.25}{151} + 0\right) = 0.922 < 1.0 \quad \text{(OK)}$$

Answer   This member satisfies the AISC Specification.

6.3   MOMENT AMPLIFICATION

The foregoing approach to the analysis of members subjected to both bending and axial load is satisfactory so long as the axial load is not too large. The presence of the axial load produces secondary moments, and unless the axial load is relatively small, these additional moments must be accounted for. For an explanation, refer to Figure 6.3, which shows a beam–column with an axial load and a transverse uniform load. At an arbitrary point O, there is a bending moment caused by the uniform load and an additional moment P_y, caused by the axial load acting at an eccentricity from the longitudinal axis of the member. This secondary moment is largest where the deflection is largest—in this case, at the centerline, where the total moment is $wL^2/8 + P\delta$. Of course, the additional moment causes an additional deflection over and above that resulting from the transverse load. Because the total deflection cannot be found directly, this problem is nonlinear, and without knowing the deflection, we cannot compute the moment.

FIGURE 6.3

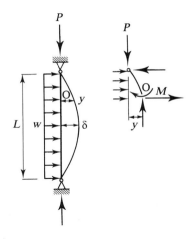

Ordinary structural analysis methods that do not take the displaced geometry into account are referred to as *first-order* methods. Iterative numerical techniques, called *second-order* methods, can be used to find the deflections and secondary moments, but these methods are impractical for manual calculations and are usually implemented with a computer program. Most current design codes and specifications, including the AISC Specification, permit the use of either a second-order analysis or the *moment amplification method.* This method entails computing the maximum bending moment resulting from flexural loading (transverse loads or member end moments) by a first-order analysis, then multiplying by a *moment amplification factor* to account for the secondary moment. An expression for this factor will now be developed.

Figure 6.4 shows a simply supported member with an axial load and an initial out-of-straightness. This initial crookedness can be approximated by

$$y_0 = e \sin \frac{\pi x}{L}$$

where e is the maximum initial displacement, occurring at midspan. For the coordinate system shown, the moment–curvature relationship can be written as

$$\frac{d^2 y}{dx^2} = -\frac{M}{EI}$$

FIGURE 6.4

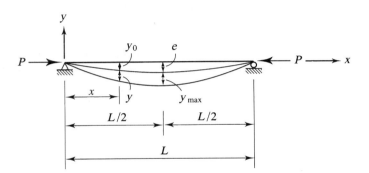

The bending moment, M, is caused by the eccentricity of the axial load, P, with respect to the axis of the member. This eccentricity consists of the initial crookedness y_0, plus additional deflection, y, resulting from bending. At any location, the moment is

$$M = P(y_0 + y)$$

Substituting this equation into the differential equation, we obtain

$$\frac{d^2y}{dx^2} = -\frac{P}{EI}\left(e \sin\frac{\pi x}{L} + y\right)$$

Rearranging gives

$$\frac{d^2y}{dx^2} + \frac{P}{EI}y = -\frac{Pe}{EI}\sin\frac{\pi x}{L}$$

which is an ordinary, nonhomogeneous differential equation. Because it is a second-order equation, there are two boundary conditions. For the support conditions shown, the boundary conditions are

$$\text{At } x = 0, \, y = 0 \text{ and at } x = L, \, y = 0$$

That is, the displacement is zero at each end. A function that satisfies both the differential equation and the boundary conditions is

$$y = B \sin\frac{\pi x}{L}$$

where B is a constant. Substituting into the differential equation, we get

$$-\frac{\pi^2}{L^2}B \sin\frac{\pi x}{L} + \frac{P}{EI}B \sin\frac{\pi x}{L} = -\frac{Pe}{EI}\sin\frac{\pi x}{L}$$

Solving for the constant gives

$$B = \frac{-\dfrac{Pe}{EI}}{\dfrac{P}{EI} - \dfrac{\pi^2}{L^2}} = \frac{-e}{1 - \dfrac{\pi^2 EI}{PL^2}} = \frac{e}{\dfrac{P_e}{P} - 1}$$

where

$$P_e = \frac{\pi^2 EI}{L^2} = \text{ the Euler buckling load}$$

$$\therefore y = B \sin\frac{\pi x}{L} = \left[\frac{e}{(P_e/P) - 1}\right]\sin\frac{\pi x}{L}$$

$$M = P(y_0 + y)$$

$$= P\left\{e \sin\frac{\pi x}{L} + \left[\frac{e}{(P_e/P) - 1}\right]\sin\frac{\pi x}{L}\right\}$$

The maximum moment occurs at $x = L/2$:

$$
\begin{aligned}
M_{max} &= P\left[e + \frac{e}{(P_e/P) - 1}\right] \\
&= Pe\left[\frac{(P_e/P) - 1 + 1}{(P_e/P) - 1}\right] \\
&= M_0\left[\frac{1}{1 - (P/P_e)}\right]
\end{aligned}
$$

where M_0 is the unamplified maximum moment. In this case, it results from initial crookedness, but in general, it can be the result of transverse loads or end moments. The moment amplification factor is therefore

$$\frac{1}{1 - (P/P_e)}$$

Because the member deflection corresponds to a buckled shape, the axial load corresponds to a failure load—that is, a load corresponding to an LRFD formulation. Therefore, the amplification factor should be written as

$$\frac{1}{1 - (P_u/P_e)} \tag{6.7}$$

where P_u is the factored axial load. The form shown in Expression 6.7 is appropriate for LRFD. For ASD, a different form, to be explained later, will be used.

As we describe later, the exact form of the AISC moment amplification factor can be slightly different from that shown in Expression 6.7.

Example 6.2   Use Expression 6.7 to compute the LRFD amplification factor for the beam–column of Example 6.1.

Solution   Since the Euler load P_e is part of an amplification factor for a *moment,* it must be computed for the axis of bending, which in this case is the *x*-axis. In terms of effective length, the Euler load can be written as

$$P_e = \frac{\pi^2 EI}{(KL)^2} = \frac{\pi^2 EI_x}{(K_x L)^2} = \frac{\pi^2 (29,000)(272)}{(1.0 \times 17 \times 12)^2} = 1871 \text{ kips}$$

From the LRFD solution to Example 6.1, $P_u = 200.4$ kips, and

$$\frac{1}{1 - (P_u/P_e)} = \frac{1}{1 - (200.4/1871)} = 1.12$$

which represents a 12% increase in bending moment. The amplified primary LRFD moment is

$$1.12 \times M_u = 1.12(107.1) = 120 \text{ ft-kips}$$

Answer   Amplification factor = 1.12.

6.4   BRACED VERSUS UNBRACED FRAMES

The AISC Specification covers moment amplification in Chapter C, "Stability Analysis and Design." Two amplification factors are used: one to account for amplification resulting from the member deflection and one to account for the effect of sway when the member is part of an unbraced frame. This approach is the same as the one used in the ACI Building Code for reinforced concrete (ACI, 2005). Figure 6.5 illustrates these two components. In Figure 6.5a, the member is restrained against sidesway, and the maximum secondary moment is $P\delta$, which is added to the maximum moment within the member. If the frame is actually unbraced, there is an additional component of the secondary moment, shown in Figure 6.5b, that is caused by sidesway. This secondary moment has a maximum value of $P\Delta$, which represents an amplification of the *end* moment.

To approximate these two effects, two amplification factors, B_1 and B_2, are used for the two types of moments. The amplified moment to be used in design is computed from the loads and moments as follows (x and y subscripts are not used here; amplified moments must be computed in the following manner for each axis about which there are moments):

$$M_r = B_1 M_{nt} + B_2 M_{\ell t}$$

(AISC Equation C2-1a)

where

   M_r = required moment strength

      = M_u for LRFD

      = M_a for ASD

FIGURE 6.5

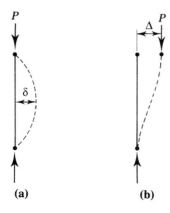

(a)   (b)

M_{nt} = maximum moment assuming that no sidesway occurs, whether the frame is actually braced or not (the subscript nt is for "no translation"). M_{nt} will be a factored load moment for LRFD and a service load moment for ASD.

$M_{\ell t}$ = maximum moment caused by sidesway (the subscript ℓt is for "lateral translation"). This moment can be caused by lateral loads or by unbalanced gravity loads. Gravity load can produce sidesway if the frame is unsymmetrical or if the gravity loads are unsymmetrically placed. $M_{\ell t}$ will be zero if the frame is actually braced. For LRFD, $M_{\ell t}$ will be a factored load moment, and for ASD, it will be a service load moment.

B_1 = amplification factor for the moments occurring in the member when it is braced against sidesway.

B_2 = amplification factor for the moments resulting from sidesway.

We cover the evaluation of B_1 and B_2 in the following sections.

6.5 MEMBERS IN BRACED FRAMES

The amplification factor given by Expression 6.7 was derived for a member braced against sidesway—that is, one whose ends cannot translate with respect to each other. Figure 6.6 shows a member of this type subjected to equal end moments producing *single-curvature bending* (bending that produces tension or compression on one side throughout the length of the member). Maximum moment amplification occurs at the center, where the deflection is largest. For equal end moments, the moment is constant throughout the length of the member, so the maximum primary moment also occurs at the center. Thus the maximum secondary moment and maximum primary moment are additive. Even if the end moments are not equal, as long as one is clockwise and the other is counterclockwise there will be single-curvature bending, and the maximum primary and secondary moments will occur near each other.

FIGURE 6.6

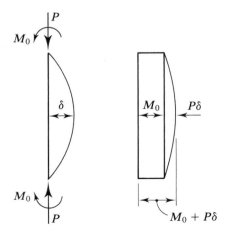

FIGURE 6.7

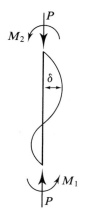

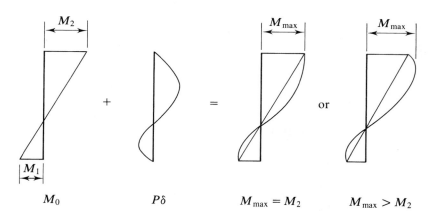

M_0 $P\delta$ $M_{max} = M_2$ $M_{max} > M_2$

That is not the case if applied end moments produce reverse-curvature bending as shown in Figure 6.7. Here the maximum primary moment is at one of the ends, and maximum moment amplification occurs between the ends. Depending on the value of the axial load P, the amplified moment can be either larger or smaller than the end moment.

The maximum moment in a beam–column therefore depends on the distribution of bending moment within the member. This distribution is accounted for by a factor, C_m, applied to the amplification factor given by Expression 6.7. The amplification factor given by Expression 6.7 was derived for the worst case, so C_m will never be greater than 1.0. The final form of the amplification factor is

$$B_1 = \frac{C_m}{1 - (\alpha P_r / P_{e1})} \geq 1 \qquad \text{(AISC Equation C2-2)}$$

where
$\quad P_r$ = required axial compressive strength
$\qquad = P_u$ for LRFD
$\qquad = P_a$ for ASD

$$\alpha = 1.00 \text{ for LRFD}$$
$$= 1.60 \text{ for ASD}$$

$$P_{e1} = \frac{\pi^2 EI}{(K_1 L)^2} \qquad \text{(AISC Equation C2-5)}$$

The required compressive strength P_r has a contribution from the $P\text{-}\Delta$ effect and is given by

$$P_r = P_{nt} + B_2 P_{\ell t} \qquad \text{(AISC Equation C2-1b)}$$

where

P_{nt} = axial load corresponding to the braced condition
$P_{\ell t}$ = axial load corresponding to the sidesway condition

As an approximation, P_r can be taken as

$$P_r = P_{nt} + P_{\ell t}$$

The moment of inertia I and the effective length factor K_1 are for the axis of bending, and $K_1 = 1.0$ unless a more accurate value is computed (AISC C2.1b). Note that throughout Chapter C of the Specification, the subscript 1 corresponds to the braced condition and the subscript 2 corresponds to the unbraced condition.

Evaluation of C_m

The factor C_m applies only to the braced condition. There are two categories of members: those with transverse loads applied between the ends and those with no transverse loads. Figure 6.8b and c illustrate these two cases (member AB is the beam–column under consideration).

FIGURE 6.8

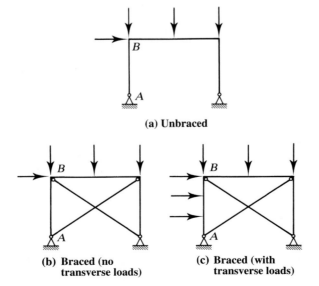

(a) Unbraced

(b) Braced (no transverse loads)

(c) Braced (with transverse loads)

FIGURE 6.9

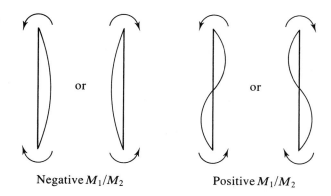

Negative M_1/M_2 Positive M_1/M_2

1. If there are no transverse loads acting on the member,

$$C_m = 0.6 - 0.4\left(\frac{M_1}{M_2}\right)$$ (AISC Equation C2-4)

M_1/M_2 is a ratio of the bending moments at the ends of the member. M_1 is the end moment that is smaller in absolute value, M_2 is the larger, and the ratio is positive for members bent in reverse curvature and negative for single-curvature bending (Figure 6.9). Reverse curvature (a positive ratio) occurs when M_1 and M_2 are both clockwise or both counterclockwise.

2. For transversely loaded members, C_m can be taken as 1.0. A more refined procedure for transversely loaded members is provided in Section C2 of the Commentary to the Specification. The reduction factor is

$$C_m = 1 + \Psi\left(\frac{\alpha P_r}{P_{e1}}\right)$$ (AISC Equation C-C2-2)

For simply supported members,

$$\Psi = \frac{\pi^2 \delta_0 EI}{M_0 L^2} - 1$$

where δ_0 is the maximum deflection resulting from transverse loading and M_0 is the maximum moment between supports resulting from the transverse loads. The factor Ψ has been evaluated for several common situations and is given in Commentary Table C-C2.1.

Example 6.3   The member shown in Figure 6.10 is part of a braced frame. Service loads are shown, and bending is about the strong axis. If A572 Grade 50 steel is used, is this member adequate? $K_x L = K_y L = 14$ feet.

FIGURE 6.10

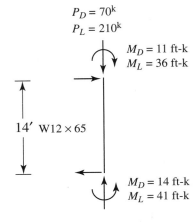

$$P_D = 70^k$$
$$P_L = 210^k$$

$M_D = 11$ ft-k
$M_L = 36$ ft-k

14' W12 × 65

$M_D = 14$ ft-k
$M_L = 41$ ft-k

LRFD Solution   The factored loads, computed from ASCE 7 load combination 2, are shown in Figure 6.11. Determine which interaction formula to apply. From the column load tables, for $KL = 1.0 \times 14 = 14$ feet, the axial compressive strength of a W12 × 65 is

$$\phi_c P_n = 685 \text{ kips}$$

$$\frac{P_u}{\phi_c P_n} = \frac{420}{685} = 0.6131 > 0.2 \qquad \therefore \text{ use Equation 6.3 (AISC Equation H1-1a)}$$

In the plane of bending,

$$P_{e1} = \frac{\pi^2 EI}{(K_1 L)^2} = \frac{\pi^2 EI_x}{(K_x L)^2} = \frac{\pi^2 (29,000)(533)}{(1.0 \times 14 \times 12)^2} = 5405 \text{ kips}$$

FIGURE 6.11

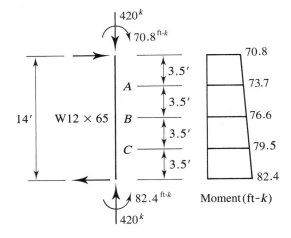

420^k

$70.8^{\text{ft-k}}$

70.8

3.5'

A — 73.7

3.5'

B — 76.6

3.5'

C — 79.5

3.5'

82.4

14'   W12 × 65

$82.4^{\text{ft-k}}$ Moment (ft-k)

420^k

$$C_m = 0.6 - 0.4\left(\frac{M_1}{M_2}\right) = 0.6 - 0.4\left(-\frac{70.8}{82.4}\right) = 0.9437$$

$$B_1 = \frac{C_m}{1-(\alpha P_r/P_{e1})} = \frac{C_m}{1-(1.00P_u/P_{e1})} = \frac{0.9437}{1-(420/5405)} = 1.023$$

From the Beam Design Charts with $C_b = 1.0$ and $L_b = 14$ feet, the moment strength is

$$\phi_b M_n = 345 \text{ ft-kips}$$

For the actual value of C_b, refer to the moment diagram of Figure 6.11:

$$C_b = \frac{12.5 M_{max}}{2.5 M_{max} + 3M_A + 4M_B + 3M_C}$$

$$= \frac{12.5(82.4)}{2.5(82.4) + 3(73.7) + 4(76.6) + 3(79.5)} = 1.060$$

$$\therefore \phi_b M_n = C_b(345) = 1.060(345) = 366 \text{ ft-kips}$$

But $\phi_b M_p = 356$ ft-kips (from the charts) < 366 ft-kips $\therefore$ use $\phi_b M_n = 356$ ft-kips
(Since a W12 $\times$ 65 is noncompact for $F_y = 50$ ksi, 356 ft-kips is the design strength based on FLB rather than full yielding of the cross section.) The factored load moments are

$$M_{nt} = 82.4 \text{ ft-kips} \quad M_{\ell t} = 0$$

From AISC Equation C2-1a, the required moment strength is
$$M_r = M_u = B_1 M_{nt} + B_2 M_{\ell t} = 1.023(82.4) + 0 = 84.30 \text{ ft-kips} = M_{ux}$$

From Equation 6.3 (AISC Equation H1-1a),

$$\frac{P_u}{\phi_c P_n} + \frac{8}{9}\left(\frac{M_{ux}}{\phi_b M_{nx}} + \frac{M_{uy}}{\phi_b M_{ny}}\right) = 0.6131 + \frac{8}{9}\left(\frac{84.30}{356} + 0\right) = 0.824 < 1.0 \qquad \text{(OK)}$$

Answer   The member is satisfactory.

ASD Solution   The service loads, computed from ASCE 7 load combination 2, are shown in Figure 6.12. Determine which interaction formula to apply. From the column load tables, for $KL = 1.0 \times 14 = 14$ feet, the axial compressive strength of a W12 $\times$ 65 is

$$\frac{P_n}{\Omega_c} = 456 \text{ kips}$$

$$\frac{P_a}{P_n/\Omega_c} = \frac{280}{456} = 0.6140 > 0.2 \quad \therefore \text{ use Equation 6.5 (AISC Equation H1-1a)}$$

FIGURE 6.12

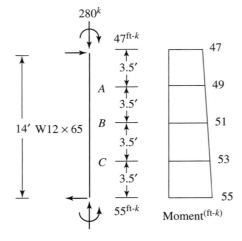

In the plane of bending,

$$P_{e1} = \frac{\pi^2 EI}{(K_1 L)^2} = \frac{\pi^2 EI_x}{(K_x L)^2} = \frac{\pi^2 (29,000)(533)}{(1.0 \times 14 \times 12)^2} = 5405 \text{ kips}$$

$$C_m = 0.6 - 0.4 \left(\frac{M_1}{M_2} \right) = 0.6 - 0.4 \left(-\frac{47}{55} \right) = 0.9418$$

$$B_1 = \frac{C_m}{1 - (\alpha P_r / P_{e1})} = \frac{C_m}{1 - (1.60 P_a / P_{e1})} = \frac{0.9418}{1 - (1.60 \times 280 / 5405)} = 1.027$$

From the Beam Design Charts with $C_b = 1.0$ and $L_b = 14$ feet, the moment strength is

$$\frac{M_n}{\Omega_b} = 230 \text{ ft-kips}$$

For the actual value of C_b, refer to the moment diagram of Figure 6.12:

$$C_b = \frac{12.5 M_{max}}{2.5 M_{max} + 3 M_A + 4 M_B + 3 M_C}$$

$$= \frac{12.5(55)}{2.5(55) + 3(49) + 4(51) + 3(53)} = 1.062$$

$$\therefore \frac{M_n}{\Omega_b} = C_b(230) = 1.062(230) = 244.3 \text{ ft-kips}$$

But $\dfrac{M_p}{\Omega_b} = 237$ ft-kips (from the charts) < 244.3 ft-kips $\therefore$ use $\dfrac{M_n}{\Omega_b} = 237$ ft-kips.

(Since a WI2 × 65 is noncompact for $F_y = 50$ ksi, 237 ft-kips is the strength based on FLB rather than full yielding of the cross section.) The unamplified moments are

$$M_{nt} = 55 \text{ ft-kips} \qquad M_{\ell t} = 0$$

From AISC Equation C2-1a, the required moment strength is

$$M_r = M_a = B_1 M_{nt} + B_2 M_{\ell t} = 1.027(55) + 0 = 56.49 \text{ ft-kips} = M_{ax}$$

From Equation 6.5 (AISC Equation H1-1a),

$$\frac{P_a}{P_n/\Omega_c} + \frac{8}{9}\left(\frac{M_{ax}}{M_{nx}/\Omega_b} + \frac{M_{ay}}{M_{ny}/\Omega_b}\right) = \frac{280}{456} + \frac{8}{9}\left(\frac{55}{237} + 0\right) = 0.820 < 1.0 \qquad \text{(OK)}$$

Answer　　The member is satisfactory.

Note that the value of C_b is nearly the same regardless of whether factored or unfactored moments are used.

Example 6.4　　The horizontal beam–column shown in Figure 6.13 is subjected to the service live loads shown. This member is laterally braced at its ends, and bending is about the x-axis. Check for compliance with the AISC Specification.

FIGURE 6.13

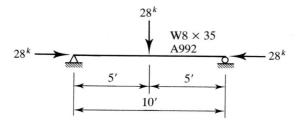

LRFD Solution　　The factored axial load is

$$P_u = 1.6(28) = 44.8 \text{ kips}$$

The factored transverse loads and bending moment are

$$Q_u = 1.6(28) = 44.8 \text{ kips}$$
$$w_u = 1.2(0.035) = 0.042 \text{ kips/ft}$$
$$M_u = \frac{44.8(10)}{4} + \frac{0.042(10)^2}{8} = 112.5 \text{ ft-kips}$$

This member is braced against sidesway, so $M_{\ell t} = 0$.

Compute the moment amplification factor. For a member braced against sidesway and transversely loaded, C_m can be taken as 1.0. A more accurate value can be found from AISC Equation C-C2-2 in the Commentary:

$$C_m = 1 + \Psi\left(\frac{\alpha P_r}{P_{e1}}\right)$$

From Commentary Table C-C2-1, $\Psi = -0.2$ for the support and loading conditions of this beam–column. For the axis of bending,

$$P_{e1} = \frac{\pi^2 EI}{(K_1 L)^2} = \frac{\pi^2 EI_x}{(K_x L)^2} = \frac{\pi^2 (29,000)(127)}{(10 \times 12)^2} = 2524 \text{ kips}$$

$$C_m = 1 + \Psi\left(\frac{\alpha P_r}{P_{e1}}\right) = 1 - 0.2\left(\frac{1.00 P_u}{P_{e1}}\right) = 1 - 0.2\left(\frac{44.8}{2524}\right) = 0.9965$$

The amplification factor is

$$B_1 = \frac{C_m}{1 - (\alpha P_r / P_{e1})} = \frac{C_m}{1 - (1.00 P_u / P_{e1})} = \frac{0.9965}{1 - (44.8/2524)} = 1.015$$

The amplified bending moment is

$$M_u = B_1 M_{nt} + B_2 M_{\ell t} = 1.015(112.5) + 0 = 114.2 \text{ ft-kips}$$

From the beam design charts, for $L_b = 10$ ft and $C_b = 1$,

$$\phi_b M_n = 123 \text{ ft-kips}$$

Because the beam weight is very small in relation to the concentrated live load, C_b may be taken from Figure 5.15c as 1.32. This value results in a design moment of

$$\phi_b M_n = 1.32(123) = 162.4 \text{ ft-kips}$$

This moment is greater than $\phi_b M_p = 130$ ft-kips, so the design strength must be limited to this value. Therefore,

$$\phi_b M_n = 130 \text{ ft-kips}$$

Check the interaction formula. From the column load tables, for $KL = 10$ ft,

$$\phi_c P_n = 358 \text{ kips}$$

$$\frac{P_u}{\phi_c P_n} + \frac{44.8}{358} = 0.1251 < 0.2 \qquad \therefore \text{ use Equation 6.4 (AISC Equation H1-1b)}$$

$$\frac{P_u}{2\phi_c P_n} + \left(\frac{M_{ux}}{\phi_b M_{nx}} + \frac{M_{uy}}{\phi_b M_{ny}}\right) = \frac{0.1251}{2} + \left(\frac{114.2}{130} + 0\right)$$

$$= 0.941 < 1.0 \text{ (OK)}$$

Answer   A W8 × 35 is adequate.

ASD Solution   The applied axial load is

$$P_a = 28 \text{ kips}$$

The applied transverse loads are

$$Q_a = 28 \text{ kips and } w_a = 0.035 \text{ kips/ft}$$

and the maximum bending moment is

$$M_{nt} = \frac{28(10)}{4} + \frac{0.035(10)^2}{8} = 70.44 \text{ ft-kips}$$

The member is braced against end translation, so $M_{\ell t} = 0$.

Compute the moment amplification factor. For a member braced against side-sway and transversely loaded, C_m can be taken as 1.0. A more accurate value can be found from AISC Equation C-C2-2 in the Commentary:

$$C_m = 1 + \Psi\left(\frac{\alpha P_r}{P_{e1}}\right)$$

From Commentary Table C-C2-1, $\Psi = -0.2$ for the support and loading conditions of this beam–column. For the axis of bending,

$$P_{e1} = \frac{\pi^2 EI}{(K_1 L)^2} = \frac{\pi^2 EI_x}{(K_x L)^2} = \frac{\pi^2 (29,000)(127)}{(10 \times 12)^2} = 2524 \text{ kips}$$

$$C_m = 1 + \Psi\left(\frac{\alpha P_r}{P_{e1}}\right) = 1 - 0.2\left(\frac{1.60 P_u}{P_{e1}}\right) = 1 - 0.2\left(\frac{1.60 \times 28}{2524}\right) = 0.9965$$

$$B_1 = \frac{C_m}{1 - (\alpha P_r/P_{e1})} = \frac{C_m}{1 - (1.60 P_a/P_{e1})} = \frac{0.9965}{1 - (1.60 \times 28/2524)} = 1.015$$

$$M_a = B_1 M_{nt} = 1.015(70.44) = 71.50 \text{ ft-kips}$$

From the Beam Design Charts with $C_b = 1.0$ and $L_b = 10$ feet, the moment strength is

$$\frac{M_n}{\Omega_b} = 82.0 \text{ ft-kips}$$

Because the beam weight is very small in relation to the concentrated live load, C_b may be taken from Figure 5.15c as 1.32. This results in an allowable moment of

$$\frac{M_n}{\Omega_b} = 1.32(82.0) = 108.2 \text{ ft-kips}$$

This result is larger than $\dfrac{M_p}{\Omega_b} = 86.6$; therefore, use $\dfrac{M_n}{\Omega_b} = \dfrac{M_p}{\Omega_b} = 86.6$ ft-kips.

Compute the axial compressive strength. From the column load tables, for $KL = 10$ ft,

$$\frac{P_n}{\Omega_c} = 238 \text{ kips}$$

Determine which interaction formula to use:

$$\frac{P_a}{P_n/\Omega_c} = \frac{28}{238} = 0.1176 < 0.2 \qquad \therefore \text{ use Equation 6.6 (AISC Equation H1-1b)}$$

$$\frac{P_a}{2P_n/\Omega_c} + \left(\frac{M_{ax}}{M_{nx}/\Omega_b} + \frac{M_{ay}}{M_{ny}/\Omega_b} \right) = \frac{0.1176}{2} + \left(\frac{71.50}{86.6} + 0 \right)$$

$$= 0.884 < 1.0 \text{ (OK)}$$

Answer   The W8 × 35 is adequate

Example 6.5   The member shown in Figure 6.14 is a W12 × 65 of A242 steel and must support the service loads and moments shown. One end is pinned, and the other is subjected to moments about both the strong and weak axes. Use $K_x = K_y = 1.0$ and investigate for compliance with the AISC Specification.

FIGURE 6.14

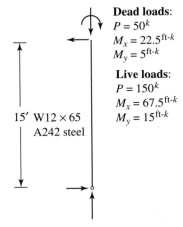

Dead loads:
$P = 50^k$
$M_x = 22.5^{\text{ft-}k}$
$M_y = 5^{\text{ft-}k}$

Live loads:
$P = 150^k$
$M_x = 67.5^{\text{ft-}k}$
$M_y = 15^{\text{ft-}k}$

15′ W12 × 65
A242 steel

Solution   First, determine the yield stress F_y. From Table 2-3, Part 2 of the *Manual,* we see that A242 steel is available in three different versions. From the dimensions and properties table in Part 1 of the *Manual,* a W12 × 65 has a flange thickness of $t_f = 0.605$ in.

This matches the thickness range corresponding to footnote 1 in Table 2-3; therefore, $F_y = 50$ ksi.

LRFD Solution Compute the factored axial load and moments.

$$P_u = 1.2(50) + 1.6(150) = 300 \text{ kips}$$
$$M_{ntx} = 1.2(22.5) + 1.6(67.5) = 135.0 \text{ ft-kips}$$
$$M_{nty} = 1.2(5) + 1.6(15) = 30.0 \text{ ft-kips}$$

Compute the strong axis bending moments.

$$C_{mx} = 0.6 - 0.4\left(\frac{M_1}{M_2}\right) = 0.6 - 0.4(0) = 0.6$$

$$P_{e1x} = \frac{\pi^2 EI}{(K_1 L)^2} = \frac{\pi^2 EI_x}{(K_x L)^2} = \frac{\pi^2 (29,000)(533)}{(1.0 \times 15 \times 12)^2} = 4708 \text{ kips}$$

$$B_{1x} = \frac{C_{mx}}{1 - (\alpha P_r/P_{e1x})} = \frac{C_{mx}}{1 - (1.00 P_u/P_{e1x})} = \frac{0.6}{1 - (300/4708)}$$
$$= 0.641 < 1.0 \quad \therefore \text{ use } B_{1x} = 1.0$$

The required moment strength is

$$M_r = M_{ux} = B_{1x} M_{ntx} + B_{2x} M_{\ell tx} = 1.0(135) + 0 = 135.0 \text{ ft-kips}$$

From the Beam Design Charts with $C_b = 1.0$ and $L_b = 15$ feet, the moment strength is

$$\phi_b M_{nx} = 340 \text{ ft-kips and } \phi_b M_{px} = 356 \text{ ft-kips}$$

From Figure 5.15g, $C_b = 1.67$ and

$$C_b \times (\phi_b M_{nx} \text{ for } C_b = 1.0) = 1.67(340) = 567.8 \text{ ft-kips}$$

This result is larger than $\phi_b M_{px}$; therefore use $\phi_b M_{nx} = \phi_b M_{px} = 356$ ft-kips. Compute the weak axis bending moments.

$$C_{my} = 0.6 - 0.4\left(\frac{M_1}{M_2}\right) = 0.6 - 0.4(0) = 0.6$$

$$P_{e1y} = \frac{\pi^2 EI}{(K_1 L)^2} = \frac{\pi^2 EI_y}{(K_y L)^2} = \frac{\pi^2 (29,000)(174)}{(1.0 \times 15 \times 12)^2} = 1537 \text{ kips}$$

$$B_{1y} = \frac{C_{my}}{1 - (\alpha P_r/P_{e1y})} = \frac{C_{my}}{1 - (1.00 P_u/P_{e1y})} = \frac{0.6}{1 - (300/1537)}$$
$$= 0.746 < 1.0 \quad \therefore \text{ use } B_{1y} = 1.0$$

The required moment strength is

$$M_r = M_{uy} = B_{1y} M_{nty} + B_{2y} M_{\ell ty} = 1.0(30) + 0 = 30 \text{ ft-kips}$$

Because the flange of this shape is noncompact (see footnote in the dimensions and properties table), the weak axis bending strength is limited by FLB (see Section 5.15 of this book and Chapter F of the AISC Specification).

$$\lambda = \frac{b_f}{2t_f} = 9.92$$

$$\lambda_p = 0.38\sqrt{\frac{E}{F_y}} = 0.38\sqrt{\frac{29,000}{50}} = 9.152$$

$$\lambda_r = 1.0\sqrt{\frac{E}{F_y}} = 1.0\sqrt{\frac{29,000}{50}} = 24.08$$

Since $\lambda_p < \lambda < \lambda_r$,

$$M_n = M_p - (M_p - 0.7F_yS_y)\left(\frac{\lambda - \lambda_p}{\lambda_r - \lambda_p}\right) \qquad \text{(AISC Equation F6-2)}$$

$$M_p = M_{py} = F_yZ_y = \frac{50(44.1)}{12} = 183.8 \text{ ft-kips}$$

$$M_n = M_{ny} = 183.8 - \left(183.8 - 0.7 \times 50 \times 29.1/12\right)\left(\frac{9.92 - 9.152}{24.08 - 9.152}\right) = 178.7 \text{ ft-kips}$$

$$\phi_b M_{ny} = 0.90(178.7) = 160.8 \text{ ft-kips}$$

The value of $\phi_b M_{ny}$ is also given in the Z_y table, listed as $\phi_b M_{py}$. For a W12 × 65, it is given as 161 ft-kips.

Determine the compressive strength. For $KL = 1.0(15) = 15$ feet, the axial compressive design strength from the column load tables is

$$\phi_c P_n = 662 \text{ kips}$$

Determine which interaction formula to use:

$$\frac{P_u}{\phi_c P_n} = \frac{300}{662} = 0.4532 > 0.2 \quad \therefore \text{ use Equation 6.3 (AISC Equation H1-1a)}$$

$$\frac{P_u}{\phi_c P_n} + \frac{8}{9}\left(\frac{M_{ux}}{\phi_b M_{nx}} + \frac{M_{uy}}{\phi_b M_{ny}}\right) = 0.4532 + \frac{8}{9}\left(\frac{135}{356} + \frac{30}{160.8}\right)$$

$$= 0.956 < 1.0 \qquad \text{(OK)}$$

Answer   The W12 × 65 is satisfactory.

ASD Solution   Compute the required axial compressive and moment strengths.

$$P_a = 50 + 150 = 200 \text{ kips}$$

$$M_{ntx} = 22.5 + 67.5 = 90.0 \text{ ft-kips}$$

$$M_{nty} = 5 + 15 = 20 \text{ ft-kips}$$

Compute the strong axis bending moments:

$$C_{mx} = 0.6 - 0.4\left(\frac{M_1}{M_2}\right) = 0.6 - 0.4(0) = 0.6$$

$$P_{e1x} = \frac{\pi^2 EI}{(K_1 L)^2} = \frac{\pi^2 EI_x}{(K_x L)^2} = \frac{\pi^2 (29,000)(533)}{(1.0 \times 15 \times 12)^2} = 4708 \text{ kips}$$

$$B_{1x} = \frac{C_{mx}}{1 - (\alpha P_r / P_{e1x})} = \frac{C_{mx}}{1 - (1.60 P_a / P_{e1x})} = \frac{0.6}{1 - (1.60 \times 200/4708)}$$

$$= 0.644 < 1.0 \therefore \text{ use } B_{1x} = 1.0$$

$$M_r = M_{ax} = B_{1x} M_{ntx} + B_{2x} M_{\ell tx} = 1.0(90) + 0 = 90 \text{ ft-kips}$$

From the Beam Design Charts with $C_b = 1.0$ and $L_b = 15$ feet, the moment strength is

$$\frac{M_{nx}}{\Omega_b} = 226 \text{ ft-kips and } \frac{M_{px}}{\Omega_b} = 237 \text{ ft-kips}$$

From Figure 5.15g, $C_b = 1.67$ and

$$C_b \times \left(\frac{M_{nx}}{\Omega_b} \text{ for } C_b = 1.0\right) = 1.67(226) = 377.4 \text{ ft-kips}$$

This result is larger than $\dfrac{M_{px}}{\Omega_b}$; therefore, use $\dfrac{M_{nx}}{\Omega_b} = \dfrac{M_{px}}{\Omega_b} = 237$ ft-kips.

Compute the weak axis bending moments.

$$C_{my} = 0.6 - 0.4\left(\frac{M_1}{M_2}\right) = 0.6 - 0.4(0) = 0.6$$

$$P_{e1y} = \frac{\pi^2 EI}{(K_1 L)^2} = \frac{\pi^2 EI_y}{(K_y L)^2} = \frac{\pi^2 (29,000)(174)}{(1.0 \times 15 \times 12)^2} = 1537 \text{ kips}$$

$$B_{1y} = \frac{C_{my}}{1 - (\alpha P_r / P_{e1y})} = \frac{C_{my}}{1 - (1.60 P_a / P_{e1y})} = \frac{0.6}{1 - (1.60 \times 200/1537)}$$

$$= 0.758 < 1.0 \therefore \text{ use } B_{1y} = 1.0$$

$$M_r = M_{ay} = B_{1y} M_{nty} + B_{2y} M_{\ell ty} = 1.0(20) + 0 = 20 \text{ ft-kips}$$

Because the flange of this shape is noncompact, the weak axis bending strength is limited by FLB (see Section 5.15 of this book and Chapter F of the AISC Specification).

$$\lambda = \frac{b_f}{2t_f} = 9.92$$

$$\lambda_p = 0.38\sqrt{\frac{E}{F_y}} = 0.38\sqrt{\frac{29{,}000}{50}} = 9.152$$

$$\lambda_r = 1.0\sqrt{\frac{E}{F_y}} = 1.0\sqrt{\frac{29{,}000}{50}} = 24.08$$

Since $\lambda_p < \lambda < \lambda_r$,

$$M_n = M_p - (M_p - 0.7F_yS_y)\left(\frac{\lambda - \lambda_p}{\lambda_r - \lambda_p}\right) \qquad \text{(AISC Equation F6-2)}$$

$$M_p = M_{py} = F_yZ_y = \frac{50(44.1)}{12} = 183.8 \text{ ft-kips}$$

$$M_n = M_{ny} = 183.8 - \left(183.8 - 0.7 \times 50 \times 29.1/12\right)\left(\frac{9.92 - 9.152}{24.08 - 9.152}\right) = 178.7 \text{ ft-kips}$$

$$\frac{M_{ny}}{\Omega_b} = \frac{178.7}{1.67} = 107.0 \text{ ft-kips}$$

The value of M_{ny}/Ω_b is also given in the Z_y table, listed as M_{py}/Ω_b. For a W12 × 65, it is given as 107 ft-kips.

Find the compressive strength. For $KL = 1.0(15) = 15$ feet, the axial compressive strength from the column load tables is

$$\frac{P_n}{\Omega_c} = 441 \text{ kips}$$

Check the interaction formula:

$$\frac{P_a}{P_n/\Omega_c} = \frac{200}{441} = 0.4535 > 0.2 \quad \therefore \text{ use Equation 6.5 (AISC Equation H1-1a)}$$

$$\frac{P_a}{P_n/\Omega_c} + \frac{8}{9}\left(\frac{M_{ax}}{M_{nx}/\Omega_b} + \frac{M_{ay}}{M_{ny}/\Omega_b}\right) = 0.4535 + \frac{8}{9}\left(\frac{90}{237} + \frac{20}{107}\right)$$

$$= 0.957 < 1.0 \quad \text{(OK)}$$

Answer   The W12 × 65 is satisfactory.

6.6   MEMBERS IN UNBRACED FRAMES

In a beam–column whose ends are free to translate, the maximum primary moment resulting from the sidesway is almost always at one end. As was illustrated in Figure 6.5, the maximum secondary moment from the sidesway is *always* at the end. As a consequence of this condition, the maximum primary and secondary moments are usually additive and there is no need for the factor C_m; in effect, $C_m = 1.0$. Even when there is a reduction, it will be slight and can be neglected. Consider the beam–column shown in Figure 6.15. Here the equal end moments are caused by the sidesway (from the horizontal load). The axial load, which partly results from loads not causing the sidesway, is carried along and amplifies the end moment. The amplification factor for the sidesway moments, B_2, is given by

$$B_2 = \frac{1}{1 - \dfrac{\alpha \Sigma P_{nt}}{\Sigma P_{e2}}} \geq 1 \qquad \text{(AISC Equation C2-3)}$$

where

$\qquad \alpha = 1.00$ for LRFD
$\qquad\quad = 1.60$ for ASD
$\quad \Sigma P_{nt} = $ sum of required load capacities for all columns in the story under consideration (factored for LRFD, unfactored for ASD)
$\quad \Sigma P_{e2} = $ sum of the Euler loads for all columns in the story under consideration

FIGURE 6.15

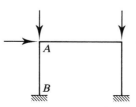

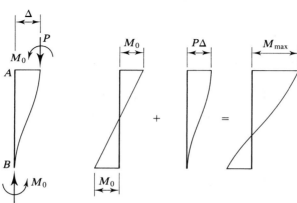

There are two forms for ΣP_{e2}, given as follows:

$$\Sigma P_{e2} = \Sigma \frac{\pi^2 EI}{(K_2 L)^2} \qquad \text{(AISC Equation C2-6a)}$$

or

$$\Sigma P_{e2} = R_M \frac{\Sigma HL}{\Delta_H} \qquad \text{(AISC Equation C2-6b)}$$

where

I = moment of inertia about the axis of bending

K_2 = effective length factor corresponding to the unbraced condition

L = story height

R_M = 1.0 for braced frames (although B_2 is not used for braced frames)

= 0.85 for unbraced frames and mixed systems

Δ_H = drift (sidesway displacement) of the story under consideration

ΣH = sum of all horizontal forces causing Δ_H

Either form of ΣP_{e2} may be used; the choice is usually one of convenience. Equation C2-6b is usually more convenient for design, where the moment of inertia is unknown, but there may be a limit on drift.

The rationale for using the summations is that B_2 applies to unbraced frames, and if sidesway is going to occur, all columns in the story must sway simultaneously. In most cases, the structure will be made up of plane frames, so ΣP_{nt} and ΣP_{e2} are for the columns within a story of the frame, and the lateral loads H are the lateral loads acting on the frame at and above the story. With Δ_H caused by ΣH, the ratio $\Sigma H / \Delta_H$ can be based on either factored or unfactored loads.

AISC Equations C2-6a and C2-6b were derived by using two different methods, but in most cases they will give almost identical results (Yura, 1988). In those cases where the two values of ΣP_{e2} would be significantly different, the axial load term of the interaction formula will be dominant, and the end results will not differ by much. As mentioned earlier, the choice is based on convenience; it depends on which terms in the equations are readily available.

In situations where M_{nt} and $M_{\ell t}$ act at two different points on the member, as in Figure 6.5, AISC Equation C2-1a will produce conservative results.

Figure 6.16 further illustrates the superposition concept. Figure 6.16a shows an unbraced frame subject to both gravity and lateral loads. The moment M_{nt} in member AB is computed by using only the gravity loads. Because of symmetry, no bracing is needed to prevent sidesway from these loads. This moment is amplified with the factor B_1 to account for the $P\delta$ effect. $M_{\ell t}$, the moment corresponding to the sway (caused by the horizontal load H), will be amplified by B_2 to account for the $P\Delta$ effect.

In Figure 6.16b, the unbraced frame supports only a vertical load. Because of the unsymmetrical placement of this load, there will be a small amount of sidesway. The moment M_{nt} is computed by considering the frame to be braced—in this case, by a fictitious horizontal support and corresponding reaction called an *artificial joint restraint* (AJR). To compute the sidesway moment, the fictitious support is removed, and a force equal to the

FIGURE 6.16

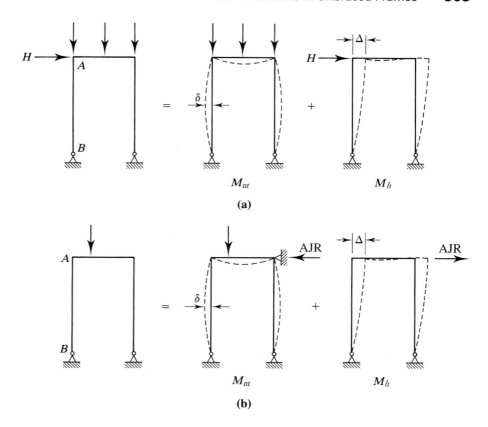

(a)

(b)

artificial joint restraint, but opposite in direction, is applied to the frame. In cases such as this one, the secondary moment $P\Delta$ will be very small, and $M_{\ell t}$ can usually be neglected.

If both lateral loads and unsymmetrical gravity loads are present, the AJR force should be added to the actual lateral loads when $M_{\ell t}$ is computed.

As an alternative to this approach, two structural analyses can be performed (Gaylord et al., 1992). In the first, the frame is assumed to be braced against sidesway. The resulting moments are the M_{nt} moments. A second analysis is performed in which the frame is assumed to be unbraced. The results of the first analysis are subtracted from the second analysis to obtain the $M_{\ell t}$ moments.

Example 6.6   A W12 × 65 of A992 steel, 15 feet long, is to be investigated for use as a column in an unbraced frame. The axial load and end moments obtained from a first-order analysis of the gravity loads (dead load and live load) are shown in Figure 6.17a. The frame is symmetrical, and the gravity loads are symmetrically placed. Figure 6.17b shows the wind load moments obtained from a first-order analysis. All bending moments are about the strong axis. Effective length factors are $K_x = 1.2$ for the sway case, $K_x = 1.0$

FIGURE 6.17

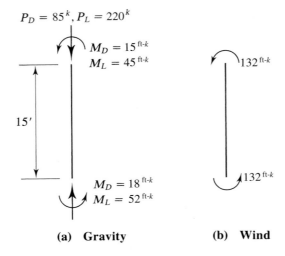

$P_D = 85^k, P_L = 220^k$

$M_D = 15^{\text{ft-}k}$
$M_L = 45^{\text{ft-}k}$

$132^{\text{ft-}k}$

15′

$M_D = 18^{\text{ft-}k}$
$M_L = 52^{\text{ft-}k}$

$132^{\text{ft-}k}$

(a) Gravity   **(b) Wind**

for the nonsway case, and $K_y = 1.0$. Determine whether this member is in compliance with the AISC Specification.

LRFD Solution   All the load combinations given in ASCE 7 involve dead load, and except for the first one, all combinations also involve live load, wind load, or both. If load types not present in this example (E, L_r, S, and R) are omitted, the load conditions can be summarized as

$$1.4D \tag{1}$$
$$1.2D + 1.6L \tag{2}$$
$$1.2D + (0.5L \text{ or } 0.8W) \tag{3}$$
$$1.2D + 1.6W + 0.5L \tag{4}$$
$$1.2D + 0.5L \tag{5}$$
$$0.9D \pm 1.6W \tag{6}$$

The dead load is less than eight times the live load, so combination (1) can be ruled out. Load combination (4) will be more critical than (3), so (3) can be eliminated. Combination (5) can be eliminated because it will be less critical than (2). Finally, combination (6) will not be as critical as (4) and can be removed from consideration, leaving only two load combinations to be investigated, (2) and (4):

$$1.2D + 1.6L \quad \text{and} \quad 1.2D + 1.6W + 0.5L$$

Figure 6.18 shows the axial loads and bending moments calculated for these two combinations.

Determine the critical axis for axial compressive strength:

$$K_y L = 15 \text{ ft}$$
$$\frac{K_x L}{r_x / r_y} = \frac{1.2(15)}{1.75} = 10.29 \text{ ft} < 15 \text{ ft} \quad \therefore \text{ use } KL = 15 \text{ ft.}$$

FIGURE 6.18

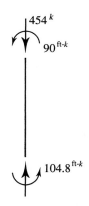

(a)   Load combination 2 (1.2D + 1.6L)

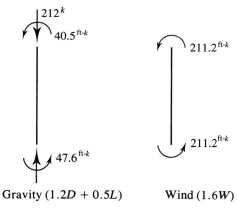

Gravity (1.2D + 0.5L)   Wind (1.6W)

(b)   Load combination 4

From the column load tables, with $KL = 15$ ft, $\phi_c P_n = 662$ kips.

For load condition 2, $P_u = 454$ kips, $M_{nt} = 104.8$ ft-kips, and $M_\ell = 0$ (because of symmetry, there are no sidesway moments). The bending factor is

$$C_m = 0.6 - 0.4\left(\frac{M_1}{M_2}\right) = 0.6 - 0.4\left(\frac{90}{104.8}\right) = 0.2565$$

For the axis of bending,

$$P_{e1} = \frac{\pi^2 EI}{(K_1 L)^2} = \frac{\pi^2 EI_x}{(K_x L)^2} = \frac{\pi^2 (29,000)(533)}{(1.0 \times 15 \times 12)^2} = 4708 \text{ kips}$$

(This case involves no sidesway, so K_x for the braced condition is used.)

FIGURE 6.19

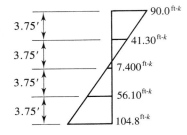

The amplification factor for nonsway moments is

$$B_1 = \frac{C_m}{1-(\alpha P_r/P_{e1})} = \frac{C_m}{1-(1.00P_u/P_{e1})} = \frac{0.2565}{1-(454/4708)} = 0.284 < 1.0 \quad \therefore \text{ use } B_1 = 1.0$$

$$M_r = M_u = B_1 M_{nt} + B_2 M_{\ell t} = 1.0(104.8) + 0 = 104.8 \text{ ft-kips}$$

From the beam design charts, with $L_b = 15$ feet,

$$\phi_b M_n = 340 \text{ ft-kips} \qquad \text{(for } C_b = 1.0)$$
$$\phi_b M_p = 356 \text{ ft-kips}$$

Figure 6.19 shows the bending moment diagram for the gravity-load moments. (The computation of C_b is based on absolute values, so a sign convention for the diagram is not necessary.) Hence,

$$\begin{aligned}
C_b &= \frac{12.5 M_{max}}{2.5 M_{max} + 3 M_A + 4 M_B + 3 M_C} \\
&= \frac{12.5(104.8)}{2.5(104.8) + 3(41.30) + 4(7.400) + 3(56.10)} = 2.24
\end{aligned}$$

For $C_b = 2.24$,

$$\phi_b M_n = 2.24(340) > \phi_b M_p = 356 \text{ kips} \qquad \therefore \text{ use } \phi_b M_n = 356 \text{ ft-kips.}$$

Determine the appropriate interaction equation:

$$\frac{P_u}{\phi_c P_n} = \frac{454}{662} = 0.6858 > 0.2 \quad \therefore \text{ use Equation 6.3 (AISC Equation H1-1a)}$$

$$\frac{P_u}{\phi_c P_n} + \frac{8}{9}\left(\frac{M_{ux}}{\phi_b M_{nx}} + \frac{M_{uy}}{\phi_b M_{ny}}\right) = 0.6858 + \frac{8}{9}\left(\frac{104.8}{356} + 0\right) = 0.947 < 1.0 \quad \text{(OK)}$$

For load condition (4), $P_u = 212$ kips, $M_{nt} = 47.6$ ft-kips, and $M_{\ell t} = 211.2$ ft-kips. For the braced condition,

$$C_m = 0.6 - 0.4\left(\frac{M_1}{M_2}\right) = 0.6 - 0.4\left(\frac{40.5}{47.6}\right) = 0.2597$$

$$P_{e1} = 4708 \text{ kips} \quad (P_{e1} \text{ is independent of the loading condition})$$

$$B_1 = \frac{C_m}{1-(\alpha P_r/P_{e1})} = \frac{C_m}{1-(1.00P_u/P_{e1})} = \frac{0.2597}{1-(212/4708)} = 0.272 < 1.0 \quad \therefore \text{ use } B_1 = 1.0$$

To compute B_2, the amplification factor for sidesway moments, the elastic critical buckling load P_{e2} must first be computed. There are not enough data available for either AISC Equation C2-6a or C2-6b, but if we assume that the ratio of applied axial load to Euler load capacity is the same for all the columns in the story as for the column under consideration, AISC Equation C2-6a can be used. The amplification factor can be written as

$$B_2 = \frac{1}{1 - \dfrac{\alpha \Sigma P_{nt}}{\Sigma P_{e2}}} \approx \frac{1}{1 - \dfrac{\alpha P_{nt}}{P_{e2}}}$$

where

$$P_{nt} = P_u = 212 \text{ kips}$$

$$P_{e2} = \frac{\pi^2 EI}{(K_2 L)^2} = \frac{\pi^2 (29,000)(533)}{(1.2 \times 15 \times 12)^2} = 3270 \text{ kips}$$

then

$$B_2 = \frac{1}{1 - \dfrac{\alpha P_{nt}}{P_{e2}}} = \frac{1}{1 - \dfrac{1.00(212)}{3270}} = 1.069$$

The total amplified moment is

$$M_r = M_u = B_1 M_{nt} + B_2 M_{\ell t} = 1.0(47.6) + 1.069(211.2) = 273.4 \text{ ft-kips}$$

Although the moments M_{nt} and $M_{\ell t}$ are different, they are distributed similarly, and C_b will be roughly the same; at any rate, they are large enough that $\phi_b M_p = 356$ ft-kips will be the design strength regardless of which moment is considered.

$$\frac{P_u}{\phi_c P_n} = \frac{212}{662} = 0.3202 > 0.2 \quad \therefore \text{ use Equation 6.3 (AISC Equation H1-1a)}$$

$$\frac{P_u}{\phi_c P_n} + \frac{8}{9}\left(\frac{M_{ux}}{\phi_b M_{nx}} + \frac{M_{uy}}{\phi_b M_{ny}}\right) = 0.3202 + \frac{8}{9}\left(\frac{273.4}{356} + 0\right) = 1.00 \quad \text{(OK)}$$

Answer   This member satisfies the AISC Specification requirements.

ASD Solution   The ASD load combinations from ASCE 7 can be reduced to the following possibilities after the elimination of all loads other than dead, live, and wind:

Load combination 2: $D + L$

Load combination 5: $D + W$

Load combination 6: $D + 0.75W + 0.75L$

FIGURE 6.20

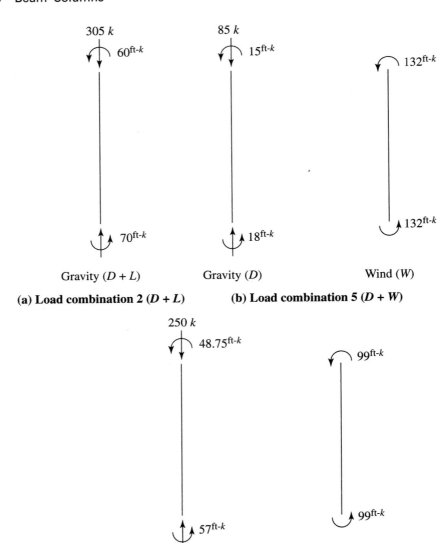

305 k

$60^{\text{ft-}k}$

$70^{\text{ft-}k}$

Gravity ($D + L$)

(a) Load combination 2 ($D + L$)

85 k

$15^{\text{ft-}k}$

$18^{\text{ft-}k}$

Gravity (D)

$132^{\text{ft-}k}$

$132^{\text{ft-}k}$

Wind (W)

(b) Load combination 5 ($D + W$)

250 k

$48.75^{\text{ft-}k}$

$57^{\text{ft-}k}$

Gravity ($D + 0.75L$)

$99^{\text{ft-}k}$

$99^{\text{ft-}k}$

Wind ($0.75W$)

(c) Load combination 6 ($D + 0.75W + 0.75L$)

Figure 6.20 shows the calculated axial load and moments for these combinations.
Load Condition 2: $P_a = 305$ kips, $M_{nt} = 70$ ft-kips, and $M_{\ell t} = 0$ (because of symmetry, there are no sidesway moments). The bending factor is

$$C_m = 0.6 - 0.4\left(\frac{M_1}{M_2}\right) = 0.6 - 0.4\left(\frac{60}{70}\right) = 0.2571$$

FIGURE 6.21

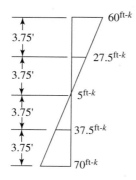

For the braced condition,

$$P_{e1} = \frac{\pi^2 EI}{(K_1 L)^2} = \frac{\pi^2 EI_x}{(K_x L)^2} = \frac{\pi^2 (29,000)(533)}{(1.0 \times 15 \times 12)^2} = 4708 \text{ kips}$$

$$B_1 = \frac{C_m}{1 - (\alpha P_r / P_{e1})} = \frac{C_m}{1 - (1.60 P_a / P_{e1})} = \frac{0.2571}{1 - (1.60 \times 305 / 4708)}$$

$$= 0.287 < 1.0 \quad \therefore \text{ use } B_1 = 1.0$$

$$M_r = M_{ax} = B_1 M_{nt} + B_2 M_{\ell t} = 1.0(70) + 0 = 70 \text{ ft-kips}$$

From the beam design charts with $L_b = 15$ ft and $C_b = 1.0$,

$$\frac{M_n}{\Omega_b} = 226 \text{ ft-kips and } \frac{M_p}{\Omega_b} = 237 \text{ ft-kips}$$

Figure 6.21 shows the bending moment diagram (absolute values shown) for the gravity-load moments to be used in the computation of C_b.

$$C_b = \frac{12.5 M_{max}}{2.5 M_{max} + 3 M_A + 4 M_B + 3 M_C}$$

$$= \frac{12.5(70)}{2.5(70) + 3(27.5) + 4(5) + 3(37.5)} = 2.244$$

$$\therefore \frac{M_n}{\Omega_b} = C_b(226) = 2.244(226) = 507 \text{ ft-kips}$$

But $\dfrac{M_p}{\Omega_b} = 237$ ft-kips < 507 ft-kips $\therefore$ use $\dfrac{M_n}{\Omega_b} = 237$ ft-kips

Compute the axial compressive strength:

$$K_y L = 15 \text{ ft}$$

$$\frac{K_x L}{r_x / r_y} = \frac{1.2(15)}{1.75} = 10.29 \text{ ft} < 15 \text{ ft} \quad \therefore \text{ use } KL = 15 \text{ ft}$$

From the column load tables, with $KL = 15$ ft, $P_n/\Omega_c = 441$ kips
Determine the appropriate interaction equation:

$$\frac{P_a}{P_n/\Omega_c} = \frac{305}{441} = 0.6916 > 0.2 \quad \therefore \text{ use Equation 6.5 (AISC Equation H1-1a)}$$

$$\frac{P_a}{P_n/\Omega_c} + \frac{8}{9}\left(\frac{M_{ax}}{M_{nx}/\Omega_b} + \frac{M_{ay}}{M_{ny}/\Omega_b}\right) = 0.6916 + \frac{8}{9}\left(\frac{70}{237} + 0\right)$$

$$= 0.954 < 1.0 \quad \text{(OK)}$$

Load Condition 5: $P_a = 85$ kips, $M_{nt} = 18$ ft-kips, and $M_{\ell t} = 132$ ft-kips. For the braced condition, the bending factor is

$$C_m = 0.6 - 0.4\left(\frac{M_1}{M_2}\right) = 0.6 - 0.4\left(\frac{15}{18}\right) = 0.2667$$

$$P_{e1} = 4708 \text{ kips} \qquad (P_{e1} \text{ is independent of the loading})$$

$$B_1 = \frac{C_m}{1-(\alpha P_r/P_{e1})} = \frac{C_m}{1-(1.60 P_a/P_{e1})} = \frac{0.2667}{1-(1.60 \times 85/4708)}$$

$$= 0.275 < 1.0 \quad \therefore \text{ use } B_1 = 1.0$$

To compute B_2, the elastic critical buckling load P_{e2} must first be computed. There are not enough data available for either AISC Equation C2-6a or C2-6b, but if we assume that the ratio of applied load to Euler load capacity is the same for all the columns in the story as for the column under consideration, AISC Equation C2-6a can be used. The amplification factor can then be written as

$$B_2 = \frac{1}{1-\dfrac{\alpha\Sigma P_{nt}}{\Sigma P_{e2}}} \approx \frac{1}{1-\dfrac{\alpha P_{nt}}{P_{e2}}}$$

where

$$P_{nt} = P_a = 85 \text{ kips}$$

$$P_{e2} = \frac{\pi^2 EI}{(K_2 L)^2} = \frac{\pi^2 (29,000)(533)}{(1.2 \times 15 \times 12)^2} = 3270 \text{ kips}$$

then

$$B_2 = \frac{1}{1-\dfrac{\alpha\Sigma P_{nt}}{\Sigma P_{e2}}} \approx \frac{1}{1-\dfrac{\alpha P_{nt}}{P_{e2}}} = \frac{1}{1-\dfrac{1.60(85)}{3270}} = 1.043$$

The total amplified moment is

$$M_r = M_{ax} = B_1 M_{nt} + B_2 M_{\ell t} = 1.0(18) + 1.043(132) = 155.7 \text{ ft-kips}$$

Although the moments M_{nt} and $M_{\ell t}$ are different, they are distributed similarly, so C_b will be approximately the same for both distributions. In each case, C_b will be large enough so that M_p/Ω_b will control; therefore, $M_n/\Omega_b = 237$ ft-kips.

$$\frac{P_a}{P_n/\Omega_c} = \frac{85}{441} = 0.1927 < 0.2 \qquad \therefore \text{ use Equation 6.6 (AISC Equation H1-1b)}$$

$$\frac{P_a}{2P_n/\Omega_c} + \left(\frac{M_{ax}}{M_{nx}/\Omega_b} + \frac{M_{ay}}{M_{ny}/\Omega_b} \right) = \frac{85}{2(441)} + \left(\frac{155.7}{237} + 0 \right)$$

$$= 0.753 < 1.0 \quad \text{(OK)}$$

Load Condition 6: $P_a = 250$ kips, $M_{nt} = 57$ ft-kips, and $M_{\ell t} = 99$ ft-kips. For the braced condition, the bending factor is

$$C_m = 0.6 - 0.4 \left(\frac{M_1}{M_2} \right) = 0.6 - 0.4 \left(\frac{48.75}{57} \right) = 0.2579$$

$$P_{e1} = 4708 \text{ kips} \qquad (P_{e1} \text{ is independent of the loading.})$$

$$B_1 = \frac{C_m}{1 - (\alpha P_r/P_{e1})} = \frac{C_m}{1 - (1.60 P_a/P_{e1})} = \frac{0.2579}{1 - (1.60 \times 250/4708)}$$

$$= 0.282 < 1.0 \quad \therefore \text{ use } B_1 = 1.0$$

$$P_{nt} = P_a = 250 \text{ kips}$$

$$P_{e2} = 3270 \text{ kips} \qquad (P_{e2} \text{ is independent of the loading condition})$$

$$B_2 \approx \frac{1}{1 - \dfrac{\alpha P_{nt}}{P_{e2}}} = \frac{1}{1 - \dfrac{1.60(250)}{3270}} = 1.139$$

The total amplified moment is

$$M_r = M_{ax} = B_1 M_{nt} + B_2 M_{\ell t} = 1.0(57) + 1.139 \, (99) = 169.8 \text{ ft-kips}$$

As before, use $M_n/\Omega_b = M_p/\Omega_b = 237$ ft-kips

$$\frac{P_a}{P_n/\Omega_c} = \frac{250}{441} = 0.5669 > 0.2 \qquad \therefore \text{ use Equation 6.5 (AISC Equation H1-1a)}$$

$$\frac{P_a}{P_n/\Omega_c} + \frac{8}{9} \left(\frac{M_{ax}}{M_{nx}/\Omega_b} + \frac{M_{ay}}{M_{ny}/\Omega_b} \right) = 0.5669 + \frac{8}{9} \left(\frac{169.8}{237} + 0 \right)$$

$$= 1.20 > 1.0 \quad \text{(N.G.)}$$

Answer   Load combination 6 controls, and the member does not satisfy the AISC Specification requirements.

6.7 DESIGN OF BEAM–COLUMNS

Because of the many variables in the interaction formulas, the design of beam–columns is essentially a trial-and-error process. A procedure developed by Aminmansour (2000) simplifies this process, especially the evaluation of trial shapes. Part 6 of the *Manual*, "Design of Members Subject to Combined Loading," contains tables based on design aids developed by Aminmansour. The procedure can be explained as follows. If we initially assume that AISC Equation H1-1a governs, then

$$\frac{P_r}{P_c} + \frac{8}{9}\left(\frac{M_{rx}}{M_{cx}} + \frac{M_{ry}}{M_{cy}}\right) \le 1.0 \qquad \text{(AISC Equation H1-1a)}$$

This can be written as

$$\left(\frac{1}{P_c}\right)P_r + \left(\frac{8}{9M_{cx}}\right)M_{rx} + \left(\frac{8}{9M_{cy}}\right)M_{ry} \le 1.0$$

or

$$pP_r + b_x M_{rx} + b_y M_{ry} \le 1.0 \qquad (6.8)$$

where

$$p = \frac{1}{P_c}$$

$$b_x = \frac{8}{9M_{cx}}$$

$$b_y = \frac{8}{9M_{cy}}$$

If AISC Equation H1-1b controls (that is, $P_r/P_c < 0.2$, or equivalently, $pP_r < 0.2$, then use

$$\frac{P_r}{2P_c} + \left(\frac{M_{rx}}{M_{cx}} + \frac{M_{ry}}{M_{cy}}\right) \le 1.0 \qquad \text{(AISC Equation H1-1b)}$$

or

$$0.5pP_r + \frac{9}{8}(b_x M_{rx} + b_y M_{ry}) \le 1.0 \qquad (6.9)$$

Table 6-1 in Part 6 of the *Manual* gives values of p, b_x, and b_y for various W shapes, in both LRFD and ASD format.

For LRFD:

$$P = \frac{1}{P_c} = \frac{1}{\phi_c P_n}$$

$$b_x = \frac{8}{9M_{cx}} = \frac{8}{9(\phi_b M_{nx})}$$

$$b_y = \frac{8}{9M_{cy}} = \frac{8}{9(\phi_b M_{ny})}$$

For ASD:

$$p = \frac{1}{P_c} = \frac{1}{P_n/\Omega_c}$$

$$b_x = \frac{8}{9M_{cx}} = \frac{8}{9(M_{nx}/\Omega_b)}$$

$$b_y = \frac{8}{9M_{cy}} = \frac{8}{9(M_{ny}/\Omega_b)}$$

Table 6-1 gives values of p, b_x, and b_y for all W shapes listed in Part 1 of the *Manual*, "Dimensions and Properties," except for those smaller than W8. The values of C_b, B_1, and B_2 must be calculated independently for use in the computation of M_r (M_u for LRFD or M_a for ASD).

The procedure for design is as follows:

1. Select a trial shape from Table 6-1.
2. Use the effective length KL to select p, and use the unbraced length L_b to select b_x (the constant b_y determines the weak axis bending strength, so it is independent of the unbraced length). The values of the constants are based on the assumption that weak axis buckling controls the axial compressive strength and that $C_b = 1.0$.
3. Compute pP_r. If this is less than or equal to 0.2, use interaction Equation 6.8. If pP_r is greater than 0.2, use Equation 6.9.
4. Evaluate the selected interaction equation with the values of p, b_x, and b_y for the trial shape.
5. If the result is not very close to 1.0, try another shape. By examining the value of each term in Equation 6.8 or 6.9, you can gain insight into which constants need to be larger or smaller.
6. Continue the process until a shape is found that gives an interaction equation result less than 1.0 and close to 1.0 (greater than 0.9).

Verification of assumptions:

- If strong axis buckling controls the compressive strength, use an effective length of

$$KL = \frac{K_x}{r_x/r_y}$$

to obtain p from Table 6-1.

- If C_b is not equal to 1.0, the value of b_x must be adjusted. Example 6.8 illustrates the procedure.

***

Example 6.7 Select a W shape of A992 steel for the beam–column of Figure 6.22. This member is part of a braced frame and is subjected to the service-load axial force and bending moments shown (the end shears are not shown). Bending is about the strong axis, and $K_x = K_y = 1.0$. Lateral support is provided only at the ends. Assume that $B_1 = 1.0$.

FIGURE 6.22

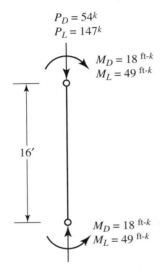

$P_D = 54^k$
$P_L = 147^k$

$M_D = 18^{\text{ft-k}}$
$M_L = 49^{\text{ft-k}}$

16′

$M_D = 18^{\text{ft-k}}$
$M_L = 49^{\text{ft-k}}$

LRFD Solution The factored axial load is

$$P_u = 1.2P_D + 1.6P_L = 1.2(54) + 1.6(147) = 300 \text{ kips}$$

The factored moment at each end is

$$M_{ntx} = 1.2M_D + 1.6M_L = 1.2(18) + 1.6(49) = 100 \text{ ft-kips}$$

Since $B_1 = 1.0$, the factored load bending moment is

$$M_{ux} = B_1 M_{ntx} = 1.0(100) = 100 \text{ ft-kips.}$$

The effective length for compression and the unbraced length for bending are the same:

$$KL = L_b = 16 \text{ ft.}$$

The bending moment is uniform over the unbraced length, so $C_b = 1.0$.
Try a W10 shape. From Table 6-1, **Try a W10 × 60**, with $p = 1.89 \times 10^{-3}$ and $b_x = 3.51 \times 10^{-3}$.

Determine which interaction equation to use:

$$pP_r = pP_u = (1.89 \times 10^{-3})(300) = 0.567 > 0.2 \quad \therefore \text{Equation 6.8 controls.}$$

$$pP_r + b_x M_{rx} + b_y M_{ry} = pP_u + b_x M_{ux} + b_y M_{uy}$$

$$= (1.89 \times 10^{-3})(300) + (3.51 \times 10^{-3})(100) + 0$$

$$= 0.918 < 1.0 \quad \text{(OK)}$$

To be sure that we have found the lightest W10, try the next lighter one, a W10 × 54, with

$$p = 2.12 \times 10^{-3} \text{ and } b_x = 3.97 \times 10^{-3}$$

From Equation 6.8,

$$pP_u + b_x M_{ux} + b_y M_{uy} = (2.12 \times 10^{-3})(300) + (3.97 \times 10^{-3})(100) + 0$$

$$= 1.03 > 1.0 \quad \text{(N.G.)}$$

Try a W12 shape. **Try a W12 × 58**, with $p = 2.00 \times 10^{-3}$ and $b_x = 3.13 \times 10^{-3}$.

$$pP_u + b_x M_{ux} + b_y M_{uy} = (2.00 \times 10^{-3})(300) + (3.13 \times 10^{-3})(100) + 0$$

$$= 0.913 < 1.0 \quad \text{(OK)}$$

Verify that Equation 6.8 is the correct one:

$$\frac{P_u}{\phi_c P_n} = pP_u = (2.00 \times 10^{-3})(300) = 0.600 > 0.2 \quad \therefore \text{Equation 6.8 controls, as assumed}$$

To be sure that we have found the lightest W12, try the next lighter one, a W12 × 53, with $p = 2.21 \times 10^{-3}$ and $b_x = 3.52 \times 10^{-3}$.

$$pP_u + b_x M_{ux} + b_y M_{uy} = (2.21 \times 10^{-3})(300) + (3.52 \times 10^{-3})(100) + 0$$

$$= 1.02 \quad \text{(N.G.)}$$

Check other categories of shapes to be sure that we have found the lightest shape. The lightest W14 that is a possibility is a W14 × 53, with $p = 2.96 \times 10^{-3}$ and $b_x = 3.51 \times 10^{-3}$.

From Equation 6.8,

$$(2.96 \times 10^{-3})(300) + (3.51 \times 10^{-3})(100) = 1.24 > 1.0 \quad \text{(N.G.)}$$

There are no other possibilities in the deeper W-shape groups.

Answer   Use a W12 × 58.

ASD Solution   The required axial load strength is

$$P_a = P_D + P_L = 54 + 147 = 201 \text{ kips}$$

The moment at each end is

$$M_{ntx} = M_D + M_L = 18 + 49 = 67 \text{ ft-kips}$$

Since $B_1 = 1.0$, The required bending moment strength is

$$M_{ax} = B_1 M_{ntx} = 1.0(67) = 67 \text{ ft-kips.}$$

The effective length for compression and the unbraced length for bending are the same:

$$KL = L_b = 16 \text{ ft.}$$

The bending moment is uniform over the unbraced length, so $C_b = 1.0$.

Try a W10 shape. From Table 6-1, **try a W10 × 60**, with $p = 2.85 \times 10^{-3}$ and $b_x = 5.27 \times 10^{-3}$.

Determine which interaction equation to use:

$$pP_r = pP_a = (2.85 \times 10^{-3})(201) = 0.573 > 0.2 \quad \therefore \text{ Equation 6.8 controls.}$$

$$\begin{aligned}
pP_r + b_x M_{rx} + b_y M_{ry} &= pP_a + b_x M_{ax} + b_y M_{ay} \\
&= (2.85 \times 10^{-3})(201) + (5.27 \times 10^{-3})(67) + 0 \\
&= 0.926 < 1.0 \quad \text{(OK)}
\end{aligned}$$

To be sure that we have found the lightest W10, try the next lighter one, a W10 × 54, with $p = 3.18 \times 10^{-3}$ and $b_x = 5.97 \times 10^{-3}$.

From Equation 6.8,

$$\begin{aligned}
pP_a + b_x M_{ax} + b_y M_{ay} &= (3.18 \times 10^{-3})(201) + (5.97 \times 10^{-3})(67) + 0 \\
&= 1.04 > 1.0 \quad \text{(N.G.)}
\end{aligned}$$

Try a W12 shape. **Try a W12 × 58**, with $p = 3.00 \times 10^{-3}$ and $b_x = 4.71 \times 10^{-3}$.

$$\begin{aligned}
pP_a + b_x M_{ax} + b_y M_{ay} &= (3.00 \times 10^{-3})(201) + (4.71 \times 10^{-3})(67) + 0 \\
&= 0.919 < 1.0 \quad \text{(OK)}
\end{aligned}$$

Verify that Equation 6.8 is the correct one:

$$\frac{P_a}{P_n/\Omega_c} = pP_a = (3.00 \times 10^{-3})(201) = 0.603 > 0.2$$

$\therefore$ Equation 6.8 controls, as assumed

To be sure that we have found the lightest W12, try the next lighter one, a W12 × 53, with $p = 3.33 \times 10^{-3}$ and $b_x = 5.29 \times 10^{-3}$.

$$\begin{aligned}
pP_a + b_x M_{ax} + b_y M_{ay} &= (3.33 \times 10^{-3})(201) + (5.29 \times 10^{-3})(67) + 0 \\
&= 1.02 \quad \text{(N.G.)}
\end{aligned}$$

Check other categories of shapes to be sure that we have found the lightest shape. The lightest W14 that is a possibility is a W14 × 53, with $p = 4.45 \times 10^{-3}$ and $b_x = 5.27 \times 10^{-3}$.

From Equation 6.8,

$$(4.45 \times 10^{-3})(201) + (5.27 \times 10^{-3})(67) = 1.25 > 1.0 \qquad \text{(N.G.)}$$

There are no other possibilities in the deeper W-shape groups.

Answer   Use a W12 × 58.

Although the absolute lightest W-shape was found in Example 6.7, in many cases a specific nominal depth, such as 12 inches, is required for architectural or other reasons. Example 6.8 illustrates the design of a beam–column of a specific depth.

Example 6.8   A structural member in a braced frame must support the following service loads and moments: an axial compressive dead load of 25 kips and a live load of 75 kips; a dead load moment of 12.5 ft-kips about the strong axis and a live load moment of 37.5 ft-kips about the strong axis; a dead load moment of 5 ft-kips about the weak axis and a live load moment of 15 ft-kips about the weak axis. The moments occur at one end; the other end is pinned. The effective length with respect to each axis is 15 feet. There are no transverse loads on the member. Use A992 steel and select a W10 shape.

LRFD Solution   The factored axial load is

$$P_u = 1.2(25) + 1.6(75) = 150 \text{ kips}$$

The factored moments are

$$M_{ntx} = 1.2(12.5) + 1.6(37.5) = 75.0 \text{ ft-kips}$$
$$M_{nty} = 1.2(5) + 1.6(15) = 30.0 \text{ ft-kips}$$

The amplification factor B_1 can be estimated as 1.0 for purposes of making a trial selection. For the two axes,

$$M_{ux} = B_{1x}M_{ntx} = 1.0(75) = 75 \text{ ft-kips}$$
$$M_{uy} = B_{1y}M_{nty} = 1.0(30) = 30 \text{ ft-kips}$$

Try a W10 × 49.   From Table 6-1, $p = 2.22 \times 10^{-3}$, $b_x = 4.35 \times 10^{-3}$, $b_y = 8.38 \times 10^{-3}$. Determine which interaction equation to use:

$$\frac{P_u}{\phi_c P_n} = pP_u = (2.22 \times 10^{-3})(150) = 0.333 > 0.2 \quad \therefore \text{ Equation 6.8 controls}$$

As a rough check (remember that B_1 has not yet been computed and C_b has not been accounted for),

$$pP_u + b_x M_{ux} + b_y M_{uy} = (2.22 \times 10^{-3})(150) + (4.35 \times 10^{-3})(75) + (8.38 \times 10^{-3})(30)$$
$$= 0.911 < 1.0 \qquad \text{(OK)}$$

It is likely that B_{1x} and B_{1y} will be equal to 1.0. In addition, the inclusion of C_b will reduce the value of b_x and the result of the interaction equation, so this shape is probably satisfactory.

Calculate B_1 for each axis:

$$C_m = 0.6 - 0.4\left(\frac{M_1}{M_2}\right) = 0.6 - 0.4\left(\frac{0}{M_2}\right) = 0.6 \quad \text{(for both axes)}$$

$$P_{elx} = \frac{\pi^2 EI_x}{(K_x L)^2} = \frac{\pi^2 (29,000)(272)}{(15 \times 12)^2} = 2403 \text{ kips}$$

$$B_{1x} = \frac{C_{mx}}{1 - \dfrac{P_u}{P_{elx}}} = \frac{0.6}{1 - \dfrac{150}{2403}} = 0.640 < 1.0 \quad \therefore B_{1x} = 1.0 \text{ as assumed}$$

$$P_{ely} = \frac{\pi^2 EI_y}{(K_y L)^2} = \frac{\pi^2 (29,000)(93.4)}{(15 \times 12)^2} = 825.1 \text{ kips}$$

$$B_{1y} = \frac{C_{my}}{1 - \dfrac{P_u}{P_{ely}}} = \frac{0.6}{1 - \dfrac{150}{825.1}} = 0.733 < 1.0 \quad \therefore B_{1y} = 1.0 \text{ as assumed}$$

From Figure 5.15g, $C_b = 1.67$. Modify b_x to account for C_b.

$$C_b \times \phi_b M_{nx} = C_b \times \frac{8}{9} \times \frac{1}{b_x} = 1.67 \times \frac{8}{9} \times \frac{1}{4.35 \times 10^{-3}} = 341.3 \text{ ft-kips}$$

From the Z_x table, $\phi_b M_{px} = 227$ ft-kips < 341.3 ft-kips $\therefore \phi_b M_{nx} = 227$ ft-kips

$$b_x = \frac{8}{9(\phi_b M_{nx})} = \frac{8}{9(227)} = 3.92 \times 10^{-3}$$

Check Equation 6.8: $p = 2.22 \times 10^{-3}$, $b_x = 3.92 \times 10^{-3}$, $b_y = 8.38 \times 10^{-3}$

$$pP_u + b_x M_{ux} + b_y M_{uy} = (2.22 \times 10^{-3})(150) + (3.92 \times 10^{-3})(75) + (8.38 \times 10^{-3})(30)$$
$$= 0.878 < 1.0 \quad \text{(OK)}$$

Try the next smaller shape. **Try a W10 × 45**, with $p = 3.01 \times 10^{-3}$, $b_x = 5.07 \times 10^{-3}$, $b_y = 11.7 \times 10^{-3}$.

$$P_{elx} = \frac{\pi^2 EI_x}{(K_x L)^2} = \frac{\pi^2 (29,000)(248)}{(15 \times 12)^2} = 2191 \text{ kips}$$

$$B_{1x} = \frac{C_{mx}}{1 - \dfrac{P_u}{P_{e1x}}} = \frac{0.6}{1 - \dfrac{150}{2191}} = 0.644 < 1.0 \quad \therefore B_{1x} = 1.0 \text{ as assumed}$$

$$P_{e1y} = \frac{\pi^2 EI_y}{(K_y L)^2} = \frac{\pi^2 (29,000)(53.4)}{(15 \times 12)^2} = 471.7 \text{ kips}$$

$$B_{1y} = \frac{C_{my}}{1 - \dfrac{P_u}{P_{e1y}}} = \frac{0.6}{1 - \dfrac{150}{471.7}} = 0.880 < 1.0 \quad \therefore B_{1y} = 1.0 \text{ as assumed}$$

$$C_b \times \phi_b M_{nx} = C_b \times \frac{8}{9} \times \frac{1}{b_x} = 1.67 \times \frac{8}{9} \times \frac{1}{5.07 \times 10^{-3}} = 292.8 \text{ ft-kips}$$

$$\phi_b M_{px} = 206 \text{ ft-kips} < 292.8 \text{ ft-kips} \quad \therefore \phi_b M_{nx} = 206 \text{ ft-kips}$$

$$b_x = \frac{8}{9(\phi_b M_{nx})} = \frac{8}{9(206)} = 4.32 \times 10^{-3}$$

Check Equation 6.8 with $p = 3.01 \times 10^{-3}$, $b_x = 4.32 \times 10^{-3}$, $b_y = 11.7 \times 10^{-3}$.

$$pP_u + b_x M_{ux} + b_y M_{uy} = (3.01 \times 10^{-3})(150) + (4.32 \times 10^{-3})(75) + (11.7 \times 10^{-3})(30)$$
$$= 1.13 > 1.0 \qquad \text{(N.G.)}$$

Answer   Use a W10 × 49.

ASD Solution   The required axial load strength is

$$P_a = P_D + P_L = 25 + 75 = 100 \text{ kips}$$

The moments at each end are

$$M_{ntx} = M_D + M_L = 12.5 + 37.5 = 50.0 \text{ ft-kips}$$
$$M_{nty} = M_D + M_L = 5 + 15 = 20.0 \text{ ft-kips}$$

If we assume that $B_1 = 1.0$, The required bending moment strengths are

$$M_{ax} = B_1 M_{ntx} = 1.0(50) = 50 \text{ ft-kips}$$
$$M_{ay} = B_1 M_{nty} = 1.0(20) = 20 \text{ ft-kips}$$

The effective length for compression and the unbraced length for bending are the same:

$$KL = L_b = 15 \text{ ft.}$$

From Table 6-1, **try a W10 × 49**, with $p = 3.34 \times 10^{-3}$, $b_x = 6.54 \times 10^{-3}$, $b_y = 12.6 \times 10^{-6}$

Determine which interaction equation to use:

$$\frac{P_a}{P_n / \Omega_c} = pP_a = (3.34 \times 10^{-3})(100) = 0.334 > 0.2 \qquad \therefore \text{ use Equation 6.8}$$

As a rough check (remember that B_1 has not yet been computed and C_b has not been accounted for),

$$pP_a + b_x M_{ax} + b_y M_{ay} = (3.34 \times 10^{-3})(100) + (6.54 \times 10^{-3})(50) + (12.6 \times 10^{-3})(20)$$
$$= 0.913 < 1.0 \quad \text{(OK)}$$

It is likely that B_{1x} and B_{1y} will be equal to 1.0. In addition, the inclusion of C_b will reduce the value of b_x and the result of the interaction equation, so this shape is probably satisfactory.

Calculate B_1 for each axis:

$$C_m = 0.6 - 0.4\left(\frac{M_1}{M_2}\right) = 0.6 - 0.4\left(\frac{0}{M_2}\right) = 0.6 \text{ (for both axes)}$$

$$P_{elx} = \frac{\pi^2 E I_x}{(K_x L)^2} = \frac{\pi^2 (29,000)(272)}{(15 \times 12)^2} = 2403 \text{ kips}$$

$$B_{1x} = \frac{C_{mx}}{1 - \dfrac{1.60 P_a}{P_{elx}}} = \frac{0.6}{1 - \dfrac{1.60(100)}{2403}} = 0.643 < 1.0 \quad \therefore B_{1x} = 1.0 \text{ as assumed}$$

$$P_{ely} = \frac{\pi^2 E I_y}{(K_y L)^2} = \frac{\pi^2 (29,000)(93.4)}{(15 \times 12)^2} = 825.1 \text{ kips}$$

$$B_{1y} = \frac{C_{my}}{1 - \dfrac{1.60 P_a}{P_{ely}}} = \frac{0.6}{1 - \dfrac{1.60(100)}{825.1}} = 0.744 < 1.0 \quad \therefore B_{1y} = 1.0 \text{ as assumed}$$

From Figure 5.15g, $C_b = 1.67$. Modify b_x to account for C_b.

$$C_b \times \frac{M_{nx}}{\Omega_b} = C_b \times \frac{8}{9} \times \frac{1}{b_x} = 1.67 \times \frac{8}{9} \times \frac{1}{6.54 \times 10^{-3}} = 227.0 \text{ ft-kips}$$

$$\frac{M_{px}}{\Omega_b} = 151 \text{ ft-kips} < 227.0 \text{ ft-kips} \qquad \therefore \text{ use } \frac{M_{nx}}{\Omega_b} = 151 \text{ ft-kips}$$

$$b_x = \frac{8}{9(M_{nx}/\Omega_b)} = \frac{8}{9(151)} = 5.89 \times 10^{-3}$$

Check Equation 6.8 with $p = 3.34 \times 10^{-3}$, $b_x = 5.89 \times 10^{-3}$, $b_y = 12.6 \times 10^{-3}$

$$pP_a + b_x M_{ax} + b_y M_{ay} = (3.34 \times 10^{-3})(100) + (5.89 \times 10^{-3})(50) + (12.6 \times 10^{-3})(20)$$
$$= 0.881 < 1.0 \quad \text{(OK)}$$

Try the next smaller shape. **Try W10 × 45**, with $p = 4.53 \times 10^{-3}$, $b_x = 7.63 \times 10^{-3}$, $b_y = 17.6 \times 10^{-3}$

$$P_{e1x} = \frac{\pi^2 EI_x}{(K_x L)^2} = \frac{\pi^2 (29,000)(248)}{(15 \times 12)^2} = 2191 \text{ kips}$$

$$B_{1x} = \frac{C_{mx}}{1 - \frac{1.60 P_a}{P_{e1x}}} = \frac{0.6}{1 - \frac{1.60(100)}{2191}} = 0.647 < 1.0 \quad \therefore B_{1x} = 1.0 \text{ as assumed}$$

$$P_{e1y} = \frac{\pi^2 EI_y}{(K_y L)^2} = \frac{\pi^2 (29,000)(53.4)}{(15 \times 12)^2} = 471.7 \text{ kips}$$

$$B_{1y} = \frac{C_{my}}{1 - \frac{1.60 P_a}{P_{e1y}}} = \frac{0.6}{1 - \frac{1.60(100)}{471.7}} = 0.908 < 1.0 \quad \therefore B_{1y} = 1.0 \text{ as assumed}$$

$$C_b \times \frac{M_{nx}}{\Omega_b} = C_b \times \frac{8}{9} \times \frac{1}{b_x} = 1.67 \times \frac{8}{9} \times \frac{1}{7.63 \times 10^{-3}} = 194.6 \text{ ft-kips}$$

$$\frac{M_{px}}{\Omega_b} = 137 \text{ ft-kips} < 194.6 \text{ ft-kips} \quad \therefore \text{ use } \frac{M_{nx}}{\Omega_b} = 137 \text{ ft-kips}$$

$$b_x = \frac{8}{9(M_{nx}/\Omega_b)} = \frac{8}{9(137)} = 6.49 \times 10^{-3}$$

Check Equation 6.8 with $p = 4.53 \times 10^{-3}$, $b_x = 6.49 \times 10^{-3}$, $b_y = 17.6 \times 10^{-3}$

$$pP_a + b_x M_{ax} + b_y M_{ay} = (4.53 \times 10^{-3})(100) + (6.49 \times 10^{-3})(50) + (17.6 \times 10^{-3})(20)$$
$$= 1.13 > 1.0 \quad \text{(N.G.)}$$

Answer   Use a W10 × 49.

The modification of b_x to account for C_b can also be done by making use of the value of b_x corresponding to $L = Lb = 0$, since that value is based on M_p. This will be illustrated in Example 6.10.

Design of Bracing

A frame can be braced to resist directly applied lateral forces or to provide stablility. The latter type, *stability bracing*, was first discussed in Chapter 4, "Compression Members." The stiffness and strength requirements for stability can be added directly to the requirements for directly applied loads.

Bracing can be classified as nodal or relative. With nodal bracing, lateral support is provided at discrete locations and does not depend on the support from other parts of the frame. Relative bracing is connected to both the point of bracing and to other parts of the frame. Diagonal or x bracing is an example of relative bracing. Relative bracing can also be defined as follows: If a lateral cut through the frame not only cuts the column to be braced but also the brace itself, the bracing is relative.

Bracing must be designed for both strength and stiffness. For relative bracing, the required strength is

$$P_{br} = 0.004P_r \qquad \text{(AISC Equation A-6-1)}$$

where

P_{br} = lateral shear force resistance to be provided by the brace
P_r = vertical load to be stabilized by the brace

In a frame, more than one column will be stabilized by a brace. Therefore, P_r will be the *total* load to be stabilized and will be the total column load acting above the brace. If more than one frame is stabilized by the brace, then the columns in all of those frames must be included.

For relative bracing, the required stiffness is

$$
\begin{aligned}
\beta_{br} &= \frac{1}{\phi}\left(\frac{2P_r}{L_b}\right) \text{ for LRFD} \\
&= \Omega\left(\frac{2P_r}{L_b}\right) \text{ for ASD}
\end{aligned}
\qquad \text{(AISC Equation A-6-2)}
$$

where

L_b = unbraced length of the column (this will be the story height unless there are intermediate braces)
$\phi = 0.75$
$\Omega = 2.00$

Example 6.9   A structure composed of three rigid frames must be stabilized by diagonal bracing in one of the frames. The braced frame is shown in Figure 6.23. The loading is the same for all frames. Use A36 steel and determine the required cross-sectional area of the bracing. Use load and resistance factor design.

FIGURE 6.23

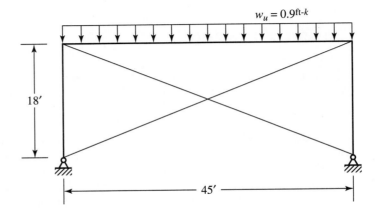

Solution   The total vertical load to be stabilized by the bracing is

$$P_r = \Sigma P_u = 0.9(45) \times 3 = 121.5 \text{ kips}$$

From AISC Equation A-6-1, the lateral shear to be resisted is

$$P_{br} = 0.004P_r = 0.004(121.5) = 0.4860 \text{ kips}$$

If we consider both braces to be tension-only members, then the entire force must be resisted by one brace. Because P_{br} is the horizontal component of the brace force (see Figure 6.24a), the brace force is

$$F = \frac{P_{br}}{\cos \theta}$$

where

$$\theta = \tan^{-1}\left(\frac{18}{45}\right) = 21.80°$$

Then,

$$F = \frac{0.4860}{\cos(21.80)} = 0.5234 \text{ kips}$$

Based on the limit state of tension yielding, the required area is

$$A = \frac{F}{0.9F_y} = \frac{0.5234}{0.9(36)} = 0.0162 \text{ in.}^2$$

From AISC Equation A-6-2, the required lateral stiffness is

$$\beta_{br} = \frac{1}{\phi}\left(\frac{2P_r}{L_b}\right) = \frac{1}{0.75}\left[\frac{2(121.5)}{(18 \times 12)}\right] = 1.500 \text{ kips/in.}$$

FIGURE 6.24

(a)

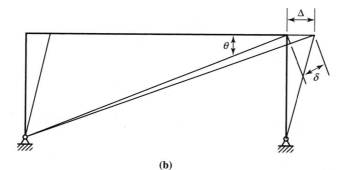

(b)

The axial stiffness of the brace is given by F/δ, where δ is the axial deformation of the brace. From Figure 6.24b,

$$\delta \approx \Delta \cos\theta$$

where Δ is the lateral displacement. The axial stiffness of the brace can be written as

$$\frac{F}{\delta} = \frac{P_{br}/\cos\theta}{\Delta\cos\theta} = \frac{P_{br}}{\Delta} \cdot \frac{1}{\cos^2\theta}$$

or

$$\frac{F}{\delta}\cos^2\theta = \frac{P_{br}}{\Delta}$$

That is,

$$\text{Axial stiffness} \times \cos^2\theta = \text{lateral stiffness}$$

From elementary mechanics of materials, the axial stiffness is AE/L, where

$$L = \text{length of brace} = (18 \times 12)/\sin(21.80) = 581.6 \text{ in.}$$

Let

$$\frac{AE}{L}\cos^2\theta = \frac{1}{\phi}\left(\frac{2P_r}{L_b}\right) = 1.500$$

$$A = \frac{1.500L}{E\cos^2\theta} = \frac{1.500(581.6)}{29,000\cos^2(21.8)} = 0.0349 \text{ in.}^2$$

The stiffness requirement controls. The results are typical in that the requirements are minimal.

Answer   Use a tension brace with a cross-sectional area of at least 0.0349 in.^{2} ■

Notes on an Allowable Strength Solution to Example 6.9

In Example 6.9, the given loads were already factored. In a more realistic situation, the service loads would be known, and the appropriate combination of these would be used for an ASD solution. Other than that, the differences are as follows:

1. The area required for strength is $A = \dfrac{F}{F_y/\Omega}$

2. The required stiffness is $\beta_{br} = \Omega\left(\dfrac{2P_r}{L_b}\right)$

Design of Unbraced Beam–Columns

The preliminary design of beam–columns in braced frames has been illustrated. The amplification factor B_1 was assumed to be equal to 1.0 for purposes of selecting a trial shape; B_1 could then be evaluated for this trial shape. In practice, B_1 will almost always be equal to 1.0. For beam–columns subject to sidesway, the amplification factor B_2 is based on several quantities that may not be known until all columns in the frame have been selected. One of these quantities is ΣP_{e2}. If AISC Equation C2-6a is used, I and K_2L may not be known for all of the members. If AISC Equation C2-6b is used, the sidesway deflection Δ_H may not be available. The following methods are suggested for evaluating B_2.

Method 1.   Assume that $B_2 = 1.0$. After a trial shape has been selected, compute

$$P_{e2} = \frac{\pi^2 EI}{(K_2 L)^2}$$

for the column under consideration, then compute B_2 from AISC Equation C2-3 by assuming that $\Sigma P_{nt}/\Sigma P_{e2}$ is the same as P_{nt}/P_{e2} for the member under consideration (as in Example 6.6).

Method 2.   Use a predetermined limit for the *drift index*, Δ_H/L, the ratio of story drift to story height. The use of a maximum permissible drift index is a serviceability requirement that is similar to a limit on beam deflection. Although no building code or other standard used in the United States contains a limit on the drift index, values of $\frac{1}{500}$ to $\frac{1}{200}$ are commonly used (Ad Hoc Committee on Serviceability, 1986). Remember that Δ_H is the drift caused by ΣH, so if the drift index is based on service loads, then the lateral loads H must also be service loads. Use of a prescribed drift index enables the designer to determine the final value of B_2 at the outset.

Example 6.10

Figure 6.25 shows a single-story, unbraced frame subjected to dead load, roof live load, and windload. The service gravity loads are shown in Figure 6.25a, and the service wind load (including an uplift, or suction, on the roof) is shown in Figure 6.25b. Use A992 steel and select a W12-shape for the columns (vertical members). Design for a drift index of $\frac{1}{400}$ based on service wind load. Bending is about the strong axis, and each column is laterally braced at the top and bottom. Use LRFD.

FIGURE 6.25

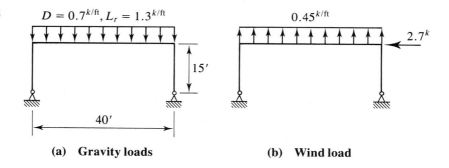

$D = 0.7^{k/ft}, L_r = 1.3^{k/ft}$

$0.45^{k/ft}$

2.7^k

15′

40′

(a)   **Gravity loads**   (b)   **Wind load**

Solution This frame is statically indeterminate to the first degree. The analysis of statically indeterminate structures is not a prerequisite for the use of this book, so the frame will not be analyzed here. The results of an approximate analysis, which is adequate for the early stages of a structural design, are summarized in Figure 6.26. The axial load and end moment are given separately for dead load, live load, wind acting on the roof, and lateral wind load. All vertical loads are symmetrically placed and contribute only to the M_{nt} moments. The lateral load produces an $M_{\ell t}$ moment.

FIGURE 6.26

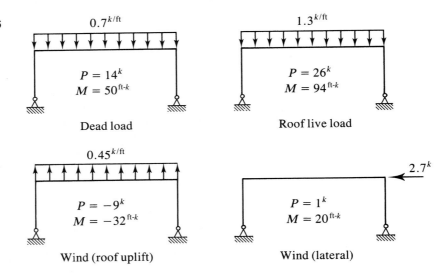

Dead load

Roof live load

Wind (roof uplift)

Wind (lateral)

Load combinations involving dead load, D, roof live load, L_r, and wind, W, are as follows.

Combination 2: $1.2D + 0.5L_r$

$P_u = 1.2(14) + 0.5(26) = 29.8$ kips

$M_{nt} = 1.2(50) + 0.5(94) = 107$ ft-kips

$M_{\ell t} = 0$

Combination 3: $1.2D + 1.6L_r + 0.8W$

$P_u = 1.2(14) + 1.6(26) + 0.8(-9 + 1) = 52.0$ kips

$M_{nt} = 1.2(50) + 1.6(94) + 0.8(-32) = 184.8$ ft-kips

$M_{\ell t} = 0.8(20) = 16.0$ ft-kips

Combination 4: $1.2D + 1.6W + 0.5L_r$

$P_u = 1.2(14) + 1.6(-9 + 1) + 0.5(26) = 17.0$ kips

$M_{nt} = 1.2(50) + 1.6(-32) + 0.5(94) = 55.8$ ft-kips

$M_{\ell t} = 1.6(20) = 32$ ft-kips

Load combination 3 will obviously govern. It produces the largest axial load and the largest total moment. (Combination 4 could not control unless B_2, the amplification factor for $M_{\ell t}$, were unrealistically large.)

For purposes of selecting a trial shape, assume that $B_1 = 1.0$. The value of B_2 can be computed using the design drift index and a value of ΣP_{e2} from AISC Equation C2-6b:

$$\Sigma P_{e2} = R_M \frac{\Sigma HL}{\Delta_H} = R_M \frac{\Sigma H}{\Delta_H/L} = 0.85\left(\frac{2.7}{1/400}\right) = 918.0$$

(The unfactored horizontal load $\Sigma H = 2.7$ kips is used because the drift index is based on the maximum drift caused by *service* loads.)

$$B_2 = \frac{1}{1 - \dfrac{\alpha \Sigma P_{nt}}{\Sigma P_{e2}}} = \frac{1}{1 - \dfrac{1.00(2 \times 52.0)}{918.0}} = 1.128$$

The required moment strength is

$$M_u = B_1 M_{nt} + B_2 M_{\ell t} = 1.0(184.8) + 1.128(16) = 202.8 \text{ ft-kips}$$

Without knowing the frame member sizes, the alignment chart for the effective length factor cannot be used. Table C-C2.2 in the Commentary to the Specification reveals that case (f) corresponds most closely to the end conditions for the sidesway case of this example and that $K_x = 2.0$.

For the braced condition, $K_x = 1.0$ will be used, and because the member is braced in the out-of-plane direction, $K_y = 1.0$ will be used. A shape can be selected by using the method given in Part 6 of the *Manual*. Since the bending moment appears to be dominant in this member, use Equation 6.9.

Try a W12 × 53. From Table 6-1, with $KL = L_b = 15$ft, $p = 2.10 \times 10^{-3}$, and $b_x = 3.45 \times 10^{-3}$.

Check to see which axis controls the axial compressive strength. Form Table 6-1, $r_x/r_y = 2.11$ and

$$\frac{K_x L}{r_x/r_y} = \frac{2.0(15)}{2.11} = 14.2 \text{ ft} < K_y L = 15 \text{ ft} \quad \therefore KL = 15 \text{ ft as assumed}$$

Verify that Equation 6.9 is the correct one:

$$\frac{P_u}{\phi_c P_n} = pP_u = (2.10 \times 10^{-3})(52) = 0.1092 < 0.2$$

$\therefore$ Equation 6.9 controls as assumed

For a bending moment that varies linearly from zero at one end to a maximum at the other, the value of C_b is 1.67 (see Figure 5.15g). Modify b_x to account for C_b.

Since $b_x = \dfrac{8}{9(\phi_b M_{nx})}$, to multiply $\phi_b M_{nx}$ by C_b, divide b_x by C_b:

$$\frac{8}{9C_b(\phi_b M_{nx})} = \frac{b_x}{C_b} = \frac{3.45 \times 10^{-3}}{1.67} = 2.07 \times 10^{-3}$$

To compare $C_b \times \phi_b M_{nx}$ with $\phi_b M_{px}$, compare this modified value of b_x with the value of b_x corresponding to $L_b = 0$, which is

$$b_x = \frac{8}{9(\phi_b M_{px})} = 3.04 \times 10^{-3}$$

Since 2.07×10^{-3} is less than 3.04×10^{-3}, $C_b \times \phi_b M_{nx}$ is greater than $\phi_b M_{px}$, so the correct modified value of b_x is 3.04×10^{-3}.
For the braced condition,

$$P_{e1} = \frac{\pi^2 E I_x}{(K_x L)^2} = \frac{\pi^2 (29{,}000)(425)}{(1.0 \times 15 \times 12)^2} = 3754 \text{ kips}$$

$$C_m = 0.6 - 0.4\left(\frac{M_1}{M_2}\right) = 0.6 - 0.4\left(\frac{0}{M_2}\right) = 0.6$$

From AISC Equation C2-2,

$$B_1 = \frac{C_m}{1 - \dfrac{\alpha P_r}{P_{e1}}} = \frac{C_m}{1 - \dfrac{1.0 P_u}{P_{e1}}} = \frac{0.6}{1 - \dfrac{52}{3754}} = 0.608 < 1.0 \quad \therefore B_1 = 1.0 \text{ as assumed}$$

Since B_1 is the value originally assumed and B_2 will not change, the previously computed value of $M_u = 202.8$ ft-kips is unchanged.
Check Equation 6.9 with $p = 2.10 \times 10^{-3}$ and $b_x = 3.04 \times 10^{-3}$.

$$0.5 p P_u + \left(\frac{9}{8}\right)(b_x M_{ux} + b_y M_{uy}) = 0.5(2.10 \times 10^{-3})(52.0)$$

$$+ \left(\frac{9}{8}\right)[(3.04 \times 10^{-3})(202.5) + 0] \qquad = 0.747 < 1.0 \quad \text{(OK)}$$

This result is significantly smaller than 1.0, so try a shape two sizes smaller.
Try a W12 × 45. From Table 6-1, $p = 3.16 \times 10^{-3}$, $b_x = 4.58 \times 10^{-3}$, and $r_x/r_y = 2.64$

$$\frac{K_x L}{r_x/r_y} = \frac{2.0(15)}{2.64} = 11.36 \text{ ft} < K_y L = 15 \text{ ft} \quad \therefore KL = 15 \text{ ft as assumed}$$

$$\frac{P_u}{\phi_c P_n} = p P_u = (3.16 \times 10^{-3})(52) = 0.1643 < 0.2$$

$\therefore$ Equation 6.9 controls as assumed

Modify b_x to account for C_b.

$$\frac{b_x}{C_b} = \frac{4.58 \times 10^{-3}}{1.67} = 2.74 \times 10^{-3}$$

This is less than b_x for $L_b = 0$, which is 3.69×10^{-3}, so use $b_x = 3.69 \times 10^{-3}$.

For the braced condition,

$$P_{e1} = \frac{\pi^2 EI_x}{(K_x L)^2} = \frac{\pi^2 (29,000)(348)}{(1.0 \times 15 \times 12)^2} = 3074 \text{ kips}$$

From AISC Equation C2-2,

$$B_1 = \frac{C_m}{1 - \dfrac{1.0 P_u}{P_{e1}}} = \frac{0.6}{1 - \dfrac{52.0}{3074}} = 0.610 < 1.0 \quad \therefore \text{ use } B_1 = 1.0$$

From Equation 6.9,

$$0.5 p P_u + \left(\frac{9}{8}\right)(b_x M_{ux} + b_y M_{uy}) = 0.5(3.16 \times 10^{-3})(52.0)$$

$$+ \left(\frac{9}{8}\right)[(3.69 \times 10^{-3})(202.5) + 0] = 0.923 < 1.0 \quad \text{(OK)}$$

Answer   Use a W12×45.

In Example 6.10, limiting the drift index was a design criterion, and there was no freedom to choose the method of computing the amplification factor B_2. If there had been no imposed drift index, a different value of B_2 could have been computed using AISC Equation C2-6a for ΣP_{e2} as follows (using properties of the W12 × 45):

$$\Sigma P_{e2x} = \Sigma \frac{\pi^2 EI}{(K_2 L)^2} = 2 \times \frac{\pi^2 EI_x}{(K_x L)^2} = \frac{2\pi^2 (29,000)(348)}{(2.0 \times 15 \times 12)^2} = 1537 \text{ kips}$$

$$B_2 = \frac{1}{1 - \dfrac{\alpha \Sigma P_{nt}}{\Sigma P_{e2}}} = \frac{1}{1 - \dfrac{1.00(2 \times 52.0)}{1537}} = 1.073$$

6.8   TRUSSES WITH TOP-CHORD LOADS BETWEEN JOINTS

If a compression member in a truss must support transverse loads between its ends, it will be subjected to bending as well as axial compression and is therefore a beam–column. This condition can occur in the top chord of a roof truss with purlins located between the joints. The top chord of an open-web steel joist must also be designed as a beam–column because an open-web steel joist must support uniformly distributed gravity loads on its top chord. To account for loadings of this nature, a truss can be modeled as an assembly of continuous chord members and pin-connected web members. The axial loads and bending moments can then be found by using a method of structural analysis, such as the stiffness method. The magnitude of the moments involved, however, does not usually warrant this degree of sophistication, and in most cases an approximate analysis will suffice. The following procedure is recommended.

1. Consider each member of the top chord to be a fixed-end beam. Use the fixed-end moment as the maximum bending moment in the member. The top chord is actually one continuous member rather than a series of individual pin-connected members, so this approximation is more accurate than treating each member as a simple beam.
2. Add the reactions from this fixed-end beam to the actual joint loads to obtain total joint loads.
3. Analyze the truss with these total joint loads acting. The resulting axial load in the top-chord member is the axial compressive load to be used in the design.

This method is represented schematically in Figure 6.27. Alternatively, the bending moments and beam reactions can be found by treating the top chord as a continuous beam with supports at the panel points.

FIGURE 6.27

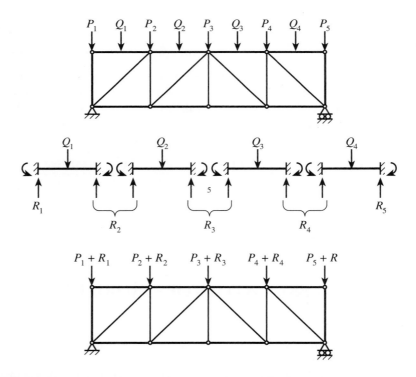

Example 6.11   The parallel-chord roof truss shown in Figure 6.28 supports purlins at the top chord panel points and midway between the panel points. The *factored* loads transmitted by the purlins are as shown. Design the top chord. Use A992 steel and select a structural tee cut from a W-shape. Use load and resistance factor design.

FIGURE 6.28

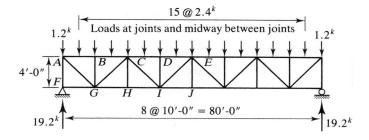

15 @ 2.4^k — Loads at joints and midway between joints

1.2^k   1.2^k

A
4'-0"   B   C D   E
F
     G   H   I   J
        8 @ 10'-0" = 80'-0"
19.2^k   19.2^k

Solution   The bending moments and panel point forces caused by the loads acting between the joints will be found by treating each top-chord member as a fixed-end beam. From Table 3-23, "Shears, Moments and Deflections," in Part 3 of the *Manual,* the fixed-end moment for each top-chord member is

$$M = M_{nt} = \frac{PL}{8} = \frac{2.4(10)}{8} = 3.0 \text{ ft-kips}$$

These end moments and the corresponding reactions are shown in Figure 6.29. When the reactions are added to the loads that are directly applied to the joints, the loading condition shown in Figure 6.29 is obtained. The maximum axial compressive force will occur in member *DE* (and in the adjacent member, to the right of the center of

FIGURE 6.29

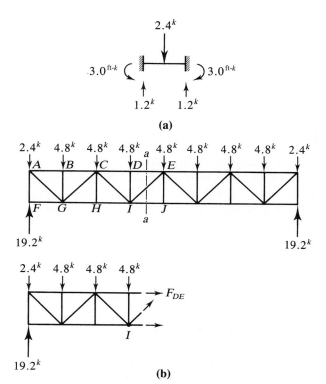

2.4^k

3.0^{ft-k} 3.0^{ft-k}

1.2^k 1.2^k

(a)

2.4^k 4.8^k 4.8^k 4.8^k 4.8^k 4.8^k 4.8^k 4.8^k 2.4^k
  A   B   C   D   E
  F   G   H   I   J

19.2^k 19.2^k

2.4^k 4.8^k 4.8^k 4.8^k

                              F_{DE}

                    I

19.2^k

(b)

the span) and can be found by considering the equilibrium of a free body of the portion of the truss to the left of section a–a:

$$\Sigma M_I = (19.2 - 2.4)(30) - 4.8(10 + 20) + F_{DE}(4) = 0$$

$$F_{DE} = -90 \text{ kips} \qquad \text{(compression)}$$

Design for an axial load of 90 kips and a bending moment of 3.0 ft-kips.

Table 6-1 in Part 6 of the *Manual* is for the design of W shapes only, and there are no corresponding tables for structural tees. An examination of the column load tables in Part 4 of the *Manual* shows that a small shape will be needed, because the axial load is small and the moment is small relative to the axial load. From the column load tables, with $K_x L = 10$ feet and $K_y L = 5$ feet, **try a W6 × 17.5**, with an axial compressive design strength of

$$\phi_c P_n = 149 \text{ kips}$$

A footnote indicates that this shape is slender for compression, but this has been accounted for in the tabulated value of the strength.

Bending is about the x axis, and the member is braced against sidesway, so

$$M_{nt} = 3.0 \text{ ft-kips}, \qquad M_{\ell t} = 0$$

Because there is a transverse load on the member, use $C_m = 1.0$ (the Commentary approach will not be used here). Compute B_1:

$$P_{e1} = \frac{\pi^2 E I_x}{(K_x L)^2} = \frac{\pi^2 (29,000)(16.0)}{(10 \times 12)^2} = 318.0 \text{ kips}$$

$$B_1 = \frac{C_m}{1 - \dfrac{\alpha P_r}{P_{e1}}} = \frac{1.0}{1 - \dfrac{1.00(90)}{318.0}} = 1.395$$

The amplified moment is

$$M_u = B_1 M_{nt} + B_2 M_{\ell t} = 1.395(3.0) + 0 = 4.185 \text{ ft-kips}$$

The nominal moment strength of structural tees with *nonslender* cross sections was introduced in Section 5.16 of this book and is covered in AISC F9. The moment strength of nonslender shapes is based on either yielding of the cross section or lateral-torsional buckling. To determine the classification of the cross section, we check the width–thickness ratios for the flange. From the dimensions and properties tables,

$$\lambda = \frac{b_f}{2t_f} = 6.31$$

$$\lambda_p = 0.38 \sqrt{\frac{E}{F_y}} = 0.38 \sqrt{\frac{29,000}{50}} = 9.152$$

$$\lambda < \lambda_p \quad \therefore \quad \text{this shape is compact}$$

(This could also be determined by observing that there is no footnote in the dimensions and properties table to indicate noncompactness for flexure.)

Check the limit state of yielding. Because the maximum moment is a fixed-end moment, the stem will be in compression, assuming that the flange is at the top. For stems in compression,

$$M_n = M_p = F_y Z \le M_y \qquad \text{(AISC Equation F9-3)}$$

Since $S_x < Z_x$, the yield moment M_y will control.t

$$M_n = M_y = F_y S_x = 50(3.23) = 161.5 \text{ in.-kips}$$

Check lateral-torsional buckling. From AISC Equation F9-5,

$$B = \pm 2.3 \left(\frac{d}{L_b} \right) \sqrt{\frac{I_y}{J}} = -2.3 \left(\frac{6.25}{10 \times 12} \right) \sqrt{\frac{12.2}{0.369}} = -0.6888$$

(The minus sign is used when the stem is in compression anywhere along the unbraced length.) From AISC Equation F9-4,

$$M_n = M_{cr} = \frac{\pi \sqrt{EI_y GJ}}{L_b} \left[B + \sqrt{1 + B^2} \right]$$

$$= \frac{\pi \sqrt{29,000(12.2)(11,200)(0.369)}}{10 \times 12} \left[-0.6888 + \sqrt{1 + (-0.6888)^2} \right]$$

$$= 526.0 \text{ in.-kips}$$

The limit state of yielding controls.

$$\phi_b M_n = 0.90(161.5) = 145.4 \text{ in.-kips} = 12.1 \text{ ft-kips}$$

Determine which interaction equation to use:

$$\frac{P_u}{\phi_c P_n} = \frac{90}{149} = 0.6040 > 0.2 \qquad \therefore \text{ use AISC Equation H1-1a}$$

$$\frac{P_u}{\phi_c P_n} + \frac{8}{9} \left(\frac{M_{ux}}{\phi_b M_{nx}} + \frac{M_{uy}}{\phi_b M_{ny}} \right) = 0.6040 + \frac{8}{9} \left(\frac{4.185}{12.1} + 0 \right)$$

$$= 0.911 < 1.0 \qquad \text{(OK)}$$

Answer   Use a WT6 × 17.5.

Problems

Note   Unless otherwise indicated, all members are laterally braced only at their ends.

Interaction Formulas

6.2-1   Determine whether the given member satisfies the appropriate AISC interaction equation. Do not consider moment amplification. The loads are 25% dead load and 75% live load. Bending is about the x-axis, and the steel is ASTM A992.

a. Use LRFD.
b. Use ASD.

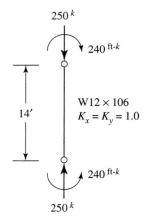

250^k

240^{ft-k}

W12 × 106
$K_x = K_y = 1.0$

14′

240^{ft-k}

250^k

FIGURE P6.2-1

6.2-2 How much service live load, in kips per foot, can be supported? The member weight is the only dead load. The axial compressive load consists of a service dead load of 10 kips and a service live load of 20 kips. Do not consider moment amplification. Bending is about the x-axis, and the steel is A992.

a. Use LRFD.
b. Use ASD.

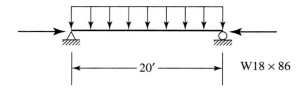

20′   W18 × 86

FIGURE P6.2-2

Members in Braced Frames

6.5-1 Compute the moment amplification factor B_1 for the member of Problem 6.2-1.

a. Use LRFD.
b. Use ASD.

6.5-2 Compute the moment amplification factor B_1 for the member of Problem 6.2-2.

a. Use LRFD.
b. Use ASD.

6.5-3 A W14 × 99 of A992 steel is used as a 14-foot-long beam–column with $K_x = 0.9$ and $K_y = 1.0$. The member is braced against sidesway, the ends are restrained, and there are transverse loads between the ends. The member is subjected to the following service loads and moments: an axial compressive load of 342 kips and a bending moment of 246 ft-kips about the strong axis. The composition of each is 33% dead load and 67% live load. Use $C_b = 1.6$ and determine whether this member satisfies the provisions of the AISC Specification.

a. Use LRFD.
b. Use ASD.

6.5-4 The member shown in Figure P6.5-4 is part of a braced frame. The load and moments are computed from service loads, and bending is about the *x*-axis (the end shears are not shown). The load and moments are 30% dead load and 70% live load. Determine whether this member satisfies the appropriate AISC interaction equation.

a. Use LRFD.
b. Use ASD.

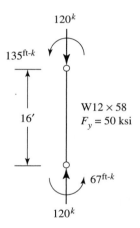

120^k

135^{ft-k}

$16'$

W12 × 58
$F_y = 50$ ksi

67^{ft-k}

120^k

FIGURE P6.5-4

6.5-5 A simply supported beam is subjected to the end couples (bending is about the strong axis) and the axial load shown in Figure P6.5-5. These moments and axial load are from service loads and consist of equal parts dead load and live load. Lateral support is provided only at the ends. Neglect the weight of the beam and investigate this member as a beam–column. Use $F_y = 50$ ksi.

a. Use LRFD.
b. Use ASD.

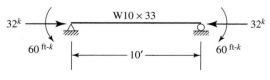

32^k — W10 × 33 — 32^k

60^{ft-k} — 10' — 60^{ft-k}

FIGURE P6.5-5

6.5-6   The beam–column in Figure P6.5-6 is braced against sidesway. Bending is about the major axis, and A992 steel is used. The given loads are service loads, consisting of 25% dead load and 75% live load. Is this member adequate?

a. Use LRFD.
b. Use ASD.

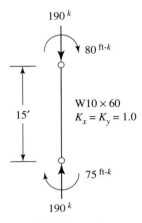

190 ^{k}

80 $^{ft-k}$

15′

W10 × 60
$K_x = K_y = 1.0$

75 $^{ft-k}$

190 ^{k}

FIGURE P6.5-6

6.5-7   The member shown in Figure P6.5-7 has lateral support at points *A*, *B*, and *C*. Bending is about the strong axis. The loads are service loads, and the uniform load includes the weight of the member. A992 steel is used. Is the member adequate?

a. Use LRFD.
b. Use ASD.

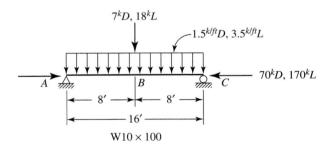

$7^kD, 18^kL$

$1.5^{k/ft}D, 3.5^{k/ft}L$

A B C $70^kD, 170^kL$

8′   8′

16′

W10 × 100

FIGURE P6.5-7

6.5-8   The member shown in Figure P6.5-8 is braced against sidesway. Bending is about the major axis. Does this member satisfy the provisions of the AISC Specification? The loads are 50% live load and 50% dead load.

a. Use LRFD.
b. Use ASD.

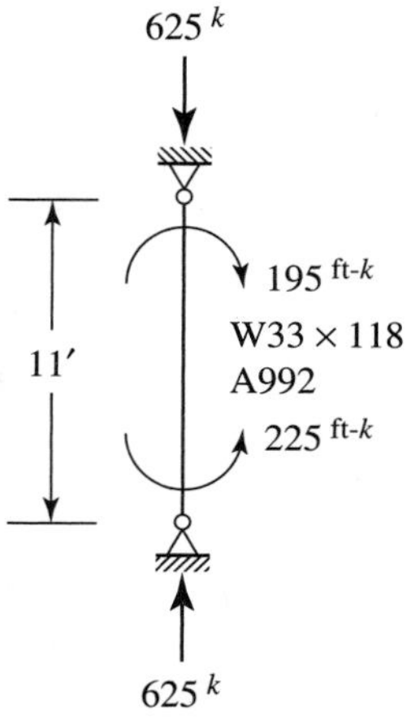

625 ^{k}

195 $^{ft\text{-}k}$

W33 × 118
A992

11′

225 $^{ft\text{-}k}$

625 ^{k}

FIGURE P6.5-8

6.5-9 The member shown in Figure P6.5-9 is a W12 × 96 of A572 Grade 50 steel and is part of a braced frame. The end moments are service load moments, and bending is about the strong axis. The end shears are not shown. If the end moments and axial load are 33% dead load and 67% live load, what is the maximum axial service load, P, that can be applied? Use $K_x = K_y = 1.0$.

a. Use LRFD.
b. Use ASD.

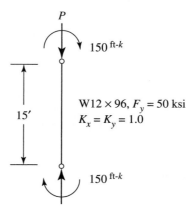

P

150 $^{ft\text{-}k}$

W12 × 96, $F_y = 50$ ksi
$K_x = K_y = 1.0$

15′

150 $^{ft\text{-}k}$

FIGURE P6.5-9

6.5-10 The loads in Figure P6.5-10 are service loads consisting of 25% dead load and 75% live load. A992 steel is used. Is this member satisfactory?

a. Use LRFD.
b. Use ASD.

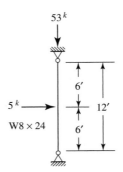

FIGURE P6.5-10

6.5-11 Evaluate the given beam–column for compliance with the AISC Specification. The member is of A992 steel and is part of a braced frame. The axial load and the end moments are unfactored, consisting of 33% dead load and 67% live load (the end shears are not shown). Use $K_x = K_y = 1.0$.

a. Use LRFD.
b. Use ASD.

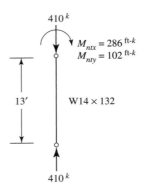

FIGURE P6.5-11

6.5-12 The member in Figure P6.5-12 is laterally supported only at its ends. If A572 Grade 50 steel is used, does the member satisfy the provisions of the AISC Specification? The loads are 50% live load and 50% dead load.

a. Use LRFD.
b. Use ASD.

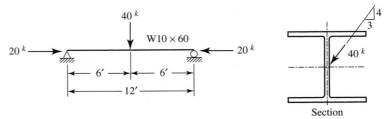

FIGURE P6.5-12

6.5-13 The fixed-end member shown in Figure P6.5-13 is a W21 × 68 of A992 steel. Bending is about the strong axis. What is the maximum permissible value of Q, a service live load? (Moments can be computed from the formulas given in Part 3 of the *Manual*.)

a. Use LRFD.
b. Use ASD.

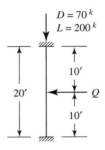

FIGURE P6.5-13

Members in Unbraced Frames

6.6-1 A W14 × 82 of 50 ksi steel is to be investigated for use as a beam–column in an unbraced frame. The length is 14 feet. First-order analyses of the frame were performed for both the sway and the nonsway cases. The factored loads and moments corresponding to one of the load combinations to be investigated are given for this member in the following table:

Type of analysis	P_u (kips)	M_{top} (ft-kips)	M_{bot} (ft-kips)
Nonsway	400	45	24
Sway	—	40	95

Bending is about the strong axis, and all moments cause double-curvature bending (all end moments are in the same direction, that is, all clockwise or all counterclockwise). The following values are also available from the results of a preliminary design:

$$\Sigma P_{e2} = 40,000 \text{ kips}, \qquad \Sigma P_u = 6,000 \text{ kips}$$

Use $K_x = 1.0$ (nonsway case), $K_x = 1.7$ (sway case), and $K_y = 1.0$. Use LRFD and determine whether this member satisfies the provisions of the AISC Specification for the given load combination.

6.6-2 A W14 × 82 of A992 steel, 16 feet long, is used as a column in an unbraced frame. The axial load and end moments obtained from a first-order analysis of the gravity loads (dead load and live load) are shown in Figure P6.6-2a. The frame is symmetric, and the gravity loads are symmetrically placed. Figure P6.6-2b shows the wind load moments obtained from a first-order analysis. All loads and moments are based on service loads, and all bending moments are about the strong axis. The effective length factors are $K_x = 0.85$ for the braced case, $K_x = 1.2$ for the unbraced case, and $K_y = 1.0$. Determine whether this member is in compliance with the AISC Specification.

a. Use LRFD.
b. Use ASD.

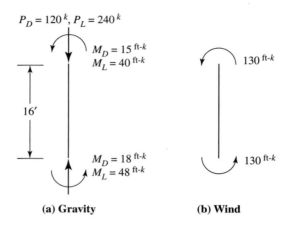

(a) Gravity　　　**(b) Wind**

FIGURE P6.6-2

Design of Beam–Columns

6.7-1　Use $F_y = 50$ ksi and select the lightest W12-shape for the beam–column shown in Figure P6.7-1. The member is part of a braced frame, and the axial load and bending moment are based on service loads consisting of 30% dead load and 70% live load (the end shears are not shown). Bending is about the strong axis, and $K_x = K_y = 1.0$.

a. Use LRFD.
b. Use ASD.

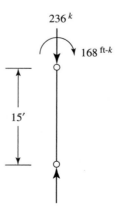

FIGURE P6.7-1

6.7-2 The beam–column in Figure P6.7-2 is part of a braced frame and is subjected to the axial load and end moments shown (the end shears are not shown). The loads and moments are based on service loads consisting of 50% dead load and 50% live load. Use A992 steel and select the lightest W10-shape. Bending is about the strong axis.

a. Use LRFD.
b. Use ASD.

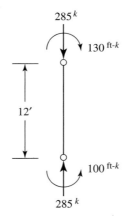

FIGURE P6.7-2

6.7-3 A member is subjected to the loads shown in Figure P6.7-3. The loads are 25% dead load and 75% live load. Bending is about the strong axis, and $K_x = K_y = 1.0$. Use A992 steel and select a W10-shape.

a. Use LRFD.
b. Use ASD.

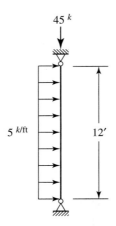

FIGURE P6.7-3

6.7-4 A member in a braced frame supports an axial compressive load and end moments that cause bending about both axes of the member. The axial load and bending moments are computed from service loads consisting of 33% dead load and 67% live load (the end shears are not shown). $K_x = K_y = 1.0$. Use A992 steel and select a W-shape.

a. Use LRFD.
b. Use ASD.

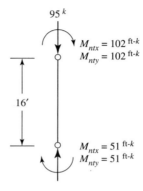

FIGURE P6.7-4

6.7-5 Select a W14 of A992 steel. Bending is about the strong axis, and there is no side-sway. Use $K_x = 0.8$ and $K_y = 1.0$.

a. Use LRFD.
b. Use ASD.

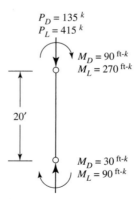

FIGURE P6.7-5

6.7-6 The member shown in Figure P6.7-6 is part of a braced frame. The axial load and end moments are based on service loads composed of equal parts dead load and live load. Select a W-shape of A992 steel.

a. Use LRFD.
b. Use ASD.

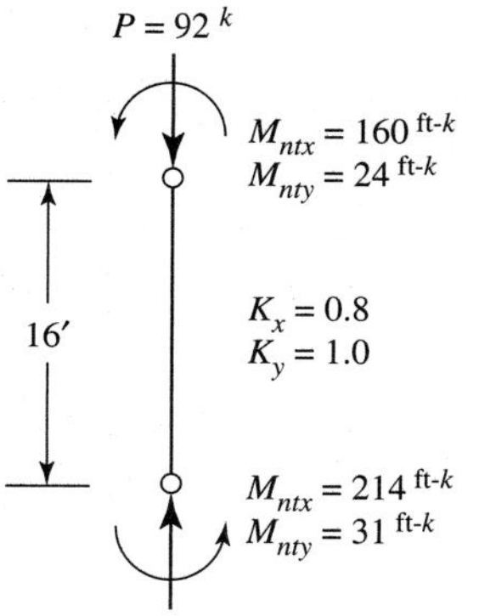

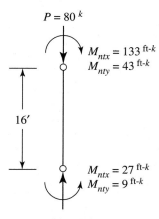

$P = 92^{\,k}$

$M_{ntx} = 160^{\,\text{ft-}k}$
$M_{nty} = 24^{\,\text{ft-}k}$

$K_x = 0.8$
$K_y = 1.0$

16′

$M_{ntx} = 214^{\,\text{ft-}k}$
$M_{nty} = 31^{\,\text{ft-}k}$

FIGURE P6.7-6

6.7-7   Use A992 steel and select a W-shape for the loads and moments shown in Figure P6.7-7. The loads and moments are unfactored and are 25% dead load and 75% live load. The member is part of a braced frame. Use $K_x = K_y = 1.0$.

    a. Use LRFD.
    b. Use ASD.

$P = 80^{\,k}$

$M_{ntx} = 133^{\,\text{ft-}k}$
$M_{nty} = 43^{\,\text{ft-}k}$

16′

$M_{ntx} = 27^{\,\text{ft-}k}$
$M_{nty} = 9^{\,\text{ft-}k}$

FIGURE P6.7-7

6.7-8   Use LRFD and select the lightest W12-shape of A992 steel to be used as a beam–column in an *unbraced* frame. The member length is 16 feet, and the effective length factors are $K_x = 1.0$ (nonsway), $K_x = 2.0$ (sway), and $K_y = 1.0$. Factored loads and moments based on first-order analyses are $P_u = 75$ kips, $M_{nt} = 270$ ft-kips, and $M_{lt} = 30$ ft-kips. Use $C_m = 0.6$ and $C_b = 1.67$. Bending is about the strong axis.

6.7-9   The single-story unbraced frame shown in Figure P6.7-9 is subjected to dead load, roof live load, and wind. The results of an approximate analysis are summarized in the figure. The axial load and end moment are given separately for dead load, roof live load, wind uplift on the roof, and lateral wind load. All vertical loads are symmetrically placed and contribute only to the M_{nt} moments. The lateral load produces

M_{lt} moments. Use A992 steel and select a W14-shape for the columns. Design for a drift index of 1/400 based on service wind load. Bending is about the strong axis, and each column is laterally braced at the top and bottom.

a. Use LRFD.

b. Use ASD.

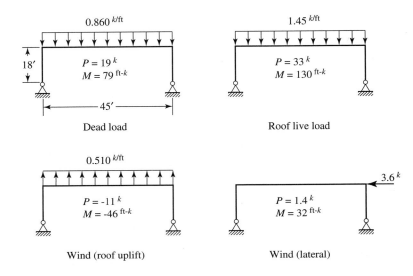

FIGURE P6.7-9

Design of Bracing

6.7-10   Use A36 steel and design single-angle diagonal bracing for the frame of Problem 6.7-9. Assume that the bracing will stabilize three frames.

a. Use LRFD.

b. Use ASD.

Trusses with Top-Chord Loads between Joints

6.8-1   Use $F_y = 50$ ksi and select a structural tee shape for the top chord of the truss shown in Figure P6.8-1. The trusses are spaced at 25 feet and are subjected to the following loads:

Purlins:	W6 × 8.5, located at joints and midway between joints
Snow:	20 psf of horizontal roof surface projection
Metal deck:	2 psf
Roofing:	4 psf
Insulation:	3 psf

a. Use LRFD.

b. Use ASD.

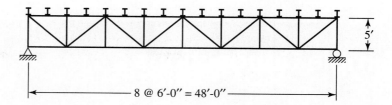

FIGURE P6.8-1

6.8-2 Use A992 steel and select a structural tee shape for the top chord of the truss shown in Figure P6.8-2. This is the truss of Example 3.15. The trusses are spaced at 25 feet on centers and support W6 × 12 purlins at the joints and midway between the joints. The other pertinent data are summarized as follows:

Metal deck:	2 psf
Built-up roof:	5 psf
Snow:	18 psf of horizontal roof projection

a. Use LRFD.
b. Use ASD.

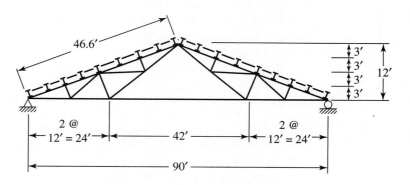

FIGURE P6.8-2

7

Simple Connections

7.1 INTRODUCTION

Connections of structural steel members are of critical importance. An inadequate connection, which can be the "weak link" in a structure, has been the cause of numerous failures. Failure of structural *members* is rare; most structural failures are the result of poorly designed or detailed connections. The problem is compounded by the confusion that sometimes exists regarding responsibility for the design of connections. In many cases, the connections are not designed by the same engineer who designs the rest of the structure, but by someone associated with the steel fabricator who furnishes the material for the project. The structural engineer responsible for the production of the design drawings, however, is responsible for the complete design, including the connections. It is therefore incumbent upon the engineer to be proficient in connection design, if only for the purpose of validating a connection designed by someone else.

Modern steel structures are connected by welding or bolting (either high-strength or "common" bolts) or by a combination of both. Until fairly recently, connections were either welded or riveted. In 1947, the Research Council of Riveted and Bolted Structural Joints was formed, and its first specification was issued in 1951. This document authorized the substitution of high-strength bolts for rivets on a one-for-one basis. Since that time, high-strength bolting has rapidly gained in popularity, and today the widespread use of high-strength bolts has rendered the rivet obsolete in civil engineering structures. There are several reasons for this change. Two relatively unskilled workers can install high-strength bolts, whereas four skilled workers were required for riveting. In addition, the riveting operation was noisy and somewhat dangerous because of the practice of tossing the heated rivet from the point of heating to the point of installation. Riveted connection design is no longer covered by the AISC Specification, but many existing structures contain riveted joints, and the analysis of these connections is required for the strength evaluation and rehabilitation of older structures. Section 5.2.6 of AISC Appendix 5, "Evaluation of Existing Structures," specifies that ASTM A502 Grade 1 rivets should be assumed unless there is evidence to the contrary. Properties of rivets can be found in the ASTM Specification (ASTM, 2005c). The analysis of riveted connections is essentially the same as for connections with common bolts; only the material properties are different.

Welding has several advantages over bolting. A welded connection is often simpler in concept and requires few, if any, holes (sometimes erection bolts may be required to hold the members in position for the welding operation). Connections that

FIGURE 7.1

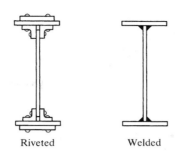

Riveted Welded

are extremely complex with fasteners can become very simple when welds are used. A case in point is the plate girder shown in Figure 7.1. Before welding became widely used, this type of built-up shape was fabricated by riveting. To attach the flange plates to the web plate, angle shapes were used to transfer load between the two elements. If cover plates were added, the finished product became even more complicated. The welded version, however, is elegant in its simplicity. On the negative side, skilled workers are required for welding, and inspection can be difficult and costly. This last disadvantage can be partially overcome by using shop welding instead of field welding whenever possible. Quality welding can be more easily ensured under the controlled conditions of a fabricating shop. When a connection is made with a combination of welds and bolts, welding can be done in the shop and bolting in the field. In the single-plate beam-to-column connection shown in Figure 7.2, the plate is shop welded to the column flange and field bolted to the beam web.

In considering the behavior of different types of connections, it is convenient to categorize them according to the type of loading. The tension member splices shown in Figure 7.3a and b subject the fasteners to forces that tend to shear the shank of the fastener. Similarly, the weld shown in Figure 7.3c must resist shearing forces.

FIGURE 7.2

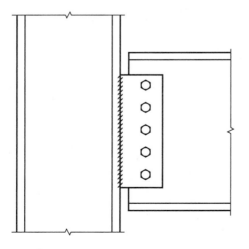
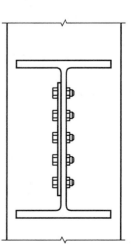

FIGURE 7.3

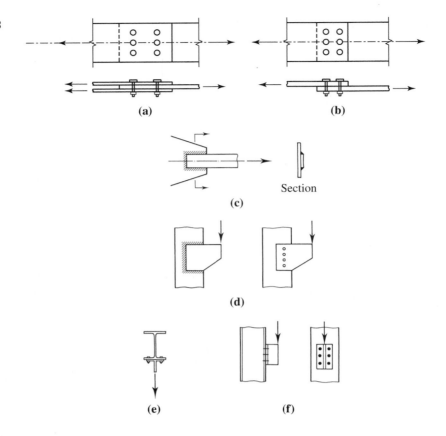

The connection of a bracket to a column flange, as in Figure 7.3d, whether by fasteners or welds, subjects the connection to shear when loaded as shown. The hanger connection shown in Figure 7.3e puts the fasteners in tension. The connection shown in Figure 7.3f produces both shear and tension in the upper row of fasteners. The strength of a fastener depends on whether it is subjected to shear or tension, or both. Welds are weak in shear and are usually assumed to fail in shear, regardless of the direction of loading.

Once the force per fastener or force per unit length of weld has been determined, it is a simple matter to evaluate the adequacy of the connection. This determination is the basis for the two major categories of connections. If the line of action of the resultant force to be resisted passes through the center of gravity of the connection, each part of the connection is assumed to resist an equal share of the load, and the connection is called a *simple* connection. In such connections, illustrated in Figure 7.3a, b, and c, each fastener or each unit length of weld will resist an equal amount of force.[*] The load capacity of the connection can then be found by multiplying the capacity of each fastener or inch of weld by the total number of fasteners or the total

---

[*]There is actually a small eccentricity in the connections of Figure 7.3b and c, but it is usually neglected.

length of weld. This chapter is devoted to simple connections. Eccentrically loaded connections, covered in Chapter 8, are those in which the line of action of the load does not act through the center of gravity of the connection. The connections shown in Figure 7.3d and f are of this type. In these cases, the load is not resisted equally by each fastener or each segment of weld, and the determination of the distribution of the load is the complicating factor in the design of this type of connection.

The AISC Specification deals with connections in Chapter J, "Design of Connections," where bolts and welds are covered.

7.2   BOLTED SHEAR CONNECTIONS: FAILURE MODES

Before considering the strength of specific grades of bolts, we need to examine the various modes of failure that are possible in connections with fasteners subjected to shear. There are two broad categories of failure: failure of the fastener and failure of the parts being connected. Consider the lap joint shown in Figure 7.4a. Failure of the fastener can be assumed to occur as shown. The average shearing stress in this case will be

$$f_v = \frac{P}{A} = \frac{P}{\pi d^2/4}$$

where P is the load acting on an individual fastener, A is the cross-sectional area of the fastener, and d is its diameter. The load can then be written as

$$P = f_v A$$

Although the loading in this case is not perfectly concentric, the eccentricity is small and can be neglected. The connection in Figure 7.4b is similar, but an analysis of free-body diagrams of portions of the fastener shank shows that each cross-sectional area is subjected to half the total load, or, equivalently, two cross sections are effective in resisting the total load. In either case, the load is $P = 2f_v A$, and this loading is called *double shear*. The bolt loading in the connection in Figure 7.4a, with only one shear

FIGURE 7.4

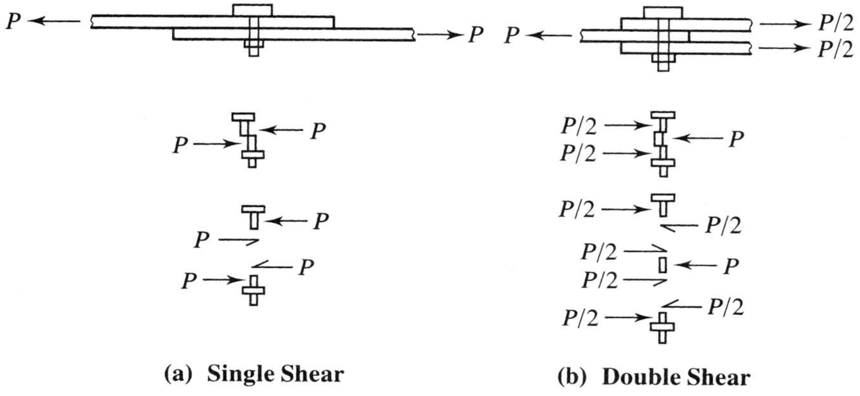

(a) **Single Shear**   (b) **Double Shear**

plane, is called *single shear*. The addition of more thicknesses of material to the connection will increase the number of shear planes and further reduce the load on each plane. However, that will also increase the length of the fastener and could subject it to bending.

Other modes of failure in shear connections involve failure of the parts being connected and fall into two general categories.

1. **Failure resulting from excessive tension, shear, or bending in the parts being connected**. If a tension member is being connected, tension on both the gross area and effective net area must be investigated. Depending on the configuration of the connection, block shear might also need to be considered. Block shear must also be examined in beam-to-column connections in which the top flange of the beam is coped. (We covered block shear in Chapters 3 and 5, and it is also described in AISC J4.3.) Depending on the type of connection and loading, connection fittings such as gusset plates and framing angles may require an analysis for shear, tension, bending, or block shear. The design of a tension member connection will usually be done in parallel with the design of the member itself because the two processes are interdependent.

2. **Failure of the connected part because of bearing exerted by the fasteners**. If the hole is slightly larger than the fastener and the fastener is assumed to be placed loosely in the hole, contact between the fastener and the connected part will exist over approximately half the circumference of the fastener when a load is applied. This condition is illustrated in Figure 7.5. The stress will vary from a maximum at A to zero at B; for simplicity, an average stress, computed as the applied force divided by the projected area of contact, is used.

Thus the bearing stress would be computed as $f_p = P/(dt)$, where P is the force applied to the fastener, d is the fastener diameter, and t is the thickness of the part subjected to the bearing. The bearing load is therefore $P = f_p dt$.

FIGURE 7.5

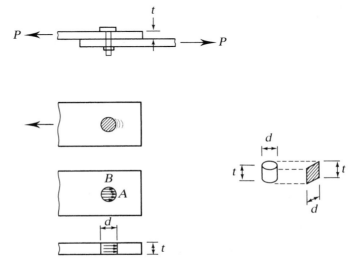

FIGURE 7.6

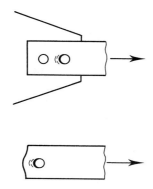

The bearing problem can be complicated by the presence of a nearby bolt or the proximity of an edge in the direction of the load, as shown in Figure 7.6. The bolt spacing and edge-distance will have an effect on the bearing strength.

7.3   BEARING STRENGTH, SPACING, AND EDGE-DISTANCE REQUIREMENTS

Bearing strength is independent of the type of fastener because the stress under consideration is on the part being connected rather than on the fastener. For this reason, bearing strength, as well as spacing and edge-distance requirements, which also are independent of the type of fastener, will be considered before bolt shear and tensile strength.

The AISC Specification provisions for bearing strength, as well as all the requirements for high-strength bolts, are based on the provisions of the specification of the Research Council on Structural Connections (RCSC, 2004). The following discussion, which is based on the commentary that accompanies the RCSC specification, explains the basis of the AISC specification equations for bearing strength.

A possible failure mode resulting from excessive bearing is shear tear-out at the end of a connected element, as shown in Figure 7.7a. If the failure surface is idealized as shown in Figure 7.7b, the failure load on one of the two surfaces is equal to the shear fracture stress times the shear area, or

$$\frac{R_n}{2} = 0.6F_uL_ct$$

where

$0.6F_u$ = shear fracture stress of the connected part
L_c = distance from edge of hole to edge of connected part
t = thickness of connected part

The total strength is

$$R_n = 2(0.6F_uL_ct) = 1.2F_uL_ct \tag{7.1}$$

FIGURE 7.7

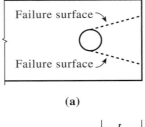

(a)

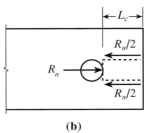

(b)

This tear-out can take place at the edge of a connected part, as shown, or between two holes in the direction of the bearing load. To prevent excessive elongation of the hole, an upper limit is placed on the bearing load given by Equation 7.1. This upper limit is proportional to the projected bearing area times the fracture stress, or

$$R_n = C \times \text{bearing area} \times F_u = CdtF_u \tag{7.2}$$

where

C = a constant
d = bolt diameter
t = thickness of the connected part

The AISC Specification uses Equation 7.1 for bearing strength, subject to an upper limit given by Equation 7.2. If excessive deformation at service load is a concern, and it usually is, C is taken as 2.4. This value corresponds to a hole elongation of about ¼ inch (RCSC, 2004). In this book, we consider deformation to be a design consideration. The nominal bearing strength of a single bolt therefore can be expressed as

$$R_n = 1.2L_c tF_u \le 2.4dtF_u \tag{AISC Equation J3-6a}$$

where

L_c = clear distance, in the direction parallel to the applied load, from the edge of the bolt hole to the edge of the adjacent hole or to the edge of the material
t = thickness of the connected part
F_u = ultimate tensile stress of the connected part (*not* the bolt)

For load and resistance factor design, the resistance factor is $\phi = 0.75$, and the design strength is

$$\phi R_n = 0.75R_n$$

FIGURE 7.8

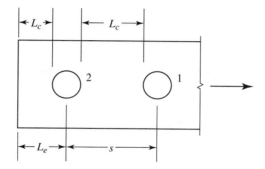

For allowable strength design, the safety factor is $\Omega = 2.00$, and the allowable strength is

$$\frac{R_n}{\Omega} = \frac{R_n}{2.00}$$

Figure 7.8 further illustrates the distance L_c. When computing the bearing strength for a bolt, use the distance from that bolt to the adjacent bolt or edge in the direction of the bearing load on the connected part. For the case shown, the bearing load would be on the left side of each hole. Thus the strength for bolt 1 is calculated with L_c measured to the edge of bolt 2, and the strength for bolt 2 is calculated with L_c measured to the edge of the connected part.

For the edge bolts, use $L_c = L_e - h/2$. For other bolts, use $L_c = s - h$, where

L_e = edge-distance to center of the hole
s = center-to-center spacing of holes
h = hole diameter

AISC Equation J3-6a is valid for standard, oversized, short-slotted and long-slotted holes with the slot parallel to the load. We use only standard holes in this book (holes $\frac{1}{16}$-inch larger than the bolt diameter). For those cases where deformation is not a design consideration, and for long-slotted holes with the slot perpendicular to the direction of the load, AISC gives other strength expressions.

When computing the distance L_c, use the actual hole diameter (which is $\frac{1}{16}$-inch larger than the bolt diameter), and do not add the $\frac{1}{16}$ inch as required in AISC D3.2 for computing the net area for tension and shear. In other words, use a hole diameter of

$$h = d + \frac{1}{16} \text{ in.}$$

not $d + \frac{1}{8}$ inch (although if $d + \frac{1}{8}$ were used, the slight error would be on the conservative side).

Spacing and Edge-Distance Requirements

To maintain clearances between bolt nuts and to provide room for wrench sockets, AISC J3.3 requires that center-to-center spacing of fasteners (in any direction) be no less than $2\frac{2}{3}d$ and preferably no less than $3d$, where d is the fastener diameter.

FIGURE 7.9

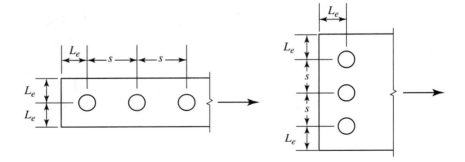

Minimum edge distances (in any direction), measured from the center of the hole, are given in AISC Table J3.4 as a function of bolt size and type of edge—sheared, rolled, or gas cut. The spacing and edge distance to be considered, denoted s and L_e, are illustrated in Figure 7.9.

Summary of Bearing Strength, Spacing, and Edge-Distance Requirements (Standard Holes)

   a. Bearing strength:

   $$R_n = 1.2L_c tF_u \leq 2.4dtF_u \qquad\qquad\text{(AISC Equation J3-6a)}$$

   b. Minimum spacing and edge distance: In any direction, both in the line of force and transverse to the line of force,

   $s \geq 2\tfrac{2}{3}d$ (preferably $3d$)
   $L_e \geq$ value from AISC Table J3.4

   For single- and double-angle shapes, the usual gage distances given in Part 1 of the *Manual* (see Section 3.6) may be used in lieu of these minimums.

Example 7.1   Check bolt spacing, edge distances, and bearing for the connection shown in Figure 7.10.

FIGURE 7.10

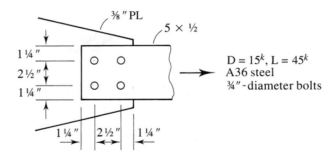

Solution   From AISC J3.3, the minimum spacing in any direction is

$$2\tfrac{2}{3}d = 2.667\left(\frac{3}{4}\right) = 2.00 \text{ in.}$$

Actual spacing = 2.50 in. > 2.00 in.   (OK)

The minimum edge distance in any direction is obtained from AISC Table J3.4. If we assume sheared edges (the worst case), the minimum edge distance is 1¼ in., so

$$\text{Actual edge distance} = 1\frac{1}{4} \text{ in.}   \text{(OK)}$$

For computation of the bearing strength, use a hole diameter of

$$h = d + \frac{1}{16} = \frac{3}{4} + \frac{1}{16} = \frac{13}{16} \text{ in.}$$

Check bearing on both the tension member and the gusset plate. For the tension member and the holes nearest the edge of the member,

$$L_c = L_e - \frac{h}{2} = 1.25 - \frac{13/16}{2} = 0.8438 \text{ in.}$$
$$R_n = 1.2L_c t F_u \le 2.4 dt F_u$$
$$1.2L_c t F_u = 1.2(0.8438)\left(\frac{1}{2}\right)(58) = 29.36 \text{ kips}$$

Check upper limit:

$$2.4 dt F_u = 2.4\left(\frac{3}{4}\right)\left(\frac{1}{2}\right)(58) = 52.20 \text{ kips}$$

29.36 kips < 52.20 kips   ∴ use $R_n = 29.36$ kips/bolt

(This result means that L_c is small enough so that it must be accounted for.)
For the other holes,

$$L_c = s - h = 2.5 - \frac{13}{16} = 1.688 \text{ in.}$$
$$R_n = 1.2L_c t F_u \le 2.4 dt F_u$$
$$1.2L_c t F_u = 1.2(1.688)\left(\frac{1}{2}\right)(58) = 58.74 \text{ kips}$$

Upper limit (the upper limit is independent of L_c and is the same for all bolts):

$$2.4 dt F_u = 52.20 \text{ kips} < 58.74 \text{ kips}   \therefore \text{ use } R_n = 52.20 \text{ kips/bolt}$$

(This result means that L_c is large enough so that it does not need to be accounted for. Hole deformation controls.)

The bearing strength for the tension member is

$$R_n = 2(29.36) + 2(52.20) = 163.1 \text{ kips}$$

For the gusset plate and the holes nearest the edge of the plate,

$$L_c = L_e - \frac{h}{2} = 1.25 - \frac{13/16}{2} = 0.8438 \text{ in.}$$

$$R_n = 1.2L_c tF_u \le 2.4dtF_u$$

$$1.2L_c tF_u = 1.2(0.8438)\left(\frac{3}{8}\right)(58) = 22.02 \text{ kips}$$

$$\text{Upper limit} = 2.4dtF_u = 2.4\left(\frac{3}{4}\right)\left(\frac{3}{8}\right)(58)$$

$$= 39.15 \text{ kips} > 22.02 \text{ kips} \qquad \therefore \text{ use } R_n = 22.02 \text{ kips/bolt}$$

For the other holes,

$$L_c = s - h = 2.5 - \frac{13}{16} = 1.688 \text{ in.}$$

$$R_n = 1.2L_c tF_u \le 2.4dtF_u$$

$$1.2L_c tF_u = 1.2(1.688)\left(\frac{3}{8}\right)(58) = 44.06 \text{ kips}$$

$$\text{Upper limit} = 2.4dtF_u = 39.15 \text{ kips} < 44.06 \text{ kips} \qquad \therefore \text{ use } R_n = 39.15 \text{ kips/bolt}$$

The bearing strength for the gusset plate is

$$R_n = 2(22.02) + 2(39.15) = 122.3 \text{ kips}$$

The gusset plate controls. The nominal bearing strength for the connection is therefore

$$R_n = 122.3 \text{ kips}$$

LRFD Solution   The design strength is $\phi R_n = 0.75(122.3) = 91.7$ kips.

The required strength is

$$R_u = 1.2D + 1.6L = 1.2(15) + 1.6(45) = 90.0 \text{ kips} < 91.7 \text{ kips} \qquad \text{(OK)}$$

ASD Solution   The allowable strength is $\dfrac{R_n}{\Omega} = \dfrac{122.3}{2.00} = 61.2$ kips.

The required strength is

$$R_a = D + L = 15 + 45 = 60 \text{ kips} < 61.2 \text{ kips} \qquad \text{(OK)}$$

Answer   Bearing strength, spacing, and edge-distance requirements are satisfied.   ■

The bolt spacing and edge distances in Example 7.1 are the same for both the tension member and the gusset plate. In addition, the same material is used. Only the thicknesses are different, so the gusset plate will control. In cases such as this one, only the thinner component need be checked. If there is a combination of differences, such as different thicknesses, edge distances, and grades of steel, both the tension member and the gusset plate should be checked.

7.4 SHEAR STRENGTH

While bearing strength is independent of the type of fastener, shear strength is not. In Section 7.2, we saw that the shear load on a bolt is

$$P = f_v A_b$$

where f_v is the shearing stress on the cross-sectional area of the bolt and A_b is the cross-sectional area. When the stress is at its limit, the shear load is the nominal strength, given by

$$R_n = F_{nv} A_b$$

where

F_{nv} = nominal shear stress
A_b = cross-sectional area of the unthreaded part of the bolt (also known as the *nominal bolt area* or *nominal body area*)

The nominal shearing stress depends on the type of bolt material. Structural bolts are available in two general categories: common bolts and high-strength bolts. Common bolts, also known as *unfinished bolts*, are designated as ASTM A307. High-strength bolts are available in two grades: ASTM A325 and ASTM A490. A490 bolts have a higher ultimate tensile strength than A325 bolts and are assigned a higher nominal strength. They were introduced long after A325 bolts had been in general use, primarily for use with high-strength steels (Bethlehem Steel, 1969). A490 bolts are more expensive than A325 bolts, but usually fewer are required. The usual approach is to determine the number of A325 bolts needed in the connection, and if too many are required, use A490 bolts.

The chief distinction between A307 bolts and high-strength bolts, other than the ultimate stress, is that the high-strength bolts can be tightened to produce a predictable tension in the bolt, which can be relied on to produce a calculable clamping force. Although A307 bolts are adequate for many applications, they are rarely used today.

The AISC provisions for high-strength bolts are based in part on the provisions of the specification of the Research Council on Structural Connections (RCSC, 2004). The nominal shear stress is taken as half of the ultimate tensile strength of the bolt. If the threads are in the plane of shear, a reduced (or net) area should be used. The Specification accounts for this by using 80% of the nominal bolt area. Instead of reducing the bolt area, a factor of 0.80 is applied to the nominal stress, resulting in a different nominal stress when the threads are in the shear plane. In this way, the nominal bolt area can be used for both cases. For example, the ultimate tensile

TABLE 7.1

Fastener	Nominal Shear Strength $R_n = F_{nv}A_b$
A307	$24A_b$
A325, threads in plane of shear	$48A_b$
A325, threads not in plane of shear	$60A_b$
A490, threads in plane of shear	$60A_b$
A490, threads not in plane of shear	$75A_b$

strength of an A325 bolt is 120 ksi, so the nominal shear stress with the threads not in the shear plane is

$$F_{nv} = 0.5(120) = 60 \text{ ksi}$$

If the threads are in the shear plane,

$$F_{nv} = 0.8(60) = 48 \text{ ksi}$$

The nominal shear stress of ASTM A307 bolts is based on the assumption that the threads will always be in the plane of shear. The shear strengths of A307, A325, and A490 bolts are summarized in Table 7.1. The values in Table 7.1 are also given in AISC Table J3.2.

AISC Table J3.2 refers to threads in a plane of shear as "*not* excluded from shear planes" and refers to threads not in a plane of shear as "excluded from shear planes." The first category, threads included in the shear plane, is sometimes referred to as connection type "N," and an A325 bolt of this type can be denoted as an A325-N bolt. The designation "X" can be used to indicate that threads are excluded from the plane of shear—for example, an A325-X bolt.

Although it is sometimes possible to determine in advance whether bolt threads will be in the plane of shear, it may depend on such things as which side of the connection the bolt is installed from. When it is not known whether the threads are in the plane of shear, assume that they are and use the lower shear strength. (In most cases, when the higher strength corresponding to threads not in shear is used, some limit state other than bolt shear will control the joint design.) For LRFD, the resistance factor is 0.75, and the design strength is

$$\phi R_n = 0.75 F_{nv}A_b$$

For ASD, the safety factor is 2.00, and the allowable strength is

$$\frac{R_n}{\Omega} = \frac{F_{nv}A_b}{2.00}$$

Example 7.2 Determine the strength of the connection shown in Figure 7.11, based on bearing and shear, for the following bolts:

 a. A307
 b. A325, threads in the plane of shear
 c. A325, threads not in the plane of shear

FIGURE 7.11

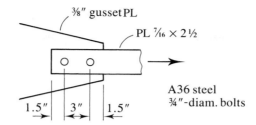

⅜″ gusset PL

PL ⁷⁄₁₆ × 2 ½

A36 steel
¾″-diam. bolts

1.5″   3″   1.5″

Solution   The connection can be classified as a simple connection, and each fastener can be considered to resist an equal share of the load. Because the bearing strength will be the same for parts a, b, and c, it will be calculated first.

Since the edge distances are the same for both the tension member and the gusset plate, the bearing strength of the gusset plate will control because it is thinner than the tension member. For bearing strength computation, use a hole diameter of

$$h = d + \frac{1}{16} = \frac{3}{4} + \frac{1}{16} = \frac{13}{16} \text{ in.}$$

For the hole nearest the edge of the gusset plate,

$$L_c = L_e - \frac{h}{2} = 1.5 - \frac{13/16}{2} = 1.094 \text{ in.}$$

$$R_n = 1.2 L_c t F_u = 1.2(1.094)\left(\frac{3}{8}\right)(58) = 28.55 \text{ kips}$$

subject to a maximum of

$$2.4 d t F_u = 2.4\left(\frac{3}{4}\right)\left(\frac{3}{8}\right)(58)$$

$$= 39.15 \text{ kips} > 28.55 \text{ kips} \quad \therefore \text{ use } R_n = 28.55 \text{ kips for this bolt}$$

For the other hole,

$$L_c = s - h = 3 - \frac{13}{16} = 2.188 \text{ in.}$$

$$R_n = 1.2 L_c t F_u = 1.2(2.188)\left(\frac{3}{8}\right)(58) = 57.11 \text{ kips}$$

The upper limit is

$$2.4 d t F_u = 39.15 \text{ kips} < 57.11 \text{ kips} \quad \therefore \text{ use } R_n = 39.15 \text{ kips for this bolt}$$

Note that the upper limit is independent of L_c and is the same for all bolts.

The nominal bearing strength for the connection is

$$R_n = 28.55 + 39.15 = 67.70 \text{ kips}$$

This value will be compared with the shear strength of each of the three types of bolts.

a. The bolts in this connection are subject to single shear, and the nominal strength of one bolt is

$$R_n = F_{nv}A_b$$

The nominal bolt area is

$$A_b = \frac{\pi d^2}{4} = \frac{\pi (3/4)^2}{4} = 0.4418 \text{ in.}^2$$

For A307 bolts, the nominal shear stress is $F_{nv} = 24$ ksi, and

$$R_n = 24(0.4418) = 10.60 \text{ kips}$$

For two bolts,

$$R_n = 2(10.60) = 21.20 \text{ kips}$$

The shear strength controls since it is less than the bearing strength.

Answer   For LRFD, the design strength is $\phi R_n = 0.75(21.20) = 15.9$ kips.

For ASD, the allowable strength is $\dfrac{R_n}{\Omega} = \dfrac{21.20}{2.00} = 10.6$ kips.

b. For A325 bolts with the threads in the plane of shear, the nominal shear stress is $F_{nv} = 48$ ksi, and

$$R_n = F_{nv}A_b = 48(0.4418) = 21.21 \text{ kips}$$

For two bolts,

$$R_n = 2(21.21) = 42.42 \text{ kips}$$

The shear strength controls since it is less than the bearing strength.

Answer   For LRFD, the design strength is $\phi R_n = 0.75(42.42) = 31.8$ kips.

For ASD, the allowable strength is $\dfrac{R_n}{\Omega} = \dfrac{42.42}{2.00} = 21.2$ kips.

c. For A325 bolts with the threads not in the plane of shear, the nominal shear stress is $F_{nv} = 60$ ksi, and

$$R_n = F_{nv}A_b = 60(0.4418) = 26.51 \text{ kips}$$

For two bolts,

$$R_n = 2(26.51) = 53.02 \text{ kips}$$

As in parts a and b, the shear strength controls.

Answer   For LRFD, the design strength is $\phi R_n = 0.75(53.02) = 39.8$ kips.

For ASD, the allowable strength is $\dfrac{R_n}{\Omega} = \dfrac{53.02}{2.00} = 26.5$ kips.

All spacing and edge-distance requirements are satisfied. For a sheared edge, the minimum edge distance required by AISC Table J3.4 is $1\frac{1}{2}$ inches, and this requirement is satisfied in both the longitudinal and transverse directions. The bolt spacing s is 3 inches, which is greater than $2\frac{2}{3}d = 2.667(\frac{3}{4}) = 2$ in.

◾

Note that some other limit state that has not been checked, such as tension on the net area of the bar, may govern the strength of the connection of Example 7.2.

Example 7.3   A plate $\frac{3}{8} \times 6$ is used as a tension member to resist a service dead load of 12 kips and a service live load of 33 kips. This member will be connected to a $\frac{3}{8}$-inch gusset plate with $\frac{3}{4}$-inch-diameter A325 bolts. A36 steel is used for both the tension member and the gusset plate. Assume that the bearing strength is adequate, and determine the number of bolts required based on bolt shear.

LRFD Solution   The factored load is

$$P_u = 1.2D + 1.6L = 1.2(12) + 1.6(33) = 67.2 \text{ kips}$$

Compute the capacity of one bolt. It is not known whether the bolt threads are in the plane of shear, so conservatively assume that they are. The nominal shear strength is therefore

$$R_n = F_{nv}A_b = 48(0.4418) = 21.21 \text{ kips}$$

and the design strength is

$$\phi R_n = 0.75\,(21.21) = 15.91 \text{ kips per bolt}$$

The number of bolts required is

$$\frac{67.2 \text{ kips}}{15.91 \text{ kips/bolt}} = 4.22 \text{ bolts}$$

Answer   Use five $\frac{3}{4}$-inch-diameter A325 bolts.

ASD Solution   The total load is

$$P_a = D + L = 12 + 33 = 45 \text{ kips}$$

Compute the capacity of one bolt. It is not known whether the bolt threads are in the plane of shear, so conservatively assume that they are. The nominal shear strength is therefore

$$R_n = F_{nv}A_b = 48(0.4418) = 21.21 \text{ kips}$$

and the allowable strength is

$$\frac{R_n}{\Omega} = \frac{21.21}{2.00} = 10.61 \text{ kips per bolt}$$

The number of bolts required is

$$\frac{45 \text{ kips}}{10.61 \text{ kips/bolt}} = 4.24 \text{ bolts}$$

Answer   Use five $\frac{3}{4}$-inch-diameter A325 bolts.

◾

In Example 7.3, the bearing strength was assumed to be adequate. In an actual design situation, the spacing and edge distances would be selected once the required number of bolts was determined, and then the bearing strength could be checked. If the bearing strength was inadequate, the spacing and edge distances could be changed, or more bolts could be used. Subsequent examples will illustrate this procedure.

7.5 INSTALLATION OF HIGH-STRENGTH BOLTS

In certain cases, A325 and A490 bolts are installed to such a degree of tightness that they are subjected to extremely large tensile forces. For example, the initial tension in a ⅝-inch-diameter, A325 bolt can be as high as 19 kips. A complete list of minimum tension values, for those connections in which a minimum tension is required, is given in AISC Table J3.1, "Minimum Bolt Pretension." Each value is equal to 70% of the minimum tensile strength of the bolt. The purpose of such a large tensile force is to achieve the clamping force illustrated in Figure 7.12. Such bolts are said to be *fully tensioned*.

As a nut is turned and advanced along the threads of a bolt, the connected parts undergo compression and the bolt elongates. The free-body diagrams in Figure 7.12a show that the total compressive force acting on the connected part is numerically

FIGURE 7.12

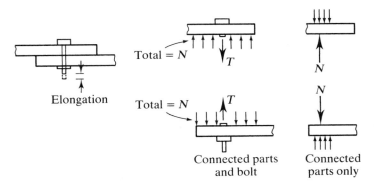

(a) **No external loads**

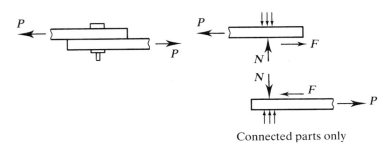

(b) **External load applied**

equal to the tension in the bolt. If an external load P is applied, a friction force will develop between the connected parts. The maximum possible value of this force is

$$F = \mu N$$

where μ is the coefficient of static friction between the connected parts, and N is the normal compressive force acting on the inner surfaces. The value of μ will depend on the surface condition of the steel—for example, whether it is painted or whether rust is present. Thus each bolt in the connection is capable of resisting a load of $P = F$, even if the bolt shank does not bear on the connected part. As long as this frictional force is not exceeded, there is no bearing or shear. If P is greater than F and slippage occurs, shear and bearing will then exist and will affect the capacity of the connection.

How is this high tension accurately achieved? There are currently four authorized procedures for the installation of high-strength bolts (RCSC, 2004).

1. **Turn-of-the-nut method**. This procedure is based on the load-deformation characteristics of the fastener and the connected parts. One full turn of a nut corresponds to a fixed length of travel along the bolt threads, which can be correlated to the elongation of the bolt. The stress–strain relationship for the bolt material can then be used to compute the tension in the bolt.

   For any size and type of bolt, therefore, the number of turns of the nut required to produce a given tensile force can be computed. Table 8.2 in the high-strength bolt specification (RCSC, 2004) gives the required nut rotation for various bolt sizes in terms of the ratio of length to diameter. The specified rotation is from a snug position, with *snug* being defined as the tightness after a few impacts with an impact wrench or after the full effort of one worker with an ordinary spud wrench, with all parts of the connection in firm contact. In spite of all the uncertainties and variables involved, the turn-of-the-nut method has proved to be reliable and surprisingly accurate.

2. **Calibrated wrench tightening**. Torque wrenches are used for this purpose. The torque required to attain a specified tension in a bolt of a given size and grade is determined by tightening this bolt in a tension-indicating device. This calibration must be done daily during construction for bolts of each size and grade.

3. **Twist-off-type bolts**. These are specially designed A325 and A490 bolts whose tips twist off when the proper tension has been achieved. Special wrenches are required for their installation. Inspection of this type of bolt installation is particularly easy.

4. **Direct tension indicators**. The most common of these devices is a washer with protrusions on its surface. When the bolt is tightened, the protrusions are compressed in proportion to the tension in the bolt. A prescribed amount of deformation can be established for any bolt, and when that amount has been achieved, the bolt will have the proper tension. The deformation can be determined by measuring the gap between the nut or bolt head and the undeformed part of the washer surface. Inspection of the bolt installation is also simplified when this type of direct tension indicator is used, as only a feeler gage is required.

Not all high-strength bolts need to be tightened to the fully tensioned condition. AISC J3.1 permits the bolts in some connections to be snug tight. These include bearing-type connections (see Section 7.6 of this book), most tension connections (Section 7.8), and most combined shear and tension connections (Section 7.9). AISC J1.10 describes the connections for which fully tensioned bolts are *required*.

7.6   SLIP-CRITICAL AND BEARING-TYPE CONNECTIONS

A connection with high-strength bolts is classified as either a *slip-critical* connection or a *bearing-type* connection. A slip-critical connection is one in which no slippage is permitted—that is, the friction force must not be exceeded. In a bearing-type connection, slip is acceptable, and shear and bearing actually occur. In some types of structures, notably bridges, the load on connections can undergo many cycles of reversal. In such cases, fatigue of the fasteners can become critical if the connection is allowed to slip with each reversal, and a slip-critical connection is advisable. In most structures, however, slip is perfectly acceptable, and a bearing-type connection is adequate. (A307 bolts are used only in bearing-type connections.) Proper installation and achievement of the prescribed initial tension is necessary for slip-critical connections. In bearing-type connections, the only practical requirement for the installation of the bolts is that they be tensioned enough so that the surfaces of contact in the connection firmly bear on one another. This installation produces the snug-tight condition referred to earlier in the discussion of the turn-of-the-nut method.

As discussed earlier, the resistance to slip will be a function of the product of the coefficient of static friction and the normal force between the connected parts. This relationship is reflected in the provisions of the AISC Specification. The nominal slip resistance of a bolt is given by

$$R_n = \mu D_u h_{sc} T_b N_s \qquad\qquad \text{(AISC Equation J3-4)}$$

where

μ = mean slip coefficient (coefficient of static friction) = 0.35 for Class A surfaces

D_u = ratio of mean actual bolt pretension to the specified minimum pretension. This is to be taken as 1.13 unless another factor can be justified.

h_{sc} = hole factor = 1.0 for standard holes

T_b = minimum fastener tension from AISC Table J3.1

N_s = number of slip planes (shear planes)

A Class A surface is one with clean mill scale (mill scale is an iron oxide that forms on the steel when it is produced). The Specification covers other surfaces, but in this book we conservatively use Class A surfaces, which are assigned the smallest slip coefficient. Although the Specification covers hole sizes other than standard, in this book we consider only standard holes, so the hole factor h_{sc} is always equal to 1.0.

Slip can be considered either a serviceability limit state or a strength limit state. AISC J3.8 specifies different resistance factors and safety factors for these two different conditions. If slip is treated as a serviceability limit state, then

$$\phi = 1.0 \quad \text{and} \quad \Omega = 1.50$$

If slip is treated as a strength limit state,

$$\phi = 0.85 \quad \text{and} \quad \Omega = 1.76$$

Slip-critical connections with standard holes *must* be designed for the serviceability condition. Since only standard holes are used in this book, we will always use $\phi = 1.0$ and $\Omega = 1.50$. Shear and bearing must also be checked in slip-critical connections (AISC J3.8).

Example 7.4   The connection shown in Figure 7.13a uses ¾-inch-diameter, A325 bolts with the threads in the shear plane. No slip is permitted. Both the tension member and the gusset plate are of A36 steel. Determine the strength of the connection.

Solution   Both the design strength (LRFD) and the allowable strength (ASD) will be computed. For efficiency, the nominal strength for each limit state will be computed before specializing the solution for LRFD and ASD.

Shear strength: For one bolt,

$$A_b = \frac{\pi(3/4)^2}{4} = 0.4418 \text{ in.}^2$$

$$R_n = F_{nv}A_b = 48(0.4418) = 21.21 \text{ kips/bolt}$$

For four bolts,

$$R_n = 4(21.21) = 84.84 \text{ kips}$$

FIGURE 7.13

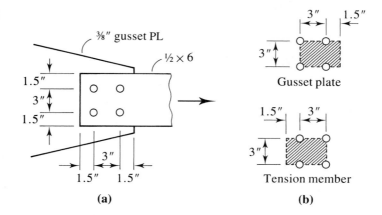

(a)   (b)

Slip-critical strength: Because no slippage is permitted, this connection is classified as slip-critical (and we will treat slip as a serviceability limit state). From AISC Table J3-1, the minimum bolt tension is $T_b = 28$ kips. From AISC Equation J3-4,

$$R_n = \mu D_u h_{sc} T_b N_s = 0.35(1.13)(1.0)(28)(1.0) = 11.07 \text{ kips/bolt}$$

For four bolts,

$$R_n = 4(11.07) = 44.28 \text{ kips}$$

Bearing strength: Since both edge distances are the same, and the gusset plate is thinner than the tension member, the gusset plate thickness of $\frac{3}{8}$ inch will be used. For bearing strength computation, use a hole diameter of

$$h = d + \frac{1}{16} = \frac{3}{4} + \frac{1}{16} = \frac{13}{16} \text{ in.}$$

For the holes nearest the edge of the gusset plate,

$$L_c = L_e - \frac{h}{2} = 1.5 - \frac{13/16}{2} = 1.094 \text{ in.}$$

$$R_n = 1.2 L_c t F_u = 1.2(1.094)\left(\frac{3}{8}\right)(58) = 28.55 \text{ kips}$$

$$\text{Upper limit} = 2.4 d t F_u = 2.4\left(\frac{3}{4}\right)\left(\frac{3}{8}\right)(58)$$

$$= 39.15 \text{ kips} > 28.55 \text{ kips} \quad \therefore \text{ use } R_n = 28.55 \text{ kips for this bolt}$$

For the other holes,

$$L_c = s - h = 3 - \frac{13}{16} = 2.188 \text{ in.}$$

$$R_n = 1.2 L_c t F_u = 1.2(2.188)\left(\frac{3}{8}\right)(58) = 57.11 \text{ kips}$$

$$\text{Upper limit} = 2.4 d t F_u$$

$$= 39.15 \text{ kips} < 57.11 \text{ kips} \quad \therefore \text{ use } R_n = 39.15 \text{ kips for this bolt}$$

The nominal bearing strength for the connection is

$$R_n = 2(28.55) + 2(39.15) = 135.4 \text{ kips}$$

Check the strength of the tension member.

Tension on the gross area:

$$P_n = F_y A_g = 36\left(6 \times \frac{1}{2}\right) = 108.0 \text{ kips}$$

Tension on the net area: All elements of the cross section are connected, so shear lag is not a factor and $A_e = A_n$. For the hole diameter, use

$$h = d + \frac{1}{8} = \frac{3}{4} + \frac{1}{8} = \frac{7}{8} \text{ in.}$$

The nominal strength is

$$P_n = F_u A_e = F_u t(w_g - \Sigma h) = 58 \left(\frac{1}{2} \right) \left[6 - 2 \left(\frac{7}{8} \right) \right] = 123.3 \text{ kips}$$

Block shear strength: The failure block for the gusset plate has the same dimensions as the block for the tension member except for the thickness (Figure 7.13b). The gusset plate, which is the thinner element, will control. There are two shear-failure planes:

$$A_{gv} = 2 \times \frac{3}{8}(3 + 1.5) = 3.375 \text{ in.}^2$$

Since there are 1.5 hole diameters per horizontal line of bolts,

$$A_{nv} = 2 \times \frac{3}{8} \left[3 + 1.5 - 1.5 \left(\frac{7}{8} \right) \right] = 2.391 \text{ in.}^2$$

For the tension area,

$$A_{nt} = \frac{3}{8} \left(3 - \frac{7}{8} \right) = 0.7969 \text{ in.}^2$$

Since the block shear will occur in a gusset plate, $U_{bs} = 1.0$. From AISC Equation J4-5,

$$R_n = 0.6 F_u A_{nv} + U_{bs} F_u A_{nt}$$
$$= 0.6(58)(2.391) + 1.0(58)(0.7969) = 129.4 \text{ kips}$$

with an upper limit of

$$0.6 F_y A_{gv} + U_{bs} F_u A_{nt} = 0.6(36)(3.375) + 1.0(58)(0.7969) = 119.1 \text{ kips}$$

The nominal block shear strength is therefore 119.1 kips.

Design Strength for LRFD

Bolt shear strength:

$$\phi R_n = 0.75(84.84) = 63.6 \text{ kips}$$

Slip-critical strength: Since slip is being treated as a serviceability limit state, $\phi = 1.0$.

$$\phi R_n = 1.0(44.28) = 44.3 \text{ kips}$$

Bearing strength:

$$\phi R_n = 0.75(135.4) = 102 \text{ kips}$$

Tension on the gross area:

$$\phi_t P_n = 0.90(108.0) = 97.2 \text{ kips}$$

Tension on the net area:

$$\phi_t P_n = 0.75(123.3) = 92.5 \text{ kips}$$

Block shear strength:

$$\phi R_n = 0.75(119.1) = 89.3 \text{ kips}$$

Of all the limit states investigated, the strength corresponding to slip is the smallest.

Answer   Design strength = 44.3 kips.

Allowable Strength for ASD

Bolt shear strength:

$$\frac{R_n}{\Omega} = \frac{84.84}{2.00} = 42.4 \text{ kips}$$

Slip-critical strength: Since slip is being treated as a serviceability limit state, $\Omega = 1.50$.

$$\frac{R_n}{\Omega} = \frac{44.28}{1.50} = 29.5 \text{ kips}$$

Bearing strength:

$$\frac{R_n}{\Omega} = \frac{135.4}{2.00} = 67.7 \text{ kips}$$

Tension on the gross area:

$$\frac{P_n}{\Omega_t} = \frac{108.0}{1.67} = 64.7 \text{ kips}$$

Tension on the net area:

$$\frac{P_n}{\Omega_t} = \frac{123.3}{2.00} = 61.7 \text{ kips}$$

Block shear strength:

$$\frac{R_n}{\Omega} = \frac{119.1}{2.00} = 59.6 \text{ kips}$$

Of all the limit states investigated, the strength corresponding to slip is the smallest.

Answer   Allowable strength = 29.5 kips.

7.7 DESIGN EXAMPLES

Although an elementary bolt design was illustrated in Example 7.3, most examples so far have been review or analysis. Examples 7.5–7.7 demonstrate more realistic design situations.

Example 7.5   A $\frac{5}{8}$-inch-thick tension member is connected to two $\frac{1}{4}$-inch splice plates, as shown in Figure 7.14. The loads shown are service loads. A36 steel and $\frac{5}{8}$-inch-diameter, A325 bolts will be used. If slip *is* permissible, how many bolts are required? Each bolt centerline shown represents a row of bolts in the direction of the width of the plates.

FIGURE 7.14

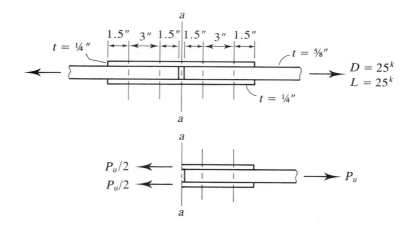

Solution   **Shear:** For shear, the nominal bolt area is

$$A_b = \frac{\pi (5/8)^2}{4} = 0.3068 \text{ in.}^2$$

Assume that the bolt threads are in the planes of shear. Then, the nominal strength for one bolt is

$$R_n = F_{nv} A_b \times 2 \text{ planes of shear} = 48(0.3068)(2) = 29.45 \text{ kips}$$

Bearing: The bearing force on the $\frac{5}{8}$-inch-thick tension member will be twice as large as the bearing force on each of the $\frac{1}{4}$-inch splice plates. Because the total load on the splice plates is the same as the load on the tension member, for the splice plates to be critical the total splice plate thickness must be less than the thickness of the tension member—and it is. Use a hole diameter of

$$h = d + \frac{1}{16} = \frac{5}{8} + \frac{1}{16} = \frac{11}{16} \text{ in.}$$

For the holes nearest the edge of the plate,

$$L_c = L_e - \frac{h}{2} = 1.5 - \frac{11/16}{2} = 1.156 \text{ in.}$$

$$R_n = 1.2L_c t F_u = 1.2(1.156)\left(\frac{1}{4} + \frac{1}{4}\right)(58) = 40.23 \text{ kips}$$

$$\text{Upper limit} = 2.4 dt F_u = 2.4\left(\frac{5}{8}\right)\left(\frac{1}{4} + \frac{1}{4}\right)(58)$$

$$= 43.51 \text{ kips} > 40.23 \text{ kips}$$

$$\therefore \text{ use } R_n = 43.51 \text{ kips for this bolt}$$

For the other holes,

$$L_c = s - h = 3 - \frac{11}{16} = 2.313 \text{ in.}$$

$$R_n = 1.2L_c t F_u = 1.2(2.313)\left(\frac{1}{4} + \frac{1}{4}\right)(58) = 80.49 \text{ kips}$$

$$\text{Upper limit} = 2.4 dt F_u = 43.51 \text{ kips} < 80.49 \text{ kips}$$

$$\therefore \text{ use } R_n = 43.51 \text{ kips for this bolt}$$

The shearing strength per bolt is smaller than both of these bearing values and therefore controls.

LRFD Solution   The design strength per bolt, based on shear, is

$$\phi R_n = 0.75(29.45) = 22.09 \text{ kips}$$

The factored load is

$$P_u = 1.2D + 1.6L = 1.2(25) + 1.6(25) = 70 \text{ kips}$$

$$\text{Number of bolts required} = \frac{\text{total load}}{\text{load per bolt}}$$

$$= \frac{70}{22.09} = 3.17 \text{ bolts}$$

Answer   Use four bolts, two per line, on each side of the splice. A total of eight bolts will be required for the connection.

ASD Solution   The allowable strength per bolt, based on shear, is

$$\frac{R_n}{\Omega} = \frac{29.45}{2.00} = 14.73 \text{ kips}$$

The total load is

$$D + L = 25 + 25 = 50 \text{ kips}$$

The number of bolts required is

$$\frac{\text{total load}}{\text{load per bolt}} = \frac{50}{14.73} = 3.39 \text{ bolts}$$

Answer   Use four bolts, two per line, on each side of the splice. A total of eight bolts will be required for the connection.

■

Example 7.6   The C8 × 18.7 shown in Figure 7.15 has been selected to resist a service dead load of 18 kips and a service live load of 54 kips. It is to be attached to a ⅜-inch gusset plate with ⅞-inch-diameter, A325 bolts. Assume that the threads are in the plane of shear and that slip of the connection is permissible. Determine the number and required layout of bolts such that the length of connection L is a minimum. A36 steel is used.

FIGURE 7.15

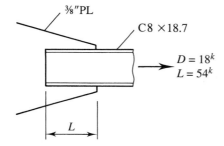

Solution   Determine the nominal capacity of a single bolt. This will be used in both the LRFD and the ASD solutions.

Shear:

$$A_b = \frac{\pi(7/8)^2}{4} = 0.6013 \text{ in.}^2$$

$$R_n = F_{nv}A_b = 48(0.6013) = 28.86 \text{ kips/bolt}$$

Bearing: The gusset plate is thinner than the web of the channel and will control. Assume that along a line parallel to the force, the length L_c is large enough so that the upper limit will control. Then

$$R_n = 2.4dtF_u = 2.4\left(\frac{7}{8}\right)\left(\frac{3}{8}\right)(58) = 45.68 \text{ kips}$$

and shear controls. The bearing strength will need to be verified once the actual bolt layout is determined.

LRFD Solution   The factored load is

$$1.2D + 1.6L = 1.2(18) + 1.6(54) = 108.0 \text{ kips}$$

FIGURE 7.16

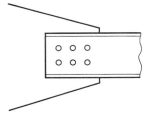

The design strength per bolt, based on shear, is

$$\phi R_n = 0.75(28.86) = 21.65 \text{ kips}$$

The number of bolts required is

$$\frac{108}{21.65} = 4.99 \text{ bolts}$$

Although five bolts will furnish enough capacity, try six bolts so that a symmetrical layout with two gage lines of three bolts each can be used, as shown in Figure 7.16. (Two gage lines are used to minimize the length of the connection.) We do not know whether the design of this tension member was based on the assumption of one line or two lines of fasteners; the tensile capacity of the channel with two lines of bolts must be checked before proceeding. For the gross area,

$$P_n = F_y A_g = 36(5.51) = 198.4 \text{ kips}$$

The design strength is

$$\phi_t P_n = 0.90(198.4) = 179 \text{ kips}$$

Tension on the effective net area:

$$A_n = 5.51 - 2\left(\frac{7}{8} + \frac{1}{8}\right)(0.487) = 4.536 \text{ in.}^2$$

The exact length of the connection is not yet known, so Equation 3.1 for U cannot be used. Assume a conservative value of $U = 0.60$.

$$A_e = A_n U = 4.536(0.60) = 2.722 \text{ in.}^2$$
$$P_n = F_u A_e = 58(2.722) = 157.9 \text{ kips}$$
$$\phi_t P_n = 0.75(157.9) = 118 \text{ kips} \quad \text{(controls)}$$

The member capacity is therefore adequate with two gage lines of bolts.

Check the spacing and edge distance transverse to the load. From AISC J3.3,

$$\text{Minimum spacing} = 2.667\left(\frac{7}{8}\right) = 2.33 \text{ in.}$$

From AISC Table J3.4,

$$\text{Minimum edge distance} = 1\tfrac{1}{8} \text{ in.}$$

A spacing of 3 inches and edge distances of 2½ inches will be used transverse to the load.

The minimum length of the connection can be established by using the minimum permissible spacing and edge distances in the longitudinal direction. The minimum spacing in any direction is $2\frac{2}{3}d = 2.33$ in. **Try 2½ in.** The minimum edge distance in any direction is $1\frac{1}{8}$ in. **Try 1⅛ in.** These distances will now be used to check the bearing strength of the connection. For the bearing strength computation, use a hole diameter of

$$h = d + \frac{1}{16} = \frac{7}{8} + \frac{1}{16} = \frac{15}{16} \text{ in.}$$

For the holes nearest the edge of the gusset plate,

$$L_c = L_e - \frac{h}{2} = 1.125 - \frac{15/16}{2} = 0.6563 \text{ in.}$$

$$R_n = 1.2L_c tF_u = 1.2(0.6563)\left(\frac{3}{8}\right)(58) = 17.13 \text{ kips}$$

$$\text{Upper limit} = 2.4dtF_u = 2.4\left(\frac{7}{8}\right)\left(\frac{3}{8}\right)(58)$$

$$= 45.68 \text{ kips} > 17.13 \text{ kips} \qquad \therefore \text{ use } R_n = 17.13 \text{ kips for this bolt}$$

For the other holes,

$$L_c = s - h = 2.5 - \frac{15}{16} = 1.563 \text{ in.}$$

$$R_n = 1.2L_c tF_u = 1.2(1.563)\left(\frac{3}{8}\right)(58) = 40.79 \text{ kips}$$

$$\text{Upper limit} = 2.4dtF_u$$

$$= 45.68 \text{ kips} > 40.79 \text{ kips} \qquad \therefore \text{ use } R_n = 40.79 \text{ kips for this bolt}$$

The total nominal bearing strength for the connection is

$$R_n = 2(17.13) + 4(40.79) = 197.4 \text{ kips}$$

The design bearing strength is

$$\phi R_n = 0.75(197.4) = 148 \text{ kips} > P_u = 108 \text{ kips} \quad \text{(OK)}$$

The tentative connection design is shown in Figure 7.17 and will now be checked for block shear in the gusset plate (the geometry of the failure block in the channel is identical, but the gusset plate is thinner).

Shear areas:

$$A_{gv} = 2 \times \frac{3}{8}(2.5 + 2.5 + 1.125) = 4.594 \text{ in.}^2$$

$$A_{nv} = 2 \times \frac{3}{8}[6.125 - 2.5(1.0)] = 2.719 \text{ in.}^2$$

Tension area:

$$A_{nt} = \frac{3}{8}(3 - 1.0) = 0.7500 \text{ in.}^2$$

FIGURE 7.17

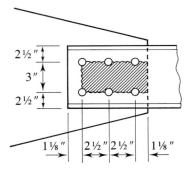

For this type of block shear, $U_{bs} = 1.0$. From AISC Equation J4-5,

$$R_n = 0.6F_u A_{nv} + U_{bs}F_u A_{nt}$$
$$= 0.6(58)(2.719) + 1.0(58)(0.7500) = 138.1 \text{ kips}$$

with an upper limit of

$$0.6F_y A_{gv} + U_{bs}F_u A_{nt} = 0.6(36)(4.594) + 1.0(58)(0.7500) = 142.7 \text{ kips}$$

The nominal block shear strength is therefore 138.1 kips, and the design strength is

$$\phi R_n = 0.75(138.1) = 104 \text{ kips} < 108 \text{ kips} \qquad \text{(N.G.)}$$

The simplest way to increase the block shear strength for this connection is to increase the shear areas by increasing the bolt spacing or the edge distance; we will increase the spacing. Although the required spacing can be determined by trial and error, it can be solved for directly, which we do here. If we assume that the upper limit in AISC Equation J4-5 does not control, the required design strength is

$$\phi R_n = 0.75(0.6F_u A_{nv} + U_{bs}F_u A_{nt})$$
$$= 0.75[0.6(58)A_{nv} + 1.0(58)(0.7500)] = 108 \text{ kips}$$

Required $A_{nv} = 2.888 \text{ in.}^2$

$$A_{nv} = \frac{3}{8}[s + s + 1.125 - 2.5(1.0)](2) = 2.888 \text{ in.}^2$$

Required $s = 2.61$ in. $\qquad \therefore$ use $s = 3$ in.

Compute the actual block shear strength.

$$A_{gv} = 2 \times \frac{3}{8}(3 + 3 + 1.125) = 5.344 \text{ in.}^2$$

$$A_{nv} = 5.344 - \frac{3}{8}(2.5 \times 1.0)(2) = 3.469 \text{ in.}^2$$

$$\phi R_n = 0.75(0.6F_u A_{nv} + U_{bs}F_u A_{nt})$$
$$= 0.75[0.6(58)(3.469) + 1.0(58)(0.7500)] = 123 \text{ kips} > 108 \text{ kips} \quad \text{(OK)}$$

FIGURE 7.18

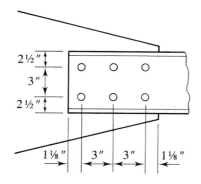

Check the upper limit:

$$\phi[0.6F_y A_{gv} + U_{bs}F_u A_{nt}] = 0.75[0.6(36)(5.344) + 1.0(58)(0.7500)]$$
$$= 119 \text{ kips} < 123 \text{ kips}$$

Therefore, the upper limit controls, but the strength is still adequate.

Using the spacing and edge distances selected, the minimum length is, therefore,

$L = 1\frac{1}{8}$ in. at the end of the channel
$\quad + 2$ spaces at 3 in.
$\quad + 1\frac{1}{8}$ in. at the end of the gusset plate
$\quad = 8\frac{1}{4}$ in. total

Answer   Use the connection detail as shown in Figure 7.18.

ASD Solution   The total load is

$$P_a = D + L = 18 + 54 = 72 \text{ kips}$$

The allowable strength per bolt, based on shear, is

$$\frac{R_n}{\Omega} = \frac{28.86}{2.00} = 14.43 \text{ kips}$$

The number of bolts required is

$$\frac{72}{14.43} = 4.99 \text{ bolts}$$

Although five bolts will furnish enough capacity, try six bolts so that a symmetrical layout with two gage lines of three bolts each can be used, as shown in Figure 7.16. (Two gage lines are used to minimize the length of the connection.) We do not know whether the design of this tension member was based on the assumption of one line or two lines of fasteners; the tensile capacity of the channel with two lines of bolts must be checked before proceeding. For the gross area,

$$P_n = F_y A_g = 36(5.51) = 198.4 \text{ kips}$$

The allowable strength is

$$\frac{P_n}{\Omega_t} = \frac{198.4}{1.67} = 119 \text{ kips}$$

Tension on the effective net area:

$$A_n = 5.51 - 2\left(\frac{7}{8} + \frac{1}{8}\right)(0.487) = 4.536 \text{ in.}^2$$

The exact length of the connection is not yet known, so Equation 3.1 for U cannot be used. Assume a conservative value of $U = 0.60$.

$$A_e = A_n U = 4.536(0.60) = 2.722 \text{ in.}^2$$
$$P_n = F_u A_e = 58(2.722) = 157.9 \text{ kips}$$
$$\frac{P_n}{\Omega_t} = \frac{157.9}{2.00} = 79.0 \text{ kips} \quad \text{(controls)}$$

The member capacity is therefore adequate with two gage lines of bolts.

From the LRFD solution, try a spacing of $s = 2\frac{1}{2}$ inches and an edge distance of $L_e = 1\frac{1}{8}$ inches. With these dimensions, the nominal bearing strength of the connection is (see the LRFD solution)

$$R_n = 197.4 \text{ kips}$$

The allowable bearing strength is

$$\frac{R_n}{\Omega} = \frac{197.4}{2.00} = 98.7 \text{ kips} > P_a = 72 \text{ kips} \quad \text{(OK)}$$

The tentative connection design is shown in Figure 7.17 and will now be checked for block shear in the gusset plate (the geometry of the failure block in the channel is identical, but the gusset plate is thinner).

Shear areas:

$$A_{gv} = 2 \times \frac{3}{8}(2.5 + 2.5 + 1.125) = 4.594 \text{ in.}^2$$

$$A_{nv} = 2 \times \frac{3}{8}[6.125 - 2.5(1.0)] = 2.719 \text{ in.}^2$$

Tension area:

$$A_{nt} = \frac{3}{8}(3 - 1.0) = 0.7500 \text{ in.}^2$$

For this type of block shear, $U_{bs} = 1.0$. From AISC Equation J4-5,

$$R_n = 0.6F_u A_{nv} + U_{bs}F_u A_{nt}$$
$$= 0.6(58)(2.719) + 1.0(58)(0.7500) = 138.1 \text{ kips}$$

with an upper limit of

$$0.6F_y A_{gv} + U_{bs}F_u A_{nt} = 0.6(36)(4.594) + 1.0(58)(0.7500) = 142.7 \text{ kips}$$

The nominal block shear strength is therefore 138.1 kips, and the allowable strength is

$$\frac{R_n}{\Omega} = \frac{138.1}{2.00} = 69.1 \text{ kips} < 72 \text{ kips} \quad (\text{N.G.})$$

The simplest way to increase the block shear strength for this connection is to increase the shear areas by increasing the bolt spacing or the edge distance; we will increase the spacing. Although the required spacing can be determined by trial and error, it can be solved for directly, which we do here. If we assume that the upper limit in AISC Equation J4-5 does not control, the required allowable strength is

$$\frac{R_n}{\Omega} = \frac{0.6F_u A_{nv} + U_{bs} F_u A_{nt}}{\Omega}$$

$$= \frac{0.6(58)A_{nv} + 1.0(58)(0.7500)}{2.00} = 72 \text{ kips}$$

Required $A_{nv} = 2.888$ in.^{2}

$$A_{nv} = \frac{3}{8}[s + s + 1.125 - 2.5(1.0)](2) = 2.888 \text{ in.}^2$$

Required $s = 2.61$ in. $\therefore$ use $s = 3$ in.

Compute the actual block shear strength.

$$A_{gv} = 2 \times \frac{3}{8}(3 + 3 + 1.125) = 5.344 \text{ in.}^2$$

$$A_{nv} = 5.344 - \frac{3}{8}(2.5 \times 1.0)(2) = 3.469 \text{ in.}^2$$

$$\frac{R_n}{\Omega} = \frac{0.6F_u A_{nv} + U_{bs} F_u A_{nt}}{\Omega}$$

$$= \frac{0.6(58)(3.469) + 1.0(58)(0.7500)}{2.00} = 82.1 \text{ kips} > 72 \text{ kips} \quad (\text{OK})$$

Check the upper limit:

$$\frac{0.6F_y A_{gv} + U_{bs} F_u A_{nt}}{\Omega} = \frac{0.6(36)(5.344) + 1.0(58)(0.7500)}{2.00}$$

$$= 79.5 \text{ kips} < 82.1 \text{ kips}$$

Therefore, the upper limit controls, but the strength is still adequate.

Using the spacing and edge distances selected, the minimum length is therefore

$L = 1\frac{1}{8}$ in. at the end of the channel
    $+ 2$ spaces at 3 in.
    $+ 1\frac{1}{8}$ in. at the end of the gusset plate
    $= 8\frac{1}{4}$ in. total

Answer   Use the connection detail as shown in Figure 7.18.

The bolt layout in Example 7.6 is symmetrical with respect to the longitudinal centroidal axis of the member. Consequently, the resultant resisting force provided by the fasteners also acts along this line, and the geometry is consistent with the definition of a simple connection. If an odd number of bolts had been required and two rows had been used, the symmetry would not exist and the connection would be eccentric. In such cases, the designer has several choices: (1) ignore the eccentricity, assuming that the effects are negligible; (2) account for the eccentricity; (3) use a staggered pattern of fasteners that would preserve the symmetry; or (4) add an extra bolt and remove the eccentricity. Most engineers would probably choose the last alternative.

Example 7.7   A 13-foot-long tension member and its connection must be designed for a service dead load of 8 kips and a service live load of 24 kips. No slip of the connection is permitted. The connection will be to a ⅜-inch-thick gusset plate, as shown in Figure 7.19. Use a single angle for the tension member. Use A325 bolts and A572 Grade 50 steel for both the tension member and the gusset plate.

LRFD Solution   The factored load to be resisted is

$$P_u = 1.2D + 1.6L = 1.2(8) + 1.6(24) = 48.0 \text{ kips}$$

Because the bolt size and layout will affect the net area of the tension member, we will begin with selection of the bolts. The strategy will be to select a bolt size for trial, determine the number required, and then try a different size if the number is too large or too small. Bolt diameters typically range from ½ inch to 1½ inches in ⅛-inch increments.

Try ⅝-inch bolts: The nominal bolt area is

$$A_b = \frac{\pi(5/8)^2}{4} = 0.3068 \text{ in.}^2$$

The shear strength is

$$\phi R_n = \phi F_{nv} A_b = 0.75(48)A_b = 0.75(48)(0.3068)$$
$$= 11.04 \text{ kips/bolt} \quad \text{(assuming that the threads are in the shear plane)}$$

No slip is permitted, so this connection is slip-critical. We will assume class A surfaces, and for a ⅝-inch-diameter A325 bolt, the minimum tension is $T_b = 19$ kips

FIGURE 7.19

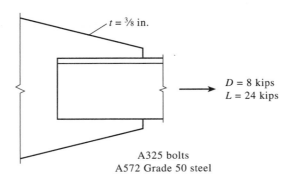

$t = ⅜$ in.

$D = 8$ kips
$L = 24$ kips

A325 bolts
A572 Grade 50 steel

FIGURE 7.20

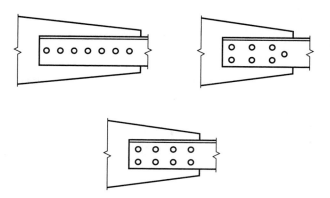

(from AISC Table J3.1). From AISC Equation J3-4, the slip-critical strength for one bolt is

$$\phi R_n = \phi(\mu D_u h_{sc} T_b N_s) = 1.0[0.35(1.13)(1.0)(19)(1)] = 7.515 \text{ kips/bolt}$$

The slip-critical strength controls. We will determine the number of bolts based on this strength and check bearing after selecting the member (because the bearing strength cannot be computed until the member thickness is known). Hence

$$\text{Number of bolts} = \frac{\text{total load}}{\text{load per bolt}} = \frac{48.0}{7.515} = 6.4 \text{ bolts}$$

A minimum of seven bolts will be required. If two rows are used, an extra bolt could be added to maintain symmetry. Figure 7.20 shows several potential bolt layouts. Although any of these arrangements could be used, the connection length can be decreased by using a larger bolt size and fewer bolts.

Try ⅞-inch bolts: The nominal bolt area is

$$A_b = \frac{\pi(7/8)^2}{4} = 0.6013 \text{ in.}^2$$

The shear strength is

$$\phi R_n = 0.75(48)A_b = 0.75(48)(0.6013)$$
$$= 21.65 \text{ kips/bolt} \qquad \text{(assuming that the threads are in the shear plane)}$$

The minimum tension for a ⅞-inch A325 bolt is $T_b = 39$ kips, so the slip-critical strength is

$$\phi R_n = \phi(\mu D_u h_{sc} T_b N_s) = 1.0[0.35(1.13)(1.0)(39)(1)]$$
$$= 15.42 \text{ kips/bolt} \quad \text{(controls)}$$

The number of ⅞-inch bolts required is

$$\frac{48.0}{15.42} = 3.1 \text{ bolts}$$

FIGURE 7.21

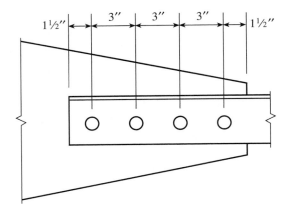

Four $\frac{7}{8}$-inch-diameter A325 bolts will be used. From AISC J3.3, the minimum spacing is

$$s = 2.667d = 2.667\left(\frac{7}{8}\right) = 2.33 \text{ in.} \quad \text{(or, preferably, } 3d = 3\left(\frac{7}{8}\right) = 2.63 \text{ in.)}$$

From AISC Table J3.4, the minimum edge distance is

$$L_e = 1.5 \text{ in.} \qquad \text{(assuming sheared edges)}$$

Try the layout shown in Figure 7.21 and select a tension member. The required gross area is

$$A_g \geq \frac{P_u}{0.90F_y} = \frac{48.0}{0.90(50)} = 1.07 \text{ in.}^2$$

and the required effective net area is

$$A_e \geq \frac{P_u}{0.75F_u} = \frac{48.0}{0.75(65)} = 0.985 \text{ in.}^2$$

The required minimum radius of gyration is

$$r_{min} = \frac{L}{300} = \frac{13(12)}{300} = 0.52 \text{ in.}$$

Try an L3½ × 2½ × ¼:

$$A_g = 1.44 \text{ in.}^2 > 1.07 \text{ in.}^2 \qquad \text{(OK)}$$
$$r_{min} = r_z = 0.541 \text{ in.} > 0.52 \text{ in.} \qquad \text{(OK)}$$

For net area computation, use a hole diameter of $\frac{7}{8} + \frac{1}{8} = 1.0$ in.

$$A_n = A_g - A_{hole} = 1.44 - \left(\frac{7}{8} + \frac{1}{8}\right)\left(\frac{1}{4}\right) = 1.19 \text{ in.}^2$$

Compute U with Equation 3.1:

$$U = 1 - \frac{\bar{x}}{L}$$

$$= 1 - \frac{0.607}{9} = 0.9326$$

where $\bar{x} = 0.607$ inch for the long leg vertical. The effective net area is

$$A_e = A_n U = 1.19(0.9326) = 1.11 \text{ in.}^2 > 0.985 \text{ in.}^2 \qquad \text{(OK)}$$

Now check the bearing strength. The edge distance for the angle is the same as the edge distance for the gusset plate and the angle is thinner than the gusset plate, so the angle thickness of ¼ inch will be used. For bearing strength computation, use a hole diameter of

$$h = d + \frac{1}{16} = \frac{7}{8} + \frac{1}{16} = \frac{15}{16} \text{ in.}$$

For the hole nearest the edge of the gusset plate,

$$L_c = L_e - \frac{h}{2} = 1.5 - \frac{15/16}{2} = 1.031 \text{ in.}$$

$$\phi R_n = \phi(1.2 L_c t F_u) = 0.75(1.2)(1.031)\left(\frac{1}{4}\right)(65) = 15.08 \text{ kips}$$

$$\text{Upper limit} = \phi(2.4 dt F_u) = 0.75(2.4)\left(\frac{7}{8}\right)\left(\frac{1}{4}\right)(65)$$

$$= 25.59 \text{ kips} > 15.08 \text{ kips}$$

$$\therefore \text{ use } \phi R_n = 15.08 \text{ kips for this bolt}$$

For the other holes,

$$L_c = s - h = 3 - \frac{15}{16} = 2.063 \text{ in.}$$

$$\phi(1.2 L_c t F_u) = 0.75(1.2)(2.063)\left(\frac{1}{4}\right)(65) = 30.17 \text{ kips}$$

$$\phi(2.4 dt F_u) = 25.59 \text{ kips} < 30.17 \text{ kips}$$

$$\therefore \text{ use } \phi R_n = 25.59 \text{ kips for this bolt}$$

The total bearing strength for the connection is

$$\phi R_n = 15.08 + 3(25.59) = 91.9 \text{ kips} > P_u = 44.8 \text{ kips} \qquad \text{(OK)}$$

Now check block shear. With the bolts placed in the long leg at the usual gage distance (see Chapter 3, Figure 3.24), the failure block is as shown in Figure 7.22. The shear areas are

$$A_{gv} = \frac{1}{4}(1.5 + 9) = 2.625 \text{ in.}^2$$

$$A_{nv} = \frac{1}{4}[1.5 + 9 - 3.5(1.0)] = 1.750 \text{ in.}^2 \quad (3.5 \text{ hole diameters})$$

FIGURE 7.22

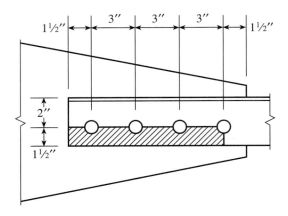

The tension area is

$$A_{nt} = \frac{1}{4}[1.5 - 0.5(1.0)] = 0.2500 \text{ in.}^2 \quad (0.5 \text{ hole diameter})$$

From AISC Equation J4-5,

$$R_n = 0.6F_u A_{nv} + U_{bs}F_u A_{nt}$$
$$= 0.6(65)(1.750) + 1.0(65)(0.2500) = 84.50 \text{ kips}$$

with an upper limit of

$$0.6F_y A_{gv} + U_{bs}F_u A_{nt} = 0.6(50)(2.625) + 1.0(65)(0.2500) = 95.00 \text{ kips}$$

The nominal block shear strength is therefore 84.50 kips, and the design strength is

$$\phi R_n = 0.75(84.50) = 63.4 \text{ kips} > 48.0 \text{ kips} \quad (OK)$$

Answer Use an L3½ × 2½ × ¼ with the long leg connected. Use four ⅞-inch-diameter, A325 bolts, as shown in Figure 7.23.

FIGURE 7.23

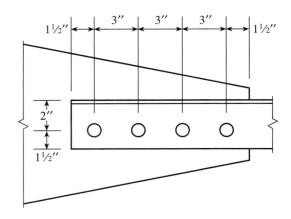

ASD Solution   The total load to be resisted is

$$P_a = D + L = 8 + 24 = 32 \text{ kips}$$

Because the bolt size and layout will affect the net area of the tension member, we will begin with selection of the bolts. The strategy will be to select a bolt size for trial, determine the number required, and then try a different size if the number is too large or too small. Bolt diameters typically range from ½ inch to 1½ inches in ⅛-inch increments.

Try ⅝-inch bolts: The nominal area is

$$A_b = \frac{\pi(5/8)^2}{4} = 0.3068 \text{ in.}^2$$

The allowable shear strength is

$$\frac{R_n}{\Omega} = \frac{F_{nv} A_b}{\Omega} = \frac{48(0.3068)}{2.00} = 7.363 \text{ kips/bolt}$$

(assuming that the threads are in the shear plane). No slip is permitted, so this connection is slip-critical. We will assume Class A surfaces, and for a ⅝-inch-diameter A325 bolt, the minimum tension is $T_b = 19$ kips (from AISC Table J3.1). From AISC Equation J3-4,

$$\frac{R_n}{\Omega} = \frac{\mu D_u h_{sc} T_b N_s}{\Omega} = \frac{[0.35(1.13)(1.0)(19)(1.0)]}{1.50} = 5.010 \text{ kips/bolt}$$

The slip-critical strength controls. We will determine the number of bolts based on this strength and check bearing after selecting the member (because the bearing strength cannot be computed until the member thickness is known). Hence the number of bolts required is

$$\frac{\text{total load}}{\text{load per bolt}} = \frac{32}{5.010} = 6.4 \text{ bolts}$$

To reduce the length of the connection without using two rows of bolts, try a larger bolt size.

Try ⅞-inch bolts: The nominal bolt area is

$$A_b = \frac{\pi(7/8)^2}{4} = 0.6013 \text{ in.}^2$$

The allowable shear strength is

$$\frac{R_n}{\Omega} = \frac{F_{nv} A_b}{\Omega} = \frac{48(0.6013)}{2.00} = 14.43 \text{ kips/bolt}$$

For a ⅞-inch-diameter A325 bolt, the minimum tension is $T_b = 39$ kips (from AISC Table J3.1).

From AISC Equation J3-4, the allowable slip-critical strength is

$$\frac{R_n}{\Omega} = \frac{\mu D h_{sc} T_b N_s}{\Omega} = \frac{[0.35(1.13)(1.0)(39)(1)]}{1.50} = 10.28 \text{ kips/bolt} \quad (\text{controls})$$

The number of bolts required is

$$\frac{32}{10.28} = 3.1 \text{ bolts}$$

Try one row of four bolts. From AISC J3.3,

$$\text{Minimum spacing } s = 2.667 \left(\frac{7}{8}\right) = 2.33 \text{ in. (or preferably, } 3d = 3\left(\frac{7}{8}\right) = 2.63 \text{ in.}$$

From AISC Table J3.4,

Minimum edge distance $L_e = 1\frac{1}{2}$ in. (assuming sheared edges)

Try the layout shown in Figure 7.21 and select a tension member.

$$\text{Required } A_g = \frac{P_a}{F_t} = \frac{P_a}{0.6F_y} = \frac{32}{0.6(50)} = 1.07 \text{ in.}^2$$

$$\text{Required } A_e = \frac{P_a}{0.5F_u} = \frac{32}{0.5(65)} = 0.985 \text{ in.}^2$$

$$\text{Required } r_{min} = \frac{L}{300} = \frac{13(12)}{300} = 0.52 \text{ in.}$$

Try an L3½ × 2½ × ¼:

$$A_g = 1.44 \text{ in.}^2 > 1.07 \text{ in.}^2 \quad (\text{OK})$$

$$r_{min} = r_z = 0.541 \text{ in.} > 0.52 \text{ in.} \quad (\text{OK})$$

$$A_n = A_g - A_{holes} = 1.44 - \left(\frac{7}{8} + \frac{1}{8}\right)\left(\frac{1}{4}\right) = 1.19 \text{ in.}^2$$

Compute U with Equation 3.1:

$$U = 1 - \frac{\overline{x}}{\ell} = 1 - \frac{0.607}{9} = 0.9326$$

$$A_e = A_n U = 1.19(0.9326) = 1.11 \text{ in.}^2 > 0.985 \text{ in.}^2 \quad (\text{OK})$$

Check the bearing strength. The edge distance for the angle is the same as the edge distance for the gusset plate and the angle is thinner than the gusset plate, so the angle thickness of ¼ inch will be used. For bearing strength computation, use a hole diameter of

$$h = d + \frac{1}{16} = \frac{7}{8} + \frac{1}{16} = \frac{15}{16} \text{ in.}$$

For the hole nearest the edge of the gusset plate,

$$L_c = L_e - \frac{h}{2} = 1.5 - \frac{15/16}{2} = 1.031 \text{ in.}$$

$$R_n = 1.2L_c tF_u = 1.2(1.031)\left(\frac{1}{4}\right)(65) = 20.1 \text{ kips}$$

$$\text{Upper limit} = 2.4dtF_u = 2.4\left(\frac{7}{8}\right)\left(\frac{1}{4}\right)(65)$$

$$= 34.13 \text{ kips} > 20.1 \text{ kips} \qquad \therefore \text{ use } R_n = 20.1 \text{ kips for this bolt}$$

For the other holes,

$$L_c = s - h = 3 - \frac{15}{16} = 2.063 \text{ in.}$$

$$R_n = 1.2L_c tF_u = 1.2(2.063)\left(\frac{1}{4}\right)(65) = 40.23 \text{ kips}$$

$$\text{Upper limit} = 2.4dtF_u = 34.13 \text{ kips} < 40.23 \text{ kips}$$

$$\therefore \text{ use } R_n = 34.13 \text{ kips for these bolts}$$

The total nominal bearing strength for the connection is

$$R_n = 20.1 + 3(34.13) = 122.5 \text{ kips}$$

The allowable bearing strength is

$$\frac{R_n}{\Omega} = \frac{122.5}{2.00} = 61.3 \text{ kips} > P_a = 32 \text{ kips} \qquad \text{(OK)}$$

The angle will now be checked for block shear. With the bolts placed in the long leg at the usual gage distance (Chapter 3, Figure 3.24), the failure block is as shown in Figure 7.22.

Shear areas:

$$A_{gv} = \frac{1}{4}(3+3+3+1.5) = 2.625 \text{ in.}^2$$

$$A_{nv} = \frac{1}{4}[3+3+3+1.5-3.5(1.0)] = 1.750 \text{ in.}^2$$

Tension area:

$$A_{nt} = \frac{1}{4}[1.5 - 0.5(1.0)] = 0.2500 \text{ in.}^2$$

From AISC Equation J4-5,

$$R_n = 0.6F_u A_{nv} + U_{bs}F_u A_{nt}$$

$$= 0.6(65)(1.750) + 1.0(65)(0.2500) = 84.50 \text{ kips}$$

with an upper limit of

$$0.6F_y A_{gv} + U_{bs}F_u A_{nt} = 0.6(50)(2.625) + 1.0(65)(0.2500) = 95.00 \text{ kips}$$

The nominal block shear strength is therefore 84.50 kips, and the allowable strength is

$$\frac{R_n}{\Omega} = \frac{84.50}{2.00} = 42.3 \text{ kips} > 32 \text{ kips} \quad \text{(OK)}$$

Answer   Use an **L3½ × 2½ × ¼** with the long leg connected. Use four ⅞-inch-diameter, A325 bolts, as shown in Figure 7.23. ∎

7.8   HIGH-STRENGTH BOLTS IN TENSION

When a tensile load is applied to a bolt with no initial tension, the tensile force in the bolt is equal to the applied load. If the bolt is pretensioned, however, a large part of the applied load is used to relieve the compression, or clamping forces, on the connected parts. This was determined by Kulak, Fisher, and Struik (1987) and is demonstrated here. Figure 7.24 shows a hanger connection consisting of a structural tee-shape bolted to the bottom flange of a W-shape and subjected to a tensile load. A single bolt and a portion of the connected parts will be isolated and examined both before and after loading.

Before loading, all forces are internal, and a free-body diagram of the assembly is as shown in Figure 7.25a. For simplicity, all forces will be assumed to be symmetrical with respect to the axis of the bolt, and any eccentricity will be neglected. If the connected parts are considered as separate free bodies, the forces consist of the bolt tension T_0 and the normal clamping force N_0, shown here as uniformly distributed. Equilibrium requires that T_0 equal N_0. When the external tensile load is applied, the forces on the assembly are as shown in Figure 7.25b, with F representing the total tensile force applied to one bolt (again, the actual distribution of the applied force per bolt has been idealized for simplicity). Figure 7.25c shows the forces acting on a free-body diagram of the segment of the structural tee flange and the corresponding segment of the bolt. Summing forces in the direction of the bolt axis gives

$$T = F + N$$

The application of force F will increase the bolt tension and cause it to elongate by an amount δ_b. Compression in the flange of the structural tee will be reduced, resulting in a distortion δ_{fl} in the same sense and amount as δ_b. A relationship between the applied force and the change in bolt tension can be approximated as follows.

FIGURE 7.24

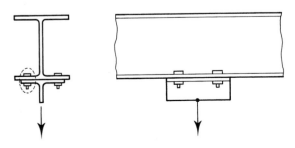

FIGURE 7.25

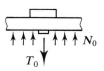

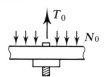

(a)

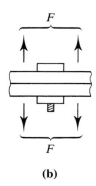

(b)

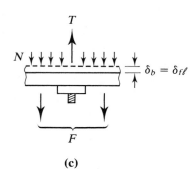

(c)

From elementary mechanics of materials, the axial deformation of an axially loaded uniform member is

$$\delta = \frac{PL}{AE} \tag{7.3}$$

where

     P = applied axial force
     L = original, undeformed length
     A = cross-sectional area
     E = modulus of elasticity

Equation 7.3 can be solved for the load:

$$P = \frac{AE\delta}{L} \tag{7.4}$$

The change in bolt force corresponding to a given axial displacement δ_b therefore is

$$\Delta T = \frac{A_b E_b \delta_b}{L_b} \tag{7.5}$$

where the subscript indicates a property or dimension of the bolt. Application of Equation 7.4 to the compression flange requires a somewhat more liberal interpretation of the load distribution in that N must be treated as if it were uniformly applied over a surface area, $A_{f\ell}$. The change in the force N is then obtained from Equation 7.4 as

$$\Delta N = \frac{A_{f\ell}E_{f\ell}\delta_{f\ell}}{L_{f\ell}} \tag{7.6}$$

where $L_{f\ell}$ is the flange thickness. As long as the connected parts (the two flanges) remain in contact, the bolt deformation, δ_b, and the flange deformation, $\delta_{f\ell}$, will be equal. Because $E_{f\ell}$ approximately equals E_b (Bickford, 1981), and $A_{f\ell}$ is much larger than A_b,

$$\frac{A_{f\ell}E_{f\ell}\delta_{f\ell}}{L_{f\ell}} \gg \frac{A_bE_b\delta_b}{L_b}$$

and therefore

$$\Delta N \gg \Delta T$$

The ratio of ΔT to ΔN is in the range of 0.05 to 0.1 (Kulak, Fisher, and Struik, 1987). Consequently, ΔT will be no greater than $0.1\Delta N$, or equivalently maximum $\Delta T/\Delta N = 0.1$, demonstrating that most of the applied load is devoted to relieving the compression of the connected parts. To estimate the magnitude of the load required to overcome completely the clamping effect and cause the parts to separate, consider the free-body diagram in Figure 7.26. When the parts have separated,

$$T = F$$

or

$$T_0 + \Delta T = F \tag{7.7}$$

At the point of impending separation, the bolt elongation and plate decompression are the same, and

$$\Delta T = \frac{A_bE_b}{L_b}\delta_b = \frac{A_bE_b}{L_b}\delta_{f\ell} \tag{7.8}$$

where $\delta_{f\ell}$ is the elongation corresponding to the release of the initial compression force N_0. From Equation 7.3,

$$\delta_{f\ell} = \frac{N_0L_{f\ell}}{A_{f\ell}E_{f\ell}}$$

FIGURE 7.26

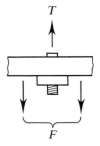

Substituting this expression for $\delta_{f\ell}$ into Equation 7.8 yields

$$\Delta T = \left(\frac{A_b E_b}{L_b}\right)\left(\frac{N_0 L_{f\ell}}{A_{f\ell} E_{f\ell}}\right) = \left(\frac{A_b E_b / L_b}{A_{f\ell} E_{f\ell} / L_{f\ell}}\right) N_0 = \left(\frac{\Delta T}{\Delta N}\right) T_0 \approx 0.1 T_0$$

From Equation 7.7,

$$T_0 + 0.1 T_0 = F \quad \text{or} \quad F = 1.1 T_0$$

Therefore, at the instant of separation, the bolt tension is approximately 10% larger than its initial value at installation. Once the connected parts separate, however, any increase in external load will be resisted entirely by a corresponding increase in bolt tension. If the bolt tension is assumed to be equal to the externally applied force (as if there were no initial tension) and the connection is loaded until the connected parts separate, the bolt tension will be underestimated by less than 10%. Although this 10% increase is theoretically possible, tests have shown that the overall strength of a connection with fasteners in tension is not affected by the installation tension (Amrine and Swanson, 2004). However, the amount of pretension was found to affect the deformation characteristics of the connection.

In summary, the tensile force in the bolt should be computed without considering any initial tension.

Prying Action

In most connections in which fasteners are subjected to tension forces, the flexibility of the connected parts can lead to deformations that increase the tension applied to the fasteners. A hanger connection of the type used in the preceding discussion is subject to this type of behavior. The additional tension is called a *prying force* and is illustrated in Figure 7.27, which shows the forces on a free body of the hanger. Before the external load is applied, the normal compressive force, N_0, is centered on the bolt. As the load is applied, if the flange is flexible enough to deform as shown, the compressive forces will migrate toward the edges of the flange. This redistribution will change the relationship between all forces, and the bolt tension will increase from B_0 to B. If the connected parts are sufficiently rigid, however, this shifting of forces will not occur, and there will be no prying action. The maximum value of the prying force, Q, will be reached when only the corners of the flange remain in contact with the other connected part. The corresponding bolt force, including the effects of prying, is B_c.

In connections of this type, bending caused by the prying action will usually control the design of the connected part. AISC J3.6 requires that prying action be included in the computation of tensile loads applied to fasteners.

A procedure for the determination of prying forces, based on research reported in *Guide to Design Criteria for Bolted and Riveted Joints* (Kulak, Fisher, and Struick, 1987), is given in the *Manual* in Part 9, "Design of Connection Elements." This method is presented here in a somewhat different form, but it gives the same results.

The method used is based on the model shown in Figure 7.28. All forces are for one fastener. Thus, T is the external tension force applied to one bolt, Q is the prying

FIGURE 7.27

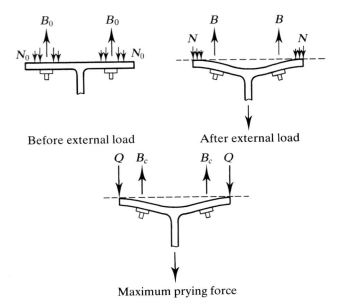

Before external load

After external load

Maximum prying force

FIGURE 7.28

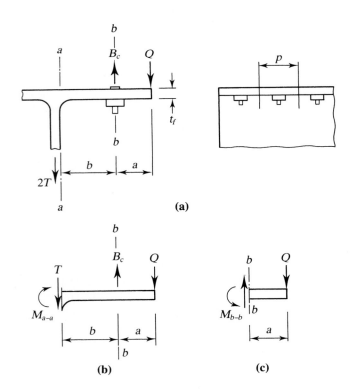

(a)

(b)

(c)

force corresponding to one bolt, and B_c is the total bolt force. The prying force has shifted to the tip of the flange and is at its maximum value.

The equations that follow are derived from consideration of equilibrium based on the free-body diagrams in Figure 7.28. From the summation of moments at b–b in Figure 7.28b,

$$Tb - M_{a-a} = Qa \tag{7.9}$$

From Figure 7.28c,

$$M_{b-b} = Qa \tag{7.10}$$

Finally, equilibrium of forces requires that

$$B_c = T + Q \tag{7.11}$$

These three equilibrium equations can be combined to obtain a single equation for the total bolt force, which includes the effects of prying action. We first define the variable α as the ratio of the moment per unit length along the bolt line to the moment per unit length at the face of the stem. For the bolt line, the length is a net length, so

$$\alpha = \frac{M_{b-b}/(p-d')}{M_{a-a}/p} = \frac{M_{b-b}}{M_{a-a}}\left(\frac{1}{1-d'/p}\right) = \frac{M_{b-b}}{\delta M_{a-a}} \tag{7.12}$$

where

p = length of flange tributary to one bolt   (see Figure 7.28a)

d' = diameter of bolt hole

$$\delta = 1 - \frac{d'}{p} = \frac{\text{net area at bolt line}}{\text{gross area at stem face}}$$

(The numerical evaluation of α will require the use of another equation, which we develop shortly.) With this notation, we can combine the three Equilibrium Equations 7.9–7.11 to obtain the total bolt force, B_c:

$$B_c = T\left[1 + \frac{\delta\alpha}{(1+\delta\alpha)}\frac{b}{a}\right] \tag{7.13}$$

At this level of loading, deformations are so large that the resultant of tensile stresses in the bolt does not coincide with the axis of the bolt. Consequently, the bolt force predicted by Equation 7.13 is conservative and does not quite agree with test results. Much better agreement is obtained if the force B_c is shifted toward the stem of the tee by an amount $d/2$, where d is the bolt diameter. The modified values of b and a are therefore defined as

$$b' = b - \frac{d}{2} \quad \text{and} \quad a' = a + \frac{d}{2}$$

(For best agreement with test results, the value of a should be no greater than $1.25b$.)

With this modification, we can write Equation 7.13 as

$$B_c = T \left[1 + \frac{\delta\alpha}{(1+\delta\alpha)} \frac{b'}{a'} \right] \tag{7.14}$$

We can evaluate α from Equation 7.14 by setting the bolt force B_c equal to the bolt tensile strength, which we denote B. Doing so results in

$$\alpha = \frac{[(B/T)-1](a'/b')}{\delta\{1-[(B/T)-1](a'/b')\}} \tag{7.15}$$

Two limit states are possible: tensile failure of the bolts and bending failure of the tee. Failure of the tee is assumed to occur when plastic hinges form at section $a\text{--}a$, the face of the stem of the tee, and at $b\text{--}b$, the bolt line, thereby creating a beam mechanism. The moment at each of these locations will equal M_p, the plastic moment capacity of a length of flange tributary to one bolt. If the absolute value of α, obtained from Equation 7.15, is less than 1.0, the moment at the bolt line is less than the moment at the face of the stem, indicating that the beam mechanism has not formed and the controlling limit state will be tensile failure of the bolt. The bolt force B_c in this case will equal the strength B. If the absolute value of α is equal to or greater than 1.0, plastic hinges have formed at both $a\text{--}a$ and $b\text{--}b$, and the controlling limit state is flexural failure of the tee flange. Since the moments at these two locations are limited to the plastic moment M_p, α should be set equal to 1.0.

The three Equilibrium Equations 7.9–7.11 can also be combined into a single equation for the required flange thickness, t_f. From Equations 7.9 and 7.10, we can write

$$Tb' - M_{a-a} = M_{b-b}$$

where b' has been substituted for b. From Equation 7.12,

$$Tb' - M_{a-a} = \delta\alpha M_{a-a} \tag{7.16}$$

$$M_{a-a} = \frac{Tb'}{(1+\delta\alpha)} \tag{7.17}$$

For LRFD, let

$$M_{a-a} = \text{design strength} = \phi_b M_p = \phi_b \left(\frac{p t_f^2 F_y}{4} \right)$$

Substituting into Equation 7.17, we get

$$\phi_b \frac{p t_f^2 F_y}{4} = \frac{Tb'}{(1+\delta\alpha)}$$

$$t_f = \sqrt{\frac{4Tb'}{\phi_b p F_y (1+\delta\alpha)}}$$

where T is the factored load per bolt.

With $\phi_b = 0.90$,

$$t_f = \sqrt{\frac{4.44Tb'}{pF_y(1+\delta\alpha)}}$$

Kulak, Fisher, and Struik (1987) recognized that this procedure is conservative when compared with test results. If the ultimate stress F_u is substituted for the yield stress F_y in the expression for flexural strength, much better agreement with test results is obtained (Thornton, 1992 and Swanson, 2002). Making this substitution, we obtain

$$\text{Required } t_f = \sqrt{\frac{4.44Tb'}{pF_u(1+\delta\alpha)}} \tag{7.18}$$

For ASD, when we substitute F_u for F_y, we get

$$M_{a-a} = \text{allowable strength} = \frac{M_p}{\Omega_b} = \frac{1}{\Omega_b}\left(\frac{pt_f^2 F_u}{4}\right)$$

With $\Omega_b = 1.67$,

$$\text{Required } t_f = \sqrt{\frac{6.68Tb'}{pF_u(1+\delta\alpha)}} \tag{7.19}$$

where T is the applied service load per bolt.

The design of connections subjected to prying is essentially a trial-and-error process. When selecting the size or number of bolts, we must make an allowance for the prying force. The selection of the tee flange thickness is more difficult in that it is a function of the bolt selection and tee dimensions. Once the trial shape has been selected and the number of bolts and their layout estimated, Equation 7.18 or 7.19 can be used to verify or disprove the choices. (If the flange thickness is adequate, the bolt strength will also be adequate.)

If the actual flange thickness is different from the required value, the actual values of α and B_c will be different from those previously calculated. If the actual bolt force, which includes the prying force Q, is desired, α will need to be recomputed as follows.

First, combine Equations 7.9 and 7.10, using b' instead of b:

$$M_{b-b} = Tb' - M_{a-a}$$

From Equation 7.12,

$$\alpha = \frac{M_{b-b}}{\delta M_{a-a}}$$

$$= \frac{Tb' - M_{a-a}}{\delta M_{a-a}} = \frac{Tb'/M_{a-a} - 1}{\delta}$$

For LRFD, set

$$M_{a-a} = \phi_b M_p = 0.90\left(\frac{pt_f^2 F_u}{4}\right)$$

then

$$\alpha = \frac{\dfrac{Tb'}{0.90\,pt_f^2 F_u/4} - 1}{\delta} = \frac{1}{\delta}\left(\frac{4.44Tb'}{pt_f^2 F_u} - 1\right) \tag{7.20}$$

where t_f is the actual flange thickness. If the calculated $\alpha < 0$, use $\alpha = 1.0$. The total bolt force can then be found from Equation 7.14.

For ASD, set M_{a-a} equal to the allowable moment:

$$M_{a-a} = \frac{M_p}{\Omega_b} = \frac{1}{1.67}\left(\frac{pt_f^2 F_u}{4}\right)$$

Then

$$\alpha = \frac{\dfrac{Tb'}{(pt_f^2 F_u/4)/1.67} - 1}{\delta} = \frac{1}{\delta}\left(\frac{6.68Tb'}{pt_f^2 F_u} - 1\right) \tag{7.21}$$

where t_f is the actual flange thickness. If $\alpha < 0$, use $\alpha = 1.0$. Find the total bolt force from Equation 7.14.

Although the prying analysis presented here is for tee sections, with a slight modification it can be used for double angles. For b, use the distance from the bolt centerline to the mid-thickness of the angle leg, rather than to the face of the leg.

Example 7.8   An 8-inch long WT10.5×66 is attached to the bottom flange of a beam as shown in Figure 7.29. This hanger must support a service dead load of 20 kips and a service live load of 60 kips. Determine the number of 7/8-inch-diameter, A325 bolts required and investigate the adequacy of the tee. A992 steel is used.

Solution   Compute the constants that are based on the geometry of the connection, which is shown in Figure 7.29.

$$b = \frac{(5.5 - 0.650)}{2} = 2.425 \text{ in.}$$

$$a = \frac{(12.4 - 5.5)}{2} = 3.450 \text{ in.}$$

Since $1.25b = 1.25(2.425) = 3.031$ in. < 3.450 in., use $a = 3.031$ in.

$$b' = b - \frac{d}{2} = 2.425 - \frac{7/8}{2} = 1.988 \text{ in.}$$

$$a' = a + \frac{d}{2} = 3.031 + \frac{7/8}{2} = 3.469 \text{ in.}$$

FIGURE 7.29

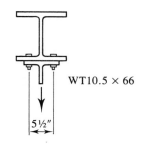

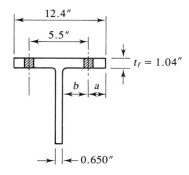

WT10.5 × 66

$$d' = d + \frac{1}{8} = \frac{7}{8} + \frac{1}{8} = 1 \text{ in.}$$

$$p = \frac{8}{2} = 4 \text{ in.}$$

$$\delta = 1 - \frac{d'}{p} = 1 - \frac{1}{4} = 0.75$$

The bolt cross-sectional area will also be needed in subsequent calculations:

$$A_b = \frac{\pi(7/8)^2}{4} = 0.6013 \text{ in.}^2$$

LRFD Solution   The design strength of one bolt is

$$B = \phi R_n = \phi F_t A_b = 0.75(90.0)(0.6013) = 40.59 \text{ kips}$$

The total factored load is

$$1.2D + 1.6L = 1.2(20) + 1.6(60) = 120 \text{ kips}$$

and the number of bolts required (without considering prying action) is $120/40.59 = 2.96$. A minimum of four bolts will be needed to maintain symmetry, so **try four bolts**. The factored external load per bolt, excluding prying force, is $T = 120/4 = 30$ kips.

Compute α:

$$\frac{B}{T} - 1 = \frac{40.59}{30} - 1 = 0.353$$

From Equation 7.15,

$$\alpha = \frac{[(B/T-1)](a'/b')}{\delta\{1-[(B/T)-1](a'/b')\}} = \frac{0.353(3.469/1.988)}{0.75[1-0.353(3.469/1.988)]} = 2.139$$

Because $|\alpha| > 1.0$, use $\alpha = 1.0$. From Equation 7.18,

$$\text{Required } t_f = \sqrt{\frac{4.44Tb'}{pF_u(1+\delta\alpha)}} = \sqrt{\frac{4.44(30)(1.988)}{4(65)[1+0.75(1.0)]}}$$
$$= 0.763 \text{ in.} < 1.04 \text{ in.} \quad \text{(OK)}$$

Both the number of bolts selected and the flange thickness are adequate, and no further computations are required. To illustrate the procedure, however, we compute the prying force, using Equations 7.20 and 7.14. From Equation 7.20,

$$\alpha = \frac{1}{\delta}\left(\frac{4.44Tb'}{pt_f^2 F_u} - 1\right) = \frac{1}{0.75}\left(\frac{4.448(30)(1.988)}{4(1.04)^2(65)} - 1\right)$$
$$= -0.07556 \quad \therefore \text{ use } \alpha = 1.0$$

From Equation 7.14, the total bolt force, including prying, is

$$B_c = T\left[1 + \frac{\delta\alpha}{(1+\delta\alpha)}\frac{b'}{a'}\right] = 30\left[1 + \frac{0.75(1.0)}{(1+0.75\times1.0)}\left(\frac{1.988}{3.469}\right)\right] = 37.37 \text{ kips}$$

The prying force is

$$Q = B_c - T = 37.37 - 30 = 7.37 \text{ kips}$$

Answer   A WT10.5 × 66 is satisfactory. Use four ⅞-inch-diameter A325 bolts.

ASD Solution   The allowable tensile strength of one bolt is

$$B = \frac{R_n}{\Omega} = \frac{F_t A_b}{\Omega} = \frac{90.0(0.6013)}{2.00} = 27.06 \text{ kips}$$

The total applied load is

$$D + L = 20 + 60 = 80 \text{ kips}$$

and the number of bolts required (without considering prying action) is $80/27.06 = 2.96$. A minimum of four bolts will be needed to maintain symmetry, so **try four bolts**. The external load per bolt, excluding prying force, is $T = 80/4 = 20$ kips.

Compute α:

$$\frac{B}{T} - 1 = \frac{27.06}{20} - 1 = 0.353$$

$$\alpha = \frac{[(B/T-1)](a'/b')}{\delta\{1-[(B/T)-1](a'/b')\}} = \frac{0.353(3.469/1.988)}{0.75[1-0.353(3.469/1.988)]} = 2.139$$

Since $|\alpha| > 1.0$, use $\alpha = 1.0$. From Equation 7.19,

$$\text{Required } t_f = \sqrt{\frac{6.68Tb'}{pF_u(1+\delta\alpha)}} = \sqrt{\frac{6.68(20)1.988}{4(65)[1+0.75(1.0)]}}$$

$$= 0.764 \text{ in.} < 1.04 \text{ in.}   \text{(OK)}$$

Determine the prying force (this is not required). From Equation 7.21,

$$\alpha = \frac{1}{\delta}\left(\frac{6.68Tb'}{pt_f^2 F_u} - 1\right) = \frac{1}{0.75}\left(\frac{6.68(20)(1.988)}{4(1.04)^2(65)} - 1\right)$$

$$= -0.07406   \therefore \text{ use } \alpha = 1.0$$

$$B_c = T\left[1 + \frac{\delta\alpha}{(1+\delta\alpha)}\frac{b'}{a'}\right] = 20\left[1 + \frac{0.75(1.0)}{(1+0.75\times1.0)}\left(\frac{1.988}{3.469}\right)\right] = 24.91 \text{ kips}$$

The prying force is

$$Q = B_c - T = 24.91 - 20 = 4.91 \text{ kips}$$

Answer   A WT10.5 × 66 is satisfactory. Use four ⅞-inch-diameter, A325 bolts.   ■

If the flange thickness in Example 7.8 had proved to be inadequate, the alternatives would include trying a larger tee-shape or using more bolts to reduce T, the external load per bolt. The prying force in Example 7.8 adds approximately 25% to the externally applied load. Neglect of this additional tension could have serious consequences. You should be alert for situations in which it may arise and, when feasible, minimize it by avoiding overly flexible connection elements.

7.9   COMBINED SHEAR AND TENSION IN FASTENERS

In most of the situations in which a bolt is subjected to both shear and tension, the connection is loaded eccentrically and falls within the realm of Chapter 8. However, in some simple connections the fasteners are in a state of combined loading. Figure 7.30

FIGURE 7.30

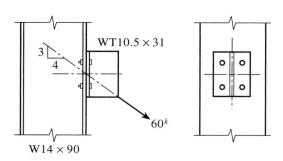

FIGURE 7.31

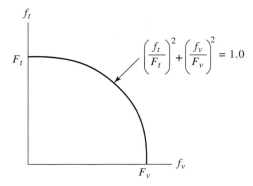

shows a structural tee segment connected to the flange of a column for the purpose of attaching a bracing member. This bracing member is oriented in such a way that the line of action of the member force passes through the center of gravity of the connection. The vertical component of the load will put the fasteners in shear, and the horizontal component will cause tension (with the possible inclusion of prying forces). Since the line of action of the load acts through the center of gravity of the connection, each fastener can be assumed to take an equal share of each component.

As in other cases of combined loading, an interaction formula approach can be used. The shear and tensile strengths for bearing-type bolts are based on test results (Chesson et al., 1965) that can be represented by the elliptical interaction curve shown in Figure 7.31. The equation of this curve can be expressed in a general way as

$$\left(\frac{\text{required tensile strength}}{\text{available tensile strength}}\right)^2 + \left(\frac{\text{required shear strength}}{\text{available shear strength}}\right)^2 = 1.0 \qquad (7.22)$$

where the strengths can be expressed as forces or stresses and in either LRFD or ASD format. If stresses are used, Equation 7.22 becomes

$$\left(\frac{f_t}{F_t}\right)^2 + \left(\frac{f_v}{F_v}\right)^2 = 1.0 \qquad (7.23)$$

where

f_t = required tensile strength (stress)
F_t = available tensile strength (stress)
f_v = required shear strength (stress)
F_v = available shear strength (stress)

An acceptable combination of shear and tension is one that lies under this curve. This fact leads to the requirement that

$$\left(\frac{f_t}{F_t}\right)^2 + \left(\frac{f_v}{F_v}\right)^2 \leq 1.0$$

FIGURE 7.32

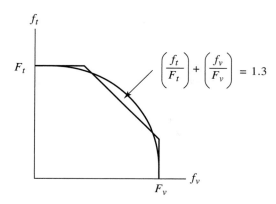

$$\left(\frac{f_t}{F_t}\right)+\left(\frac{f_v}{F_v}\right)=1.3$$

The AISC Specification approximates the elliptical curve with three straight line segments as shown in Figure 7.32. The equation of the sloping line is given by

$$\left(\frac{f_t}{F_t}\right)+\left(\frac{f_v}{F_v}\right)=1.3 \tag{7.24}$$

To avoid going above the line,

$$\left(\frac{f_t}{F_t}\right)+\left(\frac{f_v}{F_v}\right)\leq1.3$$

If Equation 7.24 is solved for the required tensile strength f_t, we obtain, for a given f_v,

$$f_t = 1.3F_t - \frac{f_v}{F_v}F_t \tag{7.25}$$

Let

available strength $= \Phi \times$ nominal strength

or

$$\text{nominal strength} = \frac{\text{available strength}}{\Phi}$$

where

$\Phi = \phi$ for LRFD

$\quad = \dfrac{1}{\Omega}$ for ASD

If f_t is viewed as the available tensile strength in the presence of shear, then from Equation 7.25, the corresponding nominal strength is

$$\frac{f_t}{\Phi} = 1.3\frac{F_t}{\Phi} - \frac{F_t}{\Phi F_v}f_v$$

or

$$F'_{nt} = 1.3F_{nt} - \frac{F_{nt}}{F_v} f_v$$

or

$$F'_{nt} = 1.3F_{nt} - \frac{F_{nt}}{\Phi F_{nv}} f_v \qquad (7.26)$$

where

F'_{nt} = nominal tensile stress in the presence of shear
F_{nt} = nominal tensile stress in the absence of shear
F_{nv} = nominal shear stress in the absence of tension
f_v = required shear stress

Note that F'_{nt} must not exceed F_{nt}, and f_v must not exceed F_{nv}. The nominal tensile strength is then

$$R_n = F'_{nt} A_b \qquad \text{(AISC Equation J3-2)}$$

Equation 7.26 will now be presented in the two design formats.

LRFD

$$\Phi = \phi$$

and

$$F'_{nt} = 1.3F_{nt} - \frac{F_{nt}}{\phi F_{nv}} f_v \leq F_{nt} \qquad \text{(AISC Equation J3-3a)}$$

where $\phi = 0.75$.

ASD

$$\Phi = \frac{1}{\Omega}$$

and

$$F'_{nt} = 1.3F_{nt} - \frac{\Omega F_{nt}}{F_{nv}} f_v \leq F_{nt} \qquad \text{(AISC Equation J3-3b)}$$

where $\Omega = 2.00$.

The Commentary to AISC J3.7 gives alternative interaction equations based on the elliptical solution. Either these alternative equations or the elliptical equation itself, Equation 7.23, may be used in lieu of AISC Equations J3-3a and J3-3b. In this book, we use AISC Equations J3-3a and J3-3b.

In slip-critical connections subject to both shear and tension, interaction of shear and tension need not be investigated. However, the effect of the applied tensile force is to relieve some of the clamping force, thereby reducing the available friction force. The AISC Specification reduces the slip-critical strength for this case.

(This reduction is not made for certain types of eccentric connections; that will be covered in Chapter 8.) The reduction is made by multiplying the slip-critical strength by a factor k_s as follows:

For LRFD,

$$k_s = 1 - \frac{T_u}{D_u T_b N_b}$$

(AISC Equation J3-5a)

For ASD,

$$k_s = 1 - \frac{1.5 T_a}{D_u T_b N_b}$$

(AISC Equation J3-5b)

where
$\quad T_u$ = total factored tensile load on the connection
$\quad T_a$ = total service tensile load on the connection
$\quad D_u$ = ratio of mean bolt pretension to specified minimum pretension; default value is 1.13
$\quad T_b$ = prescribed initial bolt tension from AISC Table J3.1
$\quad N_b$ = number of bolts in the connection

The AISC Specification approach to the analysis of bolted connections loaded in both shear and tension can be summarized as follows:

Bearing-type connections:
1. Check shear and bearing against the usual strengths.
2. Check tension against the reduced tensile strength using AISC Equation J3-3a (LRFD) or J3-3b (ASD).

Slip-critical connections:
1. Check tension, shear, and bearing against the usual strengths.
2. Check the slip-critical load against the reduced slip-critical strength.

Example 7.9   A WT10.5 × 31 is used as a bracket to transmit a 60-kip service load to a W14 × 90 column, as previously shown in Figure 7.30. The load consists of 15 kips dead load and 45 kips live load. Four ⅞-inch-diameter A325 bolts are used. The column is of A992 steel, and the bracket is A36. Assume all spacing and edge-distance requirements are satisfied, including those necessary for the use of the maximum nominal strength in bearing (i.e., $2.4 dt F_u$), and determine the adequacy of the bolts for the following types of connections: (a) bearing-type connection with the threads in shear and (b) slip-critical connection with the threads in shear.

Solution   (The following values are used in both the LRFD and ASD solutions.)

Compute the nominal bearing strength (flange of tee controls).

$$R_n = 2.4 dt F_u = 2.4 \left(\frac{7}{8} \right)(0.615)(58) = 74.91 \text{ kips}$$

Nominal shear strength:

$$A_b = \frac{\pi(7/8)^2}{4} = 0.6013 \text{ in.}^2$$

$$R_n = F_{nv}A_b = 48(0.6013) = 28.9 \text{ kips}$$

LRFD Solution $P_u = 1.2D + 1.6L = 1.2(15) + 1.6(45) = 90 \text{ kips}$

a. The total shear/bearing load is

$$V_u = \frac{3}{5}(90) = 54 \text{ kips}$$

The shear/bearing force per bolt is

$$V_{u\,bolt} = \frac{54}{4} = 13.5 \text{ kips}$$

The design bearing strength is

$$\phi R_n = 0.75(74.91) = 56.2 \text{ kips} > 13.5 \text{ kips} \qquad (\text{OK})$$

The design shear strength is

$$\phi R_n = 0.75(28.9) = 21.7 \text{ kips} > 13.5 \text{ kips} \qquad (\text{OK})$$

The total tension load is

$$T_u = \frac{4}{5}(90) = 72 \text{ kips}$$

The tensile force per bolt is

$$T_{u\,bolt} = \frac{72}{4} = 18 \text{ kips}$$

To determine the available tensile strength, use AISC Equation J3-3a:

$$F'_{nt} = 1.3F_{nt} - \frac{F_{nt}}{\phi F_{nv}} f_v \le F_{nt}$$

where
F_{nt} = nominal tensile stress in the absence of shear = 90 ksi
F_{nv} = nominal shear stress in the absence of tension = 48 ksi

$$f_v = \frac{V_{u\,bolt}}{A_b} = \frac{13.5}{0.6013} = 22.45 \text{ ksi}$$

Then

$$F'_{nt} = 1.3(90) - \frac{90}{0.75(48)}(22.45) = 60.88 \text{ ksi} < 90 \text{ ksi}$$

The nominal tensile strength is

$$R_n = F'_{nt}A_b = 60.88(0.6013) = 36.61 \text{ kips}$$

and the available tensile strength is

$$\phi R_n = 0.75(36.61) = 27.5 \text{ kips} > 18 \text{ kips} \qquad \text{(OK)}$$

Answer The connection is adequate as a bearing-type connection. (In order not to obscure the combined loading features of this example, prying action has not been included in the analysis.)

b. From Part a, the shear, bearing, and tensile strengths are satisfactory. From AISC Equation J3-4, the slip-critical strength is

$$R_n = \mu D_u h_{sc} T_b N_s$$

From AISC Table J3.1, the prescribed tension for a ⅞-inch-diameter A325 bolt is

$$T_b = 39 \text{ kips}$$

If we assume Class A surfaces, the slip coefficient is $\mu = 0.35$, and for four bolts,

$$R_n = \mu D_u h_{sc} T_b N_s \times 4 = 0.35(1.13)(1.0)(39)(1) \times 4 = 61.70 \text{ kips}$$
$$\phi R_n = 1.0(61.70) = 61.70 \text{ kips}$$

Since there is a tensile load on the bolts, the slip-critical strength must be reduced by a factor of

$$k_s = 1 - \frac{T_u}{D_u T_b N_b} = 1 - \frac{72}{1.13(39)(4)} = 0.5916$$

The reduced strength is therefore

$$k_s(61.70) = 0.5916(61.70) = 36.5 \text{ kips} < 54 \text{ kips} \qquad \text{(N.G.)}$$

Answer The connection is inadequate as a slip-critical connection.

ASD Solution $P_a = D + L = 15 + 45 = 60 \text{ kips}$

a. The total shear/bearing load is

$$V_a = \frac{3}{5}(60) = 36 \text{ kips}$$

The shear/bearing force per bolt is

$$V_{a \text{ bolt}} = \frac{36}{4} = 9.0 \text{ kips}$$

The allowable bearing strength is

$$\frac{R_n}{\Omega} = \frac{74.91}{2.00} = 37.5 \text{ kips} > 9.0 \text{ kips} \qquad \text{(OK)}$$

The allowable shear strength is

$$\frac{R_n}{\Omega} = \frac{28.9}{2.00} = 14.5 \text{ kips} > 9.0 \text{ kips} \qquad \text{(OK)}$$

The total tension load is

$$T_a = \frac{4}{5}(60) = 48 \text{ kips}$$

The tensile force per bolt is

$$T_{a \text{ bolt}} = \frac{48}{4} = 12 \text{ kips}$$

To determine the available tensile strength, use AISC Equation J3-3b:

$$F'_{nt} = 1.3F_{nt} - \frac{\Omega F_{nt}}{F_{nv}} f_v \leq F_{nt}$$

where
F'_{nt} = nominal tensile stress in the absence of shear = 90 ksi
F_{nv} = nominal shear stress in the absence of tension = 48 ksi

$$f_v = \frac{V_{a \text{ bolt}}}{A_b} = \frac{9.0}{0.6013} = 14.97 \text{ ksi}$$

Then

$$F'_{nt} = 1.3(90) - \frac{2.00(90)}{48}(14.97) = 60.86 \text{ ksi} < 90 \text{ ksi}$$

The nominal tensile strength is

$$R_n = F'_{nt}A_b = 60.86(0.6013) = 36.60 \text{ kips}$$

and the available tensile strength is

$$\frac{R_n}{\Omega} = \frac{36.60}{2.00} = 18.3 \text{ kips} > 12 \text{ kips} \qquad \text{(OK)}$$

Answer   The connection is adequate as a bearing-type connection. (In order not to obscure the combined loading features of this example, prying action has not been included in the analysis.)

b. From Part a, the shear, bearing, and tensile strengths are satisfactory. From AISC Equation J3-4, the slip-critical strength is

$$R_n = \mu D_u h_{sc} T_b N_s$$

From AISC Table J3.1, the prescribed tension for a $\frac{7}{8}$-inch-diameter A325 bolt is

$$T_b = 39 \text{ kips}$$

If we assume Class A surfaces, the slip coefficient is $\mu = 0.35$, and for four bolts,

$$R_n = \mu D_u h_{sc} T_b N_s \times 4 = 0.35(1.13)(1.0)(39)(1) \times 4 = 61.70 \text{ kip}$$

$$\frac{R_n}{\Omega} = \frac{61.70}{1.50} = 41.13 \text{ kips}$$

Since there is a tensile load on the bolts, the slip-critical strength must be reduced by a factor of

$$k_s = 1 - \frac{1.5 T_a}{D_u T_b N_b} = 1 - \frac{1.5(48)}{1.13(39)(4)} = 0.5916$$

The reduced strength is therefore

$$k_s(41.13) = 0.5916(41.13) = 24.3 \text{ kips} < 36 \text{ kips} \qquad \text{(N.G.)}$$

Answer   The connection is inadequate as a slip-critical connection.   ■

Connections with fasteners in shear and tension can be designed by trial, but a more direct procedure can be used if one assumes that the design is controlled by the strength that is reduced. If the assumption turns out to be correct, no iteration is required. This technique will be illustrated in the following example.

Example 7.10   A concentrically loaded connection is subjected to a service-load shear force of 50 kips and a service-load tensile force of 100 kips. The loads are 25% dead load and 75% live load. The fasteners will be in single shear, and bearing strength will be controlled by a $\frac{5}{16}$-inch-thick connected part of A36 steel. Assume that all spacing and edge distances are satisfactory, including those that permit the maximum nominal bearing strength of $2.4 dt F_u$ to be used. Determine the required number of $\frac{3}{4}$-inch-diameter A325 bolts for the following cases: (a) a bearing-type connection with threads in the plane of shear and (b) a slip-critical connection with threads in the plane of shear. All contact surfaces have clean mill scale.

Consider this to be a preliminary design so that no consideration of prying action is necessary.

LRFD Solution   Factored load shear = $1.2[0.25(50)] + 1.6[0.75(50)] = 75$ kips
Factored load tension = $1.2[0.25(100)] + 1.6[0.75(100)] = 150$ kips

a. For the bearing-type connection with threads in the shear plane, assume that tension controls:

$$F'_{nt} = 1.3F_{nt} - \frac{F_{nt}}{\phi F_{nv}} f_v \leq F_{nt}$$

$$= 1.3(90) - \frac{90}{0.75(48)} f_v \leq 90$$

$$= 117 - 2.5f_v \leq 90$$

$$\phi F'_{nt} = 0.75(117 - 2.5f_v) \leq 0.75(90)$$

$$= 87.75 - 1.875f_v \leq 67.5$$

Let

$$\phi F'_{nt} = \frac{150}{\Sigma A_b} \quad \text{and} \quad f_v = \frac{75}{\Sigma A_b}$$

where ΣA_b is the total bolt area. Substituting and solving for ΣA_b, we get

$$\frac{150}{\Sigma A_b} = 87.75 - 1.875\left(\frac{75}{\Sigma A_b}\right)$$

$$150 = 87.75\Sigma A_b - 1.875(75)$$

$$\Sigma A_b = 3.312 \text{ in.}^2$$

The area of one bolt is

$$A_b = \frac{\pi(3/4)^2}{4} = 0.4418 \text{ in.}^2$$

and the number of bolts required is

$$N_b = \frac{\Sigma A_b}{A_b} = \frac{3.312}{0.4418} = 7.50$$

Try eight bolts. First, check the upper limit on F'_{nt}:

$$f_v = \frac{75}{\Sigma A_b} = \frac{75}{8(0.4418)} = 21.22 \text{ ksi}$$

$$F'_{nt} = 117 - 2.5f_v = 117 - 2.5(21.22) = 64.0 \text{ ksi} < 90 \text{ ksi} \quad \text{(OK)}$$

Check shear:

$$\phi R_n = \phi F_{nv} A_b \times N_b = 0.75(48)(0.4418)(8)$$

$$= 127 \text{ kips} > 75 \text{ kips} \quad \text{(OK)}$$

Check bearing:

$$\phi R_n = \phi(2.4dtF_u) \times 8 \text{ bolts}$$

$$= 0.75(2.4)\left(\frac{3}{4}\right)\left(\frac{5}{16}\right)(58)(8) = 196 \text{ kips} > 75 \text{ kips} \quad \text{(OK)}$$

Answer   Use eight bolts.

b. Slip-critical connection. Assume that the slip load controls. The reduced slip-critical strength is

$$k_s \phi R_n$$

where

$$k_s = 1 - \frac{T_u}{D_u T_b N_b} = 1 - \frac{150}{1.13(28)N_b} = 1 - \frac{4.741}{N_b}$$

where $T_b = 28$ kips (from AISC Table J3.1).

For one bolt,

$$\phi R_n = \phi(\mu D_u h_{sc} T_b N_s) = 1.0(0.35)(1.13)(1.0)(28)(1.0) = 11.07 \text{ kips}$$

Setting the total factored-load shear load to the reduced slip-critical strength for N_b bolts,

$$75 = N_b\left(1 - \frac{4.741}{N_b}\right)(11.07)$$

$$= 11.07N_b - 52.48$$

$$N_b = 11.5$$

Since eight bolts are adequate for shear, bearing, and tension (with a *reduced* tensile strength), these limit states will not have to be checked.

Answer   Use twelve ¾-inch-diameter, A325 bolts.

ASD Solution   Applied service-load shear = 50 kips

Applied service-load tension = 100 kips

a. For the bearing-type connection with threads in the shear plane, assume that tension controls:

$$F'_{nt} = 1.3F_{nt} - \frac{\Omega F_{nt}}{F_{nv}} f_v \leq F_{nt}$$

$$= 1.3(90) - \frac{2.00(90)}{48} f_v \leq 90$$

$$= 117 - 3.75 f_v \leq 90$$

$$\frac{F'_{nt}}{\Omega} = \frac{(117 - 3.75 f_v)}{2.00} \leq \frac{90}{2.00}$$

$$= 58.5 - 1.875 f_v \leq 45$$

Let

$$\frac{F'_{nt}}{\Omega} = \frac{100}{\Sigma A_b} \quad \text{and} \quad f_v = \frac{50}{\Sigma A_b}$$

where ΣA_b is the total bolt area. Substituting and solving for ΣA_b, we get

$$\frac{100}{\Sigma A_b} = 58.5 - 1.875 \left(\frac{50}{\Sigma A_b} \right)$$

$$100 = 58.5 \Sigma A_b - 93.75$$

$$\Sigma A_b = 3.312 \text{ in.}^2$$

The area of one bolt is

$$A_b = \frac{\pi (3/4)^2}{4} = 0.4418 \text{ in.}^2$$

And the number of bolts required is

$$N_b = \frac{\Sigma A_b}{A_b} = \frac{3.312}{0.4418} = 7.50$$

Try eight bolts. First, check the upper limit on F'_{nt}:

$$f_v = \frac{50}{\Sigma A_b} = \frac{50}{8(0.4418)} = 14.15 \text{ ksi}$$

$$F'_{nt} = 117 - 3.75 f_v = 117 - 3.75(14.15) = 63.9 \text{ ksi} < 90 \text{ ksi} \quad \text{(OK)}$$

Check shear. The nominal shear stress for one bolt is

$$R_n = F_{nv} A_b = 48(0.4418) = 21.21 \text{ kips}$$

and the allowable strength for eight bolts is

$$\frac{R_n}{\Omega} \times 8 = \frac{21.21}{2.00} \times 8 = 84.8 \text{ kips} > 50 \text{ kips} \quad \text{(OK)}$$

Check bearing:

$$\frac{R_n}{\Omega} = \frac{2.4 dt F_u}{\Omega} \times 8 \text{ bolts}$$

$$= \frac{2.4(3/4)((5/16))(58)}{2.00} \times 8 = 131 \text{ kips} > 50 \text{ kips} \quad \text{(OK)}$$

Answer   Use eight bolts.

b. Slip-critical connection: assume that the slip load controls. The reduced slip-critical strength is

$$k_s \frac{R_n}{\Omega}$$

where

$$k_s = 1 - \frac{1.5T_a}{D_u T_b N_b} = 1 - \frac{1.5(100)}{1.13(28)N_b} = 1 - \frac{4.741}{N_b}$$

where $T_b = 28$ kips (from AISC Table J3.1).

For one bolt,

$$\frac{R_n}{\Omega} = \frac{\mu D_u h_{sc} T_b N_s}{\Omega} = \frac{0.35(1.13)(1.0)(28)(1.0)}{1.50} = 7.383 \text{ kips}$$

Setting the total shear load to the reduced slip-critical strength for N_b bolts,

$$50 = N_b \left(1 - \frac{4.741}{N_b}\right)(7.383)$$

$$= 7.383 N_b - 35.00$$

$$N_b = 11.5$$

From Part a, the connection is adequate in shear, bearing, and tension with eight bolts, so it will be adequate if more bolts are used.

Answer   Use 12 bolts.   ◼

7.10   WELDED CONNECTIONS

Structural welding is a process whereby the parts to be connected are heated and fused, with supplementary molten metal added to the joint. For example, the tension member lap joint shown in Figure 7.33 can be constructed by welding across the ends of both connected parts. A relatively small depth of material will become molten, and upon cooling, the structural steel and the weld metal will act as one continuous part where they are joined. The additional metal, sometimes referred to as *filler metal*, is deposited from a special electrode, which is part of an electrical circuit that includes the connected part, or *base metal*. In the shielded metal arc welding (SMAW) process, shown schematically in Figure 7.34, current arcs across a gap between the electrode and base metal, heating the connected parts and depositing part of the electrode into the molten base metal. A special coating on the electrode vaporizes and forms a protective gaseous

FIGURE 7.33

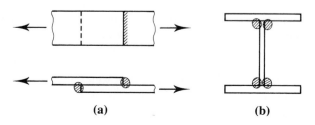

(a)   (b)

FIGURE 7.34

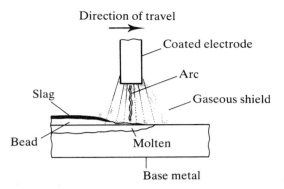

shield, preventing the molten weld metal from oxidizing before it solidifies. The electrode is moved across the joint, and a weld bead is deposited, its size depending on the rate of travel of the electrode. As the weld cools, impurities rise to the surface, forming a coating called *slag* that must be removed before the member is painted or another pass is made with the electrode.

Shielded metal arc welding is normally done manually and is the process universally used for field welds. For shop welding, an automatic or semiautomatic process is usually used. Foremost among these processes is submerged arc welding (SAW). In this process, the end of the electrode and the arc are submerged in a granular flux that melts and forms a gaseous shield. There is more penetration into the base metal than with shielded metal arc welding, and higher strength results. Other commonly used processes for shop welding include gas shielded metal arc, flux cored arc, and electroslag welding.

Quality control of welded connections is particularly difficult, because defects below the surface, or even minor flaws at the surface, will escape visual detection. Welders must be properly certified, and for critical work, special inspection techniques such as radiography or ultrasonic testing must be used.

The two most common types of welds are the *fillet weld* and the *groove weld*. The lap joint illustrated in Figure 7.33a is made with fillet welds, which are defined as those placed in a corner formed by two parts in contact. Fillet welds can also be used in a tee joint, as shown in Figure 7.33b. Groove welds are those deposited in a gap, or groove, between two parts to be connected. They are most frequently used for butt, tee, and corner joints. In most cases, one or both of the connected parts will have beveled edges, called *prepared edges,* as shown in Figure 7.35a, although relatively thin material can be groove welded with no edge preparation. The welds shown in Figure 7.35a are complete penetration welds and can be made from one side, sometimes with the aid of a backing bar. Partial penetration groove welds can be made from one or both sides, with or without edge preparation (Figure 7.35b).

Figure 7.36 shows the plug or slot weld, which sometimes is used when more weld is needed than length of edge is available. A circular or slotted hole is cut in one of the parts to be connected and is filled with the weld metal.

Of the two major types of welds, fillet welds are the most common and are considered here in some detail. The design of complete penetration groove welds is a trivial exercise in that the weld will have the same strength as the base metal and the

FIGURE 7.35

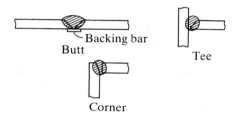

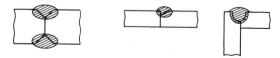

(a) Complete penetration groove welds

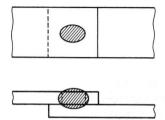

(b) Partial penetration groove welds

FIGURE 7.36

connected parts can be treated as completely continuous at the joint. The strength of a partial penetration groove weld will depend on the amount of penetration; once that has been determined, the design procedure will be essentially the same as that for a fillet weld.

7.11   FILLET WELDS

The design and analysis of fillet welds is based on the assumption that the cross section of the weld is a 45° right triangle, as shown in Figure 7.37. Any reinforcement (buildup outside the hypotenuse of the triangle) or penetration is neglected. The size of a fillet weld is denoted w and is the length of one of the two equal sides of this idealized cross section. Standard weld sizes are specified in increments of $1/16$ inch. Although a length of weld can be loaded in any direction in shear, compression, or tension, a fillet weld is weakest in shear and is always assumed to fail in this mode. Specifically, failure is assumed to occur in shear on a plane through the throat of the weld. For fillet welds made with the shielded metal arc process, the throat is the perpendicular distance from the corner, or root, of the weld to the hypotenuse and is equal to 0.707 times the size of the weld. (The effective throat thickness for a weld made with the submerged arc welding process is larger. In this book, we conservatively

FIGURE 7.37

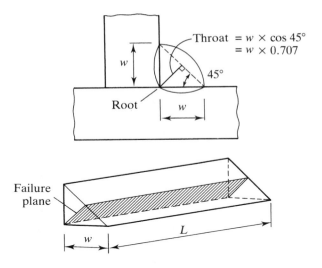

assume that the shielded metal arc welding process is used.) Thus, for a given length of weld L subjected to a load of P, the critical shearing stress is

$$f_v = \frac{P}{0.707wL}$$

where w is the weld size.

If the weld ultimate shearing stress, F_W, is used in this equation, the nominal load capacity of the weld can be written as

$$R_n = 0.707wLF_W$$

The strength of a fillet weld depends on the weld metal used—that is, it is a function of the type of electrode. The strength of the electrode is defined as its ultimate tensile strength, with strengths of 60, 70, 80, 90, 100, 110, and 120 kips per square inch available for the shielded metal arc welding process. The standard notation for specifying an electrode is the letter E followed by two or three digits indicating the tensile strength in kips per square inch and two digits specifying the type of coating. As strength is the property of primary concern to the design engineer, the last two digits are usually represented by XX, and a typical designation would be E70XX or just E70, indicating an electrode with an ultimate tensile strength of 70 ksi. Electrodes should be selected to match the base metal. For the commonly used grades of steel, only two electrodes need be considered:

Use E70XX electrodes with steels that have a yield stress less than 60 ksi.

Use E80XX electrodes with steels that have a yield stress of 60 ksi or 65 ksi.

Most of the AISC Specification provisions for welds have been taken from the *Structural Welding Code* of the American Welding Society (AWS, 2004). Exceptions are listed in AISC J2. The AWS Code should be used for criteria not covered in the AISC Specification.

FIGURE 7.38

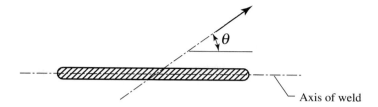

Axis of weld

The design strengths of welds are given in AISC Table J2.5. The ultimate shearing stress F_W in a fillet weld is 0.6 times the tensile strength of the weld metal, denoted F_{EXX}. The nominal stress is therefore

$$F_W = 0.60F_{EXX}$$

AISC Section J2.4a presents an alternative fillet weld strength that accounts for the direction of the load. If the angle between the direction of the load and the axis of the weld is denoted θ (see Figure 7.38), the nominal fillet weld strength is

$$F_W = 0.60F_{EXX}\left(1.0 + 0.50 \sin^{1.5} \theta\right) \qquad \text{(AISC Equation J2-5)}$$

Table 7.2 shows the strength for several values of θ. As Table 7.2 shows, if the weld axis is parallel to the load, the basic strength given by $F_W = 0.60F_{EXX}$ is correct, but when the weld is perpendicular to the load, the true strength is 50% higher.

For simple (that is, concentrically loaded) welded connections with both longitudinal and transverse welds, AISC J2.4c specifies that the larger nominal strength from the following two options be used:

1. Use the basic weld strength, $F_W = 0.6F_{EXX}$, for both the longitudinal and the transverse welds:

$$R_n = R_{w\ell} + R_{wt} \qquad \text{(AISC Equation J2-9a)}$$

   where

   $R_{w\ell}$ = strength of the longitudinal welds = $0.6F_{EXX}$
   R_{wt} = strength of the transverse weld = $0.6F_{EXX}$

2. Use the 50% increase for the transverse weld, but reduce the basic strength by 15% for the longitudinal welds. That is, use $F_W = 0.85(0.6F_{EXX})$ for the longitudinal welds and $F_W = 1.5(0.6F_{EXX})$ for the transverse welds:

$$R_n = 0.85R_{w\ell} + 1.5R_{wt} \qquad \text{(AISC Equation J2-9b)}$$

TABLE 7.2

Direction of load (θ)	$F_W = 0.60F_{EXX} (1.0 + 0.50 \sin^{1.5}\theta)$
0°	$0.60F_{EXX} (1.0)$
15°	$0.60F_{EXX} (1.066)$
30°	$0.60F_{EXX} (1.177)$
45°	$0.60F_{EXX} (1.297)$
60°	$0.60F_{EXX} (1.403)$
75°	$0.60F_{EXX} (1.475)$
90°	$0.60F_{EXX} (1.5)$

FIGURE 7.39

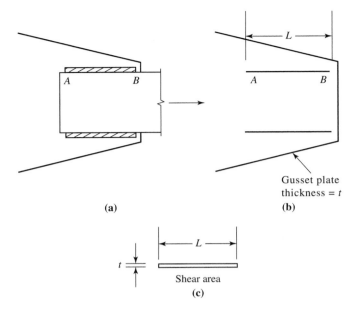

(a) (b)

Gusset plate
thickness = t

Shear area
(c)

Because AISC permits the *larger* of the two options to be used, it is permissible to use *either* and be conservative at worst. In this book, however, we will use the AISC specified approach and check both options.

For LRFD, the design strength of a fillet weld is ϕR_n, where $\phi = 0.75$. For ASD, the allowable strength is R_n/Ω, where $\Omega = 2.00$.

An additional requirement is that the shear on the base metal cannot exceed the shear strength of the base metal. This means that we cannot use a weld shear strength larger than the base metal shear strength, so the base metal shear strength is an upper limit on the weld shear strength. This requirement can be explained by an examination of the welded connection shown in Figure 7.39a. Although both the gusset plate and the tension member plate are subject to shear, we will examine the shear on the gusset plate adjacent to the weld AB. The shear would occur along line AB (Figure 7.39b) and subject an area of tL to shear (Figure 7.39c). The shear strength of the weld AB cannot exceed the shear strength of the base metal corresponding to an area tL.

The nominal shear strength of the base metal will be based on either the limit state of yielding or the limit state of rupture (fracture). As in previously discussed provisions, the shear yield and ultimate stresses are taken as 0.6 times the tensile yield and ultimate stresses. AISC J4.2 gives the shear yield strength as

$$R_n = 0.6F_y A_g \qquad \text{(AISC Equation J4-3)}$$

where A_g is the gross area of the shear surface and is equal to tL. For LRFD, the resistance factor is $\phi = 1.00$, and for ASD, the safety factor is $\Omega = 1.50$. The shear rupture strength is

$$R_n = 0.6F_u A_{nv} \qquad \text{(AISC Equation J4-4)}$$

The strength of a fillet weld can now be summarized for both load and resistance factor design and for allowable strength design.

LRFD Equations

Weld shear strength:

$$\phi R_n = 0.75(0.707wLF_W) \tag{7.27}$$

Base metal shear strength:

$$\phi R_n = \min[1.0(0.6F_y tL),\ 0.75(0.6F_u tL)] \tag{7.28}$$

It is frequently more convenient to work with the strength per unit length, in which case $L = 1$ and Equation 7.27 and 7.28 become

Weld shear strength:

$$\phi R_n = 0.75(0.707wF_W) \text{ for a one-inch length} \tag{7.29}$$

Base metal shear strength:

$$\phi R_n = \min[1.0(0.6F_y t),\ 0.75(0.6F_u t)] \text{ for a one-inch length} \tag{7.30}$$

ASD Equations

Weld shear strength: $\dfrac{R_n}{\Omega} = \dfrac{0.707wLF_W}{2.00}$ $\tag{7.31}$

Base metal shear strength: $\dfrac{R_n}{\Omega} = \min\left[\dfrac{0.6F_y tL}{1.50}, \dfrac{0.6F_u tL}{2.00}\right]$ $\tag{7.32}$

If the strength per unit length is used, $L = 1$ and Equations 7.31 and 7.32 become

Weld shear strength: $\dfrac{R_n}{\Omega} = \dfrac{0.707wF_W}{2.00}$ for a one-inch length $\tag{7.33}$

Base metal shear strength: $\dfrac{R_n}{\Omega} = \min\left[\dfrac{0.6F_y t}{1.50}, \dfrac{0.6F_u t}{2.00}\right]$ for a one-inch length

$$\tag{7.34}$$

Example 7.11   A flat bar used as a tension member is connected to a gusset plate, as shown in Figure 7.40. The welds are $\tfrac{3}{16}$-inch fillet welds made with E70XX electrodes. The connected parts are of A36 steel. Assume that the tensile strength of the member is adequate and determine the available strength of the welded connection.

Solution   Because the welds are placed symmetrically about the axis of the member, this connection qualifies as a simple connection, and there is no additional load due to eccentricity. Since both weld segments are parallel to the applied load, $\theta = 0°$ and the basic strength of the weld is $F_W = 0.60F_{EXX}$. The nominal load capacity per inch of weld is

$$R_n = 0.707wF_w = 0.707\left(\frac{3}{16}\right)(0.6 \times 70) = 5.568 \text{ kips/in.}$$

FIGURE 7.40

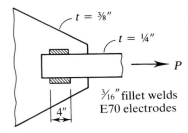

$t = \frac{3}{8}''$

$t = \frac{1}{4}''$

P

$\frac{3}{16}''$ fillet welds
E70 electrodes

4"

LRFD Solution

The design strength of the weld is

$$\phi R_n = 0.75(5.568) = 4.176 \text{ kips/in.}$$

Check the base metal shear. Since both components are of the same type of steel, the smaller thickness will control. The shear yield strength is

$$\phi R_n = \phi(0.6F_y t) = 1.00(0.6)(36)\left(\frac{1}{4}\right) = 5.4 \text{ kips/in.}$$

The shear rupture strength is

$$\phi R_n = \phi(0.6F_u t) = 0.75(0.6)(58)\left(\frac{1}{4}\right) = 6.525 \text{ kips/in.}$$

The base metal shear strength is therefore 5.4 kips/in., and the weld shear strength controls. For the connection,

$$\phi R_n = 4.176 \text{ kips/in.} \times (4 + 4) \text{ in.} = 33.4 \text{ kips}$$

Answer

Design strength of the weld is 33.4 kips.

ASD Solution

The allowable strength of the weld is

$$\frac{R_n}{\Omega} = \frac{5.568}{2.00} = 2.784 \text{ kips/in.}$$

Check the base metal shear. Since both components are of the same type of steel, the smaller thickness will control. The shear yield strength is

$$\frac{R_n}{\Omega} = \frac{0.6F_y t}{1.50} = \frac{0.6(36)(1/4)}{1.50} = 3.6 \text{ kips/in.}$$

The shear rupture strength is

$$\frac{R_n}{\Omega} = \frac{0.6F_u t}{2.00} = \frac{0.6(58)(1/4)}{2.00} = 4.35 \text{ kips/in.}$$

The base metal shear strength is therefore 3.6 kips/in., and the weld shear strength controls. For the connection,

$$\frac{R_n}{\Omega} = 2.784 \text{ kips/in.} \times (4 + 4) \text{ in.} = 22.3 \text{ kips}$$

Answer   Allowable strength of the weld is 22.3 kips.

Example 7.12   If the connection of Example 7.11 includes both the 4-inch-long longitudinal welds shown in Figure 7.40 and an additional 4-inch-long transverse weld at the end of the member, what is the available strength of the connection?

LRFD Solution   From Example 7.11, the weld shear design strength is

$$\phi R_n = 4.176 \text{ kips/in.}$$

Also from Example 7.11, the base metal shear design strength is

$$\phi R_n = 5.4 \text{ kips/in.}$$

So the weld strength of 4.176 kips/in. controls.
To determine the strength of the connection, we will investigate the two options given in AISC J2.4c.

1. Use the basic weld strength for both the longitudinal and transverse welds.

   $$\phi R_n = 4.176 (4 + 4 + 4) = 50.1 \text{ kips}$$

2. Use 0.85 times the basic weld strength for the longitudinal welds and 1.5 times the basic weld strength for the transverse weld.

   $$\phi R_n = 0.85 (4.176)(4 + 4) + 1.5(4.176)(4) = 53.5 \text{ kips}$$

The larger value controls.

Answer   Design strength of the weld is 53.5 kips.

ASD Solution   From Example 7.11, the allowable weld shear strength is 2.784 kips/in., and the allowable base metal shear strength is 3.6 kips/in. The weld strength therefore controls.
To determine the strength of the connection, we will investigate the two options given in AISC J2.4c.

1. Use the basic weld strength for both the longitudinal and transverse welds.

   $$\frac{R_n}{\Omega} = 2.784(4 + 4 + 4) = 33.4 \text{ kips}$$

2. Use 0.85 times the basic weld strength for the longitudinal welds and 1.5 times the basic weld strength for the transverse weld.

$$\frac{R_n}{\Omega} = 0.85(2.784)(4+4) + 1.5(2.784)(4) = 35.6 \text{ kips}$$

The larger value controls.

Answer   Allowable strength of the weld is 35.6 kips.   ■

When E70 electrodes are used, and this will usually be the case, computation of the weld shear strength can be simplified. The strength per unit length can be computed for a $\frac{1}{16}$-inch increment of weld size (and fillet welds are specified to the nearest $\frac{1}{16}$ inch).

LRFD: The weld design shear strength from Equation 7.29 is

$$\phi R_n = 0.75(0.707wF_W) = 0.75(0.707)\left(\frac{1}{16}\right)(0.6 \times 70) = 1.392 \text{ kips/in.}$$

Using this constant, the design shear strength of the $\frac{3}{16}$-inch fillet weld in Example 7.11 is

$$\phi R_n = 1.392 \times 3 \text{ sixteenths} = 4.176 \text{ kips/in.}$$

The base metal shear strength expression can also be somewhat simplified. The shear yield design strength per unit length is

$$\phi R_n = 1.0(0.6F_y t) = 0.6 \ F_y t \qquad \text{for a one-inch length} \qquad (7.35)$$

and the base metal shear rupture design strength per unit length is

$$\phi R_n = 0.75(0.6F_u t) = 0.45 \ F_u t \qquad \text{for a one-inch length} \qquad (7.36)$$

ASD: The weld allowable shear strength from Equation 7.31 is

$$\frac{R_n}{\Omega} = \frac{0.707wF_W}{2.00} = \frac{0.707(1/16)(0.6 \times 70)}{2.00} = 0.9279 \text{ kips/in.}$$

Using this constant, the allowable shear strength of the $\frac{3}{16}$-inch fillet weld in Example 7.11 is

$$\phi R_n = 0.9279 \times 3 \text{ sixteenths} = 2.784 \text{ kips/in.}$$

The base metal shear strength expression can also be simplified. The allowable shear yield strength per unit length is

$$\frac{R_n}{\Omega} = \frac{0.6F_y t}{1.50} = 0.4F_y t \qquad \text{for a one-inch length} \qquad (7.37)$$

and the base metal allowable shear rupture strength per unit length is

$$\frac{R_n}{\Omega} = \frac{0.6F_u t}{2.00} = 0.3F_u t \qquad \text{for a one-inch length} \tag{7.38}$$

Example 7.13　A connection of the type used in Example 7.11 must resist a service dead load of 9 kips and a service live load of 18 kips. What total length of ¼-inch fillet weld, E70XX electrode, is required? Assume that both connected parts are ⅜ inch thick.

LRFD Solution　$P_u = 1.2D + 1.6L = 1.2(9) + 1.6(18) = 39.6$ kips

The shear strength of the weld per inch of length is

$$1.392 \times 4 \text{ sixteenths} = 5.568 \text{ kips/in.}$$

The shear yield strength of the base metal is

$$0.6F_y t = 0.6(36)\left(\frac{3}{8}\right) = 8.1 \text{ kips/in.}$$

and the shear rupture strength of the base metal is

$$0.45F_u t = 0.45(58)\left(\frac{3}{8}\right) = 9.788 \text{ kips/in.}$$

The weld strength of 5.568 kips/in. governs.

The total length required is

$$\frac{39.6 \text{ kips}}{5.568 \text{ kips/in.}} = 7.11 \text{ in.}$$

Answer　Use 8 inches total, 4 inches on each side.

ASD Solution　$P_a = D + L = 9 + 18 = 27$ kips

The shear strength of the weld per inch of length is

$$0.9279 \times 4 \text{ sixteenths} = 3.712 \text{ kips/in.}$$

The allowable shear yield strength of the base metal is

$$0.4F_y t = 0.4(36)\left(\frac{3}{8}\right) = 5.4 \text{ kips/in.}$$

and the allowable shear rupture strength of the base metal is

$$0.3F_u t = 0.3(58)\left(\frac{3}{8}\right) = 6.525 \text{ kips/in.}$$

The weld strength of 3.712 kips/in. governs.

The total length required is

$$\frac{27 \text{ kips}}{3.712 \text{ kips/in.}} = 7.27 \text{ in.}$$

Answer Use 8 inches total, 4 inches on each side. ∎

Practical design of welded connections requires a consideration of such details as maximum and minimum weld sizes and lengths. The requirements for fillet welds are found in AISC J2.2b and are summarized here.

Minimum Size

The minimum size permitted is a function of the thickness of the *thinner* connected part and is given in AISC Table J2.4. This requirement is taken directly from the American Welding Society *Structural Welding Code* (AWS, 2004).

Maximum Size

Along the edge of a part less than $\frac{1}{4}$ inch thick, the maximum fillet weld size is equal to the thickness of the part. For parts $\frac{1}{4}$ inch or more in thickness, the maximum size is $t - \frac{1}{16}$ inch, where t is the thickness of the part.

For fillet welds other than those along edges (as in Figure 7.41), there is no maximum size specified. In these and all other cases, the maximum size to be used in strength computation would be that limited by the base metal shear strength.

FIGURE 7.41

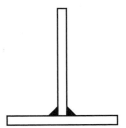

Minimum Length

The minimum permissible length of a fillet weld is four times its size. This limitation is certainly not severe, but if this length is not available, a shorter length can be used if the effective size of the weld is taken as one-fourth its length. Flat bar tension-member connections of the type shown in Figure 7.42, similar to those in the preceding examples, are in the special case category for shear lag in welded connections, covered in Chapter 3. The length of the welds in this case may not be less than the distance between them—that is, $L \geq W$.

FIGURE 7.42

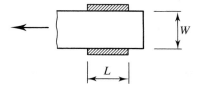

Maximum Length

AISC does not impose a limit on the length of welds, but for *end-loaded* welds, there are some restrictions. End-loaded welds are longitudinal welds at the end of an axially loaded member. If the length exceeds 100 times the weld size, a reduced effective length is used in the computation of strength. The effective length is obtained by multiplying the actual length by a factor β, where

$$\beta = 1.2 - 0.002(L/w) \le 1.0 \qquad \text{(AISC Equation J2-1)}$$

L = actual length of weld

w = weld size

If the length is larger than 300 times the weld size, use $\beta = 0.60$.

End Returns

When a weld extends to the end of a member, it is sometimes continued around the corner, as shown in Figure 7.43. The primary reason for this continuation, called an *end return*, is to ensure that the weld size is maintained over the full length of the weld. The AISC Specification does not require end returns.[*]

Small welds are generally cheaper than large welds. The maximum size that can be made with a single pass of the electrode is approximately $\frac{5}{16}$ inch, and multiple passes will add to the cost. In addition, for a given load capacity, although a small weld must be made longer, a larger and shorter weld will require more volume of weld metal. Reducing the volume of weld metal will also minimize heat buildup and residual stresses.

FIGURE 7.43

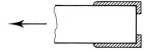

Example 7.14   A plate $\frac{1}{2} \times 4$ of A36 steel is used as a tension member to carry a service dead load of 6 kips and a service live load of 18 kips. It is to be attached to a $\frac{3}{8}$-inch gusset plate, as shown in Figure 7.44. Design a welded connection.

---

[*]If end returns are used in certain types of beam-to-column connections, AISC Commentary J2.2b places restrictions on their length.

FIGURE 7.44

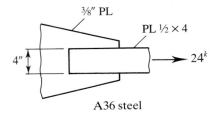

A36 steel

Solution   The base metal in this connection is A36 steel, so E70XX electrodes will be used. No restrictions on connection length have been stipulated, so weld length will not be limited, and the smallest permissible size will be used:

$$\text{Minimum size} = \frac{3}{16} \text{ in.}$$

(AISC Table J2.4)

LRFD Solution   Try a $\frac{3}{16}$-in. fillet weld, using E70XX electrodes. The design strength per inch is

$$1.392 \times 3 \text{ sixteenths} = 4.176 \text{ kips/in.}$$

The shear yield strength of the base metal is

$$0.6F_y t = 0.6(36)\left(\frac{3}{8}\right) = 8.1 \text{ kips/in.}$$

and the shear rupture strength of the base metal is

$$0.45F_u t = 0.45(58)\left(\frac{3}{8}\right) = 9.788 \text{ kips/in.}$$

The weld strength of 4.176 kips/in. governs. The factored load is

$$P_u = 1.2D + 1.6L = 1.2(6) + 1.6(18) = 36 \text{ kips}$$

and

$$\text{Required length} = \frac{36}{4.176} = 8.62 \text{ in.}$$

$$\text{Minimum length} = 4w = 4\left(\frac{3}{16}\right) = 0.75 \text{ in.} < 8.62 \text{ in.} \qquad \text{(OK)}$$

Use two 4.5-in.-long side welds for a total length of $2 \times 4.5 = 9$ in. For this type of connection, the length of the side welds must be at least equal to the transverse distance between them, or 4 inches in this case. The provided length of 4.5 inches will therefore be adequate.

Answer   Use a $\frac{3}{16}$-inch fillet weld with E70XX electrodes, with a total length of 9 inches, as shown in Figure 7.45.

ASD Solution   Try a $\frac{3}{16}$-in. fillet weld, using E70XX electrodes. The allowable strength per inch is

$$0.9279 \times 3 \text{ sixteenths} = 2.784 \text{ kips/in.}$$

FIGURE 7.45

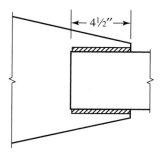

The allowable shear yield strength of the base metal is

$$0.4F_yt = 0.4(36)\left(\frac{3}{8}\right) = 5.4 \text{ kips/in.}$$

and the allowable shear rupture strength of the base metal is

$$0.3F_ut = 0.3(58)\left(\frac{3}{8}\right) = 6.525 \text{ kips/in.}$$

The weld strength of 2.784 kips/in. governs. The load to be resisted is

$$P_a = D + L = 6 + 18 = 24 \text{ kips}$$

and

$$\text{Required length} = \frac{24}{2.784} = 8.62 \text{ in.}$$

$$\text{Minimum length} = 4w = 4\left(\frac{3}{16}\right) = 0.75 \text{ in.} < 8.62 \text{ in.} \qquad \text{(OK)}$$

Use two 4.5-in.-long side welds for a total length of $2 \times 4.5 = 9$ in. For this type of connection, the length of the side welds must be at least equal to the transverse distance between them, or 4 inches in this case. The provided length of 4.5 inches will therefore be adequate.

Answer   Use a $\frac{3}{16}$-inch fillet weld with E70XX electrodes, with a total length of 9 inches, as shown in Figure 7.45.

■

Weld Symbols

Welds are specified on design drawings by standard symbols, which provide a convenient method for describing the required weld configuration. Details are given in Part 8 of the *Manual,* "Design Considerations for Welds," and so are not fully covered here. In this book, we provide only a brief introduction to the standard symbols for fillet welds. The following discussion refers to the symbols shown in Figure 7.46.

FIGURE 7.46

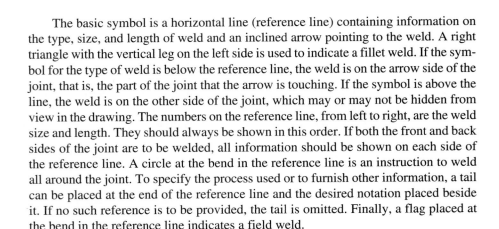

The basic symbol is a horizontal line (reference line) containing information on the type, size, and length of weld and an inclined arrow pointing to the weld. A right triangle with the vertical leg on the left side is used to indicate a fillet weld. If the symbol for the type of weld is below the reference line, the weld is on the arrow side of the joint, that is, the part of the joint that the arrow is touching. If the symbol is above the line, the weld is on the other side of the joint, which may or may not be hidden from view in the drawing. The numbers on the reference line, from left to right, are the weld size and length. They should always be shown in this order. If both the front and back sides of the joint are to be welded, all information should be shown on each side of the reference line. A circle at the bend in the reference line is an instruction to weld all around the joint. To specify the process used or to furnish other information, a tail can be placed at the end of the reference line and the desired notation placed beside it. If no such reference is to be provided, the tail is omitted. Finally, a flag placed at the bend in the reference line indicates a field weld.

Example 7.15   A plate $\frac{1}{2} \times 8$ of A36 steel is used as a tension member and is to be connected to a $\frac{3}{8}$-inch-thick gusset plate, as shown in Figure 7.47. The length of the connection cannot exceed 8 inches, and all welding must be done on the near side. Design a weld to develop the full tensile capacity of the member.

FIGURE 7.47

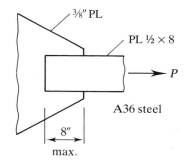

⅜″ PL

PL ½ × 8

P

A36 steel

8″
max.

LRFD Solution

The design strength of the member based on its gross area is

$$\phi_t P_n = 0.90 F_y A_g = 0.90(36)\left(\frac{1}{2}\right)(8) = 129.6 \text{ kips}$$

Compute the design strength of the member based on its effective area. For a flat bar connection, if the welds are along the sides only, $A_e = A_g U$. If there is also a transverse weld at the end, then $A_e = A_g$. Assuming the latter, we have

$$\phi_t P_n = 0.75 F_u A_e = 0.75(58)\left(\frac{1}{2}\right)(8) = 174.0 \text{ kips}$$

Design for a factored load of 129.6 kips and use E70 electrodes.

From AISC Table J2.4, the minimum weld size is ³⁄₁₆ inch. Because of the length constraint, however, try a slightly larger weld. **Try a ¼-inch E70 fillet weld:**

Design strength per inch of weld = 1.392 × 4 sixteenths = 5.568 kips/in.

The base metal shear yield strength is

$$0.6 F_y t = 0.6(36)\left(\frac{3}{8}\right) = 8.1 \text{ kips/in.}$$

and the base metal shear rupture strength is

$$0.45 F_u t = 0.45(58)\left(\frac{3}{8}\right) = 9.788 \text{ kips/in.}$$

The weld strength of 5.568 kips/in. governs. Both longitudinal and transverse welds will be used. To determine the required length of the longitudinal welds, we investigate the two options specified in AISC J2.4c. First, assuming the same strength for both the longitudinal and transverse welds,

$$\text{Total required length of weld} = \frac{129.6}{5.568} = 23.28 \text{ in.}$$

$$\text{Length of longitudinal welds} = \frac{23.28 - 8}{2} = 7.64 \text{ in.}$$

For the second option, the strength of the longitudinal welds is

$$0.85(5.568) = 4.733 \text{ kips/in.}$$

and the strength of the transverse weld is

$$1.5(5.568) = 8.352 \text{ kips/in.}$$

The load to be carried by the longitudinal welds is

$$129.6 - 8(8.352) = 62.78 \text{ kips}$$

so the required length of the longitudinal welds is

$$\frac{62.78}{2(4.733)} = 6.63 \text{ in.}$$

The second option requires shorter longitudinal welds. Try an 8-inch transverse weld and two 7-inch longitudinal welds. Check the block shear strength of the gusset plate.

$$A_{gv} = A_{nv} = 2 \times \frac{3}{8}(7) = 5.25 \text{ in.}^2$$

$$A_{nt} = \frac{3}{8}(8) = 3.0 \text{ in.}^2$$

From AISC Equation J4-5,

$$R_n = 0.6F_u A_{nv} + U_{bs}F_u A_{nt}$$
$$= 0.6(58)(5.25) + 1.0(58)(3.0) = 356.7 \text{ kips}$$

with an upper limit of

$$0.6F_y A_{gv} + U_{bs}F_u A_{nt} = 0.6(36)(5.25) + 1.0(58)(3.0) = 287.4 \text{ kips (controls)}$$

The design strength is

$$\phi R_n = 0.75(287.4) = 216 \text{ kips} > 129.6 \text{ kips} \qquad \text{(OK)}$$

Answer   Use the weld shown in Figure 7.48.

FIGURE 7.48

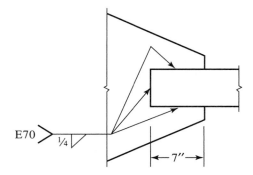

E70

$\frac{1}{4}$

7″

ASD Solution The allowable strength of the member based on its gross area is

$$\frac{P_n}{\Omega_t} = \frac{F_y A_g}{1.67} = \frac{36(1/2)(8)}{1.67} = 86.23 \text{ kips}$$

Compute the allowable strength based on the effective area. For a flat bar connection, if the welds are along the sides only, $A_e = A_g U$. If there is also a transverse weld at the end, then $A_e = A_g$. Assuming the latter, we have

$$\frac{P_n}{\Omega_t} = \frac{F_u A_e}{2.00} = \frac{58(1/2)(8)}{2.00} = 116.0 \text{ kips}$$

Design for a load of 86.23 kips and use E70 electrodes. From AISC Table J2.4, the minimum weld size is $\frac{3}{16}$ inch. Because of the length constraint, a slightly larger size will be tried.

Try a ¼-inch E70 fillet weld.

Allowable strength per inch of weld = 0.9279×4 sixteenths = 3.712 kips/in.

The allowable base metal shear yield strength is

$$0.4 F_y t = 0.4(36)\left(\frac{3}{8}\right) = 5.4 \text{ kips/in.}$$

and the allowable base metal shear rupture strength is

$$0.3 F_u t = 0.3(58)\left(\frac{3}{8}\right) = 6.525 \text{ kips/in.}$$

The weld strength of 3.712 kips/in. governs. Both longitudinal and transverse welds will be used. To determine the required length of the longitudinal welds, we investigate the two options specified in AISC J2.4c. First, assuming the same strength for both the longitudinal and transverse welds,

$$\text{Total required length of weld} = \frac{86.23}{3.712} = 23.23 \text{ in.}$$

$$\text{Length of longitudinal welds} = \frac{23.23 - 8}{2} = 7.62 \text{ in.}$$

For the second option, the strength of the longitudinal welds is

$$0.85(3.712) = 3.155 \text{ kips/in.}$$

and the strength of the transverse weld is

$$1.5(3.712) = 5.568 \text{ kips/in.}$$

The load to be carried by the longitudinal welds is

$$86.23 - 8(5.568) = 41.69 \text{ kips}$$

so the required length of the longitudinal welds is

$$\frac{41.69}{2(3.155)} = 6.61 \text{ in.}$$

The second option requires shorter longitudinal welds. Try an 8-inch transverse weld and two 7-inch longitudinal welds. Check the block shear strength of the gusset plate.

$$A_{gv} = A_{nv} = 2 \times \frac{3}{8}(7) = 5.25 \text{ in.}^2$$

$$A_{nt} = \frac{3}{8}(8) = 3.0 \text{ in.}^2$$

From AISC Equation J4-5,

$$R_n = 0.6F_u A_{nv} + U_{bs} F_u A_{nt}$$
$$= 0.6(58)(5.25) + 1.0\,(58)(3.0) = 356.7 \text{ kips}$$

with an upper limit of

$$0.6F_y A_{gv} + U_{bs} F_u A_{nt} = 0.6(36)(5.25) + 1.0(58)(3.0) = 287.4 \text{ kips} \quad \text{(controls)}$$

and the allowable strength is

$$\frac{R_n}{\Omega} = \frac{287.4}{2.00} = 144 \text{ kips} > 86.23 \text{ kips} \qquad \text{(OK)}$$

Answer   Use the weld shown in Figure 7.48.

Problems

Bearing Strength, Spacing, and Edge-Distance Requirements

7.3-1   The tension member is a PL $\frac{1}{2} \times 6$. It is connected to a $\frac{3}{8}$-inch-thick gusset plate with $\frac{7}{8}$-inch-diameter bolts. Both components are of A36 steel.

a. Check all spacing and edge-distance requirements.
b. Compute the nominal strength in bearing.

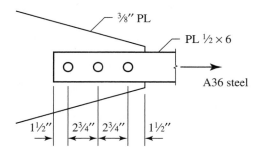

FIGURE P7.3-1

7.3-2 The tension member shown in Figure P7.3-2 is a PL ½ × 5½ of A242 steel. It is connected to a ⅜-inch-thick gusset plate (also A242 steel) with ¾-inch-diameter bolts.

a. Check all spacing and edge-distance requirements.
b. Compute the nominal strength in bearing.

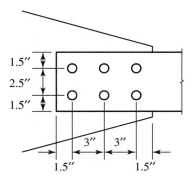

FIGURE P7.3-2

Shear Strength

7.4-1 A C8×18.7 is to be used as a tension member. The channel is bolted to a ⅜-inch gusset plate with ⅞-inch-diameter, A307 bolts. The tension member is A572 Grade 50 steel and the gusset plate is A36.

a. Check all spacing and edge-distance requirements.
b. Compute the design strength based on shear and bearing
c. Compute the allowable strength based on shear and bearing.

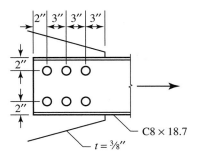

FIGURE P7.4-1

7.4-2 A ½-inch-thick tension member is spliced with two ¼-inch-thick splice plates as shown in Figure P7.4-2. The bolts are ⅞-inch-diameter, A325 and all steel is A36.

a. Check all spacing and edge-distance requirements.
b. Compute the nominal strength based on shear and bearing.

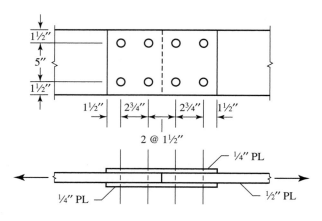

FIGURE P7.4-2

7.4-3 Determine the number of ¾-inch-diameter, A325 bolts required, based on shear and bearing, along line *a–b* in Figure P7.4-3. The given loads are service loads. A36 steel is used. Assume that the bearing strength is controlled by the upper limit of $2.4dtF_u$.

a. Use LRFD.
b. Use ASD.

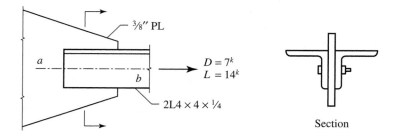

FIGURE P7.4-3

7.4-4 The splice plates shown in Figure P7.4-4 are ¼-inch-thick. How many ⅞-inch-diameter, A325 bolts are required? The given load is a service load consisting of 25% dead load and 75% live load. A36 steel is used.

a. Use LRFD.
b. Use ASD.

FIGURE P7.4-4

7.4-5 The tension member is an L6×3½×⁵⁄₁₆. It is connected to a ⁵⁄₁₆-inch-thick gusset plate with ¾-inch-diameter, A325 bolts. Both the tension member and the gusset plate are of A36 steel. What is the total *service* load that can be supported, based on bolt shear and bearing, if the ratio of live load to dead load is 2.0? The bolt threads are in the plane of shear.

a. Use LRFD.
b. Use ASD.

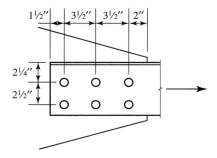

FIGURE P7.4-5

7.4-6   A double-angle tension member, $2L4 \times 3 \times \frac{1}{2}$ LLBB, is connected to a $\frac{3}{8}$-inch-thick gusset plate with $\frac{7}{8}$-inch-diameter, A325 bolts as shown in Figure P7.4-6. Both the tension member and the gusset plate are of A36 steel. Does the connection have enough capacity based on shear and bearing? It is not known whether the bolt threads will be in shear.

a. Use LRFD.
b. Use ASD.

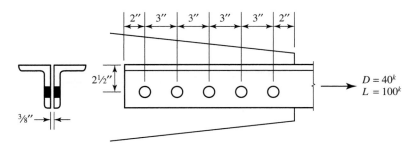

FIGURE P7.4-6

Slip-Critical and Bearing-Type Connections

7.6-1   A double-angle shape, $2L6 \times 6 \times \frac{5}{8}$, is connected to a $\frac{5}{8}$-inch gusset plate as shown in Figure P7.6-1. Determine the maximum total service load that can be applied if the ratio of dead load to live load is 8.5. The bolts are $\frac{7}{8}$-inch-diameter, A325 bearing-type bolts. A572 Grade 50 steel is used for the angle, and A36 steel is used for the gusset plate.

a. Use LRFD.
b. Use ASD.

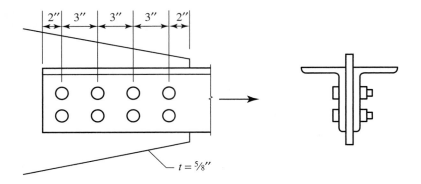

FIGURE P7.6-1

7.6-2 Determine the total number of $\frac{7}{8}$-inch, A325 bearing-type bolts required for the tension splice in Figure P7.6-2. The threads are not in the shear planes.

a. Use LRFD.
b. Use ASD.

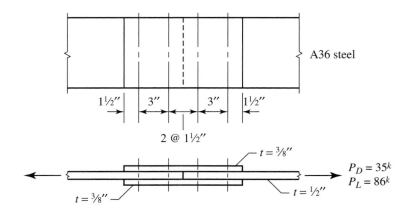

FIGURE P7.6-2

7.6-3 A WT 7×19 of A572 Grade 50 steel is used as a tension member. It will be connected to a $\frac{3}{8}$-inch-thick gusset plate, also of A572 Grade 50 steel, with $\frac{7}{8}$-inch-diameter bolts. The connection is through the flange of the tee and is a bearing-type connection. The connection must resist a service dead load of 45 kips and a service live load of 90 kips. Assume that the nominal bearing strength will be $2.4dtF_u$ and answer the following questions:

a. How many A307 bolts are required?
b. How many A325 bolts are required?
c. How many A490 bolts are required?
d. If the relative costs of A307, A325, and A490 bolts are in the ratio 1.0:1.7:2.6, which type of bolt will be the most economical for this connection?

7.6-4   a. Prepare a table showing values of A325 bolt shear strength and slip-critical strength for bolt diameters of ½ inch to 1½ inches in increments of ⅛ inch. Assume Class A surfaces and that threads are in the plane of shear. Your table should have the following form:

Bolt diameter (in.)	Single-shear design strength, ϕR_n (kips)	Slip-critical design strength, one slip plane, ϕR_n (kips)	Single-shear allowable strength, R_n/Ω (kips)	Slip-critical allowable strength, one slip plane, R_n/Ω (kips)
½	7.07	4.75	4.71	3.16
.	.	.	.	.
.	.	.	.	.
.	.	.	.	.

b. What conclusions can you draw from this table?

7.6-5   A plate ½ × 6½ of A36 steel is used as a tension member as shown in Figure P7.6-5. The gusset plate is ⅝-inch-thick and is also of A36 steel. The bolts are 1⅛-inch-diameter, A325. No slip is permitted. Using a ratio of live load-to-dead load of 3.0, determine the maximum service load, P, that can be applied. *Investigate all possible failure modes.*

a. Use LRFD.
b. Use ASD.

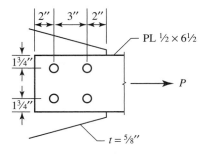

FIGURE P7.6-5

7.6-6   The tension member shown in Figure P7.6-6 is an L6×3½×½. It is connected with 1⅛-inch-diameter, A325 slip-critical bolts to a ⅜-inch-thick gusset plate. It must resist a service dead load of 20 kips, a service live load of 60 kips, and a service wind load of 20 kips. The length is 9 feet and all structural steel is A36. Are the member and its connection satisfactory?

a. Use LRFD.
b. Use ASD.

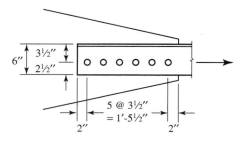

FIGURE P7.6-6

Design

7.7-1 A C9 × 20 is used as a tension member and is connected to a ½-inch gusset plate as shown in Figure P7.7-1. A242 steel is used for the tension member and A36 steel is used for the gusset plate. The member has been designed to resist a service dead load of 40 kips and a service live load of 80 kips. If the connection is to be slip critical, how many 1⅜-inch-diameter, A325 bolts are needed? Show a sketch with a possible layout. Assume that the member tensile and block shear strengths will be adequate.

a. Use LRFD.
b. Use ASD.

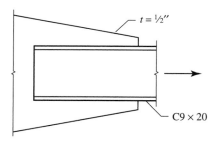

FIGURE P7.7-1

7.7-2 Design a single-angle tension member and a bolted connection for the following conditions:

- dead load = 50 kips, live load = 100 kips, and wind load = 45 kips
- A325 bolts, no slip permitted
- ⅜-inch-thick gusset plate
- A36 steel for both the tension member and the gusset plate
- length = 20 feet

Provide a complete sketch showing all information needed for the fabrication of the connection.

a. Use LRFD.
b. Use ASD.

7.7-3   Design the tension member and its connection for the following conditions:

- length = 15 feet
- the connection will be to a ⅜-inch-thick gusset plate
- all structural steel is A36
- the connection will be bolted. Slip is not permitted
- service dead load = 45 kips and service live load = 105 kips

A tension member will be a double-angle section with unequal legs, long legs back-to-back.

Provide a complete sketch showing all information needed for the fabrication of the connection.

a. Use LRFD.
b. Use ASD.

High-Strength Bolts in Tension

7.8-1   Investigate both the tee and the bolts in the hanger connection in Figure P7.8-1. Include the effects of prying.

a. Use LRFD.
b. Use ASD.

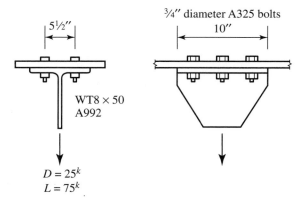

FIGURE P7.8-1

7.8-2   Determine the adequacy of the hanger connection in Figure P7.8-2 Account for prying action.

a. Use LRFD.
b. Use ASD.

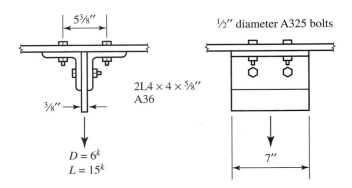

FIGURE P7.8-2

Combined Shear and Tension in Fasteners

7.9-1   A bracket must support the service loads shown in Figure P7.9-1, which act through the center of gravity of the connection. The connection to the column flange is with eight $\frac{7}{8}$-inch-diameter, A325 bearing-type bolts. A992 steel is used for all components. Is the connection adequate?

a. Use LRFD.
b. Use ASD.

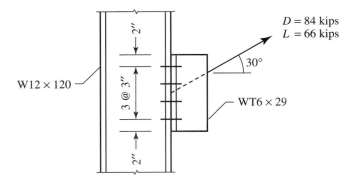

FIGURE P7.9-1

7.9-2   A structural tee bracket is attached to a column flange with six bolts as shown in Figure P7.9-2. All structural steel is A992. Check this connection for compliance with the AISC Specification. Assume that the bearing strength is controlled by the upper limit of $2.4dtF_u$.

a. Use LRFD.
b. Use ASD.

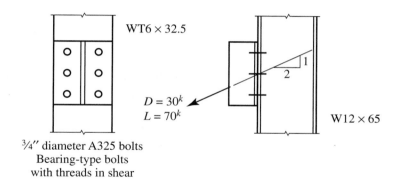

FIGURE P7.9-2

7.9-3   In the connection of Figure P7.9-3, how many ⅞-inch-diameter A325 bearing-type bolts are needed? The 80-kip load is a service load, consisting of 20 kips dead load and 60 kips live load. Assume that the bolt threads are in the plane of shear and bearing strength is controlled by the upper limit of $2.4dtF_u$.

a. Use LRFD.
b. Use ASD.

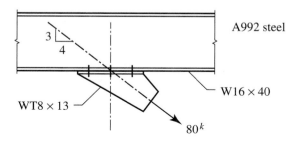

FIGURE P7.9-3

7.9-4   A double-angle tension member is attached to a ⅞-inch gusset plate, which in turn is connected to a column flange via another pair of angles as shown in Figure P7.9-4. The given load is a service load consisting of 25% dead load and 75% live load. All connections are to be made with ⅞-inch-diameter, A490 slip-critical bolts. Assume that the threads are in shear. Determine the required number of bolts and show their location on a sketch. The column is of A992 steel and the angles and plate are A36.

a. Use LRFD.
b. Use ASD.

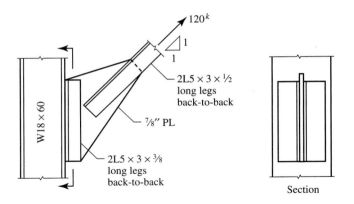

FIGURE P7.9-4

7.9-5 A bracket cut from a W12 × 120 is connected to a W12 × 120 column flange with 12 A325 bearing-type bolts as shown in Figure P7.9-5. A992 steel is used. The line of action of the load passes through the center of gravity of the connection. What size bolt is required?

a. Use LRFD.
b. Use ASD.

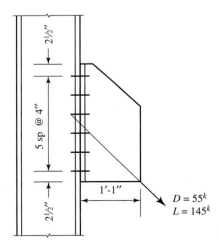

FIGURE P7.9-5

Fillet Welds

7.11-1 Determine the maximum service load that can be applied if the live load-to-dead load ratio is 2.5. Investigate all limit states. The tension member is of A572 Grade 50 steel and the gusset plate is A36. The weld is a $^3/_{16}$-inch fillet weld with E70 electrodes.

a. Use LRFD.
b. Use ASD.

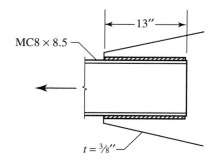

FIGURE P7.11-1

7.11-2 Determine the maximum service load that can be applied if the live load-to-dead load ratio is 3.0. Investigate all limit states. All structural steel is A36 and the weld is a ¼-inch fillet weld with E70 electrodes. Note that the tension member is a double-angle shape, and both of the angles are welded, as shown in Figure P7.11-2.

a. Use LRFD.
b. Use ASD.

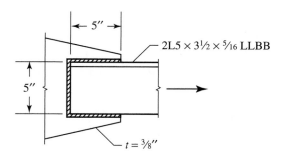

FIGURE P7.11-2

7.11-3 Determine the maximum service load, P, that can be applied if the live load-to-dead load ratio is 2.0. Each component is a PL ¾ × 7 of A242 steel. The weld is a ½-inch fillet weld, E70 electrode.

a. Use LRFD.
b. Use ASD.

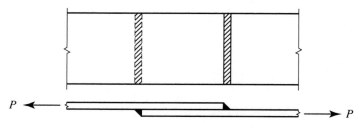

FIGURE P7.11-3

7.11-4   A tension member splice is made with ¼-inch E70 fillet welds as shown in Figure P7.11-4. Each side of the splice is welded as shown. The inner member is a PL ½ × 6 and each outer member is a PL ⁵⁄₁₆ × 3. All steel is A36. Determine the maximum service load, P, that can be applied if the load is equal parts dead load and live load.

a. Use LRFD.
b. Use ASD.

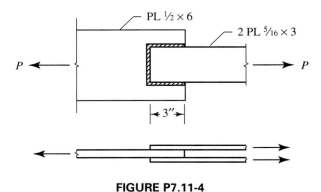

PL ½ × 6

2 PL ⁵⁄₁₆ × 3

P

P

3″

FIGURE P7.11-4

7.11-5   Design a welded connection. The given loads are service loads. Use $F_y = 50$ ksi for the angle tension member and $F_y = 36$ ksi for the gusset plate. Show your results on a sketch, complete with dimensions.

a. Use LRFD.
b. Use ASD.

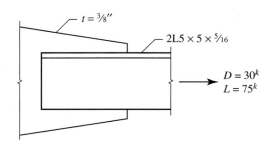

$t = ³⁄₈″$

2L5 × 5 × ⁵⁄₁₆

$D = 30^k$
$L = 75^k$

FIGURE P7.11-5

7.11-6   Design a welded connection for the conditions of Problem 7.4-3. Show your results on a sketch, complete with dimensions.

a. Use LRFD.
b. Use ASD.

7.11-7 Design a welded connection for an MC 9 × 23.9 of A572 Grade 50 steel connected to a ³⁄₈-inch-thick gusset plate. The gusset plate is A36 steel. Show your results on a sketch, complete with dimensions.

   a. Use LRFD.
   b. Use ASD.

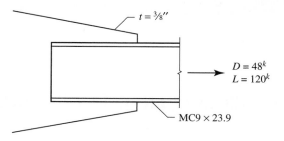

FIGURE P7.11-7

7.11-8 Design a welded connection to resist the available strength of the tension member in Figure P7.11-8 All steel is A36. Show your results on a sketch, complete with dimensions.

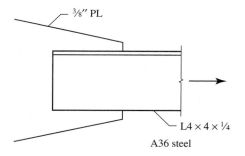

FIGURE P7.11-8

7.11-9 Select a double-angle tension member and design a welded connection to resist a dead load of 12 kips and a live load of 36 kips. The member will be 16 feet long and will be connected to a ⁵⁄₈-inch-thick gusset plate. Use A36 steel for both the tension member and the gusset plate. Show your results on a sketch, complete with dimensions.

7.11-10 Design a tension member and its connection for the following conditions:

   • The tension member will be an American Standard Channel.
   • Length = 17.5 feet.
   • The web of the channel will be welded to a ³⁄₈-inch-thick gusset plate.
   • The tension member will be A572 Grade 50 and the gusset plate will be A36.
   • Service dead load = 54 kips, service live load = 80 kips, and wind load = 75 kips.

   Show your results on a sketch, complete with dimensions.

8 Eccentric Connections

8.1 EXAMPLES OF ECCENTRIC CONNECTIONS

An eccentric connection is one in which the resultant of the applied loads does not pass through the center of gravity of the fasteners or welds. If the connection has a plane of symmetry, the centroid of the shear area of the fasteners or welds may be used as the reference point, and the perpendicular distance from the line of action of the load to the centroid is called the *eccentricity*. Although a majority of connections are probably loaded eccentrically, in many cases the eccentricity is small and may be neglected.

The *framed beam* connection shown in Figure 8.1a is a typical eccentric connection. This connection, in either bolted or welded form, is commonly used to connect beams to columns. Although the eccentricities in this type of connection are small and can sometimes be neglected, they do exist and are used here for illustration. There are actually two different connections involved: the attachment of the beam to the framing angles and the attachment of the angles to the column. These connections illustrate the two basic categories of eccentric connections: those causing only shear in the fasteners or welds and those causing both shear and tension.

If the beam and angles are considered separately from the column, as shown in Figure 8.1b, it is clear that the reaction R acts at an eccentricity e from the centroid of the areas of the fasteners in the beam web. These fasteners are thus subjected to both a shearing force and a couple that lies in the plane of the connection and causes torsional shearing stress.

If the column and the angles are isolated from the beam, as shown in Figure 8.1c, it is clear that the fasteners in the column flange are subjected to the reaction R acting at an eccentricity e from the plane of the fasteners, producing the same couple as before. In this case, however, the load is not in the plane of the fasteners, so the couple will tend to put the upper part of the connection in tension and compress the lower part. The fasteners at the top of the connection will therefore be subjected to both shear and tension.

Although we used a bolted connection here for illustration, welded connections can be similarly categorized as either shear only or shear plus tension.

Available strengths (maximum reaction capacities) for various framed beam connections are given in Tables 10-1 through 10-11 in Part 10 of the *Manual,* "Design of Simple Shear Connections."

FIGURE 8.1

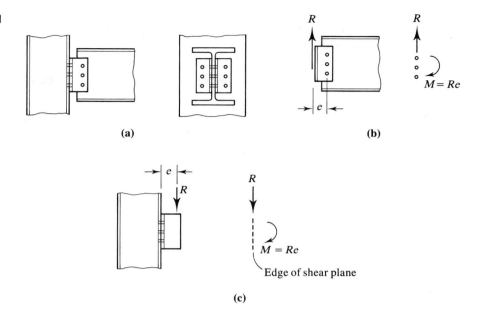

(a)   (b)

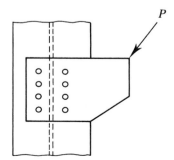

(c)

8.2   ECCENTRIC BOLTED CONNECTIONS: SHEAR ONLY

The column bracket connection shown in Figure 8.2 is an example of a bolted connection subjected to eccentric shear. Two approaches exist for the solution of this problem: the traditional elastic analysis and the more accurate (but more complex) ultimate strength analysis. Both will be illustrated.

Elastic Analysis

In Figure 8.3a, the fastener shear areas and the load are shown separate from the column and bracket plate. The eccentric load P can be replaced with the same load acting at the centroid plus the couple, $M = Pe$, where e is the eccentricity. If this

FIGURE 8.2

FIGURE 8.3

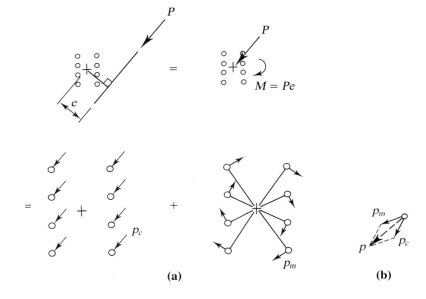

(a)   (b)

replacement is made, the load will be concentric, and each fastener can be assumed to resist an equal share of the load, given by $p_c = P/n$, where n is the number of fasteners. The fastener forces resulting from the couple can be found by considering the shearing stress in the fasteners to be the result of torsion of a cross section made up of the cross-sectional areas of the fasteners. If such an assumption is made, the shearing stress in each fastener can be found from the torsion formula

$$f_v = \frac{Md}{J} \tag{8.1}$$

where

d = distance from the centroid of the area to the point where the stress is being computed

J = polar moment of inertia of the area about the centroid

and the stress f_v is perpendicular to d. Although the torsion formula is applicable only to right circular cylinders, its use here is conservative, yielding stresses that are somewhat larger than the actual stresses.

If the parallel-axis theorem is used and the polar moment of inertia of each circular area about its own centroid is neglected, J for the total area can be approximated as

$$J = \Sigma \, Ad^2 = A \, \Sigma \, d^2$$

provided all fasteners have the same area, A. Equation 8.1 can then be written as

$$f_v = \frac{Md}{A \, \Sigma \, d^2}$$

and the shear force in each fastener caused by the couple is

$$p_m = Af_v = A \frac{Md}{A \Sigma d^2} = \frac{Md}{\Sigma d^2}$$

The two components of shear force thus determined can be added vectorially to obtain the resultant force, p, as shown in Figure 8.3b, where the lower right-hand fastener is used as an example. When the largest resultant is determined, the fastener size is selected so as to resist this force. The critical fastener cannot always be found by inspection, and several force calculations may be necessary.

It is generally more convenient to work with rectangular components of forces. For each fastener, the horizontal and vertical components of force resulting from direct shear are

$$p_{cx} = \frac{P_x}{n} \text{ and } p_{cy} = \frac{P_y}{n}$$

where P_x and P_y are the x- and y-components of the total connection load, P, as shown in Figure 8.4. The horizontal and vertical components caused by the eccentricity can be found as follows. In terms of the x- and y-coordinates of the centers of the fastener areas,

$$\Sigma d^2 = \Sigma(x^2 + y^2)$$

where the origin of the coordinate system is at the centroid of the total fastener shear area. The x-component of p_m is

$$p_{mx} = \frac{y}{d} p_m = \frac{y}{d} \frac{Md}{\Sigma d^2} = \frac{y}{d} \frac{Md}{\Sigma(x^2 + y^2)} = \frac{My}{\Sigma(x^2 + y^2)}$$

Similarly,

$$p_{my} = \frac{Mx}{\Sigma(x^2 + y^2)}$$

and the total fastener force is

$$p = \sqrt{(\Sigma p_x)^2 + (\Sigma p_y)^2}$$

FIGURE 8.4

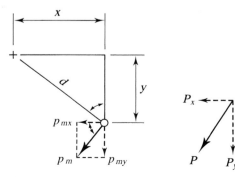

where

$$\Sigma\, p_x = p_{cx} + p_{mx}$$
$$\Sigma\, p_y = p_{cy} + p_{my}$$

If P, the load applied to the connection, is a factored load, then force p on the fastener is the factored load to be resisted in shear and bearing—that is, the required design strength. If P is a service load, then p will be the required allowable strength of the fastener.

Example 8.1 Determine the critical fastener force in the bracket connection shown in Figure 8.5.

Solution The centroid of the fastener group can be found by using a horizontal axis through the lower row and applying the principal of moments:

$$\bar{y} = \frac{2(5) + 2(8) + 2(11)}{8} = 6 \text{ in.}$$

The horizontal and vertical components of the load are

$$P_x = \frac{1}{\sqrt{5}}(50) = 22.36 \text{ kips} \leftarrow \quad \text{and} \quad p_y = \frac{2}{\sqrt{5}}(50) = 44.72 \text{ kips} \downarrow$$

Referring to Figure 8.6a, we can compute the moment of the load about the centroid:

$$M = 44.72(12 + 2.75) - 22.36(14 - 6) = 480.7 \text{ in.-kips} \quad \text{(clockwise)}$$

FIGURE 8.5

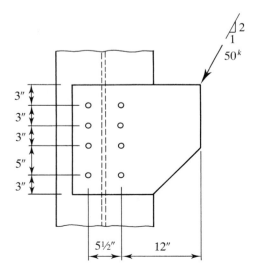

FIGURE 8.6

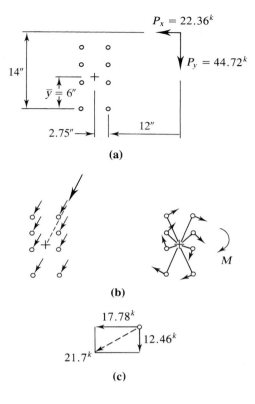

(a)

(b)

(c)

Figure 8.6b shows the directions of all component bolt forces and the relative magnitudes of the components caused by the couple. Using these directions and relative magnitudes as a guide and bearing in mind that forces add by the parallelogram law, we can conclude that the lower right-hand fastener will have the largest resultant force.

The horizontal and vertical components of force in each bolt resulting from the concentric load are

$$p_{cx} = \frac{22.36}{8} = 2.795 \text{ kips} \leftarrow \quad \text{and} \quad p_{cy} = \frac{44.72}{8} = 5.590 \text{ kips} \downarrow$$

For the couple,

$$\Sigma(x^2 + y^2) = 8(2.75)^2 + 2[(6)^2 + (1)^2 + (2)^2 + (5)^2] = 192.5 \text{ in.}^2$$

$$p_{mx} = \frac{My}{\Sigma(x^2 + y^2)} = \frac{480.7(6)}{192.5} = 14.98 \text{ kips} \leftarrow$$

$$p_{my} = \frac{Mx}{\Sigma(x^2 + y^2)} = \frac{480.7(2.75)}{192.5} = 6.867 \text{ kips} \downarrow$$

$$\Sigma p_x = 2.795 + 14.98 = 17.78 \text{ kips} \leftarrow$$

$$\Sigma p_y = 5.590 + 6.867 = 12.46 \text{ kips} \downarrow$$

$$p = \sqrt{(17.78)^2 + (12.46)^2} = 21.7 \text{ kips} \quad \text{(see Figure 8.6c)}$$

Answer The critical fastener force is 21.7 kips. Inspection of the magnitudes and directions of the horizontal and vertical components of the forces confirms the earlier conclusion that the fastener selected is indeed the critical one.

Ultimate Strength Analysis

The foregoing procedure is relatively easy to apply but is inaccurate—on the conservative side. The major flaw in the analysis is the implied assumption that the fastener load–deformation relationship is linear and that the yield stress is not exceeded. Experimental evidence shows that this is not the case and that individual fasteners do not have a well-defined shear yield stress. The procedure to be described here determines the ultimate strength of the connection by using an experimentally determined nonlinear load–deformation relationship for the individual fasteners.

The experimental study reported in Crawford and Kulak (1971) used ¾-inch-diameter A325 bearing-type bolts and A36 steel plates, but the results can be used with little error for A325 bolts of different sizes and steels of other grades. The procedure gives conservative results when used with slip-critical bolts and with A490 bolts (AISC, 1994).

The bolt force R corresponding to a deformation Δ is

$$R = R_{ult}(1 - e^{-\mu\Delta})^\lambda$$
$$= 74(1 - e^{-10\Delta})^{0.55} \tag{8.2}$$

where

R_{ult} = bolt shear force at failure = 74 kips
e = base of natural logarithms
μ = a regression coefficient = 10
λ = a regression coefficient = 0.55

The ultimate strength of the connection is based on the following assumptions:

1. At failure, the fastener group rotates about an instantaneous center (IC).
2. The deformation of each fastener is proportional to its distance from the IC and acts perpendicularly to the radius of rotation.
3. The capacity of the connection is reached when the ultimate strength of the fastener farthest from the IC is reached. (Figure 8.7 shows the bolt forces as resisting forces acting to oppose the applied load.)
4. The connected parts remain rigid.

As a consequence of the second assumption, the deformation of an individual fastener is

$$\Delta = \frac{r}{r_{max}}\Delta_{max} = \frac{r}{r_{max}}(0.34)$$

where

r = distance from the IC to the fastener
r_{max} = distance to the farthest fastener
Δ_{max} = deformation of the farthest fastener at ultimate = 0.34 in. (determined experimentally)

FIGURE 8.7

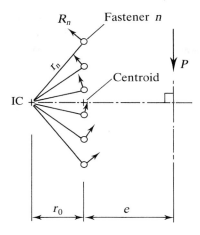

As with the elastic analysis, it is more convenient to work with rectangular components of forces, or

$$R_y = \frac{x}{r} R \text{ and } R_x = \frac{y}{r} R$$

where x and y are the horizontal and vertical distances from the instantaneous center to the fastener. At the instant of failure, equilibrium must be maintained, and the following three equations of equilibrium will be applied to the fastener group (refer to Figure 8.7):

$$\Sigma F_x = \sum_{n=1}^{m} (R_x)_n - P_x = 0 \tag{8.3}$$

$$M_{IC} = P(r_0 + e) - \sum_{n=1}^{m} (r_n \times R_n) = 0 \tag{8.4}$$

and

$$\Sigma F_y = \sum_{n=1}^{m} (R_y)_n - P_y = 0 \tag{8.5}$$

where the subscript n identifies an individual fastener and m is the total number of fasteners. The general procedure is to assume the location of the instantaneous center, then determine if the corresponding value of P satisfies the equilibrium equations. If so, this location is correct and P is the capacity of the connection. The specific procedure is as follows.

1. Assume a value for r_0.
2. Solve for P from Equation 8.4.
3. Substitute r_0 and P into Equations 8.3 and 8.5.
4. If these equations are satisfied within an acceptable tolerance, the analysis is complete. Otherwise, a new trial value of r_0 must be selected and the process repeated.

For the usual case of vertical loading, Equation 8.3 will automatically be satisfied. For simplicity and without loss of generality, we consider only this case. Even with this assumption, however, the computations for even the most trivial problems are overwhelming, and computer assistance is needed. Part (b) of Example 8.2 was worked with the aid of standard spreadsheet software.

Example 8.2

The bracket connection shown in Figure 8.8 must support an eccentric load consisting of 9 kips of dead load and 27 kips of live load. The connection was designed to have two vertical rows of four bolts, but one bolt was inadvertently omitted. If ⅞-inch-diameter A325 bearing-type bolts are used, is the connection adequate? Assume that the bolt threads are in the plane of shear. Use A36 steel for the bracket, A992 steel for the W6 × 25, and perform the following analyses: (a) elastic analysis, (b) ultimate strength analysis.

FIGURE 8.8

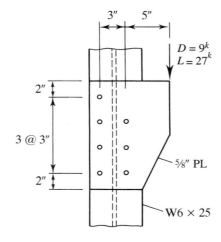

Solution

Compute the bolt shear strength.

$$A_b = \frac{\pi(7/8)^2}{4} = 0.6013 \text{ in.}^2$$

$$R_n = F_{nv}A_b = 48(0.6013) = 28.86 \text{ kips}$$

For the bearing strength, use a hole diameter of

$$h = d + \frac{1}{16} = \frac{7}{8} + \frac{1}{16} = \frac{15}{16} \text{ in.}$$

For the holes nearest the edge, use

$$L_c = L_e - \frac{h}{2} = 2 - \frac{15/16}{2} = 1.531 \text{ in.}$$

To determine which component has the smaller bearing strength, compare the values of tF_u (other variables are the same). For the plate,

$$tF_u = \left(\frac{5}{8}\right)(58) = 36.25 \text{ kips/in.}$$

For the W6 × 25,

$$tF_u = t_f F_u = 0.455(65) = 29.58 \text{ kips/in.} < 36.25 \text{ kips/in.}$$

The strength of the W6 × 25 will control.

$$R_n = 1.2 L_c t F_u = 1.2(1.531)(0.455)(65) = 54.34 \text{ kips}$$

$$\text{Upper limit} = 2.4 d t F_u = 2.4 \left(\frac{7}{8}\right)(0.455)(65)$$

$$= 62.11 \text{ kips} > 54.34 \text{ kips} \quad \therefore \text{ use } R_n = 54.34 \text{ kips for this bolt}$$

For the other holes, use $s = 3$ in. Then,

$$L_c = s - h = 3 - \frac{15}{16} = 2.063 \text{ in.}$$

$$R_n = 1.2 \, L_c t F_u = 1.2(2.063)(0.455)(65) = 73.22 \text{ kips}$$

$$2.4 d t F_u = 62.11 \text{ kips} < 73.22 \text{ kips} \quad \therefore \text{ use } R_n = 62.11 \text{ kips for these bolts}$$

Both bearing values are larger than the bolt shear strength, so the nominal shear strength of $R_n = 28.86$ kips controls.

a. **Elastic analysis**. For an x-y coordinate system with the origin at the center of the lower left bolt (Figure 8.9),

$$\bar{y} = \frac{2(3) + 2(6) + 1(9)}{7} = 3.857 \text{ in.}$$

$$\bar{x} = \frac{3(3)}{7} = 1.286 \text{ in.}$$

$$\Sigma(x^2 + y^2) = 4(1.286)^2 + 3(1.714)^2 + 2(3.857)^2 + 2(0.857)^2$$
$$+ \, 2(2.143)^2 + 1(5.143)^2 = 82.29 \text{ in.}^2$$

$$e = 3 + 5 - 1.286 = 6.714 \text{ in.}$$

LFRD Solution

$$P_u = 1.2D + 1.6L = 1.2(9) + 1.6(27) = 54 \text{ kips}$$

$$M = Pe = 54(6.714) = 362.6 \text{ in.-kips} \quad \text{(clockwise)}$$

$$p_{cy} = \frac{54}{7} = 7.714 \text{ kips} \downarrow \qquad p_{cx} = 0$$

FIGURE 8.9

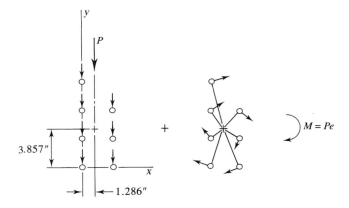

From the directions and relative magnitudes shown in Figure 8.9, the lower right bolt is judged to be critical, so

$$p_{mx} = \frac{My}{\Sigma(x^2 + y^2)} = \frac{362.6(3.857)}{82.29} = 17.00 \text{ kips} \leftarrow$$

$$p_{my} = \frac{Mx}{\Sigma(x^2 + y^2)} = \frac{362.6(1.714)}{82.29} = 7.553 \text{ kips} \downarrow$$

$$\Sigma p_x = 17.00 \text{ kips}$$

$$\Sigma p_y = 7.714 + 7.553 = 15.27 \text{ kips}$$

$$p = \sqrt{(17.00)^2 + (15.27)^2} = 22.9 \text{ kips}$$

The bolt design shear strength is

$$\phi R_n = 0.75(28.86) = 21.7 \text{ kips} < 22.9 \text{ kips} \qquad \text{(N.G.)}$$

Answer   The connection is unsatisfactory by elastic analysis.

ASD Solution

$$P_a = D + L = 9 + 27 = 36 \text{ kips}$$

$$M = Pe = 36(6.714) = 241.7 \text{ in.-kips} \quad \text{(clockwise)}$$

$$P_{cy} = \frac{36}{7} = 5.143 \text{ kips} \downarrow \qquad P_{cx} = 0$$

Check the lower right bolt.

$$p_{mx} = \frac{My}{\Sigma(x^2 + y^2)} = \frac{241.7(3.857)}{82.29} = 11.33 \text{ kips} \leftarrow$$

$$p_{my} = \frac{Mx}{\Sigma(x^2 + y^2)} = \frac{241.7(1.714)}{82.29} = 5.034 \text{ kips} \downarrow$$

$$\Sigma p_x = 11.33 \text{ kips}$$

$$\Sigma p_y = 5.143 + 5.034 = 10.18 \text{ kips}$$

$$p = \sqrt{(11.33)^2 + (10.18)^2} = 15.23 \text{ kips}$$

The bolt allowable shear strength is

$$\frac{R_n}{\Omega} = \frac{28.86}{2.00} = 14.4 \text{ kips} < 15.23 \text{ kips} \qquad \text{(N.G.)}$$

Answer   The connection is unsatisfactory by elastic analysis.

b. **Ultimate strength analysis**. This will be performed with the aid of standard spreadsheet software. The results of the final trial value of $r_o = 1.57104$ inches are given in Table 8.1. The coordinate system and bolt numbering scheme are shown in Figure 8.10. (Values shown in the table have been rounded to three decimal places for presentation purposes.)

TABLE 8.1

| Fastener | Origin at Bolt 1 | | Origin at IC | | | | | | |
	x'	y'	x	y	r	Δ	R	rR	Ry
1	0.000	0.000	0.285	−3.857	3.868	0.255	70.774	273.731	5.221
2	3.000	0.000	3.285	−3.857	5.067	0.334	72.553	367.598	47.045
3	0.000	3.000	0.285	−0.857	0.903	0.060	47.649	43.046	15.050
4	3.000	3.000	3.285	−0.857	3.395	0.224	69.563	236.188	67.310
5	0.000	6.000	0.285	2.143	2.162	0.143	63.631	137.555	8.398
6	3.000	6.000	3.285	2.143	3.922	0.259	70.891	278.061	59.377
7	0.000	9.000	0.285	5.143	5.151	0.340	72.631	374.107	4.023
Sum								1710.287	206.424

FIGURE 8.10

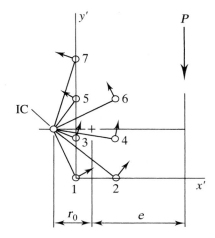

From Equation 8.4,

$$P(r_0 + e) = \Sigma \, rR$$

$$P = \frac{\Sigma \, rR}{r_0 + e} = \frac{1710.29}{1.57104 + 6.71429} = 206.424 \text{ kips}$$

where e has been expressed to six significant figures for consistency.

From Equation 8.5,

$$\Sigma \, F_y = \Sigma \, R_y - P = 206.424 - 206.424 = 0.000$$

The applied load has no horizontal component, so Equation 8.3 is automatically satisfied.

The load of 206.424 kips just computed is the failure load for the connection and is based on the critical fastener reaching its ultimate load capacity. If the connection failure load is multiplied by the ratio of fastener available strength to the fastener ultimate strength of 74 kips (Crawford and Kulak, 1971), we obtain the connection capacity.

The nominal strength of one bolt (based on shear) is $R_n = 28.86$ kips. The nominal strength of the connection is

$$P\left(\frac{R_n}{74}\right) = 206.4\left(\frac{28.86}{74}\right) = 80.50 \text{ kips}$$

LRFD Solution The design strength of the connection is

$$0.75(80.50) = 60.4 \text{ kips} > 54 \text{ kips} \qquad \text{(OK)}$$

Answer The connection is satisfactory by ultimate strength analysis.

ASD Solution The allowable strength of the connection is

$$\frac{80.50}{2.00} = 40.3 \text{ kips} > 36 \text{ kips} \qquad \text{(OK)}$$

Answer The connection is satisfactory by ultimate strength analysis.

Tables 7-7 through 7-14 in Part 7 of the *Manual,* "Design Considerations for Bolts," give coefficients for the design or analysis of common configurations of fastener groups subjected to eccentric loads. For each arrangement of fasteners considered, the tables give a value for C, the ratio of connection available strength to fastener available strength. To obtain a safe connection load, this constant must be multiplied by the available strength of the particular fastener used. For eccentric connections not included in the tables, the elastic method. which is conservative, may be used. Of course, a computer program or spreadsheet software may also be used to perform an ultimate strength analysis.

Example 8.3   Use the tables in Part 7 of the *Manual* to determine the available strength based on bolt shear, for the connection shown in Figure 8.11. The bolts are ¾-inch A325 bearing-type with the threads in the plane of shear. The bolts are in single shear.

FIGURE 8.11

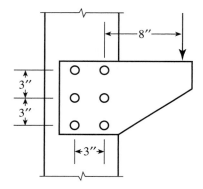

Solution   This connection corresponds to the connections in Table 7-8, for Angle = 0°. The eccentricity is

$$e_x = 8 + 1.5 = 9.5 \text{ in.}$$

The number of bolts per vertical row is

$$n = 3$$

From Table 7-8,

$$C = 1.53 \quad \text{by interpolation}$$

The nominal strength of a ¾-inch-diameter bolt in single shear is

$$r_n = F_{nv}A_b = 48(0.4418) = 21.21 \text{ kips}$$

(Here we use r_n for the nominal strength of a single bolt and R_n for the strength of the connection.)

The nominal strength of the connection is

$$R_n = Cr_n = 1.53(21.21) = 32.45 \text{ kips}$$

Answer   For LRFD, the available strength of the connection is $\phi R_n = 0.75(32.45) = 24.3$ kips.

For ASD, the available strength of the connection is $R_n/\Omega = 32.45/2.00 = 16.2$ kips.

8.3 ECCENTRIC BOLTED CONNECTIONS: SHEAR PLUS TENSION

In a connection such as the one for the tee stub bracket of Figure 8.12, an eccentric load creates a couple that will increase the tension in the upper row of fasteners and decrease it in the lower row. If the fasteners are bolts with no initial tension, the upper bolts will be put into tension and the lower ones will not be affected. Regardless of the type of fastener, each one will receive an equal share of the *shear* load.

If the fasteners are pretensioned high-strength bolts, the contact surface between the column flange and the bracket flange will be uniformly compressed before the external load is applied. The bearing pressure will equal the total bolt tension divided by the area of contact. As the load P is gradually applied, the compression at the top will be relieved and the compression at the bottom will increase, as shown in Figure 8.13a. When the compression at the top has been completely overcome, the components will separate, and the couple Pe will be resisted by tensile bolt forces and compression on

FIGURE 8.12

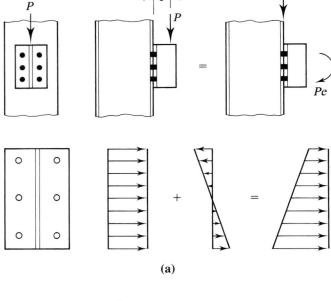

FIGURE 8.13

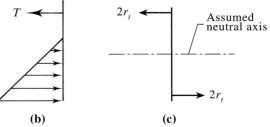

the remaining surface of contact, as shown in Figure 8.13b. As the ultimate load is approached, the forces in the bolts will approach their ultimate tensile strengths.

A conservative, simplified method will be used here. The neutral axis of the connection is assumed to pass through the centroid of the bolt areas. Bolts above this axis are subjected to tension, and bolts below the axis are assumed to be subjected to compressive forces, as shown in Figure 8.13c. Each bolt is assumed to have reached an ultimate value of r_t. Since there are two bolts at each level, each force is shown as $2r_t$. The resultant of the tensile and compressive forces is a couple that equals the resisting moment of the connection. The moment of this couple can be found by summing moments of the bolt forces about any convenient axis, such as the neutral axis. When the resisting moment is equated to the applied moment, the resulting equation can be solved for the unknown bolt tensile force r_t. (This method is the same as Case II in Part 7 of the *Manual*.)

Example 8.4   A beam-to-column connection is made with a structural tee as shown in Figure 8.14. Eight ¾-inch-diameter, A325, fully tightened bearing-type bolts are used to attach the flange of the tee to the column flange. Investigate the adequacy of this connection (the tee-to-column connection) if it is subjected to a service dead load of 20 kips and a service live load of 40 kips at an eccentricity of 2.75 inches. Assume that the bolt threads are in the plane of shear. All structural steel is A992.

FIGURE 8.14

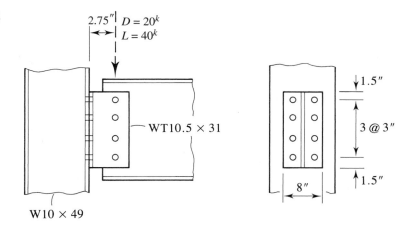

Solution   Determine the bearing and shear nominal strengths. For the strength in bearing, use a hole diameter of

$$h = d + \frac{1}{16} = \frac{3}{4} + \frac{1}{16} = \frac{13}{16} \text{ in.}$$

For the hole nearest the edge, use $L_e = 1.5$ in. Then,

$$L_c = L_e - \frac{h}{2} = 1.5 - \frac{13/16}{2} = 1.094 \text{ in.}$$

$$R_n = 1.2L_c t F_u = 1.2(1.094)(0.560)(65) = 47.79 \text{ kips}$$

$$\text{Upper limit} = 2.4 d t F_u = 2.4\left(\frac{3}{4}\right)(0.560)(65)$$

$$= 65.52 \text{ kips} > 47.79 \text{ kips}$$

$\therefore$ for this bolt, $R_n = 47.79$ kips

For the other holes, use $s = 3$ in. Then,

$$L_c = s - h = 3 - \frac{13}{16} = 2.188 \text{ in.}$$

$$R_n = 1.2L_c t F_u = 1.2(2.188)(0.560)(65) = 95.57 \text{ kips}$$

$$2.4 d t F_u = 65.52 \text{ kips} < 95.57 \text{ kips}$$

$\therefore$ for the these bolts, $R_n = 65.52$ kips

For the strength in shear,

$$A_b = \frac{\pi(3/4)^2}{4} = 0.4418 \text{ in.}^2$$

$$R_n = F_{nv} A_b = 48(0.4418) = 21.21 \text{ kips}$$

Since the nominal shear strength is less than the nominal bearing strength of any bolt, the shear strength controls.

LRFD Solution   The factored load is

$$P_u = 1.2D + 1.6L = 1.2(20) + 1.6(40) = 88 \text{ kips}$$

and the shear/bearing load per bolt is $88/8 = 11$ kips. The shear design strength per bolt is

$$\phi R_n = 0.75(21.21) = 15.9 \text{ kips} > 11 \text{ kips} \qquad \text{(OK)}$$

Compute the tensile force per bolt and then check the tension–shear interaction. Because of symmetry, the centroid of the connection is at middepth. Figure 8.15 shows the bolt areas and the distribution of bolt tensile forces.

FIGURE 8.15

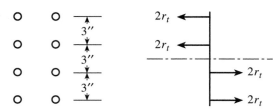

The moment of the resisting couple is found by summing moments about the neutral axis:

$$\Sigma M_{NA} = 2r_t(4.5 + 1.5 + 1.5 + 4.5) = 24r_t$$

The applied moment is.

$$M_u = P_u e = 88(2.75) = 242 \text{ in.-kips}$$

Equating the resisting and applied moments, we get

$$24r_t = 242, \quad \text{or} \quad r_t = 10.08 \text{ kips}$$

The factored load shear stress is

$$f_v = \frac{11}{0.4418} = 24.90 \text{ ksi}$$

and from AISC Equation J3-3a, the nominal tensile stress is

$$F'_{nt} = 1.3F_{nt} - \frac{F_{nt}}{\phi F_{nv}} f_v \leq F_{nt}$$

$$= 1.3(90) - \frac{90}{0.75(48)}(24.90) = 54.75 \text{ ksi} < 90 \text{ ksi}$$

The design tensile strength is

$$\phi R_n = 0.75F'_{nt}A_b = 0.75(54.75)(0.4418) = 18.1 \text{ kips} > 10.08 \text{ kips} \qquad \text{(OK)}$$

Answer   The connection is satisfactory.

ASD Solution   The total applied load is

$$P_a = D + L = 20 + 40 = 60 \text{ kips}$$

and the shear/bearing load per bolt is $60/8 = 7.5$ kips. The allowable shear strength per bolt is

$$\frac{R_n}{\Omega} = \frac{21.21}{2.00} = 10.61 \text{ kips} > 7.5 \text{ kips} \qquad \text{(OK)}$$

Compute the tensile force per bolt, then check the tension–shear interactions. The applied moment is

$$M_a = P_a e = 60(2.75) = 165 \text{ in.-kips}$$

The moment of the resisting couple is found by summing moments about the neutral axis. From Figure 8.15,

$$\Sigma M_{NA} = 2r_t(4.5 + 1.5 + 1.5 + 4.5) = 24r_t$$

Equating the resisting and applied moments, we get

$$24r_t = 165 \quad \text{or} \quad r_t = 6.875 \text{ kips}$$

The shearing stress is

$$f_v = \frac{7.5}{0.4418} = 16.98 \text{ ksi}$$

and from AISC Equation J3-3b, the nominal tensile stress is

$$F'_{nt} = 1.3F_{nt} - \frac{\Omega F_{nt}}{F_{nv}} f_v \leq F_{nt}$$

$$= 1.3(90) - \frac{2.00(90)}{48}(16.98) = 53.33 \text{ ksi} < 90 \text{ ksi}$$

The allowable tensile strength is

$$\frac{R_n}{\Omega} = \frac{F'_{nt} A_b}{\Omega} = \frac{53.33(0.4418)}{2.00} = 11.8 \text{ kips} > 6.875 \text{ kips} \qquad \text{(OK)}$$

Answer   The connection is satisfactory.

When bolts in slip-critical connections are subjected to tension, the slip-critical strength is ordinarily reduced by the factor given in AISC J3.9 (see Section 7.9). The reason is that the clamping effect, and hence the friction force, is reduced. In a connection of the type just considered, however, there is additional compression on the lower part of the connection that increases the friction, thereby compensating for the reduction in the upper part of the connection. For this reason, the slip-critical strength should *not* be reduced in this type of connection.

8.4 ECCENTRIC WELDED CONNECTIONS: SHEAR ONLY

Eccentric welded connections are analyzed in much the same way as bolted connections, except that unit lengths of weld replace individual fasteners in the computations. As in the case of eccentric bolted connections loaded in shear, welded shear connections can be investigated by either elastic or ultimate strength methods.

Elastic Analysis

The load on the bracket shown in Figure 8.16a may be considered to act in the plane of the weld—that is, the plane of the throat. If this slight approximation is made, the load will be resisted by the area of weld shown in Figure 8.16b. Computations are simplified, however, if a unit throat dimension is used. The calculated load can then be multiplied by 0.707 times the weld size to obtain the actual load.

FIGURE 8.16

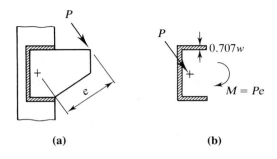

(a)   (b)

An eccentric load in the plane of the weld subjects the weld to both direct shear and torsional shear. Since all elements of the weld receive an equal portion of the direct shear, the direct shear stress is

$$f_1 = \frac{P}{L}$$

where L is the total length of the weld and numerically equals the shear area, because a unit throat size has been assumed. If rectangular components are used,

$$f_{1x} = \frac{P_x}{L} \quad \text{and} \quad f_{1y} = \frac{P_y}{L}$$

where P_x and P_y are the x- and y-components of the applied load. The shearing stress caused by the couple is found with the torsion formula

$$f_2 = \frac{Md}{J}$$

where

d = distance from the centroid of the shear area to the point where the stress is being computed

J = polar moment of inertia of that area

Figure 8.17 shows this stress at the upper right-hand corner of the given weld. In terms of rectangular components,

$$f_{2x} = \frac{My}{J} \quad \text{and} \quad f_{2y} = \frac{Mx}{J}$$

FIGURE 8.17

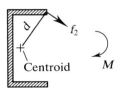

Also,

$$J = \int_A r^2\,dA = \int_A (x^2 + y^2)\,dA = \int_A x^2\,dA + \int_A y^2\,dA = I_y + I_x$$

where I_x and I_y are the rectangular moments of inertia of the shear area. Once all rectangular components have been found, they can be added vectorially to obtain the resultant shearing stress at the point of interest, or

$$f_v = \sqrt{(\Sigma\,f_x)^2 + (\Sigma\,f_y)^2}$$

As with bolted connections, the critical location for this resultant stress can usually be determined from an inspection of the relative magnitudes and directions of the direct and torsional shearing stress components.

Because a unit width of weld is used, the computations for centroid and moment of inertia are the same as for a line. In this book, we treat all weld segments as line segments, which we assume to be the same length as the edge of the connected part that they are adjacent to. Furthermore, we neglect the moment of inertia of a line segment about the axis coinciding with the line.

Example 8.5   Determine the size of weld required for the bracket connection in Figure 8.18. The service dead load is 10 kips, and the service live load is 30 kips. A36 steel is used for the bracket, and A992 steel is used for the column.

FIGURE 8.18

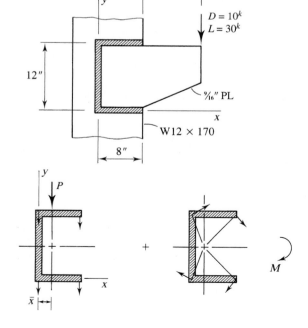

LRFD Solution

$$P_u = 1.2D + 1.6L = 1.2(10) + 1.6(30) = 60 \text{ kips}$$

The eccentric load may be replaced by a concentric load and a couple, as shown in Figure 8.18. The direct shearing stress, in kips per inch, is the same for all segments of the weld and is equal to

$$f_{1y} = \frac{60}{8 + 12 + 8} = \frac{60}{28} = 2.143 \text{ kips/in.}$$

Before computing the torsional component of shearing stress, the location of the centroid of the weld shear area must be determined. From the principle of moments with summation of moments about the y-axis,

$$\bar{x}(28) = 8(4)(2) \qquad \text{or} \qquad \bar{x} = 2.286 \text{ in.}$$

The eccentricity e is $10 + 8 - 2.286 = 15.71$ in., and the torsional moment is

$$M = Pe = 60(15.71) = 942.6 \text{ in.-kips}$$

If the moment of inertia of each horizontal weld about its own centroidal axis is neglected, the moment of inertia of the total weld area about its horizontal centroidal axis is

$$I_x = \frac{1}{12}(1)(12)^3 + 2(8)(6)^2 = 720.0 \text{ in.}^4$$

Similarly,

$$I_y = 2\left[\frac{1}{12}(1)(8)^3 + 8(4 - 2.286)^2\right] + 12(2.286)^2 = 195.0 \text{ in.}^4$$

and

$$J = I_x + I_y = 720.0 + 195.0 = 915.0 \text{ in.}^4$$

Figure 8.18 shows the directions of both components of stress at each corner of the connection. By inspection, either the upper right-hand corner or the lower right-hand corner may be taken as the critical location. If the lower right-hand corner is selected,

$$f_{2x} = \frac{My}{J} = \frac{942.6(6)}{915.0} = 6.181 \text{ kips/in.}$$

$$f_{2y} = \frac{Mx}{J} = \frac{942.6(8 - 2.286)}{915.0} = 5.886 \text{ kips/in.}$$

$$f_v = \sqrt{(6.181)^2 + (2.143 + 5.886)^2} = 10.13 \text{ kips/in.}$$

Check the strength of the base metal. The bracket is the thinner of the connected parts and controls. From Equation 7.35, the base metal shear *yield* strength per unit length is

$$\phi R_n = 0.6F_y t = 0.6(36)\left(\frac{9}{16}\right) = 12.2 \text{ kips/in.}$$

From Equation 7.36, the base metal shear *rupture* strength per unit length is

$$\phi R_n = 0.45 F_u t = 0.45(58)\left(\frac{9}{16}\right) = 14.7 \text{ kips/in.}$$

The base metal shear strength is therefore 12.2 kips/in. > 10.13kips/in. (OK)

From Equation 7.29, the weld strength per inch is

$$\phi R_n = \phi(0.707 w F_W)$$

The matching electrode for A36 steel is E70. Because the load direction varies on each weld segment, the weld shear strength varies, but for simplicity, we will conservatively use $F_w = 0.6 F_{EXX}$ for the entire weld. The required weld size is therefore

$$w = \frac{\phi R_n}{\phi(0.707) F_W} = \frac{10.13}{0.75(0.707)(0.6 \times 70)} = 0.455 \text{ in.}$$

Alternatively, for E70 electrodes, $\phi R_n = 1.392$ kips/in. per sixteenth of an inch in size. The required size in sixteenths is therefore

$$\frac{10.13}{1.392} = 7.3 \text{ sixteenths} \quad \text{use} \quad \frac{8}{16} \text{ in.} = \frac{1}{2} \text{ in.}$$

Answer   Use a ½-inch fillet weld, E70 electrode.

ASD Solution   The total load is $P_a = D + L = 10 + 30 = 40$ kips.

The eccentric load may be replaced by a concentric load and a couple, as shown in Figure 8.18. The direct shearing stress, in kips per inch, is the same for all segments of the weld and is equal to

$$f_{1y} = \frac{40}{8+12+8} = \frac{40}{28} = 1.429 \text{ kips/in.}$$

To locate the centroid of the weld shear area, use the principle of moments with summation of moments about the y-axis.

$$\bar{x}(28) = 8(4)(2) \quad \text{or} \quad \bar{x} = 2.286 \text{ in.}$$

The eccentricity e is $10 + 8 - 2.286 = 15.71$ in., and the torsional moment is

$$M = Pe = 40(15.71) = 628.4 \text{ in.-kips}$$

If the moment of inertia of each horizontal weld about its own centroidal axis is neglected, the moment of inertia of the total weld area about its horizontal centroidal axis is

$$I_x = \frac{1}{12}(1)(12)^3 + 2(8)(6)^2 = 720.0 \text{ in.}^4$$

Similarly,

$$I_y = 2\left[\frac{1}{12}(1)(8)^3 + 8(4 - 2.286)^2\right] + 12(2.286)^2 = 195.0 \text{ in.}^4$$

and

$$J = I_x + I_y = 720.0 + 195.0 = 915.0 \text{ in.}^4$$

Figure 8.18 shows the directions of both components of stress at each corner of the connection. By inspection, either the upper right-hand corner or the lower right-hand corner may be taken as the critical location. If the lower right-hand corner is selected,

$$f_{2x} = \frac{My}{J} = \frac{628.4(6)}{915.0} = 4.121 \text{ kips/in.}$$

$$f_{2y} = \frac{Mx}{J} = \frac{628.4(8 - 2.286)}{915.0} = 3.924 \text{ kips/in.}$$

$$f_v = \sqrt{(4.121)^2 + (1.429 + 3.924)^2} = 6.756 \text{ kips/in.}$$

Check the strength of the base metal. The bracket is the thinner of the connected parts and controls. From Equation 7.37, the base metal shear *yield* strength per unit length is

$$\frac{R_n}{\Omega} = 0.4F_y t = 0.4(36)\left(\frac{9}{16}\right) = 8.10 \text{ kips/in.}$$

From Equation 7.38, the base metal shear *rupture* strength per unit length is

$$\frac{R_n}{\Omega} = 0.3F_u t = 0.3(58)\left(\frac{9}{16}\right) = 9.79 \text{ kips/in.}$$

The base metal shear strength is therefore 8.10 kips/in. > 6.756 kips/in.   (OK)

From Equation 7.30, the weld strength per inch is

$$\frac{R_n}{\Omega} = \frac{0.707wF_W}{\Omega}$$

The matching electrode for A36 steel is E70. Because the load direction varies on each weld segment, the weld shear strength varies, but for simplicity, we will conservatively use $F_W = 0.6F_{EXX}$ for the entire weld. The required weld size is, therefore,

$$w = \frac{\Omega(R_n/\Omega)}{0.707F_W} = \frac{\Omega(f_v)}{0.707F_W} = \frac{2.00(6.756)}{0.707(0.6 \times 70)} = 0.455 \text{ in.}   \therefore \text{ use } \frac{1}{2} \text{ in.}$$

Alternatively, for E70 electrodes, $R_n/\Omega = 0.9279$ kips/in. per sixteenth of an inch in size. The required size in sixteenths is, therefore,

$$\frac{6.756}{0.9279} = 7.3 \text{ sixteenths}   \text{use } \frac{8}{16} \text{ in.} = \frac{1}{2} \text{ in.}$$

Answer   Use a ½-inch fillet weld, E70 electrode.

Ultimate Strength Analysis

Eccentric welded shear connections may be safely designed by elastic methods, but the factor of safety will be larger than necessary and will vary from connection to connection (Butler, Pal, and Kulak, 1972). This type of analysis suffers from some of the same shortcomings as the elastic method for eccentric bolted connections, including the assumption of a linear load-deformation relationship for the weld. Another source of error is the assumption that the strength of the weld is independent of the direction of the applied load. An ultimate strength approach, based on the relationships in AISC J2.4b, is presented in Part 8 of the *Manual* and is summarized here. It is based on research by Butler et al. (1972) and Kulak and Timler (1984) and closely parallels the method developed for eccentric bolted connections by Crawford and Kulak (1971).

Instead of considering individual fasteners, we treat the continuous weld as an assembly of discrete segments. At failure, the applied connection load is resisted by forces in each element, with each force acting perpendicular to the radius constructed from an instantaneous center of rotation to the centroid of the segment, as shown in Figure 8.19. This concept is essentially the same as that used for the fasteners. However, determining which element has the maximum deformation and computing the force in each element at failure is more difficult, because unlike the bolted case, the weld strength is a function of the direction of the load on the element. To determine the critical element, first compute the deformation of each element at maximum stress:

$$\Delta_m = 0.209(\theta + 2)^{-0.32} w$$

where

Δ_m = deformation of the element at maximum stress
θ = angle that the resisting force makes with the axis of the weld segment (see Figure 8.19)
w = weld leg size

Next, compute Δ_m/r for each element, where r is the radius from the IC to the centroid of the element. The element with the smallest Δ_m/r is the critical element, that is, the one that reaches its ultimate capacity first. For this element, the ultimate (fracture) deformation is

$$\Delta_u = 1.087(\theta + 6)^{-0.65} w \le 0.17w$$

FIGURE 8.19

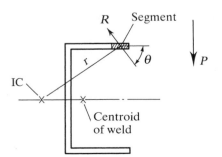

and the radius is r_{crit}. The deformation of each of the other elements is

$$\Delta = r \frac{\Delta_u}{r_{crit}}$$

The stress in each element is then

$$F_w = 0.60 F_{EXX} \left(1 + 0.5 \sin^{1.5} \theta\right)\left[p(1.9 - 0.9p)\right]^{0.3}$$

where

F_{EXX} = weld electrode strength

$p = \dfrac{\Delta}{\Delta_m}$ for the element

The force in each element is $F_w A_w$, where A_w is the weld throat area.

The preceding computations are based on an assumed location of the instantaneous center of rotation. If it is the actual location, the equations of equilibrium will be satisfied. The remaining details are the same as for a bolted connection.

1. Solve for the load capacity from the equation

    $$\Sigma M_{IC} = 0$$

    where IC is the instantaneous center.
2. If the two force equilibrium equations are satisfied, the assumed location of the instantaneous center and the load found in Step 1 are correct; otherwise, assume a new location and repeat the entire process.

The absolute necessity for the use of a computer is obvious. Computer solutions for various common configurations of eccentric welded shear connections are given in tabular form in Part 8 of the *Manual*. Tables 8-4 through 8-11 give available strength coefficients for various common combinations of horizontal and vertical weld segments based on an ultimate strength analysis. These tables may be used for either design or analysis and will cover almost any situation you are likely to encounter. For those connections not covered by the tables, the more conservative elastic method may be used.

Example 8.6   Determine the weld size required for the connection in Example 8.5, based on ultimate strength considerations. Use the tables for eccentrically loaded weld groups given in Part 8 of the *Manual*.

Solution   The weld of Example 8.5 is the same type as the one shown in Table 8-8 (angle $= 0°$), and the loading is similar. The following geometric constants are required for entry into the table:

$$a = \frac{a\ell}{\ell} = \frac{e}{\ell} = \frac{15.7}{12} = 1.3$$

$$k = \frac{k\ell}{\ell} = \frac{8}{12} = 0.67$$

By interpolation in Table 8-8 for $a = 1.3$,

$$C = 1.52 \quad \text{for} \quad k = 0.6 \quad \text{and} \quad C = 1.73 \quad \text{for} \quad k = 0.7$$

Interpolating between these two values for $k = 0.67$ gives

$$C = 1.67$$

For E70XX electrodes, $C_1 = 1.0$.

LRFD Solution From Table 8-8, the nominal strength of the connection is given by

$$R_n = CC_1 D\ell$$

For LRFD,

$$\phi R_n = P_u$$

so

$$\frac{P_u}{\phi} = CC_1 D\ell$$

and the required value of D is

$$D = \frac{P_u}{\phi CC_1 \ell} = \frac{60}{0.75(1.67)(1.0)(12)} = 3.99 \text{ sixteenths}$$

The required weld size is therefore

$$\frac{3.99}{16} = 0.249 \text{ in.} \quad \text{(versus 0.455 inch required in Example 8.5)}$$

Answer Use a ¼-inch fillet weld, E70 electrode.

ASD Solution From Table 8-8, the nominal strength of the connection is given by

$$R_n = CC_1 D\ell$$

For ASD,

$$\frac{R_n}{\Omega} = P_a$$

so

$$\Omega P_a = CC_1 D\ell$$

and the required value of D is

$$D = \frac{\Omega P_a}{CC_1 \ell} = \frac{2.00(40)}{1.67(1.0)(12)} = 3.99 \text{ sixteenths}$$

The required weld size is, therefore,

$$\frac{3.99}{16} = 0.249 \text{ in.} \quad \text{(versus 0.455 inch required in Example 8.5)}$$

Answer Use a ¼-inch fillet weld, E70 electrode.

Special Provision for Axially Loaded Members

When a structural member is axially loaded, the stress is uniform over the cross section and the resultant force may be considered to act along the *gravity axis*, which is a longitudinal axis through the centroid. For the member to be concentrically loaded at its ends, the resultant resisting force furnished by the connection must also act along this axis. If the member has a symmetrical cross section, this result can be accomplished by placing the welds or bolts symmetrically. If the member is one with an unsymmetrical cross section, such as the double-angle section in Figure 8.20, a symmetrical placement of welds or bolts will result in an eccentrically loaded connection, with a couple of *Te,* as shown in Figure 8.20b.

AISC J1.7 permits this eccentricity to be neglected in statically loaded members. When the member is subjected to fatigue caused by repeated loading or reversal of stress, the eccentricity must be eliminated by an appropriate placement of the welds or bolts. (Of course, this solution may be used even if the member is subjected to static loads only.) The correct placement can be determined by applying the force and moment equilibrium equations. For the welded connection shown in Figure 8.21, the first equation can be obtained by summing moments about the lower longitudinal weld:

$$\Sigma M_{L_2} = Tc - P_3 \frac{L_3}{2} - P_1 L_3 = 0$$

FIGURE 8.20

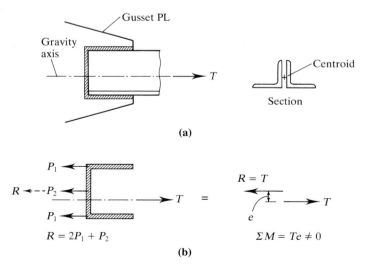

FIGURE 8.21

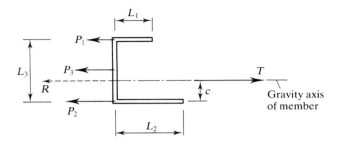

This equation can be solved for P_1, the required resisting force in the upper longitudinal weld. This value can then be substituted into the force equilibrium equation:

$$\Sigma F = T - P_1 - P_2 - P_3 = 0$$

This equation can be solved for P_2, the required resisting force in the lower longitudinal weld. For any size weld, the lengths L_1 and L_2 can then be determined. We illustrate this procedure, known as *balancing the welds*, in Example 8.7.

Example 8.7   A tension member consists of a double-angle section, 2L5 × 3 × ½ LLBB (long legs placed back-to-back). The angles are attached to a ⅜-inch-thick gusset plate. All steel is A36. Design a welded connection, balanced to eliminate eccentricity, that will resist the full tensile capacity of the member.

Solution   The nominal strength of the member based on the gross section is

$$P_n = F_y A_g = 36(7.51) = 270.4 \text{ kips}$$

The nominal strength based on the net section requires a value of U, but the length of the connection is not yet known, so U cannot be computed from Equation 3.1. We will use an estimated value of 0.80, and, if necessary, revise the solution after the connection length is known.

$$A_e = A_g U = 7.51(0.80) = 6.008 \text{ in.}^2$$
$$P_n = F_u A_e = 58(6.008) = 348.5 \text{ kips}$$

For A36 steel, the appropriate electrode is E70XX, and

Minimum weld size $= \dfrac{3}{16}$ in.   (AISC Table J2.4)

Maximum size $= \dfrac{1}{2} - \dfrac{1}{16} = \dfrac{7}{16}$ in.   (AISC J2.2b)

LRFD Solution   Compute the required design strength. For yielding of the gross section,

$$\phi_t P_n = 0.90(270.4) = 243.4 \text{ kips}$$

For fracture of the net section,

$$\phi_t P_n = 0.75(348.5) = 261.4 \text{ kips}$$

Yielding of the gross section controls. For one angle, the required design strength is

$$\frac{243.4}{2} = 121.7 \text{ kips}$$

Try a ⁵⁄₁₆-inch fillet weld:

Capacity per inch of length $= \phi R_n = 1.392D = 1.392(5) = 6.960 \text{ kips/in.}$

where D is the weld size in sixteenths of an inch.

Check the base metal shear strength. The gusset plate is the thinner of the connected parts and controls. From Equation 7.35, the base metal shear yield strength per unit length is

$$\phi R_n = 0.6 F_y t = 0.6(36)\left(\frac{3}{8}\right) = 8.100 \text{ kips/in.}$$

From Equation 7.36, the base metal shear rupture strength per unit length is

$$\phi R_n = 0.45 F_u t = 0.45(58)\left(\frac{3}{8}\right) = 9.788 \text{ kips/in.}$$

The base metal shear strength is therefore 8.100 kips/in., and the weld strength of 6.960 kips/in. controls.

Refer to Figure 8.22. Since there are both transverse and longitudinal welds, we will try both of the options given in AISC J2.4c. First, we will use the basic electrode strength of $0.6 F_{EXX}$ for both the longitudinal and transverse welds (this corresponds to $\phi R_n = 6.960$ kips/in.) The capacity of the weld across the end of the angle is

$$P_3 = 6.960(5) = 34.80 \text{ kips}$$

Summing moments about an axis along the bottom, we get

$$\sum M_{L_2} = 121.7(3.26) - 34.80\left(\frac{5}{2}\right) - P_1(5) = 0, \quad P_1 = 61.95 \text{ kips}$$

$$\sum F = 121.7 - 61.95 - 34.80 - P_2 = 0, \quad P_2 = 24.95 \text{ kips}$$

$$L_1 = \frac{P_1}{6.960} = \frac{61.95}{6.960} = 8.90 \text{ in.} \quad \therefore \text{ use 9 in.}$$

$$L_2 = \frac{P_2}{6.960} = \frac{24.95}{6.960} = 3.59 \text{ in.} \quad \therefore \text{ use 4 in.}$$

FIGURE 8.22

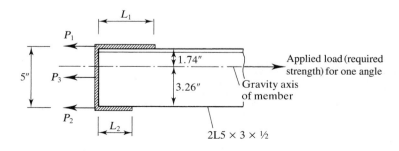

We will now examine the second option given in AISC J2.4c, in which we use 150% of the basic strength for the transverse weld and 85% of the basic strength for the longitudinal welds. For the transverse weld, use

$$\phi R_n = 1.5 \times 6.960 = 10.44 \text{ kips/in.}$$

and for the longitudinal welds,

$$\phi R_n = 0.85 \times 6.960 = 5.916 \text{ kips/in.}$$

The capacity of the weld across the end of the angle is

$$P_3 = 10.44(5) = 52.20 \text{ kips}$$

Summing moments about an axis along the bottom, we get

$$\sum M_{L_2} = 121.7(3.26) - 52.20\left(\frac{5}{2}\right) - P_1(5) = 0, \quad P_1 = 53.25 \text{ kips}$$

$$\sum F = 121.7 - 53.25 - 52.20 - P_2 = 0, \quad P_2 = 16.25 \text{ kips}$$

$$L_1 = \frac{P_1}{5.916} = \frac{53.25}{5.916} = 9.00 \text{ in.} \quad \therefore \text{ use 9 in.}$$

$$L_2 = \frac{P_2}{5.916} = \frac{16.25}{5.916} = 2.75 \text{ in.} \quad \therefore \text{ use 3 in.}$$

The second option results in a slight savings and will be used.

Verify the assumption on U to be sure that the strength of the member is governed by yielding. From Equation 3.1,

$$U = 1 - \frac{\overline{x}}{\ell} = 1 - \frac{0.746}{9} = 0.9171$$

This is larger than the value of 0.80 that was initially used, so the strength based on fracture will be larger than originally calculated, and the strength will be governed by yielding as assumed.

Answer   Use the weld shown in Figure 8.23.

FIGURE 8.23

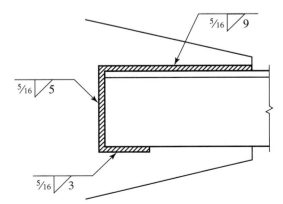

ASD Solution   Compute the required allowable strength. For yielding of the gross section,

$$\frac{P_n}{\Omega_t} = \frac{270.4}{1.67} = 161.9 \text{ kips}$$

For fracture of the net section,

$$\frac{P_n}{\Omega_t} = \frac{348.5}{2.00} = 174.3 \text{ kips}$$

Yielding of the gross section controls. For one angle, the required allowable strength is

$$\frac{161.9}{2} = 80.95 \text{ kips}$$

Try a ⁵⁄₁₆-inch fillet weld:

Capacity per inch of length $= \dfrac{R_n}{\Omega} = 0.9279D = 0.9279(5) = 4.640 \text{ kips/in.}$

where D is the weld size in sixteenths of an inch.

Check the base metal shear strength. The gusset plate is the thinner of the connected parts and controls. From Equation 7.37, the base metal shear yield strength per unit length is

$$\frac{R_n}{\Omega} = 0.4F_y t = 0.4(36)\left(\frac{3}{8}\right) = 5.400 \text{ kips/in.}$$

From Equation 7.38, the base metal shear rupture strength per unit length is

$$\frac{R_n}{\Omega} = 0.3F_u t = 0.3(58)\left(\frac{3}{8}\right) = 6.525 \text{ kips/in.}$$

The base metal shear strength is therefore 5.400 kips/in., and the weld strength of 4.640 kips/in. controls.

Refer to Figure 8.22. Since there are both transverse and longitudinal welds, we will try both of the options given in AISC J2.4c. First, we will use the basic electrode strength of $0.6F_{EXX}$ for both the longitudinal and transverse welds (this corresponds to $R_n/\Omega = 4.640$ kips/in.) The capacity of the weld across the end of the angle is

$$P_3 = 4.640(5) = 23.20 \text{ kips}$$

Summing moments about an axis along the bottom, we get

$$\sum M_{L_2} = 80.95(3.26) - 23.20\left(\frac{5}{2}\right) - P_1(5) = 0, \quad P_1 = 41.18 \text{ kips}$$

$$\sum F = 80.95 - 41.18 - 23.20 - P_2 = 0, \quad P_2 = 16.57 \text{ kips}$$

$$L_1 = \frac{P_1}{4.640} = \frac{41.18}{4.640} = 8.88 \text{ in.} \quad \therefore \text{ use 9 in.}$$

$$L_2 = \frac{P_2}{4.640} = \frac{16.57}{4.640} = 3.57 \text{ in.} \quad \therefore \text{ use 4 in.}$$

We will now examine the second option given in AISC J2.4c, in which we use 150% of the basic strength for the transverse weld and 85% of the basic strength for the longitudinal welds. For the transverse weld, use

$$\frac{R_n}{\Omega} = 1.5 \times 4.640 = 6.960 \text{ kips/in.}$$

and for the longitudinal welds,

$$\frac{R_n}{\Omega} = 0.85 \times 4.640 = 3.944 \text{ kips/in.}$$

The capacity of the weld across the end of the angle is

$$P_3 = 6.960(5) = 34.80 \text{ kips}$$

Summing moments about an axis along the bottom, we get

$$\sum M_{L_2} = 80.95(3.26) - 34.80\left(\frac{5}{2}\right) - P_1(5) = 0, \quad P_1 = 35.38 \text{ kips}$$

$$\sum F = 80.95 - 35.38 - 34.80 - P_2 = 0, \quad P_2 = 10.77 \text{ kips}$$

$$L_1 = \frac{P_1}{3.944} = \frac{35.38}{3.944} = 8.97 \text{ in.} \quad \therefore \text{ use 9 in.}$$

$$L_2 = \frac{P_2}{3.944} = \frac{10.77}{3.944} = 2.73 \text{ in.} \quad \therefore \text{ use 3 in.}$$

The second option results in a slight savings and will be used.

Verify the assumption on U to be sure that the strength of the member is government by yielding. From Equation 3.1,

$$U = 1 - \frac{\overline{x}}{\ell} = 1 - \frac{0.746}{9} = 0.9171$$

This is larger than the value of 0.80 that was initially used, so the strength based on fracture will be larger than originally calculated, and the strength will be governed by yielding as assumed.

Answer   Use the weld shown in Figure 8.23.

8.5   ECCENTRIC WELDED CONNECTIONS: SHEAR PLUS TENSION

Many eccentric connections, particularly beam-to-column connections, place the welds in tension as well as shear. Two such connections are illustrated in Figure 8.24.

The seated beam connection consists primarily of a short length of angle that serves as a "shelf" to support the beam. The welds attaching this angle to the column

FIGURE 8.24

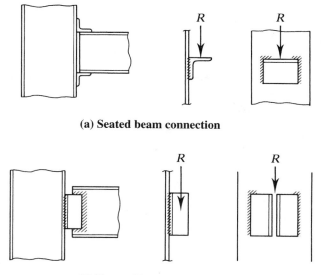

(a) Seated beam connection

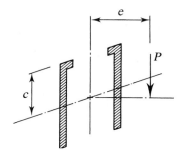

(b) Framed beam connection

must resist the moment caused by the eccentricity of the reaction as well as the beam reaction in direct shear. The angle connecting the top flange provides torsional stability to the beam at its end and does not assist in supporting the reaction. It may be attached to the beam web instead of the top flange. The beam-to-angle connections can be made with either welds or bolts and will not carry any calculated load.

The framed beam connection, which is very common, subjects the vertical angle-to-column welds to the same type of load as the seated beam connection. The beam-to-angle part of the connection is also eccentric, but the load is in the plane of shear, so there is no tension. Both the seated and the framed connections have their bolted counterparts.

In each of the connections discussed, the vertical welds on the column flange are loaded as shown in Figure 8.25. As with the bolted connection in Section 8.3, the eccentric load P can be replaced by a concentric load P and a couple $M = Pe$. The shearing stress is

$$f_v = \frac{P}{A}$$

FIGURE 8.25

where A is the total throat area of the weld. The maximum tensile stress can be computed from the flexure formula

$$f_t = \frac{Mc}{I}$$

where I is the moment of inertia about the centroidal axis of the area consisting of the total throat area of the weld, and c is the distance from the centroidal axis to the farthest point on the tension side. The maximum resultant stress can be found by adding these two components vectorially:

$$f_r = \sqrt{f_v^2 + f_t^2}$$

For units of kips and inches, this stress will be in kips per square inch. If a unit throat size is used in the computations, the same numerical value can also be expressed as kips per linear inch. If f_r is derived from factored loads, it may be compared with the design strength of a unit length of weld. Although this procedure assumes elastic behavior, it will be conservative when used in an LRFD context.

Example 8.8

An L6 × 4 × ½ is used in a seated beam connection, as shown in Figure 8.26. It must support a service load reaction of 5 kips dead load and 10 kips live load. The angles are A36 and the column is A992. E70XX electrodes are to be used. What size fillet welds are required for the connection to the column flange?

FIGURE 8.26

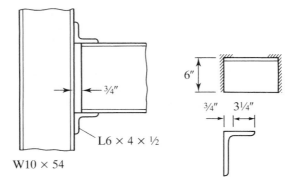

W10 × 54

Solution

As in previous design examples, a unit throat size will be used in the calculations. Although an end return is usually used for a weld of this type, conservatively for simplicity it will be neglected in the following calculations.

For the beam setback of ¾ inch, the beam is supported by 3.25 inches of the 4-inch outstanding leg of the angle. If the reaction is assumed to act through the center of this contact length, the eccentricity of the reaction with respect to the weld is

$$e = 0.75 + \frac{3.25}{2} = 2.375 \text{ in.}$$

FIGURE 8.27

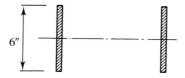

For the assumed weld configuration in Figure 8.27,

$$I = \frac{2(1)(6)^3}{12} = 36 \text{ in.}^4, \quad c = \frac{6}{2} = 3 \text{ in.}$$

LRFD Solution   The factored-load reaction is $P_u = 1.2D + 1.6L = 1.2(5) + 1.6(10) = 22$ kips.

$$M_u = P_u e = 22(2.375) = 52.25 \text{ ft-kips}$$

$$f_t = \frac{M_u c}{I} = \frac{52.25(3)}{36} = 4.354 \text{ kips/in.}$$

$$f_v = \frac{P_u}{A} = \frac{22}{2(1)(6)} = 1.833 \text{ kips/in.}$$

$$f_r = \sqrt{f_t^2 + f_v^2} = \sqrt{(4.354)^2 + (1.833)^2} = 4.724 \text{ kips/in.}$$

The required weld size w can be found by equating f_r to the weld capacity per inch of length:

$$f_r = 1.392D$$

$$4.724 = 1.392D, \quad D = 3.394$$

where D is the weld size in sixteenths of an inch for E70 electrodes. The required size is therefore

$$w = \frac{4}{16} = \frac{1}{4} \text{ in.}$$

From AISC Table J2.4,

$$\text{Minimum weld size} = \frac{3}{16} \text{ in.}$$

From AISC J2.2b,

$$\text{Maximum size} = \frac{1}{2} - \frac{1}{16} = \frac{7}{16} \text{ in.}$$

Try $w = \frac{1}{4}$ inch.

Check the shear capacity of the base metal (the angle controls):

Applied direct shear $= f_v = 1.833$ kips/in.

From Equation 7.35, the shear yield strength of the angle leg is

$$\phi R_n = 0.6F_y t = 0.6(36)\left(\frac{1}{2}\right) = 10.8 \text{ kips/in.}$$

From Equation 7.36, the shear rupture strength is

$$\phi R_n = 0.45 F_u t = 0.45(58)\left(\frac{1}{2}\right) = 13.1 \text{ kips/in.}$$

The base metal shear strength is therefore 10.8 kips/in. > 1.833 kips/in. (OK)

Answer   Use a ¼-inch fillet weld, E70XX electrodes.

ASD Solution   The total reaction is $P_a = D + L = 5 + 10 = 15$ kips

$$M_a = P_a e = 15(2.375) = 35.63 \text{ ft-kips}$$

$$f_t = \frac{M_a c}{I} = \frac{35.63(3)}{36} = 2.969 \text{ kips/in.}$$

$$f_v = \frac{P_a}{A} = \frac{15}{2(1)(6)} = 1.250 \text{ kips/in.}$$

$$f_r = \sqrt{f_t^2 + f_v^2} = \sqrt{(2.969)^2 + (1.250)^2} = 3.221 \text{ kips/in.}$$

To find the required weld size, equate f_r to the weld capacity per inch of length:

$$f_r = 0.9279 D$$
$$3.221 = 0.9279 D, \quad D = 3.471$$

where D is the weld size in sixteenths of an inch for E70 electrodes. The required size is therefore

$$w = \frac{4}{16} = \frac{1}{4} \text{ in.}$$

From AISC Table J2.4,

$$\text{Minimum weld size} = \frac{3}{16} \text{ in.}$$

From AISC J2.2b,

$$\text{Maximum size} = \frac{1}{2} - \frac{1}{16} = \frac{7}{16} \text{ in.}$$

Try $w = \frac{1}{4}$ inch.

Check the shear capacity of the base metal (the angle controls):

Applied direct shear $= f_v = 1.250$ kips/in.

From Equation 7.37, the shear yield strength of the angle leg is

$$\frac{R_n}{\Omega} = 0.4 F_y t = 0.4(36)\left(\frac{1}{2}\right) = 7.20 \text{ kips/in.}$$

From Equation 7.38, the shear rupture strength is

$$\frac{R_n}{\Omega} = 0.3 F_u t = 0.3(58)\left(\frac{1}{2}\right) = 8.70 \text{ kips/in.}$$

The base metal shear strength is therefore 7.20 kips/in. > 1.250 kips/in.   (OK)

Answer   Use a ¼-inch fillet weld, E70XX electrodes.

Example 8.9   A welded framed beam connection is shown in Figure 8.28. The framing angles are $4 \times 3 \times \frac{1}{2}$, and the column is a W12 × 72. A36 steel is used for the angles, and A992 steel is used for the W-shapes. The welds are $\frac{3}{8}$-inch fillet welds made with E70XX electrodes. Determine the maximum available beam reaction as limited by the welds at the column flange.

FIGURE 8.28

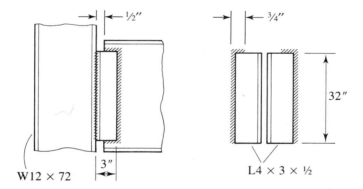

W12 × 72
L4 × 3 × ½

Solution   The beam reaction will be assumed to act through the center of gravity of the connection to the framing angles. The eccentricity of the load with respect to the welds at the column flange will therefore be the distance from this center of gravity to the column flange. For a unit throat size and the weld shown in Figure 8.29a,

$$\bar{x} = \frac{2(2.5)(1.25)}{32 + 2(2.5)} = 0.1689 \text{ in.} \quad \text{and} \quad e = 3 - 0.1689 = 2.831 \text{ in.}$$

The moment on the column flange welds is

$$M = Re = 2.831R \text{ in.-kips}$$

where R is the beam reaction in kips.

From the dimensions given in Figure 8.29b, the properties of the column flange welds can be computed as

$$\bar{y} = \frac{32(16)}{32 + 0.75} = 15.63 \text{ in.}$$

$$I = \frac{1(32)^3}{12} + 32(16 - 15.63)^2 + 0.75(15.63)^2 = 2918 \text{ in.}^4$$

FIGURE 8.29

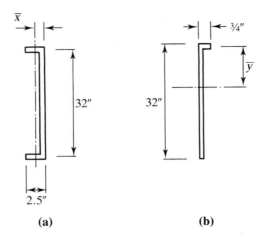

(a)   (b)

For the two welds,

$$I = 2(2918) = 5836 \text{ in.}^4$$

$$f_t = \frac{Mc}{I} = \frac{2.831R(15.63)}{5836} = 0.007582R \text{ kips/in.}$$

$$f_v = \frac{R}{A} = \frac{R}{2(32 + 0.75)} = 0.01527R \text{ kips/in.}$$

$$f_r = \sqrt{(0.007582R)^2 + (0.01527R)^2} = 0.01705R \text{ kips/in.}$$

LRFD Solution   Let

$$0.01705R_u = 1.392 \times 6$$

where R_u is the factored load reaction and 6 is the weld size in sixteenths of an inch. Solving for R_u, we get

$$R_u = 489.9 \text{ kips}$$

Check the shear capacity of the base metal (the angle controls). From Equation 7.35, the shear yield strength of the angle leg is

$$\phi R_n = 0.6F_y t = 0.6(36)\left(\frac{1}{2}\right) = 10.8 \text{ kips/in.}$$

From Equation 7.36, the shear rupture strength is

$$\phi R_n = 0.45F_u t = 0.45(58)\left(\frac{1}{2}\right) = 13.1 \text{ kips/in.}$$

The direct shear to be resisted by one angle is

$$\frac{R_u}{A} = \frac{489.9}{2(32.75)} = 7.48 \text{ kips/in.} < 10.8 \text{ kips/in.} \quad (\text{OK})$$

Answer   The maximum factored load reaction = 490 kips.

ASD Solution   Let

$$0.01705R_a = 0.9279 \times 6$$

where R_a is the service load reaction and 6 is the weld size in sixteenths of an inch. Solving for R_a, we get

$$R_a = 326.5 \text{ kips}$$

Check the shear capacity of the base metal (the angle controls). From Equation 7.37, the shear yield strength of the angle leg is

$$\frac{R_n}{\Omega} = 0.4F_y t = 0.4(36)\left(\frac{1}{2}\right) = 7.20 \text{ kips/in.}$$

From Equation 7.38, the shear rupture strength is

$$\frac{R_n}{\Omega} = 0.3F_u t = 0.3(58)\left(\frac{1}{2}\right) = 8.70 \text{ kips/in.}$$

The direct shear to be resisted by one angle is

$$\frac{R_a}{A} = \frac{326.5}{2(32.75)} = 4.99 \text{ kips/in.} < 7.20 \text{ kips/in.} \qquad \text{(OK)}$$

Answer   The maximum service load reaction = 327 kips.

8.6   MOMENT-RESISTING CONNECTIONS

Although most connections appear to be capable of transmitting moment, most are not, and special measures must be taken to make a connection moment-resisting. Consider the beam connection shown in Figure 8.30. This connection is sometimes referred to as a *shear connection*, because it can transmit shear but virtually no moment.

FIGURE 8.30

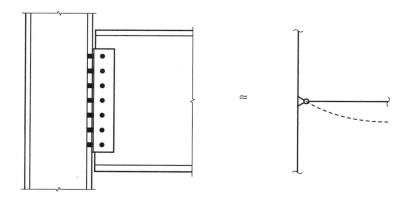

FIGURE 8.31

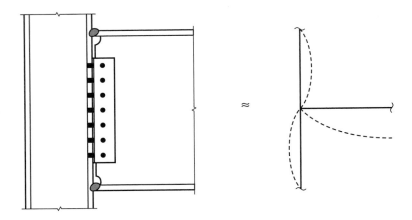

It can be treated as a simple support. In a beam connection, any moment transfer takes place mostly through the flanges, in the form of a couple. This couple consists of a compressive force in one flange and a tensile force in the other. In a shear connection, there is no connection of the flanges, and the web connection is designed to be flexible enough to allow some relative rotation of the members at the joint. Only a very small rotation is necessary in order for a connection to be treated as pinned.

The connection in Figure 8.31 is the same as the one in Figure 8.30, except that the beam flanges are welded to the column flange. A connection of this type can transfer both shear and moment. The shear is transmitted mostly through the web connection, as in the shear connection, and the moment is transferred through the flanges. This connection is treated as a rigid connection and can be modeled as shown.

Although connections are usually treated as either simple or rigid, the reality is that most connections fall somewhere in between and can be accurately described as *partially restrained* or *semirigid*. The distinction between different types of connections can be made by examining the degree of relative rotation that can take place between the connected members.

Figure 8.32 shows a *moment-rotation curve* for a connection. This graph shows the relationship between moment transferred at a joint and the corresponding rotation of the connection. Moment-rotation curves can be constructed for specific connections,

FIGURE 8.32

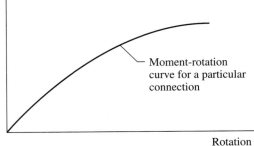

Moment

Moment-rotation curve for a particular connection

Rotation

FIGURE 8.33

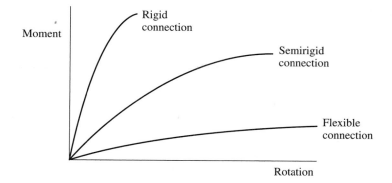

and the relationship can be determined experimentally or analytically. Moment-rotation curves for three different connections are shown in Figure 8.33. Connections that are designed to be fully restrained (or rigid), actually permit some rotation; otherwise, the curve would just be represented by the vertical axis. Connections designed to be moment-free (or flexible), have some moment restraint. A perfectly flexible connection would be represented by the horizontal axis.

Figure 8.34 shows the moment-rotation curve for a beam connection and includes the moment-rotation relationship for the beam as well as the connection. The straight line in the graph is called the *beam line*, or *load line*. It can be constructed as follows. If the end of the beam were fully restrained, the rotation would be zero. The fixed-end moment, caused by the actual load on the beam, is plotted on the moment axis (rotation = 0). If the end were simply supported (pinned end), the moment would be zero. The end rotation corresponding to a simple support and the actual loading is

FIGURE 8.34

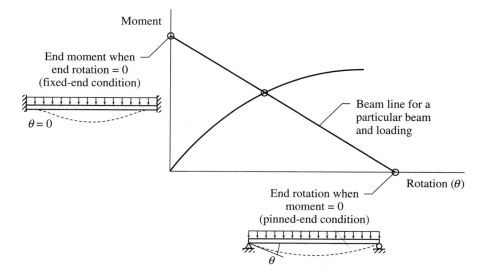

FIGURE 8.35

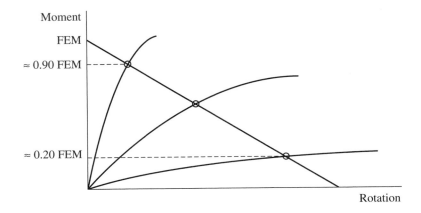

then plotted on the rotation axis (moment = 0). The line connecting the two points is the beam line, and points on the line represent different degrees of end restraint. The curved line in the graph is the moment-curvature relation for the connection that is used with the beam. The intersection of these curves gives the moment and rotation for this particular combination of beam, connection, and loading.

In Figure 8.35, a beam line is superimposed on the three moment-rotation curves of Figure 8.33. Although a connection designed as rigid has a theoretical moment capacity equal to the fixed-end moment (FEM) of the beam, it will actually have a moment resistance of about 90% of the fixed-end moment. The moment and rotation for this design is represented by the intersection of the moment-rotation curve for the connection and the beam line. Similarly, a connection designed as pinned (simply supported; no moment) would actually be capable of transmitting a moment of about 20% of the fixed-end moment, with a rotation of approximately 80% of the simple support rotation. The design moment for a partially restrained connection corresponds to the intersection of the beam line with the actual moment-rotation curve for the partially restrained connection.

The advantage of a partially-restrained connection is that it can equalize the negative and positive moments within a span. Figure 8.36a shows a uniformly loaded beam with simple supports, along with the corresponding moment diagram. Figure 8.36b shows the same beam and loading with fixed supports. Regardless of the support conditions, whether simple, fixed, or something in between, the same static moment of $wL^2/8$ will have to be resisted. The effect of partially restrained connections is to shift the moment diagram as shown in Figure 8.36c. This will increase the positive moment but decrease the negative moment, which is the maximum moment in the beam, thereby potentially resulting in a lighter beam.

One of the difficulties in designing a partially restrained system is obtaining an accurate moment-rotation relationship for the connection, whether bolted or welded. Such relationships have been developed, and this is an area of ongoing research (Christopher and Bjorhovde, 1999). Another drawback is the requirement for a structural frame analysis that incorporates the partial joint restraint. Chapter 11 of the *Manual*, "Design of Flexible Moment Connections," presents a simplified alternative to partially restrained connections.

FIGURE 8.36

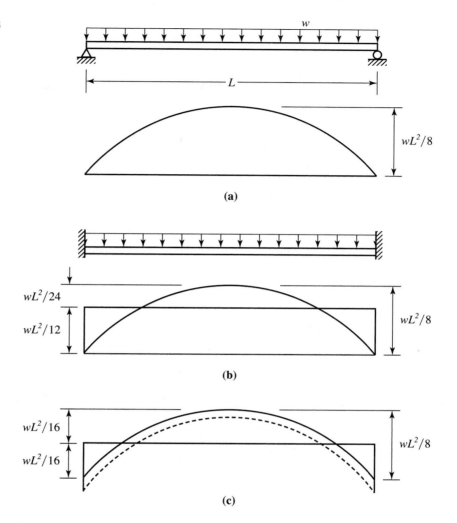

(a)

(b)

(c)

The AISC Specification defines three categories of connections in Section B3.6," Design of Connections." They are the three types we have just covered:

- Simple (flexible connection)
- FR—Fully Restrained (rigid connection)
- PR—Partially Restrained (semirigid connection)

In the present chapter of this book, we will consider only fully restrained moment connections designed to resist a specific value of moment.

Several examples of commonly used moment connections are illustrated in Figure 8.37. As a general rule, most of the moment transfer is through the beam flanges, and most of the moment capacity is developed there. The connection in Figure 8.37a (the same type of connection as in Figure 8.31) typifies this concept. The plate connecting the beam web to the column is shop welded to the column and field bolted to

FIGURE 8.37

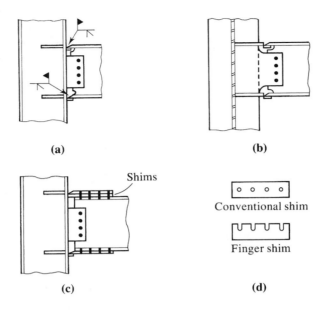

(a)   (b)
Shims
Conventional shim
Finger shim
(c)   (d)

the beam. With this arrangement, the beam is conveniently held in position so that the flanges can be field welded to the column. The plate connection is designed to resist only shear and takes care of the beam reaction. Complete penetration groove welds connect the beam flanges to the column and can transfer a moment equal to the moment capacity of the beam flanges. This will constitute most of the moment capacity of the beam, but a small amount of restraint will also be provided by the plate connection. (Because of strain hardening, the full plastic moment capacity of the beam can actually be developed through the flanges.) Making the flange connection requires that a small portion of the beam web be removed and a "backing bar" used at each flange to permit all welding to be done from the top. When the flange welds cool, they will shrink, typically about ⅛ inch. The resulting longitudinal displacement can be accounted for by using slotted bolt holes and tightening the bolts after the welds have cooled. The connection illustrated also uses column stiffeners, which are not always required (see Section 8.7).

The moment connection of Figure 8.37a also illustrates a recommended connection design practice: Whenever possible, welding should be done in the fabricating shop, and bolting should be done in the field. Shop welding is less expensive and can be more closely controlled.

In most beam-to-column moment connections, the members are part of a plane frame and are oriented as shown in Figure 8.37a — that is, with the webs in the plane of the frame so that bending of each member is about its major axis. When a beam must frame into the web of a column rather than its flange (for example, in a space frame), the connection shown in Figure 8.37b can be used. This connection is similar to the one shown in Figure 8.37a but requires the use of column stiffeners to make the connections to the beam flanges.

Although the connection shown in Figure 8.37a is simple in concept, its execution requires close tolerances. If the beam is shorter than anticipated, the gap between the column and the beam flange may cause difficulties in welding, even when a backing bar is used. The three-plate connection shown in Figure 8.37c does not have this handicap, and it has the additional advantage of being completely field bolted. The flange plates and the web plate are all shop welded to the column flange and field bolted to the beam. To provide for variation in the beam depth, the distance between flange plates is made larger than the nominal depth of the beam, usually by about ⅜ inch. This gap is filled at the top flange during erection with *shims,* which are thin strips of steel used for adjusting the fit at joints. Shims may be one of two types: either conventional or "finger" shims, which can be inserted after the bolts are in place, as shown in Figure 8.37d. In regions where seismic forces are large, the connection shown in Figure 8.37a requires special design procedures (FEMA, 2000).

Example 8.10 illustrates the design of a three-plate moment connection, including the requirements for connecting elements, which are covered by AISC J4.

Example 8.10   Design a three-plate moment connection of the type shown in Figure 8.38 for the connection of a W21 × 50 beam to the flange of a W14 × 99 column. Assume a beam setback of ½ inch. The connection must transfer the following service load effects: a dead-load moment of 35 ft-kips, a live-load moment of 105 ft-kips, a dead-load shear of 5.5 kips, and a live-load shear of 16.5 kips. All plates are to be shop welded to the column with E70XX electrodes and field bolted to the beam with A325 bearing-type bolts. A36 steel is used for the plates, and A992 steel is used for the beam and column.

FIGURE 8.38

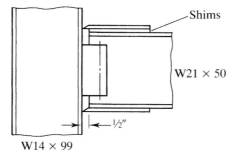

W21 × 50

½"

W14 × 99

LRFD Solution
$$M_u = 1.2\,M_D + 1.6\,M_L = 1.2(35) + 1.6(105) = 210.0 \text{ ft-kips}$$
$$V_u = 1.2\,V_D + 1.6\,V_L = 1.2(5.5) + 1.6(16.5) = 33.0 \text{ kips}$$

For the web plate, **try ¾-inch-diameter bolts**. Neglect eccentricity and assume that the threads are in the plane of shear. The nominal shear strength of one bolt is

$$R_n = F_{nv}A_b = 48(0.4418) = 21.21 \text{ kips}$$

and the design strength is

$$\phi R_n = 0.75(21.21) = 15.91 \text{ kips}$$

$$\text{Number of bolts required } = \frac{33}{15.91} = 2.07$$

Try three blots. The minimum spacing is $2\frac{2}{3}d = 2.667(3/4) = 2.0$ in. From AISC Table J3.4, the minimum edge distance is $1\frac{1}{4}$ inches. Try the layout shown in Figure 8.39 and determine the plate thickness required for bearing. Use a hole diameter of

$$h = d + \frac{1}{16} = \frac{3}{4} + \frac{1}{16} = \frac{13}{16} \text{ in.}$$

For the hole nearest the edge,

$$L_c = L_e - \frac{h}{2} = 1.5 - \frac{13/16}{2} = 1.094 \text{ in.}$$

The nominal bearing strength is

$$R_n = 1.2L_c t F_u = 1.2(1.094)t(58) = 76.14t$$

subject to an upper limit of

$$2.4 dt F_u = 2.4\left(\frac{3}{4}\right)t(58) = 104.4t$$

Use $R_n = 76.14t$ and

$$\phi R_n = 0.75(76.14t) = 57.11t \text{ kips/bolt}$$

For the other holes,

$$L_c = s - h = 3 - \frac{13}{16} = 2.188 \text{ in.}$$
$$R_n = 1.2L_c t F_u = 1.2(2.188)t(58) = 152.3t$$
$$2.4 dt F_u = 104.4t \text{ (controls)}$$
$$\phi R_n = 0.75(104.4t) = 78.30t \text{ kips/bolt}$$

FIGURE 8.39

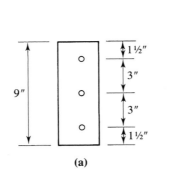
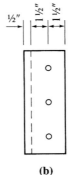

(a)   (b)

To find the required plate thickness, equate the total bearing strength to the applied load:

$$57.11t + 2(78.30t) = 33, \quad t = 0.154 \text{ in.}$$

(The beam web has a thickness of $t_w = 0.380$ in. > 0.154 in., and since F_y for the beam is larger than F_y for the plate, the bearing strength of the beam web will be adequate.)

Other limit states for connection plates (in addition to bolt bearing) can be found in AISC J4, "Affected Elements of Members and Connecting Elements." To determine the plate thickness required for the vertical shear, consider both yielding of the gross section and rupture of the net section. For shear yielding, from AISC Equation J4-3,

$$R_n = 0.60F_yA_g = 0.60(36)(8.5t) = 183.6t$$
$$\phi R_n = 1.00(183.6t) = 183.6t$$

For shear rupture, the net area is

$$A_{nv} = \left[8.5 - 3\left(\frac{3}{4} + \frac{1}{8} \right) \right] t = 5.875t$$

From AISC Equation J4-4,

$$R_n = 0.6F_uA_{nv} = 0.6(58)(5.875t) = 204.5t$$
$$\phi R_n = 0.75(204.5t) = 153.4t$$

Shear rupture controls. Let

$$153.4t = 33, \quad t = 0.215 \text{ in.}$$

The largest required thickness is for the limit state of shear rupture. **Try $t = \frac{1}{4}$ in**.

For the connection of the shear plate to the column flange, the required strength per inch is

$$\frac{V_u}{L} = \frac{33}{9} = 3.667 \text{ kips/in.}$$

From Equation 7.35, the base metal shear yield strength per unit length is

$$\phi R_n = 0.6F_yt = 0.6(36)\left(\frac{1}{4} \right) = 5.400 \text{ kips/in.}$$

From Equation 7.36, the base metal shear rupture strength per unit length is

$$\phi R_n = 0.45F_ut = 0.45(58)\left(\frac{1}{4} \right) = 6.525 \text{ kips/in.}$$

The base metal shear strength is therefore 5.400 kips/in., which is greater than the required strength of 3.667 kips/in. For welds on both sides of the plate, the required strength per weld is $3.667/2 = 1.834$ kips/in. To determine the required weld size, let

$$1.392D = 1.834, \quad D = 1.32 \text{ sixteenths of an inch}$$

The minimum weld size from AISC Table J2.4 is $\frac{1}{8}$ inch, based on the thinner connected part (the shear plate). There is no maximum size requirement for this type of connection (one where the weld is not along an edge). **Use a $\frac{1}{8}$-inch fillet weld on each side of the plate**.

The minimum width of the plate can be determined from a consideration of edge distances. The load being resisted (the beam reaction) is vertical, so the horizontal edge distance need only conform to the clearance requirements of AISC Table J3.4. If we assume a sheared edge, the minimum edge distance is $1\frac{1}{4}$ inch.

With a beam setback of $\frac{1}{2}$ inch and edge distances of $1\frac{1}{2}$ inches, as shown in Figure 8.39b, the width of the plate is

$$0.5 + 2(1.5) = 3.5 \text{ in.}$$

Try a plate $\frac{1}{4} \times 3\frac{1}{2}$. Check block shear. The shear areas are

$$A_{gv} = \frac{1}{4}(3 + 3 + 1.5) = 1.875 \text{ in.}^2$$

$$A_{nv} = \frac{1}{4}\left[3 + 3 + 1.5 - 2.5\left(\frac{3}{4} + \frac{1}{8}\right)\right] = 1.328 \text{ in.}^2$$

The tension area is

$$A_{nt} = \frac{1}{4}\left[1.5 - 0.5\left(\frac{3}{4} + \frac{1}{8}\right)\right] = 0.2656 \text{ in.}^2$$

For this type of block shear, $U_{bs} = 1.0$. From AISC Equation J4-5,

$$R_n = 0.6F_u A_{nv} + U_{bs}F_u A_{nt}$$
$$= 0.6(58)(1.328) + 1.0(58)(0.2656) = 61.62 \text{ kips}$$

with an upper limit of

$$0.6F_y A_{gv} + U_{bs}F_u A_{nt} = 0.6(36)(1.875) + 1.0(58)(0.2656) = 55.90 \text{ kips} < 61.62 \text{ kips}$$
$$\phi R_n = 0.75(55.90) = 41.9 \text{ kips} > 33 \text{ kips} \quad \text{(OK)}$$

Use a plate $\frac{1}{4} \times 3\frac{1}{2}$.

For the flange connection, first select the bolts. From Figure 8.40, the force at the interface between the beam flange and the plate is

$$H = \frac{M}{d} = \frac{210(12)}{20.8} = 121.2 \text{ kips}$$

where d is the depth of the beam. Although the moment arm of the couple is actually the distance from center of flange plate to center of flange plate, the plate thickness is not yet known, and use of the beam depth is conservative.

FIGURE 8.40

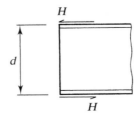

Try ¾-inch A325 bolts. (Since ¾-inch-diameter bolts were selected for the shear connection, try the same size here.) If bolt shear controls, the number of bolts required is

$$\frac{121.2}{15.91} = 7.62 \quad \therefore \quad \text{use 8 bolts (4 pairs)}$$

Use edge distances of 1½ inches, spacings of 3 inches, and determine the minimum plate thickness required for bearing. Use a hole diameter of

$$h = d + \frac{1}{16} = \frac{3}{4} + \frac{1}{16} = \frac{13}{16} \text{ in.}$$

For the hole nearest the edge,

$$L_c = L_e - \frac{h}{2} = 1.5 - \frac{13/16}{2} = 1.094 \text{ in.}$$

The nominal bearing strength is

$$R_n = 1.2 L_c t F_u = 1.2(1.094)t(58) = 76.14t$$

subject to an upper limit of

$$2.4 dt F_u = 2.4\left(\frac{3}{4}\right)t(58) = 104.4t$$

Use $R_n = 76.14t$ and

$$\phi R_n = 0.75(76.14t) = 57.11t \text{ kips/bolt}$$

For the other holes,

$$L_c = s - h = 3 - \frac{13}{16} = 2.188 \text{ in.}$$
$$R_n = 1.2 L_c t F_u = 1.2(2.188)t(58) = 152.3t$$
$$2.4 dt F_u = 104.4t \text{ (controls)}$$
$$\phi R_n = 0.75(104.4t) = 78.30t \text{ kips/bolt}$$

To find the required thickness, equate the total bearing strength to the applied load:

$$2(57.11t) + 6(78.30t) = 121.2, \quad t = 0.208 \text{ in.}$$

The flange plate will be designed as a tension-connecting element (the top plate) then checked for compression (the bottom plate). The minimum cross section required for tension on the gross and net areas will now be determined. From AISC Equation J4-1,

$$R_n = F_y A_g$$
$$\phi R_n = 0.90 F_y A_g$$

$$\text{Required } A_g = \frac{\phi R_n}{0.90 F_y} = \frac{H}{0.90 F_y} = \frac{121.2}{0.90(36)} = 3.741 \text{ in.}^2$$

From AISC Equation J4-2,

$$R_n = F_u A_e$$

$$\phi R_n = 0.75 F_u A_e$$

$$\text{Required } A_e = \frac{\phi R_n}{0.75 F_u} = \frac{H}{0.75 F_u} = \frac{121.2}{0.75(58)} = 2.786 \text{ in.}^2$$

Try a plate width of w_g = 6.5 in. (equal to the beam flange width). Compute the thickness needed to satisfy the gross area requirement.

$$A_g = 6.5t = 3.741 \text{ in.}^2 \quad \text{or} \quad t = 0.576 \text{ in.}$$

Compute the thickness needed to satisfy the net area requirement.

$$A_e = A_n = t w_n = t(w_g - \Sigma d_{\text{hole}}) = t\left[6.5 - 2\left(\frac{7}{8}\right)\right] = 4.750t$$

Let

$$4.750t = 2.786 \text{ in.}^2 \quad \text{or} \quad t = 0.587 \text{ in.} \quad \text{(controls)}$$

This thickness is also greater than that required for bearing, so it will be the minimum acceptable thickness. **Try a plate $\frac{5}{8} \times 6\frac{1}{2}$.**

Check compression. Assume that the plate acts as a fixed-end compression member between fasteners, with $L = 3$ in. and $K = 0.65$.

$$r = \sqrt{\frac{I}{A}} = \sqrt{\frac{6.5(5/8)^3/12}{6.5(5/8)}} = 0.1804 \text{ in.}$$

$$\frac{KL}{r} = \frac{0.65(3)}{0.1804} = 10.81$$

From AISC J4.4, for compression elements with $KL/r < 25$, the nominal strength is

$$P_n = F_y A_g \qquad \qquad \text{(AISC Equation J4-6)}$$

and for LRFD, $\phi = 0.9$

$$\phi P_n = 0.9 F_y A_g = 0.9(36)\left(\frac{5}{8} \times 6.5\right) = 132 \text{ kips} > 121.2 \text{ kips} \quad \text{(OK)}$$

Check block shear in the plate. In the transverse direction, use bolt edge distances of $1\frac{1}{2}$ inches and a spacing of $3\frac{1}{2}$ inches. This places the bolts at the "workable gage" location in the beam flange (see Part 1 of the *Manual*). Figure 8.41 shows the bolt layout and two possible block shear failure modes. The shear areas for both cases are

$$A_{gv} = \frac{5}{8}(3+3+3+1.5) \times 2 = 13.13 \text{ in.}^2$$

$$A_{nv} = \frac{5}{8}\left[3+3+3+1.5-3.5\left(\frac{3}{4}+\frac{1}{8}\right)\right] \times 2 = 9.297 \text{ in.}^2$$

FIGURE 8.41

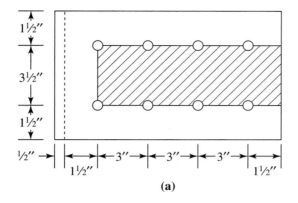

(a)

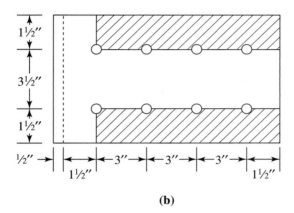

(b)

If a transverse tension area between the bolts is considered (Figure 8.41a), the width is 3.5 in. If two outer blocks are considered (Figure 8.41b), the total tension width is $2(1.5) = 3.0$ in. This will result in the smallest block shear strength.

$$A_{nt} = \frac{5}{8}\left[1.5 - 0.5\left(\frac{3}{4} + \frac{1}{8}\right)\right] \times 2 = 1.328 \text{ in.}^2$$

$$R_n = 0.6F_u A_{nv} + U_{bs} F_u A_{nt}$$
$$= 0.6(58)(9.279) + 1.0\,(58)(1.328) = 399.9 \text{ kips}$$

Upper limit $= 0.6F_y A_{gv} + U_{bs} F_u A_{nt} = 0.6(36)(13.13) + 1.0(58)(1.328) = 360.6$ kips

$$\phi R_n = 0.75(360.6) = 271 \text{ kips} > 121 \text{ kips}   \text{(OK)}$$

Check block shear in the beam flange. The bolt spacing and edge distance are the same as for the plate.

$$A_{gv} = 13.13\left(\frac{0.535}{5/8}\right) = 11.24 \text{ in.}^2$$

$$A_{nv} = 9.297\left(\frac{0.535}{5/8}\right) = 7.958 \text{ in.}^2$$

$$A_{nt} = 1.328\left(\frac{0.535}{5/8}\right) = 1.137 \text{ in.}^2$$

$$R_n = 0.6F_u A_{nv} + U_{bs}F_u A_{nt} = 0.6(65)(7.958) + 1.0(65)(1.137)$$
$$= 384.3 \text{ kips}$$

$$0.6F_y A_{gv} + U_{bs}F_u A_{nt} = 0.6(50)(11.24) + 1.0(65)(1.137) = 411.1 \text{ kips} > 384.3 \text{ kips}$$

Since 384.3 kips > 360.6 kips, block shear in the plate controls.

Use a plate $\frac{5}{8} \times 6\frac{1}{2}$. Part of the beam flange area will be lost because of the bolt holes. Use the provisions of AISC F13.1 to determine whether we need to account for this loss. The gross area of one flange is

$$A_{fg} = t_f b_f = 0.535(6.53) = 3.494 \text{ in.}^2$$

The effective hole diameter is

$$d_h = \frac{3}{4} + \frac{1}{8} = \frac{7}{8} \text{ in.}$$

and the net flange area is

$$A_{fn} = A_{fg} - t_f \sum d_h = 3.494 - 0.535\left(2 \times \frac{7}{8}\right) = 2.558 \text{ in.}^2$$

$$F_u A_{fn} = 65(2.558) = 166.3 \text{ kips}$$

Determine Y_t. For A992 steel, the maximum F_y/F_u ratio is 0.85. Since this is greater than 0.8, use $Y_t = 1.1$.

$$Y_t F_y A_{fg} = 1.1(50)(3.494) = 192.2 \text{ kips}$$

Since $F_u A_{fn} < Y_t F_y A_{fg}$, the holes must be accounted for. From AISC Equation F13-1,

$$M_n = \frac{F_u A_{fn}}{A_{fg}} S_x = \frac{166.3}{3.494}(94.5) = 4498 \text{ in.-kips}$$

$$\phi_b M_n = 0.90(4498) = 4048 \text{ in.-kips} = 337 \text{ ft-kips} > 210 \text{ ft-kips} \quad \text{(OK)}$$

Answer Use the connection shown in Figure 8.42 (column stiffener requirements will be considered in Section 8.7).[*]

ASD Solution (Portions of this solution will be taken from the LRFD solution)

$$M_a = M_D + M_L = 35 + 105 = 140 \text{ ft-kips}$$
$$V_a = V_D + V_L = 5.5 + 16.5 = 22 \text{ kips}$$

---

[*]Figure 8.42 also shows the symbol for a bevel groove weld, used here for the beam flange plate-to-column connection

FIGURE 8.42

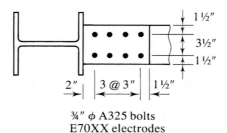

¾" φ A325 bolts
E70XX electrodes

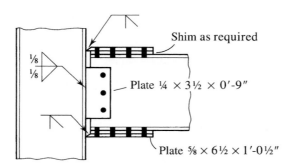

Shim as required

Plate ¼ × 3½ × 0'-9"

Plate ⅝ × 6½ × 1'-0½"

For the web plate, **try ¾-inch-diameter bolts**. Neglect eccentricity and assume that the threads are in the plane of shear. The nominal shear strength of one bolt is

$$R_n = F_{nv}A_b = 48(0.4418) = 21.21 \text{ kips}$$

and the allowable strength is

$$\frac{R_n}{\Omega} = \frac{21.21}{2.00} = 10.61 \text{ kips}$$

$$\text{Number of bolts required} = \frac{22}{10.61} = 2.07$$

Try three bolts. The minimum spacing is $2\frac{2}{3}d = 2.667(3/4) = 2.0$ in. From AISC Table J3.4, the minimum edge distance is $1\frac{1}{4}$ inches. Try the layout shown in Figure 8.39 and determine the plate thickness required for bearing. From the LRFD solution, for the hole nearest the edge,

$$R_n = 76.14t$$

$$\frac{R_n}{\Omega} = \frac{76.14t}{2.00} = 38.07t \text{ kips/bolt}$$

For the other holes,

$$R_n = 104.4t$$

$$\frac{R_n}{\Omega} = \frac{104.4t}{2.00} = 52.2t \text{ kips/bolt}$$

To find the required thickness, equate the total bearing strength to the applied load:

$$38.07t + 2(52.2t) = 22, \quad t = 0.154 \text{ in.}$$

(The beam web has a thickness of $t_w = 0.380$ in. > 0.154 in., and since F_y for the beam is larger than F_y for the plate, the bearing strength of the beam web will be adequate.)

Other limit states for connection plates (in addition to bolt bearing) can be found in AISC J4, "Affected Elements of Members and Connecting Elements." To determine the plate thickness required for the vertical shear, consider both yielding of the gross section and rupture of the net section. For shear yielding, from AISC Equation J4-3,

$$R_n = 0.60F_y A_g = 0.60(36)(8.5t) = 183.6t$$

$$\frac{R_n}{\Omega} = \frac{183.6t}{1.50} = 122.4t$$

For shear rupture, the net area is

$$A_{nv} = \left[8.5 - 3\left(\frac{3}{4} + \frac{1}{8}\right)\right]t = 5.875t$$

From AISC Equation J4-4,

$$R_n = 0.6F_u A_{nv} = 0.6(58)(5.875t) = 204.5t$$

$$\frac{R_n}{\Omega} = \frac{204.5t}{2.00} = 102.2t$$

Shear rupture controls. Let

$$102.2t = 22, \quad t = 0.215 \text{ in.}$$

The largest required thickness is therefore required by the limit state of shear rupture. **Try $t = \frac{1}{4}$ in.**

For the connection of the shear plate to the column flange, the required strength per inch is

$$\frac{V_a}{L} = \frac{22}{9} = 2.444 \text{ kips/in.}$$

From Equation 7.37, the base metal shear yield strength per unit length is

$$\frac{R_n}{\Omega} = 0.4F_y t = 0.4(36)\left(\frac{1}{4}\right) = 3.6 \text{ kips/in.}$$

From Equation 7.38, the base metal shear rupture strength per unit length is

$$\frac{R_n}{\Omega} = 0.3F_u t = 0.3(58)\left(\frac{1}{4}\right) = 4.35 \text{ kips/in.}$$

The base metal shear strength is therefore 3.6 kips/in., which is greater than the required weld strength of 2.444 kip/in. For welds on both sides of the plate, the required strength per weld is 2.444/2 = 1.222 kips/in. To determine the required weld size, let

$$0.9279D = 1.294, \quad D = 1.40 \text{ sixteenths of an inch}$$

The minimum weld size from AISC Table J2.4 is $\frac{1}{8}$ inch, based on the thinner connected part (the shear plate). There is no maximum size requirement for this type of connection (one where the weld is not along an edge). **Use a $\frac{1}{8}$-inch fillet weld on each side of the plate**.

The minimum width of the plate can be determined from a consideration of edge distances. The load being resisted (the beam reaction) is vertical, so the horizontal edge distance need only conform to the clearance requirements of AISC Table J3.4. If we assume a sheared edge, the minimum edge distance is $1\frac{1}{4}$ inch.

With a beam setback of $\frac{1}{2}$ inch and edge distances of $1\frac{1}{2}$ inches, as shown in Figure 8.39b, the width of the plate is

$$0.5 + 2(1.5) = 3.5 \text{ in.}$$

Try a plate $\frac{1}{4} \times 3\frac{1}{2}$. Check block shear. From the LRFD solution,

$$R_n = 55.90 \text{ kips}$$

$$\frac{R_n}{\Omega} = \frac{55.90}{2.00} = 28.0 \text{ kips} > 22 \text{ kips} \quad \text{(OK)}$$

Use a plate $\frac{1}{4} \times 3\frac{1}{2}$.

For the flange connection, first select the bolts. From Figure 8.40, the force at the interface between the beam flange and the plate is

$$H = \frac{M}{d} = \frac{140(12)}{20.8} = 80.77 \text{ kips}$$

where d is the depth of the beam. Although the moment arm of the couple is actually the distance from center of flange plate to center of flange plate, the plate thickness is not yet known, and use of the beam depth is conservative.

Try $\frac{3}{4}$-inch A325 bolts. (Since $\frac{3}{4}$-inch-diameter bolts were selected for the shear connection, try the same size here). If bolt shear controls, the number of bolts required is

$$\frac{80.77}{10.61} = 7.61 \text{ bolts} \qquad \text{Use 8 bolts (4 pairs)}$$

Use edge distances of $1\frac{1}{2}$ inches, spacings of 3 inches, and determine the minimum plate thickness required for bearing. From the LRFD solution, for the hole nearest the edge,

$$R_n = 76.14t$$

$$\frac{R_n}{\Omega} = \frac{76.14t}{2.00} = 38.07t \text{ kips/bolt}$$

For the other holes,

$$R_n = 104.4t$$

$$\frac{R_n}{\Omega} = \frac{104.4t}{2.00} = 52.2t \text{ kips/bolt}$$

To find the required thickness, equate the total bearing strength to the applied load:

$$2(38.07t) + 6(52.2t) = 80.77, \qquad t = 0.208 \text{ in.}$$

The flange plate will be designed as a tension-connecting element (the top plate), then checked for compression (the bottom plate). The minimum cross section required for tension on the gross and net areas will now be determined. From AISC Equation J4-1,

$$R_n = F_y A_g$$

$$\frac{R_n}{\Omega} = \frac{F_y A_g}{1.67}$$

$$\text{Required } A_g = \frac{1.67(R_n/\Omega)}{F_y} = \frac{1.67H}{F_y} = \frac{1.67(80.77)}{36} = 3.747 \text{ in.}^2$$

From AISC Equation J4-2,

$$R_n = F_u A_e$$

$$\frac{R_n}{\Omega} = \frac{F_u A_e}{2.00}$$

$$\text{Required } A_e = \frac{2.00(R_n/\Omega)}{F_u} = \frac{2.00H}{F_u} = \frac{2.00(80.77)}{58} = 2.785 \text{ in.}^2$$

Try a plate width of $w_g = 6.5$ in. (equal to the beam flange width). Compute the thickness needed to satisfy the gross area requirement.

$$A_g = 6.5t = 3.747 \quad \text{or} \quad t = 0.576 \text{ in.}$$

Compute the thickness needed to satisfy the net area requirement.

$$A_e = A_n = tw_n = t(w_g - \Sigma d_{\text{hole}}) = t\left[6.5 - 2\left(\frac{7}{8}\right)\right] = 4.750t$$

Let

$$4.750t = 2.785 \text{ in.}^2 \quad \text{or} \quad t = 0.586 \text{ in. (controls)}$$

This thickness is also greater than that required for bearing, so it will be the minimum acceptable thickness. **Try a plate $\frac{5}{8} \times 6\frac{1}{2}$.**

Check compression. Assume that the plate acts as a fixed-end compression member between fasteners, with $L = 3$ in. and $K = 0.65$. From the LRFD solution,

$$P_n = F_y A_g$$

and for ASD, $\Omega = 1.67$.

$$\frac{P_n}{\Omega} = \frac{F_y A_g}{1.67} = \frac{36\left(\frac{5}{8} \times 6.5\right)}{1.67} = 87.6 \text{ kips} > 80.77 \text{ kips} \quad \text{(OK)}$$

Check block shear in the plate. In the transverse direction, use bolt edge distances of $1\frac{1}{2}$ inches and a spacing of $3\frac{1}{2}$ inches. This places the bolts at the "workable gage" location in the beam flange (see Part 1 of the *Manual*). Figure 8.41 shows the bolt layout and two possible block shear failure modes. From the LRFD solution,

$$R_n = 360.6 \text{ kips}$$

$$\frac{R_n}{\Omega} = \frac{360.6}{2.00} = 180 \text{ kips} > 80.77 \text{ kips} \qquad \text{(OK)}$$

Check block shear in the beam flange. The bolt spacing and edge distance are the same as for the plate. From the LRFD solution,

$$R_n = 384.3 \text{ kips}$$

Since 384.3 kips > 360.6 kips, block shear in the plate controls.

Use a plate ⅝ × 6½. Part of the beam flange area will be lost because of the bolt holes. Use the provisions of AISC F13.1 to determine whether we need to account for this loss. From the LRFD solution, the reduced nominal moment strength is

$$M_n = 4498 \text{ in.-kips}$$

$$\frac{M_n}{\Omega} = \frac{4498}{1.67} = 2693 \text{ in.-kips} = 224 \text{ ft-kips} > 140 \text{ ft-kips} \qquad \text{(OK)}$$

Answer   Use the connection shown in Figure 8.42 (column stiffener requirements are considered in Section 8.7).   ■

8.7   COLUMN STIFFENERS AND OTHER REINFORCEMENT

Since most of the moment transferred from the beam to the column in a rigid connection takes the form of a couple consisting of the tensile and compressive forces in the beam flanges, the application of these relatively large concentrated forces may require reinforcement of the column. For negative moment, as would be the case with gravity loading, these forces are directed as shown in Figure 8.43, with the top flange of the beam delivering a tensile force to the column and the bottom flange delivering a compressive force.

FIGURE 8.43

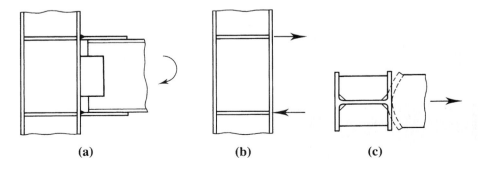

(a)   (b)   (c)

Both forces are transmitted to the column web, with compression being more critical because of the stability problem. The tensile load at the top can distort the column flange (exaggerated in Figure 8.43c), creating additional load on the weld connecting the beam flange to the column flange. A stiffener of the type shown can provide anchorage for the column flange. Obviously, this stiffener must be welded to both web and flange. If the applied moment never changes direction, the stiffener resisting the compressive load (the bottom stiffener in this illustration) can be fitted to bear on the flange and need not be welded to it.

AISC Specification Requirements

The AISC Requirements for column web reinforcement are covered in Section J10, "Flanges and Webs with Concentrated Forces." For the most part, these provisions are based on theoretical analyses that have been modified to fit test results. If the applied load transmitted by the beam flange or flange plate exceeds the available strength for any of the limit states considered, stiffeners must be used.

Local Bending of the Column Flange. To avoid a local bending failure of the column flange, the tensile load from the beam flange must not exceed the available strength. The nominal strength is

$$R_n = 6.25t_f^2 F_{yf} \qquad \text{(AISC Equation J10-1)}$$

where

t_f = thickness of the column flange
F_{yf} = yield stress of the column flange

For LRFD, the design strength is ϕR_n, where $\phi = 0.90$. For ASD, the allowable strength is R_n/Ω, where $\Omega = 1.67$.

If the tensile load is applied at a distance less than $10t_f$ from the end of the column, reduce the strength given by AISC Equation J10-1 by 50%.

Local Web Yielding. For the limit state of local web yielding in compression, when the load is applied at a distance from the end of the column that is more than the depth of the column,

$$R_n = (5k + N)F_{yw}t_w \qquad \text{(AISC Equation J10-2)}$$

When the load is applied at a distance from the end of the column that is less than the depth of the column,

$$R_n = (2.5k + N)F_{yw}t_w \qquad \text{(AISC Equation J10-3)}$$

where

k = distance from the outer flange surface of the column to the toe of the fillet in the column web
N = length of applied load = thickness of beam flange or flange plate
F_{yw} = yield stress of the column web
t_w = thickness of the column web

We also used AISC Equations J10-2 and J10-3 in Section 5.14 to investigate web yielding in beams subjected to concentrated loads.

For LRFD, $\phi = 1.00$. For ASD, $\Omega = 1.50$.

Web Crippling. To prevent web crippling when the compressive load is delivered to one flange only, as in the case of an exterior column with a beam connected to one side, the applied load must not exceed the available strength. (We also addressed web crippling in Section 5.14.) When the load is applied at a distance of least $d/2$ from the end of the column,

$$R_n = 0.80t_w{}^2 \left[1 + 3\left(\frac{N}{d}\right)\left(\frac{t_w}{t_f}\right)^{1.5}\right]\sqrt{\frac{EF_{yw}t_f}{t_w}} \qquad \text{(AISC Equation J10-4)}$$

where d = total column depth.

If the load is applied at a distance less than $d/2$ from the end of the column,

$$R_n = 0.40t_w^2 \left[1 + 3\left(\frac{N}{d}\right)\left(\frac{t_w}{t_f}\right)^{1.5}\right]\sqrt{\frac{EF_{yw}t_f}{t_w}} \qquad \left(\text{for } \frac{N}{d} \le 0.2\right)$$

$$\text{(AISC Equation J10-5a)}$$

or

$$R_n = 0.40t_w^2 \left[1 + \left(\frac{4N}{d} - 0.2\right)\left(\frac{t_w}{t_f}\right)^{1.5}\right]\sqrt{\frac{EF_{yw}t_f}{t_w}} \qquad \left(\text{for } \frac{N}{d} > 0.2\right)$$

$$\text{(AISC Equation J10-5b)}$$

For LRFD, $\phi = 0.75$. For ASD, $\Omega = 2.00$.

Compression Buckling of the Web. Compression buckling of the column web must be investigated when the loads are delivered to *both* column flanges. Such loading would occur at an interior column with beams connected to both sides. The nominal strength for this limit state is

$$R_n = \frac{24t_w^3\sqrt{EF_{yw}}}{h} \qquad \text{(AISC Equation J10-8)}$$

where h = column web depth from toe of fillet to toe of fillet (Figure 8.44).

If the connection is near the end of the column (i.e., if the load is applied within a distance $d/2$ from the end), the strength given by AISC Equation J10-8 should be reduced by half.

For LRFD, $\phi = 0.90$. For ASD, $\Omega = 1.67$.

FIGURE 8.44

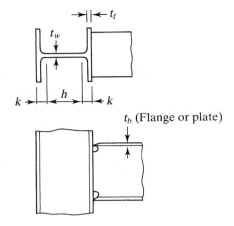

In summary, to investigate the need for column stiffeners, three limit states should be checked:

1. Local flange bending (AISC Equation J10-1).
2. Local web yielding (AISC Equation J10-2 or J10-3).
3. Web crippling *or* compression buckling of the web. (If the compressive load is applied to *one* flange only, check web crippling [AISC Equation J10-4 or J10-5]. If the compressive load is applied to *both* flanges, check compression buckling of the web [AISC Equation J10-8].)

If the required strength is greater than the available strength corresponding to any of these limit states, stiffeners must be provided.

For the stiffener available strength, use yielding as the limit state. The nominal stiffener strength will therefore be

$$F_{yst}A_{st}$$

where

F_{yst} = yield stress of the stiffener
A_{st} = area of stiffener

For LRFD, equate the design strength of the stiffener to the extra strength needed and solve for the required stiffener area:

$$\phi_{st}F_{yst}A_{st} = P_{bf} - \phi R_{n\ min}$$

$$A_{st} = \frac{P_{bf} - \phi R_{n\ min}}{\phi_{st}F_{yst}} \tag{8.6}$$

where

ϕ_{st} = 0.90 (since this is a yielding limit state)
P_{bf} = applied factored load from the beam flange or flange plate
$\phi R_{n\ min}$ = smallest of the strengths corresponding to the three limit states

For ASD, equate the allowable strength of the stiffener to the extra strength needed and solve for the required stiffener area:

$$\frac{F_{yst} A_{st}}{\Omega_{st}} = P_{bf} - (R_n/\Omega)_{\min}$$

$$A_{st} = \frac{P_{bf} - (R_n/\Omega)_{\min}}{F_{yst}/\Omega_{st}} \tag{8.7}$$

where

$\Omega_{st} = 1.67$ (since this is a yielding limit state)

$P_{bf} =$ applied service load from the beam flange or flange plate

$(R_n/\Omega)_{\min} =$ smallest of the strengths corresponding to the three limit states

AISC J10.8 gives the following guidelines for proportioning the stiffeners.

- The width of the stiffener plus half the column web thickness must be equal to at least one third of the width of the beam flange or plate delivering the force to the column, or, from Figure 8.45,

$$b + \frac{t_w}{2} \geq \frac{b_b}{3} \qquad \therefore b \geq \frac{b_b}{3} - \frac{t_w}{2}$$

- The stiffener thickness must be at least half the thickness of the beam flange or plate, or

$$t_{st} \geq \frac{t_b}{2}$$

- The stiffener thickness must be at least equal to its width divided by 15, or

$$t_{st} \geq \frac{b}{15}$$

- Full-depth stiffeners are required for the compression buckling case, but half-depth stiffeners are permitted for the other limit states. Thus full-depth stiffeners are required only when beams are connected to both sides of the column.

For any of the limit states, the decision on whether to weld the stiffener to the flange should be based on the following criteria:

- On the tension side, the stiffeners should be welded to both the web and flange.
- On the compression side, the stiffeners need only bear on the flange but may be welded.

FIGURE 8.45

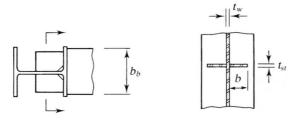

In Part 4 of the *Manual*, "Design of Compression Members," the column load tables contain constants that expedite the evaluation of the need for stiffeners. Their use is explained in the *Manual* and is not covered here.

Shear in the Column Web

The transfer of a large moment to a column can produce large shearing stresses in the column web within the boundaries of the connection; for example, region *ABCD* in Figure 8.46. This region is sometimes called the *panel zone*. The *net* moment is of concern, so if beams are connected to both sides of the column, the algebraic sum of the moments induces this web shear.

Each flange force can be taken as

$$H = \frac{M_1 + M_2}{d_m}$$

where d_m is the moment arm of the couple (for beams of equal depth on both sides of the column).

If the column shear adjacent to the panel is V and is directed as shown, the total shear force in the panel (required shear strength) is

$$F = H - V = \frac{M_1 + M_2}{d_m} - V \tag{8.8}$$

The web nominal shear strength R_n is given in AISC J10.6. It is a function of P_r, the required axial strength of the column, and P_c, the yield strength of the column.

FIGURE 8.46

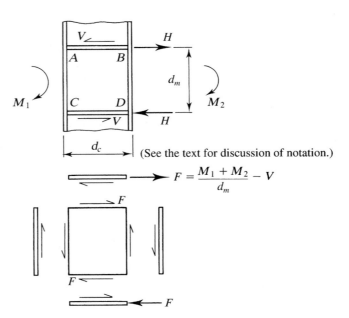

(See the text for discussion of notation.)

$$F = \frac{M_1 + M_2}{d_m} - V$$

When $P_r \le 0.4P_c$,

$$R_n = 0.60F_y d_c t_w \qquad \text{(AISC Equation J10-9)}$$

When $P_r > 0.4P_c$,

$$R_n = 0.60F_y d_c t_w \left(1.4 - \frac{P_r}{P_c} \right) \qquad \text{(AISC Equation J10-10)}$$

where

P_r = required axial strength of the column
= factored axial load for LRFD
= service axial load for ASD
P_c = yield strength of the column
= $P_y = F_y A$ for LRFD
= $0.6P_y = 0.6F_y A$ for ASD
A = cross-sectional area of column
F_y = yield stress of column
d_c = total column depth
t_w = column web thickness

For LRFD, the design shear strength is ϕR_n, where $\phi = 0.90$.

For ASD, the allowable shear strength is R_n/Ω, where $\Omega = 1.67$.

If the column web has insufficient shear strength, it must be reinforced. A *doubler plate* with sufficient thickness to make up the deficiency can be welded to the web or a pair of diagonal stiffeners can be used. Frequently, the most economical alternative is to use a column section with a thicker web.

AISC J10.6 also provides equations to be used when frame stability, including plastic deformation of the panel zone, is considered in the analysis. They are not covered here.

Example 8.11   Determine whether stiffeners or other column web reinforcement is required for the connection of Example 8.10. Assume that $V = 0$ and $P_r/P_c = 0.4$. The connection is not near the end of the column.

LRFD Solution   From Example 8.10, the flange force can be conservatively taken as

$P_{bf} = H = 121.2$ kips

Check local flange bending with AISC Equation J10-1:

$$\phi R_n = \phi(6.25t_f^2 F_{yf})$$
$$= 0.90[6.25(0.780)^2(50)] = 171 \text{ kips} > 121.2 \text{ kips} \qquad \text{(OK)}$$

Check local web yielding with AISC Equation J10-2:

$$\phi R_n = \phi[(5k + N)F_{yw}t_w]$$

$$= 1.0\left[5(1.38) + \frac{5}{8}\right](50)(0.485) = 182 \text{ kips} > 121.2 \text{ kips} \quad \text{(OK)}$$

Check web crippling with AISC Equation J10-4:

$$\phi R_n = \phi 0.80 t_w^2 \left[1 + 3\left(\frac{N}{d}\right)\left(\frac{t_w}{t_f}\right)^{1.5}\right]\sqrt{\frac{EF_{yw}t_f}{t_w}}$$

$$= 0.75(0.80)(0.485)^2\left[1 + 3\left(\frac{5/8}{14.2}\right)\left(\frac{0.485}{0.780}\right)^{1.5}\right]\sqrt{\frac{29,000(50)(0.780)}{0.485}}$$

$$= 229 \text{ kips} > 121.2 \text{ kips} \quad \text{(OK)}$$

Answer Column stiffeners are not required.

For shear in the column web, from Equation 8.8 and neglecting the thickness of shims in the computation of d_m, the factored load shear force in the column web panel zone is

$$F = \frac{(M_1 + M_2)}{d_m} - V = \frac{(M_1 + M_2)}{d_b + t_{PL}} - V$$

$$= \frac{210(12)}{20.8 + 5/8} - 0 = 118 \text{ kips}$$

Because $P_r = 0.4P_c$, use AISC Equation J10-9:

$$R_n = 0.60F_y d_c t_w = 0.60(50)(14.2)(0.485) = 206.6 \text{ kips}$$

The design strength is

$$\phi R_n = 0.90(206.6) = 186 \text{ kips} > 118 \text{ kips} \quad \text{(OK)}$$

Answer Column web reinforcement is not required.

ASD Solution From Example 8.10, the flange force can be conservatively taken as

$$P_{bf} = H = 80.77 \text{ kips}$$

Check local flange bending with AISC Equation J10-1:

$$\frac{R_n}{\Omega} = \frac{6.25 t_f^2 F_{yf}}{\Omega}$$

$$= \frac{6.25(0.780)^2(50)}{1.67} = 114 \text{ kips} > 80.77 \text{ kips} \quad \text{(OK)}$$

Check local web yielding with AISC Equation J10-2:

$$\frac{R_n}{\Omega} = \frac{(5k + N)F_{yw}t_w}{\Omega}$$

$$= \frac{\left[5(1.38) + (5/8)\right](50)(0.485)}{1.50} = 122 \text{ kips} > 80.77 \text{ kips} \qquad (\text{OK})$$

Check web crippling with AISC Equation J10-4:

$$\frac{R_n}{\Omega} = \frac{0.80t_w^2\left[1 + 3\left(\frac{N}{d}\right)\left(\frac{t_w}{t_f}\right)^{1.5}\right]\sqrt{\frac{EF_{yw}t_f}{t_w}}}{\Omega}$$

$$= \frac{0.80(0.485)^2\left[1 + 3\left(\frac{5/8}{14.2}\right)\left(\frac{0.485}{0.780}\right)^{1.5}\right]\sqrt{\frac{29{,}000(50)0.780}{0.485}}}{2.00}$$

$$= 153 \text{ kips} > 80.77 \text{ kips} \qquad (\text{OK})$$

Answer   Column stiffeners are not required.

For shear in the column web, From Equation 8.8 and neglecting the thickness of shims in the computation of d_m, the service load shear force in the column web panel zone is

$$F = \frac{(M_1 + M_2)}{d_m} - V = \frac{M}{d_b + t_{PL}} - V = \frac{140(12)}{20.8 + 5/8} - 0 = 89.8 \text{ kips}$$

From AISC Equation J10-9, the nominal strength is

$$R_n = 0.60F_y d_c t_w = 0.60(50)(14.2)(0.485) = 206.6 \text{ kips}$$

The allowable strength is

$$\frac{R_n}{\Omega} = \frac{206.6}{1.67} = 123.7 \text{ kips} > 89.8 \text{ kips} \qquad (\text{OK})$$

Answer   Column web reinforcement is not required.

Example 8.12   The beam-to-column connection shown in Figure 8.47 must transfer the following gravity load moments: a service dead-load moment of 32.5 ft-kips and a service live-load moment of 97.5 ft-kips. A992 steel is used for the beam and column, and A36 is used for the plate material. E70XX electrodes are used. Investigate column stiffener and web panel-zone reinforcement requirements. The connection is not near the end of the column. Assume that $V = 0$ and $P_r/P_c = 0$.

FIGURE 8.47

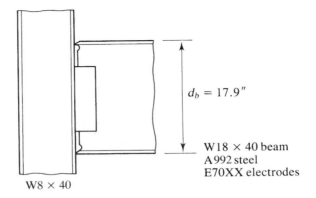

$d_b = 17.9''$

W18 × 40 beam
A992 steel
E70XX electrodes

W8 × 40

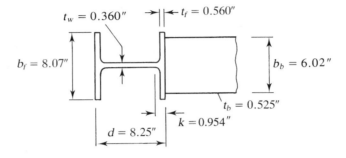

$t_w = 0.360''$ $t_f = 0.560''$

$b_f = 8.07''$ $b_b = 6.02''$

$t_b = 0.525''$

$k = 0.954''$

$d = 8.25''$

LRFD Solution The factored-load moment is

$$M_u = 1.2M_D + 1.6(M_L) = 1.2(32.5) + 1.6(97.5) = 195.0 \text{ ft-kips}$$

The flange force is

$$P_{bf} = \frac{M_u}{d_b - t_b} = \frac{195(12)}{17.9 - 0.525} = 134.7 \text{ kips}$$

To check local flange bending, use AISC Equation J10-1:

$$\phi R_n = \phi(6.25t_f^2 F_{yf}) = 0.90[6.25(0.560)^2(50)]$$
$$= 88.20 \text{ kips} < 134.7 \text{ kips} \quad \text{(N.G.)}$$

∴ stiffeners are required to prevent local flange bending

To check for local web yielding, use AISC Equation J10-2:

$$\phi R_n = \phi\big[(5k + N)F_{yw}t_w\big]$$
$$= 1.0\big\{ [5(0.954) + 0.525](50)(0.360) \big\} = 95.3 \text{ kips} < 134.7 \text{ kips} \quad \text{(N.G.)}$$

∴ stiffeners are required to prevent local web yielding

To check web crippling strength, use AISC Equation J10-4:

$$\phi R_n = \phi 0.80 t_w^2 \left[1 + 3 \left(\frac{N}{d} \right) \left(\frac{t_w}{t_f} \right)^{1.5} \right] \sqrt{\frac{EF_{yw} t_f}{t_w}}$$

$$= 0.75(0.80)(0.360)^2 \left[1 + 3 \left(\frac{0.525}{8.25} \right) \left(\frac{0.360}{0.560} \right)^{1.5} \right] \sqrt{\frac{29,000(50)(0.560)}{0.360}}$$

$$= 128.3 \text{ kips} < 134.7 \text{ kips} \quad \text{(N.G.)}$$

∴ stiffeners are required to prevent web crippling

The local flange bending strength of 88.20 kips is the smallest of the three limit states. From Equation 8.6, the required stiffener area is

$$A_{st} = \frac{P_{bf} - \phi R_{n \, \min}}{\phi_{st} F_{yst}} = \frac{134.7 - 88.20}{0.90(36)} = 1.44 \text{ in.}^2$$

The stiffener dimensions will be selected on the basis of the criteria given in AISC Section J10.8, and the area of the resulting cross section will then be checked.
The minimum width is

$$b \geq \frac{b_b}{3} - \frac{t_w}{2} = \frac{6.02}{3} - \frac{0.360}{2} = 1.83 \text{ in.}$$

If the stiffeners are not permitted to extend beyond the edges of the column flange, the maximum width is

$$b \leq \frac{8.07 - 0.360}{2} = 3.86 \text{ in.}$$

The minimum thickness is

$$\frac{t_b}{2} = \frac{0.525}{2} = 0.263 \text{ in.}$$

Try a plate ⁵⁄₁₆ × 3:

$$A_{st} = 3 \left(\frac{5}{16} \right) \times 2 \text{ stiffeners} = 1.88 \text{ in.}^2 > 1.44 \text{ in.}^2 \quad \text{(OK)}$$

Check for $t_{st} \geq b/15$:

$$\frac{b}{15} = \frac{3}{15} = 0.2 \text{ in.} < \frac{5}{16} \text{ in.} \quad \text{(OK)}$$

This connection is one-sided, so full-depth stiffeners are not required. Try a depth of

$$\frac{d}{2} = \frac{8.25}{2} = 4.125 \text{ in.} \quad \textbf{Try 4½ inches.}$$

Try two plates ⁵⁄₁₆ × 3 × 0′- 4½″ (Clip the inside corners to avoid the fillets at the column flange-to-web intersection. Clip at a 45° angle for ⅝ inch.)

For the stiffener to column web welds,

$$\text{Minimum size} = \frac{3}{16} \text{ in.} \quad \text{(AISC Table J2.4, based on plate thickness)}$$

The size required for strength is

$$w = \frac{\text{force resisted by stiffener}}{\phi(0.707LF_W)}$$

Since flange local bending controls in this example, the force to be resisted by the stiffener is

$$134.7 - 88.20 = 46.50 \text{ kips}$$

The length available for welding the stiffener to the web is

$$L = (length - clip) \times 2 \text{ sides} \times 2 \text{ stiffeners}$$
$$= \left(4.5 - \frac{5}{8}\right)(2)(2) = 15.5 \text{ in.} \quad \text{(See Figure 8.48)}$$

The required weld size is therefore

$$w = \frac{46.50}{0.75(0.707)(15.5)(0.6 \times 70)} = 0.135 \text{ in.} < \frac{3}{16} \text{ in. minimum}$$

Try w = ³⁄₁₆ in.

Check the base metal shear strength. From Equation 7.35, the shear yield strength of the web is

$$\phi R_n = 0.6F_y t \times 2 \text{ sides of stiffener} = 0.6(50)(0.360)(2) = 21.6 \text{ kips/in.}$$

(The load is shared by two thicknesses of web, one on each side of the stiffener pair.) From Equation 7.36, the shear rupture strength is

$$\phi R_n = 0.45F_u t \times 2 = 0.45(65)(0.360)(2) = 21.06 \text{ kips/in.}$$

For the stiffener shear yield strength,

$$\phi R_n = 0.6F_y t_{st} \times 2 = 0.6(36)\left(\frac{5}{16}\right)(2) = 13.5 \text{ kips/in.}$$

FIGURE 8.48

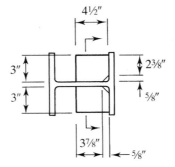

Section

The shear rupture strength is

$$\phi R_n = 0.45 F_u t_{st} \times 2 = 0.45(58)\left(\frac{5}{16}\right)(2) = 16.31 \text{ kips/in.}$$

The base metal shear strength is therefore 13.5 kips/in. The *required* strength of the weld is

$$\phi(0.707 w F_W) \times 2 \times 2 = 0.75(0.707)(0.135)(0.6 \times 70)(2)(2)$$
$$= 12.03 \text{ kips/in.} < 13.5 \text{ kips/in.} \quad \text{(OK)}$$

Answer   Use two plates $\frac{5}{16} \times 3 \times 0' - 4\frac{1}{2}''$ Clip the inside corners to avoid the fillets at the column flange-to-web intersection. Clip at a 45° angle for $\frac{5}{8}$ inch. Use a $\frac{3}{16}$-inch fillet weld for the stiffener-to-column web welds.

Stiffener-to-column flange welds: Since the beam tension flange force is an issue (for both column flange bending and web yielding), the stiffener must be welded to the column flange. Use a weld sufficient to develop the yield strength of the stiffener. Let

$$\phi[0.707 w L (0.6 F_{EXX})](1.5) = \phi_{st} F_{yst} A_{st}$$

where

$$L = (b - clip) \times 2 \text{ sides} \times 2 \text{ stiffeners}$$

and the factor of 1.5 is used because the load is perpendicular to the axis of the weld. Then,

$$w = \frac{\phi_{st} F_{yst} A_{st}}{\phi[0.707 L (0.6 F_{EXX})](1.5)}$$

$$= \frac{0.90(36)(1.88)}{0.75(0.707)[(3 - 5/8) \times 2 \times 2](0.6 \times 70)(1.5)} = \frac{60.91}{317.4} = 0.192 \text{ in.}$$

Minimum size $= \dfrac{3}{16}$ in. (AISC Table J2.4, based on plate thickness)

Answer   Use a $\frac{1}{4}$-inch fillet weld for the stiffener-to-column flange connection. (Because the applied moment is caused by gravity loads and is not reversible, the stiffeners opposite the beam compression flange can be fitted to bear on the column flange and need not be welded, but this option is not exercised here.)

Check the column web for shear. From Equation 8.8,

$$F = \frac{(M_1 + M_2)}{d_m} - V = \frac{M_u}{d_b - t_b} - V_u = \frac{195(12)}{17.9 - 0.525} - 0 = 134.7 \text{ kips}$$

From AISC Equation J10-9,

$$R_n = 0.60 F_y d_c t_w = 0.60(50)(8.25)(0.360) = 89.1 \text{ kips}$$

The design strength is

$$\phi R_n = 0.90(89.1) = 80.19 \text{ kips} < 134.7 \text{ kips} \qquad \text{(N.G.)}$$

Alternative 1: Use a Web Doubler Plate. Use AISC Equation J10-9 to find the required doubler plate thickness. Multiplying both sides by ϕ and solving for t_w gives

$$t_w = \frac{\phi R_n}{\phi(0.60 F_y d_c)}$$

Substituting the plate thickness t_d for t_w and using the yield stress of the doubler plate, we get

$$t_d = \frac{\phi R_n}{\phi(0.60 F_y d_c)} = \frac{134.7 - 80.19}{0.90(0.60)(36)(8.25)} = 0.340 \text{ in.}$$

where $134.7 - 80.19$ is the extra strength, in kips, to be furnished by the doubler plate.

The design of the welds connecting the doubler plate to the column will depend on the exact configuration of the plate, including whether the plate extends beyond the transverse stiffeners. For this and other details, refer to AISC Design Guide 13, *Stiffening of Wide-Flange Columns at Moment Connections: Wind and Seismic Applications* (Carter, 1999).

Answer Use a ⅜-inch-thick doubler plate.

Alternative 2: Use a Diagonal Stiffener. With this alternative, use full-depth horizontal stiffeners as shown in Figure 8.49.

FIGURE 8.49

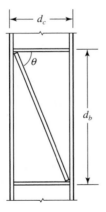

The shear force to be resisted by the web reinforcement is $134.7 - 80.19 = 54.51$ kips. If this force is taken as the horizontal component of an axial compressive force P in the stiffener,

$$P \cos \theta = 54.51 \text{ kips}$$

where

$$\theta = \tan^{-1}\left(\frac{d_b}{d_c}\right) = \tan^{-1}\left(\frac{17.9}{8.25}\right) = 65.26°$$

$$P = \frac{54.51}{\cos(65.26°)} = 130.3 \text{ kips}$$

Since the stiffener is connected along its length, we will treat it as a compression member whose effective length KL is zero. From AISC J4.4, for compression elements with $KL/r < 25$, the nominal strength is

$$P_n = F_y A_g$$

and for LRFD, $\phi = 0.90$

$$\phi P_n = 0.90(36)A_g = 32.4 A_g$$

Equating this strength to the required strength, we obtain the required area of stiffener:

$$32.4 A_g = 130.3$$
$$A_g = 4.02 \text{ in.}^2$$

Try two stiffeners, $\frac{3}{4} \times 3$, one on each side of the web:

$$A_{st} = 2(3)\left(\frac{3}{4}\right) = 4.50 \text{ in.}^2 > 4.02 \text{ in.}^2 \text{ required} \qquad \text{(OK)}$$

Check for $t_{st} \geq b/15$:

$$\frac{b}{15} = \frac{3}{15} = 0.2 \text{ in.} < \frac{3}{4} \text{ in.} \quad \text{(OK)}$$

Design the welds. The approximate length of each diagonal stiffener is

$$L_{st} = \frac{d_c}{\cos\theta} = \frac{8.25}{\cos(65.26°)} = 19.7 \text{ in.}$$

If welds are used on both sides of the stiffeners, the available length for welding is

$$L = 19.7(4) = 78.8 \text{ in.}$$

The weld size, in sixteenths of an inch, required for strength is

$$D = \frac{P}{1.392L} = \frac{130.3}{1.392(78.8)} = 1.2 \text{ sixteenths}$$

Based on the web thickness, use a minimum size of $\frac{3}{16}$ inch (AISC Table J2.4).

Because of the small size required for strength, investigate the possibility of using intermittent welds. From AISC J2.2b,

$$\text{Minimum length} = 4w = 4\left(\frac{3}{16}\right)$$
$$= 0.75 \text{ in., but not less than 1.5 in.} \qquad \text{(1.5 in. controls)}$$

For a group of four welds, the capacity is

$$4(1.392DL) = 4(1.392)(3)(1.5) = 25.06 \text{ kips}$$

Required weld capacity per inch $= \dfrac{P}{L_{st}} = \dfrac{130.3}{19.7} = 6.614$ kips/in.

Check the base metal shear strength of the column web. From Equation 7.35, the shear yield strength per unit length (considering a web thickness on each side of the stiffener pair) is

$$\phi R_n = 0.6F_y t_w \times 2 = 0.6(50)(0.360)(2) = 21.6 \text{ kips/in.}$$

From Equation 7.36, the shear rupture strength is

$$\phi R_n = 0.45F_u t_w \times 2 = 0.45(65)(0.360)(2) = 21.1 \text{ kips/in.}$$

For the stiffener, the shear yield strength is

$$\phi R_n = 0.6F_y t_{st} \times 2 \text{ stiffeners} = 0.6(36)(3/4)(2) = 32.4 \text{ kips/in.}$$

and the shear rupture strength is

$$\phi R_n = 0.45F_u t_{st} \times 2 = 0.45(58)(3/4) \times 2 = 39.15 \text{ kips/in.}$$

The controlling base metal shear strength is 21.1 kips/in., but this is greater than the weld strength of 6.614 kips/in.

$$\text{Required spacing of welds} = \frac{\text{strength of weld group (kips)}}{\text{required strength per inch (kips/in.)}} = \frac{25.06}{6.614} = 3.79 \text{ in.}$$

Answer   Use two stiffeners, ¾ × 3, one on each side of the web, and ³⁄₁₆-inch×1½-inch intermittent fillet welds spaced at 3½ inches on center, on each side of each diagonal stiffener.

ASD Solution   The service load moment is

$$M_a = M_D + M_L = 32.5 + 97.5 = 130 \text{ ft-kips}$$

The flange force is

$$P_{bf} = \frac{M_a}{d_b - t_b} = \frac{130(12)}{17.9 - 0.525} = 89.78 \text{ kips}$$

Check local flange bending with AISC Equation J10-1:

$$\frac{R_n}{\Omega} = \frac{6.25 t_f^2 F_{yf}}{\Omega}$$

$$= \frac{6.25(0.560)^2 (50)}{1.67} = 58.68 \text{ kips} < 89.78 \text{ kips} \quad \text{(N.G.)}$$

Check local web yielding with AISC Equation J10-2:

$$\frac{R_n}{\Omega} = \frac{(5k + N)F_{yw} t_w}{\Omega}$$

$$= \frac{[5(0.954) + 0.525](50)(0.360)}{1.50} = 63.54 \text{ kips} < 89.78 \text{ kips} \quad \text{(N.G.)}$$

Check web crippling with AISC Equation J10-4:

$$\frac{R_n}{\Omega} = \frac{0.80t_w^2\left[1+3\left(\frac{N}{d}\right)\left(\frac{t_w}{t_f}\right)^{1.5}\right]\sqrt{\frac{EF_{yw}t_f}{t_w}}}{\Omega}$$

$$= \frac{0.80(0.360)^2\left[1+3\left(\frac{0.525}{8.25}\right)\left(\frac{0.360}{0.560}\right)^{1.5}\right]\sqrt{\frac{29{,}000(50)0.560}{0.360}}}{2.00}$$

$$= 85.52 \text{ kips} < 89.78 \text{ kips} \quad \text{(N.G.)}$$

Stiffeners are required. The smallest strength is 58.68 kips, for the limit state of local flange bending. From Equation 8.7, the required stiffener area is

$$A_{st} = \frac{P_{bf} - (R_n/\Omega)_{\min}}{F_{yst}/\Omega} = \frac{89.78 - 58.68}{36/1.67} = 1.44 \text{ in.}^2$$

The minimum width is

$$b \geq \frac{b_b}{3} - \frac{t_w}{2} = \frac{6.02}{3} - \frac{0.360}{2} = 1.83 \text{ in.}$$

Use a maximum width of

$$b \leq \frac{8.07 - 0.360}{2} = 3.86 \text{ in.}$$

The minimum thickness is

$$\frac{t_b}{2} = \frac{0.525}{2} = 0.263 \text{ in.}$$

Try a plate ⁵⁄₁₆ × 3:

$$A_{st} = 3\left(\frac{5}{16}\right) \times 2 \text{ stiffeners} = 1.88 \text{ in.}^2 > 1.44 \text{ in.}^2 \quad \text{(OK)}$$

Check for $t_{st} \geq b/15$:

$$\frac{b}{15} = \frac{3}{15} = 0.2 \text{ in.} < \frac{5}{16} \text{ in.} \quad \text{(OK)}$$

This connection is one-sided, so full-depth stiffeners are not required. Use a depth of

$$\frac{d}{2} = \frac{8.25}{2} = 4.125 \text{ in.} \quad \text{Try } 4\tfrac{1}{2} \text{ in.}$$

Try two plates ⁵⁄₁₆ × 3 × 0′ - 4½″ (Clip the inside corners to avoid the fillets at the column flange-to-web intersection. Clip at a 45° angle for ⅝ inch.)

Stiffener-to-column web welds: The size required for strength is

$$w = \frac{\text{force resisted by stiffener}}{0.707LF_W/\Omega}$$

The force to be resisted by the stiffener $= 89.78 - 58.68 = 31.10$ kips. The length available for welding the stiffener to the web is

$$L = (length - clip) \times 2 \text{ sides} \times 2 \text{ stiffeners}$$
$$= \left(4.5 - \frac{5}{8}\right)(2)(2) = 15.5 \text{ in.} \quad \text{(see Figure 8.48)}$$

The required weld size is therefore

$$w = \frac{31.10}{0.707(15.5)(0.6 \times 70)/2.00} = 0.135 \text{ in.}$$

$$\text{Minimum size} = \frac{3}{16} \text{ in. (AISC Table J2.4, based on plate thickness)}$$

Try w = ³⁄₁₆ in.

From Equation 7.37, the base metal shear yield strength of the web is

$$\frac{R_n}{\Omega} = 0.4F_y t \times 2 \text{ sides of stiffener} = 0.4(50)(0.360)(2) = 14.4 \text{ kips/in.}$$

(The load is shared by two thicknesses of web, one on each side of the stiffener pair.) From Equation 7.38, the shear rupture strength is

$$\frac{R_n}{\Omega} = 0.3F_u t \times 2 = 0.3(65)(0.360)(2) = 14.04 \text{ kips/in.}$$

For the stiffener shear yield strength,

$$\frac{R_n}{\Omega} = 0.4F_y t \times 2 = 0.4(36)(5/16)(2) = 9.0 \text{ kips/in.}$$

The shear rupture strength of the stiffener is

$$\frac{R_n}{\Omega} = 0.3F_u t \times 2 = 0.3(58)(5/16)(2) = 10.88 \text{ kips/in.}$$

The stiffener yield strength of 9.0 kips/in. controls the base metal shear strength. The *required* strength of the weld is

$$\frac{R_n}{\Omega} = \frac{0.707wF_W}{\Omega} = \frac{0.707(0.135)(0.6 \times 70)}{2.00} \times 4$$
$$= 8.017 \text{ kips/in.} < 9.0 \text{ kips/in.} \quad \text{(OK)}$$

Answer Use two plates $5/16 \times 3 \times 0' - 4\frac{1}{2}''$ Clip the inside corners to avoid the fillets at the column flange-to-web intersection. Clip at a 45° angle for $5/8$ inch. Use a $3/16$-inch fillet weld for the stiffener-to-column web welds.

Stiffener-to-column flange welds: Since the beam tension flange force is an issue (for both column flange bending and web yielding), the stiffener must be welded to the column flange. Use a weld sufficient to develop the yield strength of the stiffener. Let

$$\frac{0.707wL(0.6F_{EXX})(1.5)}{\Omega} = \frac{F_{yst}A_{st}}{\Omega_{st}}$$

where

$$L = (b - clip) \times 2 \text{ sides} \times 2 \text{ stiffeners}$$

and the factor of 1.5 is used because the load is perpendicular to the axis of the weld. Then

$$w = \frac{F_{yst}A_{st}/\Omega_{st}}{0.707L(0.6F_{EXX})(1.5)/\Omega} = \frac{36(1.88)/1.67}{0.707[(3-5/8)\times 2\times 2](0.6\times 70)(1.5)/2.00}$$
$$= 0.192 \text{ in.}$$

Minimum size $= \dfrac{3}{16}$ in. (AISC Table J2.4, based on plate thickness)

Answer Use a $\frac{1}{4}$-inch fillet weld for the stiffener-to-column flange connection. (Because the applied moment is caused by gravity loads and is not reversible, the stiffeners opposite the beam compression flange can be fitted to bear on the column flange and need not be welded, but this option is not exercised here.)

Check shear in the column panel zone: From Equation 8.8,

$$F = \frac{M_1 + M_2}{d_m} - V = \frac{M_a}{d_b - t_b} - V_a = \frac{130(12)}{17.9 - 0.525} - 0 = 89.78 \text{ kips}$$

From AISC Equation J10-9, the nominal strength is

$$R_n = 0.60F_y d_c t_w = 0.60(50)(8.25)(0.360) = 89.1 \text{ kips}$$

The allowable strength is

$$\frac{R_n}{\Omega} = \frac{89.1}{1.67} = 53.35 \text{ kips} < 89.78 \text{ kips} \quad \text{(N.G.)}$$

Alternative 1: Use a Web Doubler Plate. Use AISC Equation J10-9 to find the required plate thickness. Dividing both sides of AISC Equation J10-9 by Ω and substituting the plate thickness t_d for t_w, we get

$$\frac{R_n}{\Omega} = \frac{0.60F_y d_c t_d}{\Omega}$$

where the left side of the equation is the required allowable strength of the doubler plate. Solving for t_d, we get

$$t_d = \frac{R_n/\Omega}{0.60F_y d_c/\Omega} = \frac{89.78 - 53.35}{0.60(36)(8.25)/1.67} = 0.341 \text{ in.}$$

where $89.78 - 53.35$ is the extra strength, in kips, to be furnished by the doubler plate.

The design of the welds connecting the doubler plate to the column will depend on the exact configuration of the plate, including whether the plate extends beyond the transverse stiffeners. For this and other details, refer to AISC Design Guide 13, *Stiffening of Wide-Flange Columns at Moment Connections: Wind and Seismic Applications* (Carter, 1999).

Answer   Use a ⅜-inch-thick doubler plate.

Alternative 2: Use a Diagonal Stiffener. With a diagonal stiffener, use full-depth horizontal stiffeners as shown in Figure 8.49.

The shear force to be resisted by the web reinforcement is $89.78 - 53.35 = 36.43$ kips. If this force is considered to be the horizontal component of an axial compressive force P in the stiffener,

$$P \cos\theta = 36.43 \text{ kips}$$

where

$$\theta = \tan^{-1}\left(\frac{d_b}{d_c}\right) = \tan^{-1}\left(\frac{17.9}{8.25}\right) = 65.26°$$

$$P = \frac{36.43}{\cos(65.26°)} = 87.05 \text{ kips}$$

Because the stiffener will be connected along its length, we will consider it to be a compression member whose effective length is zero. From AISC J4.4, for compression elements with $KL/r < 25$, the nominal strength is

$$P_n = F_y A_g$$

and for ASD, $\Omega = 1.67$, so

$$\frac{P_n}{\Omega} = \frac{36A_g}{1.67} = 21.56A_g$$

Equating this available strength to the required strength, we obtain the required area of stiffener:

$$21.56A_g = 87.05$$
$$A_g = 4.04 \text{ in.}^2$$

Try two stiffeners, ¾ × 3, one on each side of the web:

$$A_{st} = 2(3)\left(\frac{3}{4}\right) = 4.50 \text{ in.}^2 > 4.04 \text{ in.}^2 \text{ required} \quad \text{(OK)}$$

Check for $t_{st} \geq b/15$:

$$\frac{b}{15} = \frac{5}{15} = 0.2 \text{ in.} < \frac{3}{4} \text{ in.} \quad \text{(OK)}$$

Design the welds. The approximate length of each diagonal stiffener is

$$L_{st} = \frac{d_c}{\cos\theta} = \frac{8.25}{\cos(65.26°)} = 19.7 \text{ in.}$$

If welds are used on both sides of the stiffeners, the available length for welding is

$$L = 19.7(4) = 78.8 \text{ in.}$$

The weld size, in sixteenths of an inch, required for strength is

$$D = \frac{P}{0.9279L} = \frac{87.05}{0.9279(78.8)} = 1.2 \text{ sixteenths}$$

Use the minimum size of $3/16$ inch (from Table J2.4, based on the web thickness). Because of the small size required for strength, try intermittent welds. From AISC J2.2b,

$$\text{Minimum length } = 4w = 4\left(\frac{3}{16}\right) = 0.75 \text{ in., but not less than 1.5 in. (1.5 in. controls)}$$

For a group of four welds, the allowable strength is

$$4(0.9279DL) = 4(0.9279)(3)(1.5) = 16.7 \text{ kips}$$

The required weld capacity per inch of stiffener length is

$$\frac{P}{L_{st}} = \frac{87.05}{19.7} = 4.419 \text{ kips/in.}$$

Check the base metal shear strength. From Equation 7.37, the shear yield strength of the web is

$$\frac{R_n}{\Omega} = 0.4F_y t \times 2 \text{ sides of stiffener} = 0.4(50)(0.360)(2) = 14.4 \text{ kips/in.}$$

From Equation 7.38, the shear rupture strength is

$$\frac{R_n}{\Omega} = 0.3F_u t \times 2 = 0.3(65)(0.360)(2) = 14.04 \text{ kips/in.}$$

For the stiffener shear yield strength,

$$\frac{R_n}{\Omega} = 0.4F_y t \times 2 \text{ stiffeners} = 0.4(36)(3/4)(2) = 21.6 \text{ kips/in.}$$

The shear rupture strength of the stiffener is

$$\frac{R_n}{\Omega} = 0.3F_u t \times 2 = 0.3(58)(3/4)(2) = 26.1 \text{ kips/in.}$$

The shear strength of the weld controls.

$$\text{Required spacing of welds} = \frac{\text{strength of weld group (kips)}}{\text{required strength per inch (kips/in.)}} = \frac{16.7}{4.419} = 3.78 \text{ in.}$$

Answer   Use two stiffeners, $\frac{3}{4} \times 3$, one on each side of the web, and $\frac{3}{16}$-inch $\times$ $1\frac{1}{2}$-inch intermittent fillet welds spaced at $3\frac{1}{2}$ inches on center, on each side of each diagonal stiffener.

     As previously mentioned, the most economical alternative may be simply to use a larger column section rather than stiffeners or doubler plates. The labor costs associated with doubler plates and stiffeners may outweigh the additional cost of material for a larger column.

     AISC Design Guide 13 (Carter, 1999) contains detailed guidelines and examples of the design of column reinforcement.

8.8   END PLATE CONNECTIONS

The end plate connection is a popular beam-to-column and beam-to-beam connection that has been in use since the mid-1950s. *Figure 8.50* illustrates two categories of the beam-to-column connection: the simple (or shear only) connection (Type PR construction) and the rigid (moment-resisting) connection (Type FR construction). The rigid

FIGURE 8.50

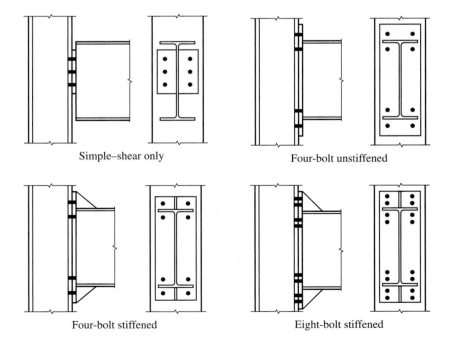

Simple–shear only   Four-bolt unstiffened

Four-bolt stiffened   Eight-bolt stiffened

version is also called an *extended end plate* connection because the plate extends beyond the beam flange. The basis of both types is a plate that is shop welded to the end of a beam and field bolted to a column or another beam. This feature is one of the chief advantages of this type of connection, the other being that ordinarily fewer bolts are required than with other types of connections, thus making erection faster. However, there is little room for error in beam length, and the end must be square. Cambering will make the fit even more critical and is not recommended. Some latitude can be provided in beam length by fabricating it short and achieving the final fit with shims.

In a simple connection, care must be taken to make the connection sufficiently flexible that enough end rotation of the beam is possible. This flexibility can be achieved if the plate is short and thin, compared to the fully restrained version of this connection. The *Steel Construction Manual,* in Part 10, "Design of Simple Shear Connections," gives guidelines for achieving this flexibility. This part of the *Manual* also presents other guidelines. Table 10-4 is a design aid that contains available strengths for various combinations of plates and bolts.

Figure 8.50 shows three versions of the rigid end plate connection: an unstiffened four-bolt connection, a stiffened four-bolt connection, and a stiffened eight-bolt connection. The number of bolts in the designation refers to the number of bolts adjacent to the tension flange in a negative moment connection. In the three rigid connections shown in Figure 8.50, the same number of bolts are used at each flange, so that the connections can be used in cases of moment reversal. Although other configurations are possible, these three are commonly used.

The design of moment-resisting end-plate connections requires determination of the bolt size, the plate thickness, and the weld details. The design of the bolts and the welds is a fairly straightforward application of traditional analysis procedures, but determination of the plate thickness relies on yield-line theory (Murray and Sumner, 2003). This approach is based on analysis of patterns of yield lines that form a collapse mechanism. This theory is similar to what is called plastic analysis for linear members such as beams, where the yielding is at one or more points along the length (see Appendix A). Yield-line theory was originally formulated for reinforced concrete slabs but is equally valid for steel plates.

Part 12 of the *Manual,* "Design of Fully Restrained (FR) Moment Connections," gives guidance for extended end-plate moment connections, but the actual design procedures, along with examples, can be found in AISC Design Guide 4 (Murray and Sumner, 2003) and AISC Design Guide 16 (Murray and Shoemaker, 2002).

Design Guide 16, *Flush and Extended Multiple Row Moment End Plate Connections,* is the more general of the two documents. It covers two approaches: "thick plate" theory, in which thick plates and small diameter bolts result, and "thin plate" theory, with thin plates and large bolts. The thin plate approach requires the incorporation of prying action, whereas the thick plate theory does not. Design Guide 4, *Extended End-Plate Moment Connections—Seismic and Wind Applications,* covers only thick plate theory and three configurations: four-bolt unstiffened, four-bolt stiffened, and eight-bolt stiffened. Design Guide 4 can be used for both static and seismic loads, but Design Guide 16 is restricted to static loading (which includes wind and *low* seismic loading). We will use the approach of Design Guide 4 in this book and consider only the four-bolt unstiffened end-plate connection.

FIGURE 8.51

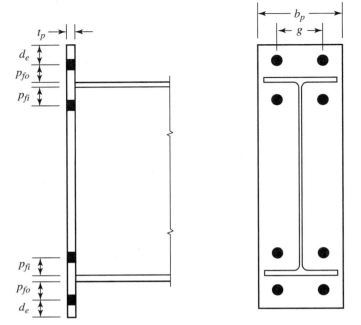

The guidelines and assumptions from Part 12 of the *Manual* can be summarized as follows (this list has been specialized for the connections covered in Design Guide 4):

- Pretensioned ASTM A325 or A490 bolts, with diameter $d_b \leq 1\frac{1}{2}$ in., should be used. Although pretensioning is required, the connection need not be treated as slip-critical.
- The end-plate yield stress should be no greater than 50 ksi.
- The following bolt pitch is recommended: $d_b + \frac{1}{2}$ in. for bolt diameters up to 1 inch and $d_b + \frac{3}{4}$ in. for larger bolts. Many fabricators, however, prefer 2 inches or $2\frac{1}{2}$ inches for all bolt sizes. Figure 8.51 shows the pitch of the outer bolts (p_{fo}) and the pitch of the inner bolts (p_{fi}).
- All of the shear is resisted by the compression-side bolts. (For negative moment, this would be the four bolts on the bottom side. By using the same bolt configuration on the top and bottom sides, the connection will work for both positive and negative moments.)
- The maximum effective end-plate width is the beam flange width plus 1 inch. Larger values can be used for the actual width, but this effective width should be used in the computations if it is smaller than the actual width.
- The tension bolt gage (g in Figure 8.51) should be no larger than the beam flange width. The "workable gage" distance given in Part 1 of the *Manual* can be used.
- If the required moment strength is less than the full moment capacity of the beam (in a nonseismic connection), design for at least 60% of the beam strength. (Neither Design Guide 4 nor Design Guide 16 lists this assumption.)

- Beam web-to-end plate welds near the tension bolts should be designed to develop the yield stress of the beam web unless the required moment is less than 60% of the beam strength.
- Only part of the beam web welds are considered effective in resisting shear. The length of web to be used is the smaller of the following:
    a. From mid-depth of the beam to the inside face of the compression flange
    b. From the inner row of tension bolts plus 2 bolt diameters to the inside face of the compression flange

Design Procedure

We will first consider the procedure for LRFD and then present a summary for ASD. The approach is the same for both, but the differences have to be considered throughout the process.

1. **Determine whether the connection moment is at least 60% of the beam moment strength**. If not, design the connection for 60% of the beam moment strength.
2. **Select a trial layout.** Select a plate width b_p and bolt locations relative to the beam flanges (p_{fi}, p_{fo}, and g).
3. **Determine the bolt diameter**. At failure, the bolts are assumed to have reached their ultimate tensile stress and have a strength of

$$P_t = F_t A_b$$

where F_t is the ultimate tensile stress of the bolt and A_b is the bolt area. Figure 8.52 shows these forces. Assuming that the moment-resisting couple is formed by these forces and the compressive force in the bottom beam flange,

$$M_u = \phi(2P_t h_0 + 2P_t h_1)$$
$$= \phi 2 P_t (h_0 + h_1)$$
$$= \phi 2 F_t A_b (h_0 + h_1)$$

$$M_u = \phi 2 F_t \left(\frac{\pi d_b^2}{4} \right)(h_0 + h_1) \tag{8.9}$$

where
M_u = required moment strength
h_0 = distance from center of beam compression flange to center of outer row of bolts on the tension side
h_1 = distance from center of compression flange to center of inner row of bolts
$\phi = 0.75$

Equation 8.9 can be solved for the required bolt diameter.

$$d_b = \sqrt{\frac{2M_u}{\pi \phi F_t (h_0 + h_1)}} \tag{8.10}$$

FIGURE 8.52

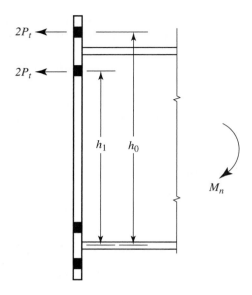

Select a bolt diameter, then using the actual bolt size, compute the actual moment strength.

$$\phi M_n = \phi[2P_t(h_0 + h_1)] \tag{8.11}$$

where
$$\phi = 0.75$$

4. **Determine the plate thickness.** The design strength for plate flexural yielding is

$$\phi_b F_y t_p^2 Y_p \tag{8.12}$$

where
$$\phi_b = 0.90$$
$$t_p = \text{end plate thickness}$$
$$Y_p = \text{yield line mechanism parameter}$$

For a four-bolt unstiffened extended end plate, the yield-line mechanism parameter is given in AISC Design Guide 4 as

$$Y_p = \frac{b_p}{2}\left[h_1\left(\frac{1}{p_{fi}} + \frac{1}{s}\right) + h_0\left(\frac{1}{p_{fo}}\right) - \frac{1}{2}\right] + \frac{2}{g}[h_1(p_{fi} + s)] \tag{8.13}$$

where

$$s = \frac{1}{2}\sqrt{b_p g} \quad (\text{if } p_{fi} > s, \text{ use } p_{fi} = s)$$

To ensure thick-plate behavior (no prying action), use 90% of the strength given by Equation 8.12 to match the moment strength provided by the bolts (ϕM_n from Equation 8.11).

$$0.90 \phi_b \, F_y t_p^2 Y_p = \phi M_n$$

$$t_p = \sqrt{\frac{1.11 \phi M_n}{\phi_b \, F_y Y_p}}$$

(8.14)

Select a practical plate thickness.

5. **Check shear in the plate**. The beam flange force is

$$F_{fu} = \frac{M_u}{d - t_{fb}}$$

where t_{fb} is the beam flange thickness. Half of this force will produce shear in the plate on each side of the flange. This will be shear on the gross area of the plate (Sections a–a and b–b in Figure 8.53) and shear on the net area (Sections c–c and d–d). For both cases, check for

$$\frac{F_{fu}}{2} \le \phi R_n$$

For shear yielding,

$$\phi = 0.90$$
$$R_n = (0.6 F_y) A_g$$
$$A_g = t_p b_p$$

FIGURE 8.53

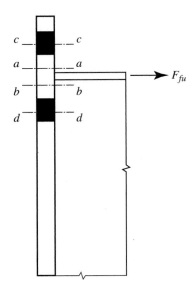

For shear rupture,

$$\phi = 0.75$$
$$R_n = (0.6\,F_u)A_n$$
$$A_n = t_p\,[b_p - 2(d_b + 1/8)]$$

Increase t_p if necessary.

6. **Check bolt shear and bearing.**
   The total beam reaction must be resisted by the four bolts on the compression side.

   $$V_u \le \phi R_n$$

   where V_u = beam end shear (reaction). For bolt shear,

   $$\phi = 0.75$$
   $$R_n = F_{nv}\,A_b \times 4 \text{ bolts}$$

   For bolt bearing,

   $$\phi = 0.75$$
   $$R_n = 1.2L_c t F_u \le 2.4 d_b t F_u \text{ per bolt}$$

   where t is the thickness of the end plate or the column flange, and F_u is the ultimate tensile strength of the end plate or the column flange. For the four bolts,

   $$R_n = 2R_n \text{ (inner bolts)} + 2R_n \text{ (outer bolts)}$$

7. **Design the welds.**
8. **Check column strength and stiffening requirements.**

Design Procedure for ASD. For allowable strength design, we can use the same nominal strength expressions as for LRFD. To obtain the design equations, we use service loads and moments rather than factored loads and moments, and we divide the nominal strengths by a safety factor rather than multiplying by a resistance factor. To convert the equations, we simply replace the u subscripts with a subscripts and replace ϕ by $1/\Omega$.

1. **Determine whether the connection moment is at least 60% of the beam moment strength.** If not, design the connection for 60% of the beam moment strength.
2. **Select a trial layout.**
3. **Determine the required bolt diameter.**

   $$d_b = \sqrt{\frac{\Omega(2M_a)}{\pi F_t(h_0 + h_1)}}$$

   where
   $$\Omega = 2.00$$

4. **Determine the required plate thickness.**

   $$t_p = \sqrt{\frac{1.11(M_n/\Omega)}{F_y Y_p/\Omega_b}}$$

where
$$M_n = 2P_t\,(h_0 + h_1)$$
$$\Omega = 2.00$$
$$\Omega_b = 1.67$$

5. **Check shear in the plate**. Check for

$$\frac{F_{fa}}{2} \le \frac{R_n}{\Omega}$$

where

$$F_{fa} = \frac{M_a}{d - t_{fb}}$$

For shear yielding,

$$\Omega = 1.67$$
$$R_n = (0.6F_y)A_g$$
$$A_g = t_p b_p$$

For shear rupture,

$$\Omega = 2.00$$
$$R_n = (0.6F_u)A_n$$
$$A_n = t_p\,[b_p - 2(d_b + 1/8)]$$

6. **Check bolt shear and bearing**.

$$V_a \le \frac{R_n}{\Omega}$$

For bolt shear,

$$\Omega = 2.00$$
$$R_n = F_{nv}A_b \times 4 \text{ bolts}$$

For bolt bearing,

$$\Omega = 2.00$$
$$R_n = 1.2L_c t F_u \le 2.4 d_b t F_u \text{ per bolt}$$

For the four bolts,

$$R_n = 2R_n \text{ (inner bolts)} + 2R_n \text{ (outer bolts)}$$

7. **Design the welds**.
8. **Check column strength and stiffening requirements**.

Example 8.13

Use LRFD and design a four-bolt end-plate connection for a W18 × 35 beam. This connection must be capable of transferring a dead-load moment of 23 ft-kips and a live-load moment of 91 ft-kips. The end shear consists of 4 kips of dead load and 18 kips of live load. Use A992 steel for the beam, A36 for the plate, E70XX electrodes, and fully tensioned A325 bolts.

Solution

$$M_u = 1.2M_D + 1.6M_L = 1.2(23) + 1.6(91) = 173.2 \text{ ft-kips}$$
$$V_u = 1.2V_D + 1.6V_L = 1.2(4) + 1.6(18) = 33.6 \text{ kips}$$

For a W18 × 35,

$$d = 17.7 \text{ in.,} \quad t_w = 0.300 \text{ in.,} \quad b_{fb} = 6.00 \text{ in.,} \quad t_{fb} = 0.425 \text{ in.,}$$
$$\text{workable gage} = 3.50 \text{ in.,}$$
$$Z_x = 66.5 \text{ in.}^3$$

Compute 60% of the flexural strength of the beam:

$$\phi_b M_p = \phi_b F_y Z_x = 0.90(50)(66.5) = 2993 \text{ in.-kips}$$
$$0.60(\phi_b M_p) = 0.60(2993) = 1796 \text{ in.-kips} = 150 \text{ ft-kips}$$
$$M_u = 173.2 \text{ ft-kips} > 150 \text{ ft-kips} \quad \therefore \text{ design for 173.2 ft-kips}$$

For the bolt pitch, try $p_{fo} = p_{fi} = 2$ in.

For the gage distance, use the workable gage distance given in Part 1 of the *Manual*.

$$g = 3.50 \text{ in.}$$

Required bolt diameter:

$$h_0 = d - \frac{t_{fb}}{2} + p_{fo} = 17.7 - \frac{0.425}{2} + 2 = 19.49 \text{ in.}$$

$$h_1 = d - \frac{t_{fb}}{2} - t_{fb} - p_{fi} = 17.7 - \frac{0.425}{2} - 0.425 - 2 = 15.06 \text{ in.}$$

$$d_{b\text{Req'd}} = \sqrt{\frac{2M_u}{\pi \phi F_t (h_0 + h_1)}} = \sqrt{\frac{2(173.2 \times 12)}{\pi(0.75)(90)(19.49 + 15.06)}} = 0.753 \text{ in.}$$

Try $d_b = \frac{7}{8}$ inch.

Moment strength based on bolt strength:

$$P_t = F_t A_b = 90(0.6013) = 54.12 \text{ kips/bolt}$$
$$M_n = 2P_t (h_0 + h_1) = 2(54.12)(19.49 + 15.06) = 3740 \text{ in.-kips}$$
$$\phi M_n = 0.75(3740) = 2805 \text{ in-kips}$$

Determine end-plate width. For the bolt hole edge distance, use AISC Table J3.4. If we assume sheared edges, then for $\frac{7}{8}$-inch diameter bolts,

$$\text{Minimum } L_e = 1\frac{1}{2} \text{ in.}$$

The minimum plate width is

$$g + 2L_e = 3.50 + 2(1.5) = 6.5 \text{ in.}$$

but no less than the beam flange width of 6.00 in.

Maximum *effective* end-plate width = $b_{fb} + 1 = 6.00 + 1 = 7.00$ in.

Try $b_p = 7$ in. Compute the required plate thickness:

$$s = \frac{1}{2}\sqrt{b_p g} = \frac{1}{2}\sqrt{7(3.5)} = 2.475 \text{ in.} > p_{fi} \quad \therefore \text{ use the original value of } p_{fi} = 2.0 \text{ in.}$$

From Equation 8.13,

$$Y_p = \frac{b_p}{2}\left[h_1\left(\frac{1}{p_{fi}} + \frac{1}{s}\right) + h_0\left(\frac{1}{p_{fo}}\right) - \frac{1}{2}\right] + \frac{2}{g}[h_1(p_{fi} + s)]$$

$$= \frac{7}{2}\left[15.06\left(\frac{1}{2} + \frac{1}{2.475}\right) + 19.49\left(\frac{1}{2}\right) - \frac{1}{2}\right] + \frac{2}{3.5}[15.06(2 + 2.475)]$$

$$= 118.5$$

From Equation 8.14,

$$\text{Required } t_p = \sqrt{\frac{1.11\phi M_n}{\phi_b F_y Y_p}} = \sqrt{\frac{1.11(2805)}{0.9(36)(118.5)}} = 0.901 \text{ in.}$$

Try $t_p = 1$ in.

Beam flange force:

$$F_{fu} = \frac{M_u}{d - t_{fb}} = \frac{173.2 \times 12}{17.7 - 0.425} = 120.3 \text{ kips}$$

$$\frac{F_{fu}}{2} = \frac{120.3}{2} = 60.2 \text{ kips}$$

The shear yield strength of the end plate is

$$\phi(0.6)F_y t_p b_p = 0.90(0.6)(36)(1)(7) = 136 \text{ kips} > 60.2 \text{ kips} \quad (\text{OK})$$

Shear rupture strength of end plate:

$$A_n = t_p\left[b_p - 2\left(d_b + \frac{1}{8}\right)\right] = (1)\left[7 - 2\left(\frac{7}{8} + \frac{1}{8}\right)\right] = 5.000 \text{ in.}^2$$

$$\phi(0.6)F_u A_n = 0.75(0.6)(58)(5.000) = 131 \text{ kips} > 60.2 \text{ kips} \quad (\text{OK})$$

Check bolt shear. The compression side bolts must be capable of resisting the entire vertical shear.

$$A_b = \frac{\pi d_b^2}{4} = \frac{\pi(7/8)^2}{4} = 0.6013 \text{ in.}^2$$

$$\phi R_n = \phi F_{nv} A_b = 0.75(48)(0.6013) = 21.65 \text{ kips/bolt}$$

For 4 bolts, $\phi R_n = 4 \times 21.65 = 86.6$ kips

$$V_u = 33.6 \text{ kips} < 86.6 \text{ kips} \quad (\text{OK})$$

Check bearing at the compression side bolts.

$$h = d + \frac{1}{16} = \frac{7}{8} + \frac{1}{16} = \frac{15}{16} \text{ in.}$$

For the outer bolts

$$L_c = p_{fo} + t_{fb} + p_{fi} - h = 2 + 0.425 + 2 - \frac{15}{16} = 3.488 \text{ in.}$$

$$\phi R_n = \phi(1.2 L_c t F_u) = 0.75(1.2)(3.488)(1)(58) = 182.1 \text{ kips}$$

The upper limit is

$$\phi(2.4 d t F_u) = 0.75(2.4)(7/8)(1)(58) = 91.35 \text{ kips} < 182.1 \text{ kips}$$

$$\therefore \text{ use } \phi R_n = 91.35 \text{ kips/bolt.}$$

Since the inner bolts are not near an edge or adjacent bolts, the outer bolts control. The total bearing strength is

$$4 \times 91.35 = 365 \text{ kips} > V_u = 33.6 \text{ kips} \quad \text{(OK)}$$

The plate length, using detailing dimensions and the notation of Figure 8.51, is

$$d + 2 p_{fo} + 2 d_e = 17 \tfrac{3}{4} + 2(2) + 2(1 \tfrac{1}{2}) = 24 \tfrac{3}{4} \text{ in.}$$

Answer Use a PL $1 \times 7 \times 2' - 0 \tfrac{3}{4}''$ and four $\tfrac{7}{8}$-inch diameter A325 fully tightened bolts at each flange.

Beam flange-to-plate weld design: The flange force is

$$F_{fu} = 120.3 \text{ kips}$$

AISC Design Guide 4 recommends that the minimum design flange force should be 60% of the flange yield strength:

$$\text{Min. } F_{fu} = 0.6 F_y (b_{fb} t_{fb}) = 0.6(50)(6.00)(0.425) = 76.5 \text{ kips} < 120.3 \text{ kips}$$

Therefore, use the actual flange force of 120.3 kips. The flange weld length is

$$b_{fb} + (b_{fb} - t_w) = 6.00 + (6.00 - 0.300) = 11.70 \text{ in.}$$

The weld strength is

$$\phi R_n = 1.392 D \times 11.70 \times 1.5$$

where D is the weld size in sixteenths of an inch, and the factor of 1.5 accounts for the direction of the load on the weld. If we equate the weld strength to the flange force,

$$1.392 D \times 11.70 \times 1.5 = 120.3, \quad D = 4.92 \text{ sixteenths}$$

From AISC Table J2.4, the minimum weld size is $\tfrac{3}{16}$ in. (based on the thickness of the flange, which is the thinner connected part).

Answer Use a $\tfrac{5}{16}$-inch fillet weld at each flange.

Beam web-to-plate weld design: To develop the yield stress in the web near the tension bolts, let

$$1.392 D \times 2 = 0.6 F_y t_w$$

for two welds, one on each side of the web. The required weld size is

$$D = \frac{0.6F_y t_w}{1.392(2)} = \frac{0.6(50)(0.300)}{1.392(2)} = 3.23 \text{ sixteenths}$$

Answer   Use a ¼-inch fillet weld on each side of the web in the tension region.
The applied shear of $V_u = 33.6$ kips must be resisted by welding a length of web equal to the smaller of the following two lengths:

1. From mid-depth to the compression flange:

$$L = \frac{d}{2} - t_{fb} = \frac{17.7}{2} - 0.425 = 8.425 \text{ in.}$$

2. From the inner row of tension bolts plus $2d_b$ to the compression flange:

$$L = d - 2t_{fb} - p_{fi} - 2d_b = 17.7 - 2(0.425) - 2.0 - 2(7/8) = 13.10 \text{ in.} > 8.425 \text{ in.}$$

Equating the weld strength to the required shear strength, we get

$$1.392D \times 8.425 \times 2 = 33.6, \quad D = 1.43 \text{ sixteenths } (w = 1/8 \text{ in.})$$

From AISC Table J2.4, the minimum weld size is ³⁄₁₆ in.

FIGURE 8.54

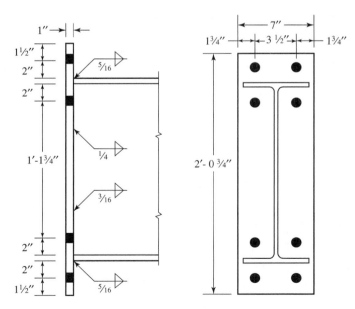

Answer   Use a ³⁄₁₆-inch fillet weld on each side of the web between mid-depth and the compression flange. This design is summarized in Figure 8.54.

Special Requirements for Columns in End-Plate Connections

When end-plate connections are used, column flange bending strength is based on yield-line theory. The equation for the required column flange thickness is the same as that for the required end-plate thickness, but the equation for the yield-line mechanism parameter is different. The required column flange thickness for LRFD is

$$t_{fc} = \sqrt{\frac{1.11\phi M_n}{\phi_b F_{yc} Y_c}}$$

where

ϕM_n = moment design strength based on bolt tension
ϕ_b = 0.90
F_{yc} = yield stress of the column
Y_c = yield-line mechanism parameter for an *unstiffened* column flange

$$= \frac{b_{fc}}{2}\left[h_1\left(\frac{1}{s}\right) + h_0\left(\frac{1}{s}\right) \right] + \frac{2}{g}\left[h_1\left(s + \frac{3c}{4} \right) + h_0\left(s + \frac{c}{4} \right) + \frac{c^2}{2} \right] + \frac{g}{2}$$

b_{fc} = column flange width

$$s = \frac{1}{2}\sqrt{b_{fc}g}$$

$$c = p_{fo} + t_{fb} + p_{fi}$$

For ASD,

$$t_{fc} = \sqrt{\frac{1.11(M_n/\Omega)}{F_{yc} Y_c/\Omega}}$$

where

Ω = 2.00
Ω_b = 1.67

If the existing column flange is not thick enough, another column shape can be selected or stiffeners can be used. If stiffeners are used, a different equation for Y_c from Design Guide 4 must be used. As a reminder, all equations given here pertain to four-bolt unstiffened end-plate connections. For other configurations, some equations may be different, and Design Guide 4 should be consulted.

The other difference for columns in end-plate connections is in the column web yielding strength. This involves a minor liberalization of AISC Equation J10-2. AISC Equation J10-2 is based on limiting the stress on a cross section of the web formed by its thickness and a length of $t_{fb} + 5k$, as shown in Figure 8.55a. As shown in Figure 8.55b, a larger area will be available when the load is transmitted through the additional thickness of the end plate. If the beam flange-to-plate welds are taken into account and the load is assumed to disperse at a slope of 1:1 through the plate, the loaded length of web will be $t_{fb} + 2w + 2t_p + 5k$. Based on experimental studies (Hendrick and Murray, 1984),

FIGURE 8.55

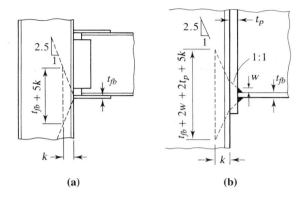

(a)   (b)

the term $5k$ can be replaced with $6k$, resulting in the following equation for web yielding strength:

$$\phi R_n = \phi\left[(6k + t_{fb} + 2w + 2t_p)F_{yw}t_w\right] \qquad (8.15)$$

where w = weld size.

If the beam is near the top of the column (that is, if the distance from the top face of the beam flange to the top of the column is less than the column depth), the result from Equation 8.15 should by reduced by half. This is in lieu of using AISC Equation J10-3.

All other requirements for column web stiffening and panel zone shear strength are the same as for other types of moment connections.

8.9 CONCLUDING REMARKS

In this chapter we emphasized the design and analysis of bolts and welds rather than connection fittings, such as framing angles and beam seats. In most cases, the provisions for bearing in bolted connections and base metal shear in welded connections will ensure adequate strength of these parts. Sometimes, however, additional shear investigation is needed. At other times, direct tension or bending must be considered.

Flexibility of the connection is another important consideration. In a shear connection (simple framing), the connecting parts must be flexible enough to permit the connection to rotate under load. Type FR connections (rigid connections), however, should be stiff enough so that relative rotation of the connected members is kept to a minimum.

This chapter is intended to be introductory only and is by no means a complete guide to the design of building connections. Blodgett (1966) is a useful source of detailed welded connection information. Although somewhat dated, it contains numerous practical suggestions. Also recommended is *Detailing for Steel Construction* (AISC, 2002), which is intended for detailers but also contains information useful for designers.

Problems

Eccentric Bolted Connections: Shear Only

8.2-1   Use an elastic analysis to determine the maximum bolt shear force in the bracket connection of Figure P8.2-1.

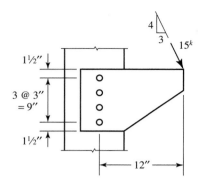

FIGURE P8.2-1

8.2-2 The bolt group shown in Figure P8.2-2 consists of ¾-inch-diameter, A325 slip-critical bolts in single shear. Assume that the bearing strength is adequate and use an elastic analysis to determine:

   a. The maximum factored load that can be applied if LRFD is used.
   b. The maximum total service load that can be applied if ASD is used.

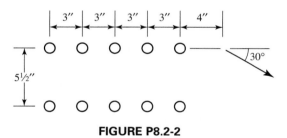

FIGURE P8.2-2

8.2-3 A plate is used as a bracket and is attached to a column flange as shown in Figure P8.2-3. Use an elastic analysis and compute the maximum bolt shear force.

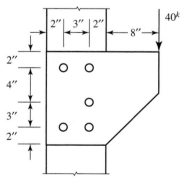

FIGURE P8.2-3

8.2-4 The fasteners in the connection of Figure P8.2-4 are placed at the usual gage distance (see Figure 3.24). What additional force is experienced as a consequence of the fasteners not being on the centroidal axis of the member?

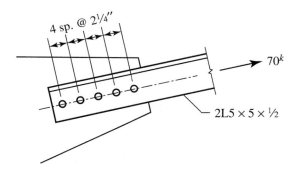

4 sp. @ $2\frac{1}{4}''$

70^k

$2L5 \times 5 \times \frac{1}{2}$

FIGURE P8.2-4

8.2-5 A plate is used as a bracket and is attached to a column flange as shown in Figure P8.2-5. Use an elastic analysis and compute the maximum bolt shear force.

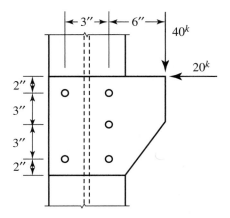

$\leftarrow 3'' \rightarrow \leftarrow 6'' \rightarrow$

40^k

20^k

$2''$
$3''$
$3''$
$2''$

FIGURE P8.2-5

8.2-6 The load on the bracket plate consists of a service dead load of 20 kips and a service live load of 35 kips. What diameter A325 bearing-type bolt is required? Use an elastic analysis and assume that the bearing strengths of the bracket and the column flange are adequate.

a. Use LRFD.
b. Use ASD.

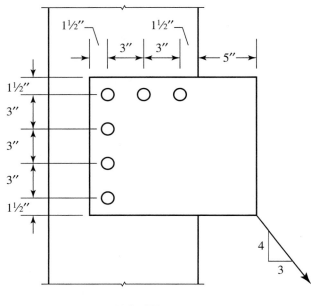

FIGURE P8.2-6

8.2-7 What diameter A325 bearing-type bolt is required? Use an elastic analysis and assume that the bearing strengths of the bracket and the column flange are adequate.

a. Use LRFD.
b. Use ASD.

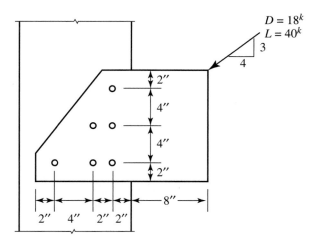

FIGURE P8.2-7

8.2-8 Use A325 slip-critical bolts and select a bolt size for the service loads shown in Figure P8.2-8. Use an elastic analysis and assume that the bearing strengths of the connected parts are adequate.

a. Use LRFD.
b. Use ASD.

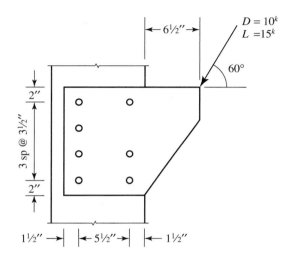

FIGURE P8.2-8

8.2-9 A325 bolts are used in the connection in Figure P8.2-9. Use an elastic analysis and determine the required size if slip is permitted. The 10-kip load consists of 2.5 kips of service dead load and 7.5 kips of service live load. All structural steel is A36.

a. Use LRFD.
b. Use ASD.

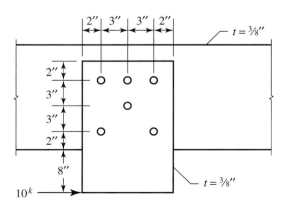

FIGURE P8.2-9

8.2-10   A plate is connected to the web of a channel as shown in Figure P8.2-10.

a. Use an elastic analysis and compute the maximum bolt force.

b. Use the tables in Part 7 of the *Manual* to find the maximum bolt force by the ulti-mate strength method. (Note that the coefficient *C* can be interpreted as the ratio of total connection load to maximum fastener force.) What is the percent differ-ence from the value obtained by the elastic analysis?

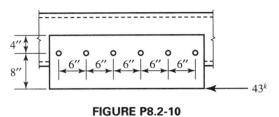

FIGURE P8.2-10

8.2-11   Solve Problem 8.2-2 by the ultimate strength method (use the tables in Part 7 of the *Manual*.)

8.2-12   The bolt group shown in Figure P8.2-12 consists of ¾-inch-diameter, A325 bearing-type bolts in single shear. Assuming that the bearing strength is adequate, use the ta-bles in Part 7 of the *Manual* to determine:

a. The maximum permissible factored load for LRFD.

b. The maximum permissible service load for ASD.

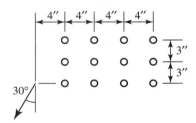

FIGURE P8.2-12

8.2-13   For the conditions of Problem 8.2-12, determine the number of bolts required per row (instead of three, as shown) if the service dead load is 40 kips and the service live load is 90 kips.

a. Use LRFD.

b. Use ASD.

Eccentric Bolted Connections: Shear Plus Tension

8.3-1   Check the adequacy of the bolts. The given loads are service loads.

a. Use LRFD.

b. Use ASD.

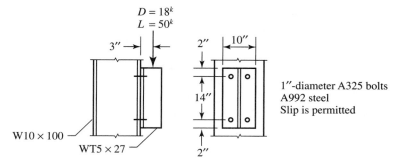

FIGURE P8.3-1

8.3-2   A beam is connected to a column with ⅞-inch-diameter, A325 bearing-type bolts, as shown in Figure P8.3-2. Eight bolts connect the tee to the column. A992 steel is used. Is the tee-to-column connection adequate?

    a. Use LRFD.
    b. Use ASD.

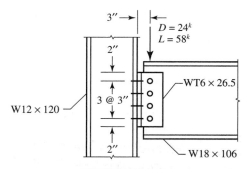

FIGURE P8.3-2

8.3-3   Check the adequacy of the bolts for the given service loads.

    a. Use LRFD.
    b. Use ASD.

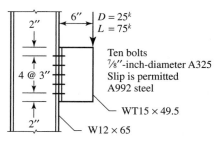

FIGURE P8.3-3

8.3-4 Check the adequacy of the bolts. The given load is a service load.

a. Use LRFD.
b. Use ASD.

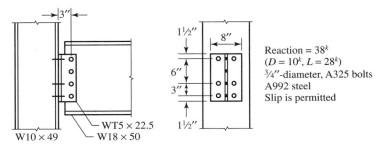

Reaction = 38^k
($D = 10^k$, $L = 28^k$)
$\frac{3}{4}$"-diameter, A325 bolts
A992 steel
Slip is permitted

FIGURE P8.3-4

8.3-5 Check the adequacy of the bolts in the connection shown in Figure P8.3-5. The load is a service load consisting of 33% dead load and 67% live load. The bolts are $\frac{7}{8}$-inch, A325 bearing-type. Assume that the connected parts have adequate bearing strength.

a. Use LRFD.
b. Use ASD.

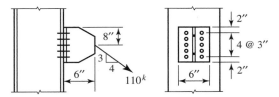

FIGURE P8.3-5

8.3-6 The flange of a portion of a WT6×20 is used as a bracket and is attached to the flange of a W14×61 column as shown in Figure P8.3-6. All steel is A992. Determine whether the bolts are adequate.

a. Use LRFD.
b. Use ASD.

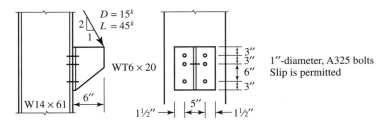

FIGURE P8.3-6

8.3-7 A beam is connected to a column with ¾-inch-diameter, A325 slip-critical bolts, as shown in Figure P8.3-7. A992 steel is used for the beam and column, and A36 steel is used for the angles. The force R is the beam reaction. Based on the strength of the 10 angle-to-column bolts, determine;

a. The maximum available factored load reaction, R_u, for LRFD.
b. The maximum available service load reaction, R_a, for ASD.

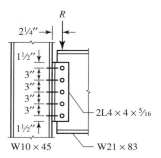

FIGURE P8.3-7

8.3-8 A bracket cut from a WT-shape is connected to a column flange with 10 A325 slip-critical bolts, as shown in Figure P8.3-8. A992 steel is used. The loads are service loads, consisting of 30% dead load and 70% live load. What size bolt is required?

a. Use LRFD.
b. Use ASD.

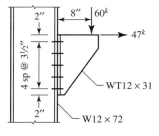

FIGURE P8.3-8

8.3-9 For the following designs, use A36 steel for the angles and A992 steel for the beam and column. Use LRFD.

a. Design a simply supported beam for the conditions shown in Figure P8.3-9. In addition to its own weight, the beam must support a service live load of 4 kips/foot. Assume continuous lateral support of the compression flange. Deflection is not a design consideration.
b. Design an all-bolted, double-angle connection. Do not consider eccentricity. Use bearing-type bolts.
c. Consider eccentricity and check the connection designed in part b. Revise the design if necessary.
d. Prepare a detailed sketch of your recommended connection.

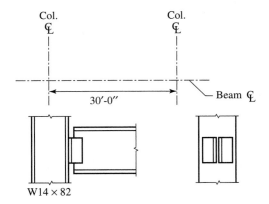

FIGURE P8.3-9

8.3-10   Same as Problem 8.3-9, but use ASD.

Eccentric Welded Connections: Shear Only

8.4-1   Use an elastic analysis and determine the maximum load in the weld (in kips per inch of length).

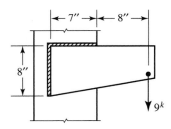

FIGURE P8.4-1

8.4-2   Use an elastic analysis and determine the maximum load in the weld (in kips per inch of length).

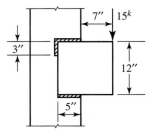

FIGURE P8.4-2

8.4-3 Use an elastic analysis and determine the maximum load per inch of weld.

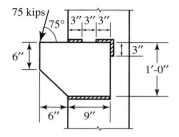

FIGURE P8.4-3

8.4-4 Use an elastic analysis and check the adequacy of the weld. Assume that the shear in the base metal is acceptable. The 10-kip load is a service load, composed of 25% dead load and 75% live load.

a. Use LRFD.
b. Use ASD.

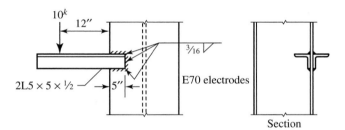

FIGURE P8.4-4

8.4-5 Use E70 electrodes and determine the required weld size. Use an elastic analysis. Assume that the base metal shear strength is adequate.

a. Use LRFD.
b. Use ASD.

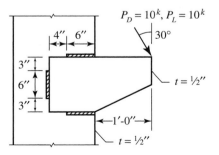

FIGURE P8.4-5

8.4-6 Check the adequacy of the weld. The 20-kip load is a service load, with a live load-to-dead load ratio of 2.0. Use an elastic analysis and assume that the shear strength of the base metal is adequate.

a. Use LRFD.
b. Use ASD.

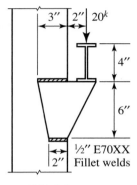

FIGURE P8.4-6

8.4-7 Use an elastic analysis and compute the extra load in the weld (in kips per inch of length) caused by the *eccentricity*.

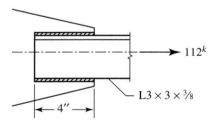

FIGURE P8.4-7

8.4-8 Use an elastic analysis and compute the extra load in the weld (in kips per inch of length) caused by the *eccentricity*.

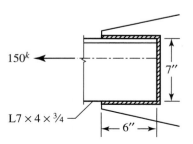

FIGURE P8.4-8

8.4-9   An L6×6×⅜ is attached to a ⅜-in.-thick gusset plate with E70 fillet welds. Design the welds to develop the available strength of the member. Use a placement of welds that will eliminate eccentricity. Assume that the strengths of the connected parts do not govern. Use A36 steel.

  a. Use LRFD.
  b. Use ASD.

8.4-10   Solve Problem 8.4-1 by the ultimate strength method (use the tables in Part 8 of the *Manual*).

8.4-11   Solve Problem 8.4-4 by the ultimate strength method (use the tables in Part 8 of the *Manual*).

8.4-12   Solve Problem 8.4-7 by the ultimate strength method (use the tables in Part 8 of the *Manual*).

8.4-13   A connection is to be made with the weld shown in Figure P8.4-13. The applied load is a service load. Use LRFD.

  a. Determine the required weld size. Use an elastic analysis.
  b. Determine the required weld size by the ultimate strength method (use the tables in Part 8 of the *Manual*).

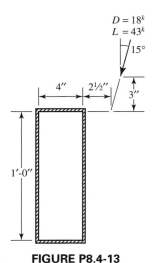

FIGURE P8.4-13

8.4-14   Same as Problem 8.4-13, but use ASD.

8.4-15   Use the elastic method and design a welded connection for an L6 × 6 × 5/16 of A36 steel connected to a ⅜-inch-thick gusset plate, also of A36 steel. The load to be resisted is a service dead load of 31 kips and a service live load of 31 kips. Use LRFD.

  a. Do not balance the welds. Show your design on a sketch.
  b. Balance the welds. Show your design on a sketch.

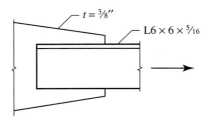

FIGURE P8.4-15

8.4-16 Same as Problem 8.4-15, but use ASD.

8.4-17 A single-angle tension member is connected to a gusset plate as shown in Figure P8.4-17. A36 steel is used for both the angle and the gusset plate.

    a. Use LRFD and the minimum size fillet weld to design a connection. Do not balance the welds.

    b. Check the design of part a, accounting for the eccentricity. Revise if necessary.

    c. Show your final design on a sketch.

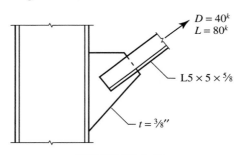

FIGURE P8.4-17

8.4-18 Same as Problem 8.4-17, but use ASD.

8.4-19 a. Use LRFD and design a welded connection for the bracket shown in Figure P8.4-19. All structural steel is A36. The horizontal 10-inch dimension is a *maximum*.

    b. State why you think your weld size and configuration are best.

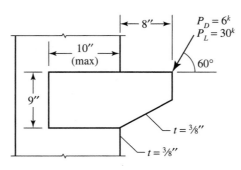

FIGURE P8.4-19

8.4-20   Same as Problem 8.4-19, but use ASD.

Eccentric Welded Connections: Shear Plus Tension

8.5-1   Determine the maximum load in the weld in kips per inch of length.

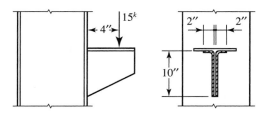

FIGURE P8.5-1

8.5-2   Determine the maximum load in the weld in kips per inch of length.

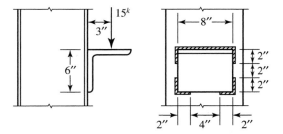

FIGURE P8.5-2

8.5-3   Use the maximum size E70 fillet weld and compute the available reaction R (as limited by the strength of the weld) that can be supported by the connection of Figure P8.5-3. The beam and column steel is A992, and the shelf angle is A36. Neglect the end returns shown at the top of the welds.

a. Use LRFD.
b. Use ASD.

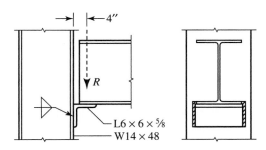

FIGURE P8.5-3

8.5-4   A bracket plate of A36 steel is welded to a W12×50 of A992 steel. Use E70 electrodes and determine the required fillet weld size. The applied load consists of 8 kips dead load and 18 kips live load.

a. Use LRFD.
b. Use ASD.

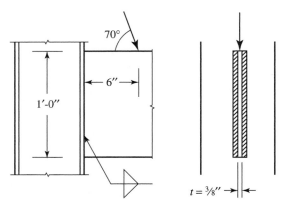

FIGURE P8.5-4

8.5-5   A WT7×41 bracket is connected to a W14×159 column with ⁵⁄₁₆-inch E70 fillet welds as shown in Figure P8.5-5. What is the maximum factored load P_u that can be supported? What is the maximum service load P_a that can be supported?

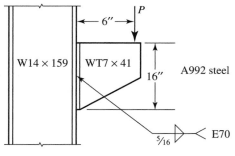

FIGURE P8.5-5

Moment-Resisting Connections

8.6-1   A W18 × 50 beam is connected to a W14 × 99 column. To transfer moment, plates are bolted to the beam flanges. The service-load moment to be transferred is 180 ft-kips, consisting of 45 ft-kips of dead-load moment and 135 ft-kips of live-load moment. The bolts are ⅞-inch-diameter A325-N, and eight bolts are used in each flange. Do these bolts have enough shear strength?

a. Use LRFD.
b. Use ASD.

8.6-2 A W16×45 beam is connected to a W10×45 column as shown in Figure P8.6-2. The structural shapes are A992, and the plates are A36 steel. Twenty ⅞-inch, A325-N bearing-type bolts are used: eight at each flange, and four in the web. The electrodes are E70. Use LRFD.

   a. Determine the available shear strength of the connection.
   b. Determine the available flexural strength.

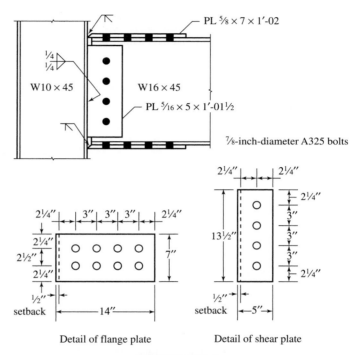

Detail of flange plate   Detail of shear plate

FIGURE P8.6-2

8.6-3 Same as problem 8.6-2, but use ASD.

8.6-4 Design a three-plate moment connection of the type shown in Problem 8.6-2 for the connection of a W18×35 beam to a W14×99 column for the following conditions: The service dead-load moment is 42 ft-kips, the service live-load moment is 104 ft-kips, the service dead-load beam reaction is 8 kips, and the service live-load beam reaction is 21 kips. Use A325 bearing-type bolts and E70 electrodes. The beam and column are of A992 steel, and the plate material is A36.

   a. Use LRFD.
   b. Use ASD.

Column Stiffeners and Other Reinforcement

8.7-1 A service-load moment of 118 ft-kips, 30% dead load and 70% live load, is applied to the connection of Problem 8.6-2. Assume that the connection is at a distance from the end of the column that is more than the depth of the column and determine whether column stiffeners are required. If so, use A36 steel and determine the required dimensions.

a. Use LRFD.
b. Use ASD.

8.7-2   Determine whether column stiffeners are required for the maximum force that can be developed in the beam flange plate, which is A36 steel. If they are, use A36 steel and specify the required dimensions. A992 steel is used for the beam and column. The connection is at a distance from the end of the column that is more than the depth of the column.

a. Use LRFD.
b. Use ASD.

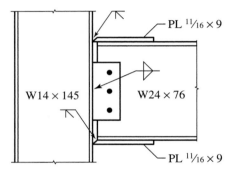

FIGURE P8.7-2

8.7-3   a. A W18×35 beam is to be connected to a W14×53 column. $F_y = 50$ ksi for both the beam and the column. The connection is at a distance from the end of the column that is more than the depth of the column. Use A36 steel for the web plate and use LRFD to design a connection similar to the one shown in Figure 8.37a for a factored moment of 220 ft-kips and a factored reaction of 45 kips. If column stiffeners are needed, use A36 steel and specify the required dimensions.

    b. If the factored column shear adjacent to the connection is $V_u = 0$, and $P_u/P_y = 0.6$, determine whether panel zone reinforcement is required. If it is, provide two alternatives: (1) a doubler plate of A36 steel and (2) diagonal stiffeners of A36 steel.

End-Plate Connections

8.8-1   Investigate the adequacy of the bolts in the given end-plate connection. The loads are service loads, consisting of 25% dead load and 75% live load. The beam and column are A992 steel, and the end plate is A36 steel.

a. Use LRFD.
b. Use ASD.

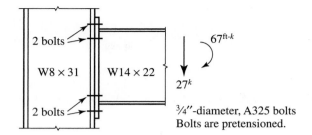

FIGURE P8.8-1

8.8-2 Investigate the adequacy of the bolts in the given end-plate connection. The loads are service loads, consisting of 25% dead load and 75% live load. The beam and column are A992 steel, and the end plate is A36 steel.

    a. Use LRFD.
    b. Use ASD.

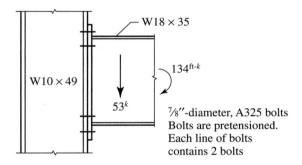

FIGURE P8.8-2

8.8-3 Design a four-bolt unstiffened end-plate connection for a W18 × 40 beam to a W8 × 40 column. Design for the full moment and shear capacities of the beam. Use A992 steel for the members and A36 for the end plate. Use A325 pretensioned bolts.

    a. Use LRFD.
    b. Use ASD.

8.8-4 Design a four-bolt unstiffened end-plate connection for a W12 × 30 beam to a W10 × 60 column. The shear consists of a 13-kip service dead load and a 34-kip service live load. The service dead-load moment is 20 ft-kips, and the service live-load moment is 48 ft-kips. Use A992 steel for the structural shapes, A36 for the end plate, and pretensioned A325 bolts.

    a. Use LRFD.
    b. Use ASD.

9
Composite Construction

9.1 INTRODUCTION

Composite construction employs structural members that are composed of two materials: structural steel and reinforced concrete. Strictly speaking, any structural member formed with two or more materials is composite. In buildings and bridges, however, that usually means structural steel and reinforced concrete, and that usually means composite beams or columns. Composite columns are being used again in some structures after a period of disuse; we cover them later in this chapter. Our coverage of beams is restricted to those that are part of a floor or roof system. Composite construction is covered in AISC Specification Chapter I, "Design of Composite Members."

Composite beams can take several forms. The earliest versions consisted of beams encased in concrete (Figure 9.1a). This was a practical alternative when the primary means of fireproofing structural steel was to encase it in concrete; the rationale was that if the concrete was there, we might as well account for its contribution to the strength of the beam. Currently, lighter and more economical methods of fireproofing are available, and encased composite beams are rarely used. Instead, composite behavior is achieved by connecting the steel beam to the reinforced concrete slab it supports, causing the two parts to act as a unit. In a floor or roof system, a portion of the slab acts with each steel beam to form a composite beam consisting of the rolled steel shape augmented by a concrete flange at the top (Figure 9.1b).

This unified behavior is possible only if horizontal slippage between the two components is prevented. That can be accomplished if the horizontal shear at the interface is resisted by connecting devices known as *shear connectors*. These devices —which can be headed studs or short lengths of small channel shapes—are welded to the top flange of the steel beam at prescribed intervals and provide the connection mechanically through anchorage in the hardened concrete (Figure 9.1c). Studs are the most commonly used type of shear connectors, and more than one can be used at each location if the flange is wide enough to accommodate them (which depends on the allowable spacing, which we consider in Section 9.4). One reason for the popularity of shear studs is their ease of installation. It is essentially a one-worker job made possible by an automatic tool that allows the operator to position the stud and weld it to the beam in one operation.

A certain number of shear connectors will be required to make a beam fully composite. Any fewer than this number will permit some slippage to occur between the

FIGURE 9.1

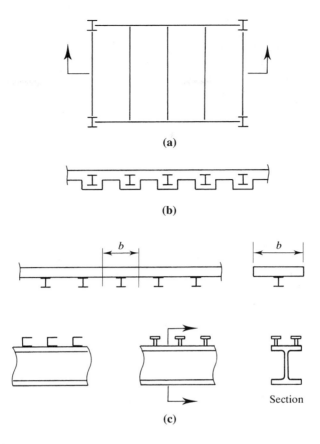

(a)

(b)

(c)

Section

steel and concrete; such a beam is said to be *partially composite*. Partially composite beams (which actually are more efficient than fully composite beams) are covered in Section 9.7.

Most composite construction in buildings utilizes formed steel deck, which serves as formwork for the concrete slab and is left in place after the concrete cures. This metal deck also contributes to the strength of the slab, the design of which we do not consider here. The deck can be used with its ribs oriented either perpendicular or parallel to the beams. In the usual floor system, the ribs will be perpendicular to the floor beams and parallel to the supporting girders. The shear studs are welded to the beams from above, through the deck. Since the studs can be placed only in the ribs, their spacing along the length of the beam is limited to multiples of the rib spacing. Figure 9.2 shows a slab with formed steel deck and the ribs perpendicular to the beam axis.

Almost all highway bridges that use steel beams are of composite construction, and composite beams are frequently the most economical alternative in buildings. Although smaller, lighter rolled steel beams can be used with composite construction, this advantage will sometimes be offset by the additional cost of the shear connectors. Even so, other advantages may make composite construction attractive. Shallower

FIGURE 9.2

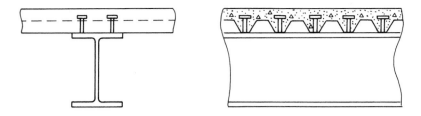

beams can be used, and deflections will be smaller than with conventional noncomposite construction.

Elastic Stresses in Composite Beams

Although the available strength of composite beams is usually based on conditions at failure, an understanding of the behavior at service loads is important for several reasons. Deflections are always investigated at service loads, and in some cases, the available strength is based on the limit state of first yield.

Flexural and shearing stresses in beams of homogeneous materials can be computed from the formulas

$$f_b = \frac{Mc}{I} \qquad \text{and} \qquad f_v = \frac{VQ}{It}$$

A composite beam is not homogeneous, however, and these formulas are not valid. To be able to use them, an artifice known as the *transformed section* is employed to "convert" the concrete into an amount of steel that has the same effect as the concrete. This procedure requires the strains in the fictitious steel to be the same as those in the concrete it replaces. Figure 9.3 shows a segment of a composite beam with stress and strain diagrams superimposed. If the slab is properly attached to the rolled steel shape, the strains will be as shown, with cross sections that are plane before bending remaining plane after bending. However, a continuous linear stress distribution as shown in part c of the figure is valid only if the beam is assumed to be homogeneous.

FIGURE 9.3

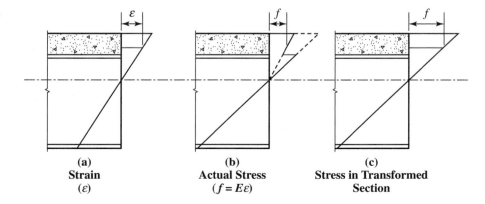

(a)	(b)	(c)
Strain	**Actual Stress**	**Stress in Transformed**
(ε)	$(f = E\varepsilon)$	**Section**

We first require that the strain in the concrete at any point be equal to the strain in any replacement steel at that point:

$$\varepsilon_c = \varepsilon_s \ \text{ or } \ \frac{f_c}{E_c} = \frac{f_s}{E_s}$$

and

$$f_s = \frac{E_s}{E_c} f_c = n f_c \tag{9.1}$$

where

E_c = modulus of elasticity of concrete

$n = \dfrac{E_s}{E_c}$ = modular ratio

AISC I2.lb gives the modulus of elasticity of concrete as*

$$E_c = w_c^{1.5} \sqrt{f_c'}$$

where

w_c = unit weight of concrete (lb/ft^3) (normal-weight concrete weighs approximately 145 lb/ft^3)

f_c' = 28-day compressive strength of concrete (kips/in.^{2})

The AISC Specification also gives a metric version of the equation for E_c.

Equation 9.1 can be interpreted as follows: n square inches of concrete are required to resist the same force as one square inch of steel. To determine the area of steel that will resist the same force as the concrete, divide the concrete area by n. That is, replace A_c by A_c/n. The result is the *transformed area*.

Consider the composite section shown in Figure 9.4a (determination of the effective flange width b when the beam is part of a floor system is discussed presently). To transform the concrete area, A_c, we must divide by n. The most convenient way to do this is to divide the width by n and leave the thickness unchanged. Doing so results in the homogeneous steel section of Figure 9.4b. To compute stresses, we locate the neutral axis of this composite shape and compute the corresponding moment of inertia. We can then compute bending stresses with the flexure formula. At the top of the steel,

$$f_{st} = \frac{My_t}{I_{tr}}$$

At the bottom of the steel,

$$f_{sb} = \frac{My_b}{I_{tr}}$$

---

*The ACI Building Code (ACI, 2005) gives the value of E_c as $w_c^{1.5}(33)\sqrt{f_c'}$, where f_c' is in pounds per square inch.

FIGURE 9.4

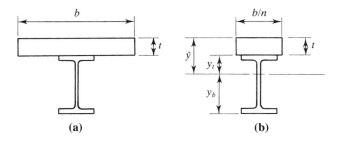

(a)   (b)

where

M = applied bending moment

I_{tr} = moment of inertia about the neutral axis (same as the centroidal axis for this homogeneous section).

y_t = distance from the neutral axis to the top of the steel

y_b = distance from the neutral axis to the bottom of the steel

The stress in the concrete may be computed in the same way, but because the material under consideration is steel, the result must be divided by n (see Equation 9.1) so that

$$\text{Maximum } f_c = \frac{M\bar{y}}{nI_{tr}}$$

where $\bar{y}$ is the distance from the neutral axis to the top of the concrete.

This procedure is valid only for a positive bending moment, with compression at the top, because concrete has negligible tensile strength.

Example 9.1   A composite beam consists of a W16 × 36 of A992 steel with a 5-inch-thick × 87-inch-wide reinforced concrete slab at the top. The strength of the concrete is $f_c' =$ 4 ksi. Determine the maximum stresses in the steel and concrete resulting from a positive bending moment of 160 ft-kips.

Solution
$$E_c = w_c^{1.5}\sqrt{f_c'} = (145)^{1.5}\sqrt{4} = 3492 \text{ ksi}$$

$$n = \frac{E_s}{E_c} = \frac{29,000}{3492} = 8.3 \quad \therefore \text{ use } n = 8$$

Since the modulus of elasticity of concrete can only be approximated, the usual practice of rounding n to the nearest whole number is sufficiently accurate. Thus,

$$\frac{b}{n} = \frac{87}{8} = 10.88 \text{ in.}$$

The transformed section is shown in Figure 9.5. Although the neutral axis is shown below the top of the steel, it is not known yet whether it lies in the steel or the concrete.

FIGURE 9.5

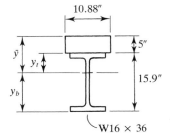

The location of the neutral axis can be found by applying the principle of moments with the axis of moments at the top of the slab. The computations are summarized in Table 9.1, and the distance from the top of the slab to the centroid is

$$\bar{y} = \frac{\Sigma Ay}{\Sigma A} = \frac{273.3}{65.00} = 4.205 \text{ in.}$$

Since this is less than 5 inches (the thickness of the slab) the neutral axis lies within the slab. Applying the parallel axis theorem and tabulating the computations in Table 9.2, we obtain the moment of inertia of the transformed section as

$$I_{tr} = 1530 \text{ in.}^4$$

The distance from the neutral axis to the top of the steel is

$$y_t = \bar{y} - t = 4.205 - 5.000 = -0.795 \text{ in.}$$

where t is the thickness of the slab. The negative sign means that the top of the steel is below the neutral axis and is therefore in tension. The stress at the top of the steel is

$$f_{st} = \frac{My_t}{I_{tr}} = \frac{(160 \times 12)(0.795)}{1530} = 0.998 \text{ ksi} \qquad \text{(tension)}$$

TABLE 9.1

Component	A	y	Ay
Concrete	54.40	2.50	136.0
W16 × 36	10.6	12.95	137.3
	65.00		273.3

TABLE 9.2

Component	A	y	$\bar{I}$	d	$\bar{I} + Ad^2$
Concrete	54.40	2.50	113.3	1.705	271.4
W16 × 36	10.6	12.95	448	8.745	1259
					1530.4

Stress at the bottom of the steel:

$$y_b = t + d - \bar{y} = 5 + 15.9 - 4.205 = 16.70 \text{ in.}$$

$$f_{sb} = \frac{My_b}{I_{tr}} = \frac{(160 \times 12)(16.70)}{1530} = 21.0 \text{ ksi} \quad \text{(tension)}$$

The stress at the top of the concrete is

$$f_c = \frac{M\bar{y}}{nI_{tr}} = \frac{(160 \times 12)(4.205)}{8(1530)} = 0.660 \text{ ksi}$$

If the concrete is assumed to have no tensile strength, the concrete below the neutral axis should be discounted. The geometry of the transformed section will then be different from what was originally assumed; to obtain an accurate result, the location of the neutral axis should be recomputed on the basis of this new geometry. Referring to Figure 9.6 and Table 9.3, we can compute the new location of the neutral axis as follows:

$$\bar{y} = \frac{\Sigma Ay}{\Sigma A} = \frac{5.44\bar{y}^2 + 137.3}{10.88\bar{y} + 10.6}$$

$$\bar{y}(10.88\bar{y} + 10.6) = 5.44\bar{y}^2 + 137.3$$

$$5.44\bar{y}^2 + 10.6\bar{y} - 137.3 = 0$$

$$\bar{y} = 4.143 \text{ in.}$$

The moment of inertia of this revised composite area is

$$I_{tr} = \frac{1}{3}(10.88)(4.143)^3 + 448 + 10.6(12.95 - 4.143)^2 = 1528 \text{ in.}^4$$

and the stresses are

$$f_{st} = \frac{(160 \times 12)(5 - 4.143)}{1528} = 1.08 \text{ ksi} \quad \text{(tension)}$$

$$f_{sb} = \frac{(160 \times 12)(5 + 15.9 - 4.143)}{1528} = 21.1 \text{ ksi} \quad \text{(tension)}$$

$$f_c = \frac{(160 \times 12)(4.143)}{8(1528)} = 0.651 \text{ ksi}$$

The difference between the two analyses is negligible, so the refinement in locating the neutral axis is not necessary.

FIGURE 9.6

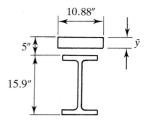

TABLE 9.3

Component	A	y	Ay
Concrete	$10.88\bar{y}$	$\bar{y}/2$	$5.44\bar{y}^2$
W16 × 36	10.6	12.95	137.3

Answer   The maximum stress in the steel is 21.1 ksi tension, and the maximum stress in the concrete is 0.651 ksi compression. ■

Flexural Strength

In most cases, the nominal flexural strength will be reached when the entire steel cross section yields and the concrete crushes in compression (for positive bending moment). The corresponding stress distribution on the composite section is called a *plastic stress distribution.* The AISC Specification provisions for flexural strength are as follows:

* For shapes with compact webs—that is, $h/t_w \le 3.76\sqrt{E/F_y}$ —the nominal strength M_n is obtained from the plastic stress distribution.
* For shapes with noncompact webs—$(h/t_w > 3.76\sqrt{E/F_y}$ — M_n is obtained from the elastic stress distribution corresponding to first yielding of the steel.
* For LRFD, the design strength is $\phi_b M_n$, where $\phi_b = 0.90$.
* For ASD, the allowable strength is M_n/Ω_b, where $\Omega_b = 1.67$.

All W, M, and S shapes tabulated in the *Manual* have compact webs (for flexure) for $F_y \le 50$ ksi, so the first condition will govern for all composite beams except those with built-up steel shapes. We consider only compact shapes in this chapter.

When a composite beam has reached the plastic limit state, the stresses will be distributed in one of the three ways shown in Figure 9.7. The concrete stress is shown as a uniform compressive stress of $0.85f_c'$, extending from the top of the slab to a depth that may be equal to or less than the total slab thickness. This distribution is the *Whitney equivalent stress distribution,* which has a resultant that matches that of the actual stress distribution (ACI, 2005). Figure 9.7a shows the distribution corresponding to full tensile yielding of the steel and partial compression of the concrete, with the plastic neutral axis (PNA) in the slab. The tensile strength of concrete is small and is discounted, so no stress is shown where tension is applied to the concrete. This condition will usually prevail when there are enough shear connectors provided to prevent slip completely—that is, to ensure full composite behavior. In Figure 9.7b, the concrete stress block extends the full depth of the slab, and the PNA is in the flange of the steel shape. Part of the flange will therefore be in compression to augment the compressive force in the slab. The third possibility, the PNA in the web, is shown in Figure 9.7c. Note that the concrete stress block need not extend the full depth of the slab for any of these three cases.

In each case shown in Figure 9.7, we can find the nominal moment capacity by computing the moment of the couple formed by the compressive and tensile resultants. This can be accomplished by summing the moments of the resultants about any

FIGURE 9.7

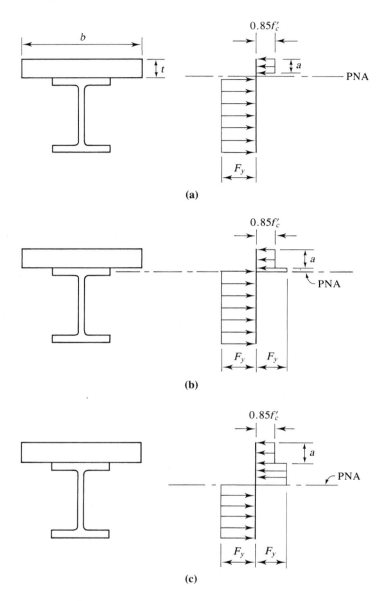

(a)

(b)

(c)

convenient point. Because of the connection of the steel shape to the concrete slab, lateral-torsional buckling is no problem once the concrete has cured and composite action has been achieved.

To determine which of the three cases governs, compute the compressive resultant as the smallest of

1. $A_s F_y$
2. $0.85 f_c' A_c$
3. ΣQ_n

where

A_s = cross-sectional area of steel shape
A_c = area of concrete
     = tb (see Figure 9.7)
ΣQ_n = total shear strength of the shear connectors

Each possibility represents a horizontal shear force at the interface between the steel and the concrete. When the first possibility controls, the steel is being fully utilized, and the stress distribution of Figure 9.7a applies. The second possibility corresponds to the concrete controlling, and the PNA will be in the steel (Figure 9.7b or c). The third case governs only when there are fewer shear connectors than required for full composite behavior, resulting in partial composite behavior. Although partial composite action can exist with either solid slabs or slabs with formed steel deck, it will be covered in Section 9.7, "Composite Beams with Formed Steel Deck."

Example 9.2   Compute the available strength of the composite beam of Example 9.1. Assume that sufficient shear connectors are provided for full composite behavior.

Solution   Determine the compressive force C in the concrete (horizontal shear force at the interface between the concrete and steel). Because there will be full composite action, this force will be the smaller of $A_s F_y$ and $0.85 f_c' A_c$:

$$A_s F_y = 10.6(50) = 530 \text{ kips}$$
$$0.85 f_c' A_c = 0.85(4)(5 \times 87) = 1479 \text{ kips}$$

The steel controls; $C = 530$ kips. This means that the full depth of the slab is not needed to develop the required compression force. The stress distribution shown in Figure 9.8 will result.

The resultant compressive force can also be expressed as

$$C = 0.85 f_c' ab$$

from which we obtain

$$a = \frac{C}{0.85 f_c' b} = \frac{530}{0.85(4)(87)} = 1.792 \text{ in.}$$

FIGURE 9.8

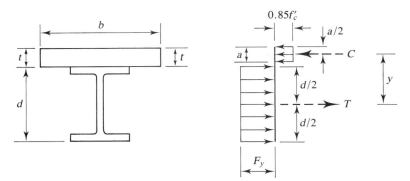

The force C will be located at the centroid of the compressive area at a depth of $a/2$ from the top of the slab. The resultant tensile force T (equal to C) will be located at the centroid of the steel area. The moment arm of the couple formed by C and T is

$$y = \frac{d}{2} + t - \frac{a}{2} = \frac{15.9}{2} + 5 - \frac{1.792}{2} = 12.05 \text{ in.}$$

The nominal strength is the moment of the couple, or

$$M_n = Cy = Ty = 530(12.05) = 6387 \text{ in.-kips} = 532.2 \text{ ft-kips}$$

Answer   For LRFD, the design strength is $\phi_b M_n = 0.90(532.2) = 479$ ft-kips.

For ASD, the allowable strength is $\dfrac{M_n}{\Omega_b} = \dfrac{532.2}{1.67} = 319$ ft-kips.

When full composite behavior exists, the plastic neutral axis will normally be within the slab, as in Example 9.2. Analysis for the case of the plastic neutral axis located within the steel section will be deferred until partial composite action has been covered.

9.2 SHORED VERSUS UNSHORED CONSTRUCTION

Until the concrete has cured and attained its design strength (at least 75% of its 28-day compressive strength, f_c'), there can be no composite behavior, and the weight of the slab must be supported by some other means. Once the concrete has cured, composite action is possible, and all subsequently applied loads will be resisted by the composite beam. If the steel shape is supported at a sufficient number of points along its length before the slab is placed, the weight of the wet concrete will be supported by these temporary *shores* rather than by the steel. Once the concrete has cured, the temporary shoring can be removed, and the weight of the slab, as well as any additional loads, will be carried by the composite beam. If shoring is not used, however, the rolled steel shape must support not only its own weight, but also the weight of the slab and its formwork during the curing period. Once composite behavior is attained, additional loads, both dead and live, will be supported by the composite beam. We now consider these different conditions in more detail.

Unshored: Before Concrete Cures

AISC I3.lc requires that when temporary shoring is not provided, the steel shape alone must have sufficient strength to resist all loads applied before the concrete attains 75% of its strength, f_c'. The flexural strength is computed in the usual way, based on Chapter F of the Specification (Chapter 5 of this book). Depending on its design, the formwork for the concrete slab may or may not provide lateral support for the steel beam. If not, the unbraced length L_b must be taken into account, and lateral-torsional buckling may control the flexural strength. If temporary shoring is not used, the steel beam may also be called on to resist incidental construction loads. To account for these loads, an additional 20 pounds per square foot is recommended (Hansell et al., 1978).

Unshored: After Concrete Cures

After composite behavior is achieved, all loads subsequently applied will be supported by the composite beam. At failure, however, all loads will be resisted by the internal couple corresponding to the stress distribution at failure. Thus the composite section must have adequate strength to resist all loads, including those applied to the steel beam before the concrete cures (except for construction loads, which will no longer be present).

Shored Construction

In shored construction, only the composite beam need be considered, because the steel shape will not be required to support anything other than its own weight.

Shear Strength

AISC I3.1b conservatively requires that all shear be resisted by the web of the steel shape, as provided for in Chapter G of the Specification (Chapter 5 of this book).

Example 9.3

A W12×50 acts compositely with a 4-inch-thick concrete slab. The effective slab width is 72 inches. Shoring is not used, and the applied bending moments are as follows: from the beam weight, M_{beam} = 13 ft-kips; from the slab weight, M_{slab} = 77 ft-kips; and from the live load, M_L = 150 ft-kips. (Do not consider any additional construction loads in this example.) Steel is A992, and f_c' = 4 ksi. Determine whether the flexural strength of this beam is adequate. Assume full composite action and assume that the formwork provides lateral support of the steel section before curing of the concrete.

Solution

Compute the nominal strength of the composite section. The compressive force C is the smaller of

$$A_s F_y = 14.6(50) = 730 \text{ kips}$$

or

$$0.85 f_c' A_c = 0.85(4)(4 \times 72) = 979.2 \text{ kips}$$

The PNA is in the concrete and C = 730 kips. From Figure 9.8, the depth of the compressive stress block is

$$a = \frac{C}{0.85 f_c' b} = \frac{730}{0.85(4)(72)} = 2.982 \text{ in.}$$

The moment arm is

$$y = \frac{d}{2} + t - \frac{a}{2} = \frac{12.2}{2} + 4 - \frac{2.982}{2} = 8.609 \text{ in.}$$

The nominal moment strength is

$$M_n = Cy = 730(8.609) = 6285 \text{ in.-kips} = 523.8 \text{ ft-kips}$$

LRFD Solution Before the concrete cures, there is only dead load (no construction load in this example). Hence load combination A4-1 controls, and the factored load moment is

$$M_u = 1.4(M_D) = 1.4(13 + 77) = 126 \text{ ft-kips}$$

From the Z_x table in Part 3 of the *Manual*,

$$\phi_b M_n = \phi_b M_p = 270 \text{ ft-kips} > 126 \text{ ft-kips} \qquad \text{(OK)}$$

After the concrete has cured, the factored load moment that must be resisted by the composite beam is

$$M_u = 1.2M_D + 1.6M_L = 1.2(13 + 77) + 1.6(150) = 348 \text{ ft-kips}$$

The design moment is

$$\phi_b M_n = 0.9(523.8) = 471 \text{ ft-kips} > 348 \text{ ft-kips} \qquad \text{(OK)}$$

Answer The beam has sufficient flexural strength.

ASD Solution Before the concrete cures, there is only dead load (there is no construction load in this example).

$$M_a = M_D = 13 + 77 = 90 \text{ ft-kips}$$

From the Z_x table in Part 3 of the *Manual*,

$$\frac{M_n}{\Omega_b} = \frac{M_p}{\Omega_b} = 179 \text{ ft-kips} > 90 \text{ ft-kips} \qquad \text{(OK)}$$

After the concrete has cured, the required moment strength is

$$M_a = M_D + M_L = 13 + 77 + 150 = 240 \text{ ft-kips}$$

$$\frac{M_n}{\Omega_b} = \frac{523.8}{1.67} = 314 \text{ ft-kips} > 204 \text{ ft-kips} \qquad \text{(OK)}$$

Answer The beam has sufficient flexural strength. ■

Obviously, shored construction is more efficient than unshored construction because the steel section is not called on to support anything other than its own weight. In some situations, the use of shoring will enable a smaller steel shape to be used.

Most composite construction is unshored, however, because the additional cost of the shores, especially the labor cost, outweighs the small savings in steel weight that may result. Consequently, we devote the remainder of this chapter to unshored composite construction.

9.3   EFFECTIVE FLANGE WIDTH

The portion of the floor slab that acts compositely with the steel beam is a function of several factors, including the span length and beam spacing. AISC I3.1a requires that the effective width of floor slab on *each side* of the beam centerline be taken as the smallest of

1. one eighth of the span length,
2. one half of the beam center-to-center spacing, or
3. the distance from the beam centerline to the edge of the slab.

The third criterion will apply to edge beams only, so for *interior* beams, the full effective width will be the smaller of one fourth the span length or the center-to-center spacing of the beams (assuming that the beams are evenly spaced).

Example 9.4   A composite floor system consists of W18 × 35 steel beams spaced at 9 feet and supporting a 4.5-inch-thick reinforced concrete slab. The span length is 30 feet. In addition to the weight of the slab, there is a 20 psf partition load and a live load of 125 psf (light manufacturing). The steel is A992, and the concrete strength is $f_c' = 4$ ksi. Investigate a typical interior beam for compliance with the AISC Specification if no temporary shores are used. Assume full lateral support during construction and an additional construction load of 20 psf. Sufficient shear connectors are provided for full composite action.

Solution   The loads and the strength of the composite section are common to both the LRFD and ASD solutions. These common parts will be presented first, followed by the LRFD solution and then the ASD solution.

Loads applied before the concrete cures:

$$\text{Slab weight} = \left(\frac{4.5}{12}\right)(150) = 56.25\,\text{psf}$$

(Although normal-weight concrete weighs 145 pcf, recall that *reinforced* concrete is assumed to weigh 150 pcf.) For a beam spacing of 9 feet, the dead load is

$$56.25(9) = 506.3\ \text{lb/ft}$$
$$+\ \text{Beam weight} = \underline{\ 35.0\ \text{lb/ft}}$$
$$541.3\ \text{lb/ft}$$

The construction load is $20(9) = 180$ lb/ft, which is treated as live load.

Loads applied after the concrete cures: After the concrete cures, the construction loads do not act, but the partition load does, and it will be treated as live load (see Example 5.13):

$$w_D = 506.3 + 35 = 541.3 \text{ lb/ft}$$

The live load is

$$w_L = (125 + 20)(9) = 1305 \text{ lb/ft}$$

Strength of the composite section: To obtain the effective flange width, use either

$$\frac{\text{Span}}{4} = \frac{30(12)}{4} = 90 \text{ in.} \quad \text{or} \quad \text{Beam spacing} = 9(12) = 108 \text{ in.}$$

Since this member is an interior beam, the third criterion is not applicable. Use $b = 90$ inches as the effective flange width. Then, as shown in Figure 9.9, the compressive force is the smaller of

$$A_s F_y = 10.3(50) = 515.0 \text{ kips}$$

or

$$0.85 f'_c A_c = 0.85(4)(4.5 \times 90) = 1377 \text{ kips}$$

Use $C = 515$ kips. From Figure 9.9,

$$a = \frac{C}{0.85 f'_c b} = \frac{515}{0.85(4)(90)} = 1.683 \text{ in.}$$

$$y = \frac{d}{2} + t - \frac{a}{2} = 8.85 + 4.5 - \frac{1.683}{2} = 12.51 \text{ in.}$$

$$M_n = Cy = 515(12.51) = 6443 \text{ in.-kips} = 536.9 \text{ ft-kips}$$

FIGURE 9.9

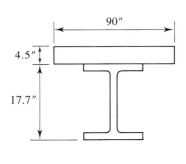
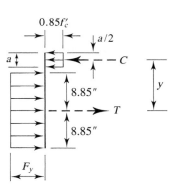

LRFD Solution   Before the concrete cures, the factored load and moment are

$$w_u = 1.2w_D + 1.6w_L = 1.2(541.3) + 1.6(180) = 937.6 \text{ lb/ft}$$

$$M_u = \frac{1}{8} w_u L^2 = \frac{1}{8}(0.9376)(30)^2 = 106 \text{ ft-kips}$$

From the Z_x table,

$$\phi_b M_n = \phi_b M_p = 249 \text{ ft-kips} > 106 \text{ ft-kips} \qquad \text{(OK)}$$

After the concrete cures, the factored load and moment are

$$w_u = 1.2 w_D + 1.6 w_L = 1.2(541.3) + 1.6(1305) = 2738 \text{ lb/ft}$$

$$M_u = \frac{1}{8}(2.738)(30)^2 = 308 \text{ ft-kips}$$

The design strength of the composite section is

$$\phi_b M_n = 0.90(536.9) = 483 \text{ ft-kips} > 308 \text{ ft-kips} \qquad \text{(OK)}$$

Check shear:

$$V_u = \frac{w_u L}{2} = \frac{2.738(30)}{2} = 41.1 \text{ kips}$$

From the Z_x table,

$$\phi_v V_n = 159 \text{ kips} > 41.1 \text{ kips} \qquad \text{(OK)}$$

Answer   The beam complies with the AISC Specification.

ASD Solution   Before the concrete cures, the load and moment are

$$w_a = w_D + w_L = 541.3 + 180 = 721.3 \text{ lb/ft}$$

$$M_a = \frac{1}{8} w_a L^2 = \frac{1}{8}(0.7213)(30)^2 = 81.2 \text{ ft-kips}$$

From the Z_x table,

$$\frac{M_n}{\Omega_b} = \frac{M_p}{\Omega_b} = 166 \text{ ft-kips} > 81.2 \text{ ft-kips} \qquad \text{(OK)}$$

After the concrete cures, the load and moment are

$$w_a = w_D + w_L = 541.3 + 1305 = 1846 \text{ lb/ft}$$

$$M_a = \frac{1}{8}(1.846)(30)^2 = 208 \text{ ft-kips}$$

The allowable strength of the composite section is

$$\frac{M_n}{\Omega_b} = \frac{536.9}{1.67} = 322 \text{ ft-kips} > 208 \text{ ft-kips} \qquad \text{(OK)}$$

Check shear:

$$V_a = \frac{w_a L}{2} = \frac{1.846(30)}{2} = 27.7 \text{ kips}$$

From the Z_x table,

$$\frac{V_n}{\Omega_v} = 106 \text{ kips} > 27.7 \text{ kips} \qquad \text{(OK)}$$

Answer   The beam complies with the AISC Specification.   ■

9.4   SHEAR CONNECTORS

As we have shown, the horizontal shear force to be transferred between the concrete and the steel is equal to the compressive force in the concrete, C. We denote this horizontal shear force V'. Thus V' is given by the smallest of $A_s F_y$, $0.85 f_c' A_c$, or ΣQ_n. If $A_s F_y$ or $0.85 f_c' A_c$ controls, full composite action will exist and the number of shear connectors required between the points of zero moment and maximum moment is

$$N_1 = \frac{V'}{Q_n} \tag{9.2}$$

where Q_n is the nominal shear strength of one connector. The N_1 connectors should be uniformly spaced within the length where they are required. The AISC Specification gives equations for the strength of both stud and channel shear connectors. As indicated at the beginning of this chapter, stud connectors are the most common, and we consider only this type. For one stud shear connector,

$$Q_n = 0.5 A_{sc} \sqrt{f_c' E_c} \leq R_g R_p A_{sc} F_u \qquad \text{(AISC Equation I3-3)}$$

where

A_{sc} = cross-sectional area of stud (in.^{2})
f_c' = 28-day compressive strength of the concrete (ksi)
E_c = modulus of elasticity of the concrete (ksi)
$R_p = R_g$ = 1.0 for solid slabs (no formed steel deck). When formed steel deck
  is used, R_p and R_g depend on the deck properties. This is considered in
  Section 9.7 of this book, "Composite Beams with Formed Steel Deck."
F_u = minimum tensile strength of stud (ksi)

For studs used as shear connectors in composite beams, the tensile strength F_u is 65 ksi (AWS, 2004). Values given by AISC Equation I3-3 are based on experimental studies. No factors are applied to Q_n (neither a resistance factor for LRFD nor a safety factor for ASD); the overall flexural resistance factor or safety factor accounts for all strength uncertainties.

Equation 9.2 gives the number of shear connectors required between the point of zero moment and the point of maximum moment. Consequently, for a simply supported, uniformly loaded beam, $2N_1$ connectors will be required, and they should be equally spaced. When concentrated loads are present, AISC I3.2d(6) requires that enough of the N_1 connectors be placed between the concentrated load and the adjacent

FIGURE 9.10

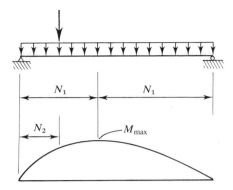

point of zero moment to develop the moment required at the load. We denote this portion N_2, and this requirement is illustrated in Figure 9.10. Note that the total number
of shear connectors is not affected by this requirement.

Miscellaneous Requirements for Headed Studs

Except as noted, the following requirements are from AISC I3.2d(6):

- Maximum diameter = 2.5 × flange thickness of steel shape (unless placed
  directly over the web)
- Minimum length = 4 × stud diameter (this requirement is from AISC I1.3)
- Minimum longitudinal spacing (center-to-center) = 6 × stud diameter
- Maximum longitudinal spacing (center-to-center) = 8 × slab thickness ≤ 36″
- Minimum transverse spacing (center-to-center) = 4 × stud diameter
- Minimum lateral cover = 1 inch (except in the ribs of formed steel deck; see
  Section 9.7)

Note that there is no minimum *vertical* cover except when formed steel deck is used.
This is covered in Section 9.7.

The AWS Structural Code (AWS, 2004) lists standard stud diameters of ½, ⅝,
¾, ⅞, and 1 inch. Matching these diameters with the minimum lengths prescribed by
AISC, we get the common stud sizes of ½ × 2, ⅝ × 2½, ¾ × 3, ⅞ × 3½, and 1 × 4
(but longer studs may be used).

Example 9.5   Design shear connectors for the floor system in Example 9.4.

Solution   Summary of data from Example 9.4:

W18 × 35, A992 steel
$f_c' = 4$ ksi
Slab thickness $t = 4.5$ in.
Span length = 30 ft

From Example 9.4, the horizontal shear force V' corresponding to full composite action is

$$V' = C = 515 \text{ kips}$$

Try ½ in. × 2-inch studs. The maximum permissible diameter is

$$2.5t_f = 2.5(0.425) = 1.063 \text{ in.} > 0.5 \text{ in.} \qquad \text{(OK)}$$

The cross-sectional area of one shear connector is

$$A_{sc} = \frac{\pi(0.5)^2}{4} = 0.1963 \text{ in.}^2$$

If we assume normal-weight concrete, the modulus of elasticity of the concrete is

$$E_c = w_c^{1.5}\sqrt{f_c'} = (145)^{1.5}\sqrt{4} = 3492 \text{ ksi}$$

From AISC Equation I3-3, the nominal shear strength of one connector is

$$Q_n = 0.5A_{sc}\sqrt{f_c'E_c} \leq R_g R_p A_{sc} F_u$$
$$= 0.5(0.1963)\sqrt{4(3492)} = 11.60 \text{ kips}$$
$$R_g R_p A_{sc} F_u = 1.0(1.0)(0.1963)(65) = 12.76 \text{ kips} > 11.60 \text{ kips}$$
$$\therefore \text{ use } Q_n = 11.60 \text{ kips}$$

and

Minimum longitudinal spacing is $6d = 6(0.5) = 3$ in.
Minimum transverse spacing is $4d = 4(0.50) = 2$ in.
Maximum longitudinal spacing is $8t = 8(4.5) = 36$ in. (upper limit is 36 in.)

The number of studs required between the end of the beam and midspan is

$$N_1 = \frac{V'}{Q_n} = \frac{515}{11.60} = 44.4$$

Use a minimum of 45 for half the beam, or 90 total. If one stud is used at each section, the required spacing will be

$$s = \frac{30(12)}{90} = 4 \text{ in.}$$

For two studs per section,

$$s = \frac{30(12)}{90/2} = 8 \text{ in.}$$

FIGURE 9.11

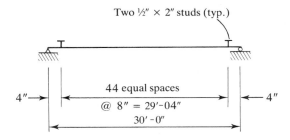

Two ½″ × 2″ studs (typ.)

4″

44 equal spaces
@ 8″ = 29′-04″

30′ -0″

4″

Either arrangement is satisfactory; either spacing will be between the lower and upper limits. The layout shown in Figure 9.11 will be recommended.

Answer   Use 90 studs, ½ in. × 2 in., placed as shown in Figure 9.11.

Note that there is no distinction between LRFD and ASD when designing shear connectors for full composite behavior. This is because the required number of studs is determined by dividing a nominal strength V' by a nominal strength Q_n, and no applied loads are involved.

9.5   DESIGN

The first step in the design of a floor system is to select the thickness of the floor slab, whether it is solid or ribbed (formed with steel deck). The thickness will be a function of the beam spacing, and several combinations of slab thickness and beam spacing may need to be investigated so that the most economical system can be found. The design of the slab is beyond the scope of this text, however, and we will assume that the slab thickness and beam spacing are known. Having made this assumption, we can take the following steps to complete the design of an unshored floor system.

1. Compute the moments acting before and after the concrete cures.
2. Select a steel shape for trial.
3. Compare the available strength of the steel shape to the required moment strength acting before the concrete cures. Account for the unbraced length if the formwork does not provide adequate lateral support. If this shape is not satisfactory, try a larger one.
4. Compute the available strength of the composite section and compare it to the *total* required moment strength. If the composite section is inadequate, select another steel shape for trial.
5. Check the shear strength of the steel shape.
6. Design the shear connectors:
   a. Compute V', the horizontal shear force at the interface between the concrete and the steel.

FIGURE 9.12

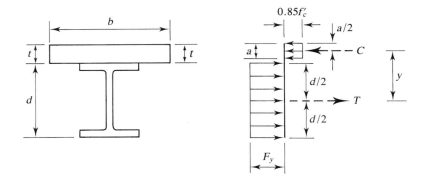

   b. Divide this force by Q_n, the shear capacity of a single connector, to obtain N_1, which, in most cases, is half the total number of shear connectors required. Using this number of connectors will provide full composite behavior. If partial composite action is desired, the number of shear connectors can be reduced (this is covered in Section 9.7).
7. Check deflections (we cover this in Section 9.6).

The major task in the trial-and-error procedure just outlined is the selection of a trial steel shape. A formula that will give the required area (or, alternatively, the required weight per foot of length) can be developed if a beam depth is assumed. Assuming full composite action and the PNA in the slab (i.e., steel controlling, the most common case for full composite action), we can write the nominal strength (refer to Figure 9.12) as

$$M_n = Ty = A_s F_y y$$

LRFD Procedure. Equate the design strength to the factored load moment and solve for A_s:

$$\phi_b M_n = \phi_b A_s F_y y = M_u$$

$$A_s = \frac{M_u}{\phi_b F_y y}$$

or

$$A_s = \frac{M_u}{\phi_b F_y \left(\dfrac{d}{2} + t - \dfrac{a}{2} \right)} \tag{9.3}$$

Equation 9.3 can also be written in terms of weight rather than area. Since one foot of length has a volume of $A_s / 144$ cubic feet and structural steel weights 490 pounds per cubic foot,

$$w = \frac{A_s}{144}(490) = 3.4 A_s \ \text{lb/ft} \quad \text{(for } A_s \text{ in square inches)}$$

From Equation 9.3, the estimated weight per foot is, therefore,

$$w = \frac{3.4M_u}{\phi_b F_y \left(\dfrac{d}{2} + t - \dfrac{a}{2}\right)} \text{ lb/ft} \tag{9.4}$$

where M_u is in in.-kips; F_y is in ksi; and d, t, and a are in inches.

ASD Procedure. Equate the allowable strength to the service load moment and solve for A_s:

$$\frac{M_n}{\Omega_b} = \frac{A_s F_y y}{\Omega_b} = M_a$$

$$A_s = \frac{\Omega_b M_a}{F_y y}$$

or

$$A_s = \frac{\Omega_b M_a}{F_y \left(\dfrac{d}{2} + t - \dfrac{a}{2}\right)} \tag{9.5}$$

Using

$$w = \frac{A_s}{144}(490) = 3.4 A_s \text{ lb/ft} \quad \text{(for } A_s \text{ in square inches)}$$

we obtain from Equation 9.5 the estimated weight per foot:

$$w = \frac{3.4\Omega_b M_a}{F_y \left(\dfrac{d}{2} + t - \dfrac{a}{2}\right)} \text{ lb/ft} \tag{9.6}$$

where M_a is in in.-kips; F_y is in ksi; and d, t, and a are in inches.

Both Equations 9.4 (LRFD) and 9.6 (ASD) require an assumed depth and an estimate of $d/2$. The stress block depth will generally be very small; consequently, an error in the estimate of $a/2$ will have only a slight effect on the estimated value of A_s. An assumed value of $a/2 = 0.5$ inch is suggested.

Example 9.6   The span length of a certain floor system is 30 feet, and the beam spacing is 10 feet center-to-center. Select a rolled steel shape and the shear connectors needed to achieve full composite behavior with a 3.5-inch-thick reinforced concrete floor slab. Superimposed loading consists of a 20 psf partition load and a 100 psf live load. Concrete strength is $f_c' = 4$ ksi, and A992 steel is to be used. Assume that the beam has full lateral support during construction and that there is a 20 psf construction load.

Solution   Loads to be supported before the concrete cures are

> Slab: $(3.5/12)(150) = 43.75$ psf
> Weight per linear foot: $43.75(10) = 437.5$ lb/ft
> Construction load: $20(10) = 200$ lb/ft

(The beam weight will be accounted for later.)
Loads to be supported after the concrete cures are

$$w_D = w_{\text{slab}} = 437.5 \text{ lb/ft}$$
$$w_L = (100 + 20)(10) = 1200 \text{ lb/ft}$$

where the 20 psf partition load is treated as a live load.

LRFD Solution   The composite section must resist a factored load and moment of

$$w_u = 1.2w_D + 1.6w_L = 1.2(437.5) + 1.6(1200) = 2445 \text{ lb/ft}$$
$$M_u = \frac{1}{8} w_u L^2 = \frac{1}{8}(2.445)(30)^2 = 275 \text{ ft-kips}$$

Try a nominal depth of $d = 16$ inches. From Equation 9.4, the estimated beam weight is

$$w = \frac{3.4 M_u}{\phi_b F_y \left(\dfrac{d}{2} + t - \dfrac{a}{2} \right)} = \frac{3.4(275 \times 12)}{0.90(50)\left(\dfrac{16}{2} + 3.5 - 0.5 \right)} = 22.7 \text{ lb/ft}$$

Try a W16 × 26. Check the unshored steel shape for loads applied before the concrete cures (the weight of the slab, the weight of the beam, and the construction load).

$$w_u = 1.2(0.4375 + 0.026) + 1.6(0.200) = 0.8762 \text{ kips/ft}$$
$$M_u = \frac{1}{8}(0.8762)(30)^2 = 98.6 \text{ ft-kips}$$

From the Z_x table,

$$\phi_b M_n = \phi_b M_p = 166 \text{ ft-kips} > 98.6 \text{ ft-kips} \qquad \text{(OK)}$$

After the concrete cures and composite behavior has been achieved,

$$w_D = w_{\text{slab}} + w_{\text{beam}} = 0.4375 + 0.026 = 0.4635 \text{ kips/ft}$$
$$w_u = 1.2w_D + 1.6w_L = 1.2(0.4635) + 1.6(1.200) = 2.476 \text{ kips/ft}$$
$$M_u = \frac{1}{8} w_u L^2 = \frac{1}{8}(2.476)(30)^2 = 279 \text{ ft-kips}$$

Before computing the design strength of the composite section, we must first determine the effective slab width. For an interior beam, the effective width is the smaller of

$$\frac{\text{Span}}{4} = \frac{30(12)}{4} = 90 \text{ in.} \quad \text{or} \quad \text{Beam spacing} = 10(12) = 120 \text{ in.}$$

Use $b = 90$ in. For full composite behavior, the compressive force in the concrete at ultimate (equal to the horizontal shear at the interface between the concrete and steel) will be the smaller of

$$A_s F_y = 7.68(50) = 384 \text{ kips}$$

or

$$0.85 f_c' A_c = 0.85(4)(90)(3.5) = 1071 \text{ kips}$$

Use $C = V' = 384$ kips. The depth of the compressive stress block in the slab is

$$a = \frac{C}{0.85 f_c' b} = \frac{384}{0.85(4)(90)} = 1.255 \text{ in.}$$

and the moment arm of the internal resisting couple is

$$y = \frac{d}{2} + t - \frac{a}{2} = \frac{15.7}{2} + 3.5 - \frac{1.255}{2} = 10.72 \text{ in.}$$

The design flexural strength is

$$\phi_b M_n = \phi_b(Cy) = 0.90(384)(10.72)$$
$$= 3705 \text{ in.-kips} = 309 \text{ ft-kips} > 279 \text{ ft-kips} \quad \text{(OK)}$$

Check the shear:

$$V_u = \frac{w_u L}{2} = \frac{2.476(30)}{2} = 37.1 \text{ kips}$$

From the Z_x table,

$$\phi_v V_n = 106 \text{ kips} > 37.1 \text{ kips} \quad \text{(OK)}$$

Answer   Use a $W16 \times 26$.

Design the shear connectors.

$$\text{Maximum diameter} = 2.5 t_f = 2.5(0.345) = 0.863 \text{ in.}$$

Try ½ in. × 2 in. studs.

$$d = 1/2 \text{ in.} < 0.863 \text{ in.} \quad \text{(OK)}$$

$$A_{sc} = \frac{\pi d^2}{4} = \frac{\pi (0.5)^2}{4} = 0.1963 \text{ in.}^2$$

For normal weight concrete,

$$E_c = w^{1.5}\sqrt{f_c'} = (145)^{1.5}\sqrt{4} = 3492 \text{ ksi}$$

From AISC Equation I3-3,

$$Q_n = 0.5A_{sc}\sqrt{f_c'E_c} \le R_g R_p A_{sc} F_u$$
$$= 0.5(0.1963)\sqrt{4(3492)} = 11.60 \text{ kips}$$
$$R_g R_p A_{sc} F_u = 1.0(1.0)(0.1963)(65)$$
$$= 12.76 \text{ kips} > 11.60 \text{ kips} \quad \therefore \text{ use } Q_n = 11.60 \text{ kips}$$

The number of studs required between the end of the beam and midspan is

$$N_1 = \frac{V'}{Q_n} = \frac{384}{11.60} = 33.1 \quad \therefore \text{ use 34 for half the beam, or 68 total}$$

and

Minimum longitudinal spacing is $6d = 6(0.5) = 3$ in.
Minimum transverse spacing is $4d = 4(0.5) = 2$ in.
Maximum longitudinal spacing is $8t = 8(3.5) = 28$ in. (< 36 in. upper limit)

If one stud is used at each section, the approximate spacing will be

$$s = \frac{30(12)}{68} = 5.3 \text{ in.}$$

This spacing is between the upper and lower limits and is therefore satisfactory.

Answer   Use the design shown in Figure 9.13.

ASD Solution   The composite section must resist a service load and moment of

$$w_a = w_D + w_L = 437.5 + 1200 = 1638 \text{ lb/ft}$$
$$M_a = \frac{1}{8}w_a L^2 = \frac{1}{8}(1.638)(30)^2 = 184.3 \text{ ft-kips}$$

FIGURE 9.13

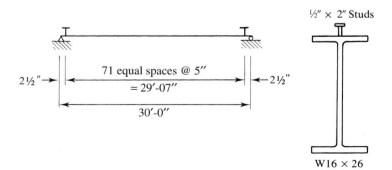

½" × 2" Studs

2½"→

71 equal spaces @ 5"
= 29'-07"

←2½"

30'-0"

W16 × 26

Try a nominal depth of $d = 16$ inches. From Equation 9.6, the estimated beam weight is

$$w = \frac{3.4\Omega_b M_a}{F_y\left(\dfrac{d}{2} + t - \dfrac{a}{2}\right)} = \frac{3.4(1.67)(184.3 \times 12)}{50\left(\dfrac{16}{2} + 3.5 - 0.5\right)} = 22.8 \text{ lb/ft}$$

Try a W16 × 26. Check the unshored steel shape for loads applied before the concrete cures (the weight of the slab, the weight of the beam, and the construction load).

$$w_a = w_{\text{slab}} + w_{\text{beam}} + w_{\text{const}} = 0.4375 + 0.026 + 0.200 = 0.6635 \text{ lb/ft}$$

$$M_a = \frac{1}{8} w_a L^2 = \frac{1}{8}(0.6635)(30)^2 = 74.6 \text{ ft-kips}$$

From the Z_x table,

$$\frac{M_n}{\Omega_b} = \frac{M_p}{\Omega_b} = 110 \text{ ft-kips} > 74.6 \text{ ft-kips} \qquad \text{(OK)}$$

After the concrete cures and composite behavior has been achieved, the load and moment are

$$w_a = w_{\text{slab}} + w_{\text{beam}} + w_{\text{part}} + w_L = 0.4375 + 0.026 + 0.200 + 1.000 = 1.664 \text{ kips/ft}$$

where the 20 psf partition load has been included as part of W_L.

$$M_a = \frac{1}{8}(1.664)(30)^2 = 187 \text{ ft-kips}$$

The effective slab width for a typical interior beam is the smaller of

$$\frac{\text{Span}}{4} = \frac{30 \times 12}{4} = 90 \text{ in.} \qquad \text{or} \qquad \text{Beam spacing} = 10 \times 12 = 120 \text{ in.}$$

Use $b = 90$ in. For full composite behavior, the compressive force in the concrete at ultimate (equal to the horizontal shear at the interface between the concrete and steel) will be the smaller of

$$A_s F_y = 7.68(50) = 384 \text{ kips}$$

or

$$0.85 f_c' A_c = 0.85(4)(3.5 \times 90) = 1071 \text{ kips}$$

Use $C = 384$ kips.

$$a = \frac{C}{0.85 f_c' b} = \frac{384}{0.85(4)(90)} = 1.255 \text{ in.}$$

$$y = \frac{d}{2} + t - \frac{a}{2} = \frac{15.7}{2} + 3.5 - \frac{1.255}{2} = 10.72 \text{ in.}$$

The allowable flexural strength is

$$\frac{M_n}{\Omega_b} = \frac{Cy}{\Omega_b} = \frac{384(10.72)}{1.67} = 2465 \text{ in.-kips} = 205 \text{ ft-kips} > 187 \text{ ft-kips} \qquad \text{(OK)}$$

Check shear:

$$V_a = \frac{w_a L}{2} = \frac{1.664(30)}{2} = 25.0 \text{ kips}$$

From the Z_x table,

$$\frac{V_n}{\Omega_v} = 70.5 \text{ kips} > 25.0 \text{ kips} \qquad \text{(OK)}$$

Answer   Use a W16 × 26.

The ASD design of the shear connectors is identical to what was done for LRFD and will not be repeated here.

Answer   Use the shear connector design shown in Figure 9.13.   ◼

9.6   DEFLECTIONS

Because of the large moment of inertia of the transformed section, deflections in composite beams are smaller than in noncomposite beams. For unshored construction, this larger moment of inertia is available only after the concrete slab has cured. Deflections caused by loads applied before the concrete cures must be computed with the moment of inertia of the steel shape. An additional complication arises if the beam is subject to sustained loading, such as the weight of partitions, after the concrete cures. In positive moment regions, the concrete will be in compression continuously and is subject to a phenomenon known as *creep*. Creep is a deformation that takes place under sustained compressive load. After the initial deformation, additional deformation will take place at a very slow rate over a long period of time. The effect on a composite beam is to increase the curvature and hence the vertical deflection. Long-term deflection can only be estimated; one technique is to use a reduced area of concrete in the transformed section so as to obtain a smaller moment of inertia and a larger computed deflection. The reduced area is computed by using $2n$ or $3n$ instead of the actual modular ratio n. For ordinary buildings, the additional creep deflection caused by sustained dead load is very small, and we will not cover it in this book. If a significant portion of the live load is considered to be sustained, then creep should be accounted for. Detailed coverage of this topic, as well as deflections caused by shrinkage of the concrete during curing, can be found in Viest et al. (1997). A method for estimating the shrinkage deflection can be found in the Commentary to the Specification.

Use of the transformed moment of inertia for computing composite beam deflections tends to underestimate the deflections by 15 to 30% (AISC Specification Commentary, Section I3.1). To counter this effect, the Commentary recommends that the calculated transformed moment of inertia be reduced by 25% (that is, use $0.75I_{tr}$).

Example 9.7 Compute deflections for the beam in Example 9.4.

Solution Summary of data from Example 9.4:

> W18 × 35, A992 steel
> Slab thickness $t = 4.5$ in. and effective width $b = 90$ in.
> $f_c' = 4$ ksi
> Dead load applied before concrete cures is $w_D = 541.3$ lb/ft (slab plus beam)
> Construction load is $w_{\text{const}} = 180$ lb/ft
> Live load is $w_L = 125(9) = 1125$ lb/ft
> Partition load is $w_{\text{part}} = 20(9) = 180$ lb/ft

Immediate deflection: For the beam plus the slab, $w = 541.3$ lb/ft and

$$\Delta_1 = \frac{5wL^4}{384EI_s} = \frac{5(0.5413/12)(30 \times 12)^4}{384(29{,}000)(510)} = 0.6670 \text{ in.}$$

The construction load is $w = 180$ lb/ft and

$$\Delta_2 = \frac{5wL^4}{384EI_s} = \frac{5(0.180/12)(30 \times 12)^4}{384(29{,}000)(510)} = 0.2218 \text{ in.}$$

The total immediate deflection is $\Delta_1 + \Delta_2 = 0.6670 + 0.2218 = 0.889$ in.

For the remaining deflections, the moment of inertia of the transformed section will be needed. For a normal-weight concrete and $f_c' = 4$ ksi, $E_c = 3492$ ksi, and the modular ratio is

$$n = \frac{E_s}{E_c} = \frac{29{,}000}{3492} = 8.3 \quad \therefore \text{ use } n = 8$$

The effective width is

$$\frac{b}{n} = \frac{90}{8} = 11.25 \text{ in.}$$

Figure 9.14 shows the corresponding transformed section. The computations for the neutral axis location and the moment of inertia are summarized in Table 9.4.

FIGURE 9.14

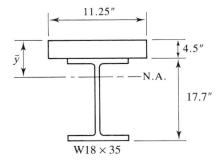

W18 × 35

TABLE 9.4

Component	A	y	Ay	$\bar{I}$	d	$\bar{I} + Ad^2$
Concrete	50.62	2.25	113.9	85.43	1.877	264
W18 × 35	10.3	13.35	137.5	510	9.223	1386
Sum	60.92		251.4			1650 in.^{4}

$$\bar{y} = \frac{\Sigma Ay}{\Sigma A} = \frac{251.4}{60.92} = 4.127\,\text{in.}$$

As recommended in the Commentary, we will reduce the transformed moment of inertia by 25% to obtain an effective moment of inertia I_{eff}.

$$I_{\text{eff}} = 0.75 I_{tr} = 0.75(1650) = 1238 \text{ in.}^4$$

The deflection caused by the partition weight is

$$\Delta_3 = \frac{5w_{\text{part}}L^4}{384EI_{\text{eff}}} = \frac{5(0.180/12)(30 \times 12)^4}{384(29,000)(1238)} = 0.09137 \text{ in.}$$

The deflection caused by the live load is

$$\Delta_4 = \frac{5w_L L^4}{384EI_{\text{eff}}} = \frac{5(1.125/12)(30 \times 12)^4}{384(29,000)(1238)} = 0.5711 \text{ in.}$$

Answer   The following is a summary of the deflections:

Immediate deflection, before composite behavior is attained:

$$\Delta_1 + \Delta_2 = 0.6670 + 0.2218 = 0.889 \text{ in.}$$

Deflection after composite behavior is attained, with partitions but no live load:

$$\Delta_1 + \Delta_3 = 0.6670 + 0.09137 = 0.758 \text{ in.}$$

Total deflection, with live load:

$$\Delta_1 + \Delta_3 + \Delta_4 = 0.6670 + 0.09137 + 0.5711 = 1.33 \text{ in.}$$

■

9.7   COMPOSITE BEAMS WITH FORMED STEEL DECK

The floor slab in many steel-framed buildings is formed on ribbed steel deck, which is left in place to become an integral part of the structure. Although there are exceptions, the ribs of the deck are usually oriented perpendicular to floor beams and

FIGURE 9.15

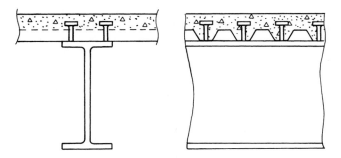

parallel to supporting girders. In Figure 9.15, the ribs are shown perpendicular to the beam. The installation of shear studs is done in the same way as without the deck; the studs are welded to the beam flange directly through the deck. The attachment of the deck to the beam can be considered to provide lateral support for the beam before the concrete has cured. The design or analysis of composite beams with formed steel deck is essentially the same as with slabs of uniform thickness, with the following exceptions.

1. The concrete in the ribs—that is, below the top of the deck—is neglected in determining section properties when the ribs are perpendicular to the beam [AISC I3.2c(2)]. When the ribs are parallel to the beam, the concrete may be included in determining section properties, and it *must* be included in computing A_c [AISC I3.2c(3)].
2. The capacity of the shear connectors will possibly be reduced.
3. Full composite behavior will not usually be possible. The reason is that the spacing of the shear connectors is limited by the spacing of the ribs, and the exact number of required connectors cannot always be used. Although partial composite design can be used without formed steel deck, it is covered here because it is almost a necessity with formed steel deck. This is not a disadvantage; in fact, it will be the most economical alternative.

Most composite beams with formed steel deck are floor beams with the deck ribs oriented perpendicular to the beam, and we limit our coverage to this case. Special requirements that apply when the ribs are parallel to the beam are presented in AISC I3.2c(3).

Reduced Capacity of Shear Connectors

Depending on the configuration of the deck, the shear strength of the studs could be reduced. This is accounted for by the factors R_g and R_p in the shear strength equation

$$Q_n = 0.5 A_{sc} \sqrt{f_c' E_c} \leq R_g R_p A_{sc} F_u$$

(AISC Equation I3-3)

Since we are limiting our coverage to formed steel deck with the ribs perpendicular to the beam, the definitions of R_g and R_p will be limited to that case:

$R_g = 1.0$ for studs welded directly to the top flange (for example, with no deck)

$\qquad = 1.0$ for one stud per rib

$\qquad = 0.85$ for two studs per rib (as in Figure 9.15)

$\qquad = 0.7$ for three or more studs per rib

$R_p = 1.0$ for studs welded directly to the top flange (for example, with no deck). If there is a *haunch,* there are other conditions to be satisfied (we do not cover haunches in this book)

$\qquad = 0.75$ for $e_{\text{mid-ht}} \geq 2$ in.

$\qquad = 0.6$ for $e_{\text{mid-ht}} < 2$ in.

$e_{\text{mid-ht}} = $ distance from mid-height of the rib to the stud, measured in the load-bearing direction (toward the point of maximum moment in a simply supported beam)

Most steel deck is manufactured with a longitudinal stiffener in the middle of the rib, so the stud must be placed on one side or the other of the stiffener. Tests have shown that placement on the side farthest from the point of maximum moment results in a higher strength. Since it is difficult to know in advance where the stud will actually be placed, it is conservative to use a value of $R_p = 0.6$. In this book, we will use $R_p = 0.6$ when formed steel deck is used.

Example 9.8

Determine the shear strength of a ½-in. × 2½-in. stud, where there are two studs per rib. The 28-day compressive strength of the concrete is $f_c' = 4$ ksi ($E_c = 3492$ ksi).

Solution

$$A_{sc} = \frac{\pi(0.5)^2}{4} = 0.1963 \text{ in.}^2$$

$$Q_n = 0.5 A_{sc} \sqrt{f_c' E_c} \leq R_g R_p A_{sc} F_u$$

$$= 0.5(0.1963)\sqrt{4(3492)} = 11.60 \text{ kips}$$

Upper limit:

$$R_g R_p A_{sc} F_u = 0.85(0.6)(0.1963)(65) = 6.51 \text{ kips}$$

Since 11.60 kips > 6.51 kips, the upper limit controls.

Answer

$Q_n = 6.51$ kips

Partial Composite Action

Partial composite action exists when there are not enough shear connectors to completely prevent slip between the concrete and steel. Neither the full strength of the concrete nor that of the steel can be developed, and the compressive force is limited to the maximum force that can be transferred across the interface between the steel and the concrete—that is, the strength of the shear connectors, ΣQ_n. Recall that C is the smallest of $A_s F_y$, $0.85 f_c' A_c$, and ΣQ_n.

With partial composite action, the plastic neutral axis (PNA) will usually fall within the steel cross section. This location will make the strength analysis somewhat more difficult than if the PNA were in the slab, but the basic principles are the same.

When an elastic analysis must be made, as when deflections are being computed, an estimate of the moment of inertia of the partially composite section must be made. A parabolic transition from I_s (for the steel shape alone) to I_{tr} (for the fully composite section) works well (Hansell et al., 1978). This approximation results in the following equation for an effective moment of inertia, presented in the Commentary to the AISC Specification:

$$I_{\text{eff}} = I_s + \sqrt{\Sigma Q_n / C_f}\,(I_{tr} - I_s) \qquad \text{(AISC Equation C-I3-3)}$$

where C_f is the compressive force in the concrete for the fully composite condition—the smaller of $A_s F_y$ and $0.85 f_c' A_c$. Since ΣQ_n is the actual compressive force for the partially composite case, the ratio $\Sigma Q_n / C_f$ is the fraction of "compositeness" that exists. If the ratio is less than 0.25, AISC Equation C-I3-3 should not be used (Hansell et al., 1978).

The steel strength will not be fully developed in a partially composite beam, so a larger shape will be required than with full composite behavior. However, fewer shear connectors will be required, and the costs of both the steel and the shear connectors (including the cost of installation) must be taken into account in any economic analysis. Whenever a fully composite beam has excess capacity, which almost always is the case, the design can be fine-tuned by eliminating some of the shear connectors, thereby creating a partially composite beam.

Miscellaneous Requirements

The following requirements are from AISC Sections I3.2c. Only those provisions not already discussed are listed.

- Maximum rib height $h_r = 3$ inches.
- Minimum average width of rib $w_r = 2$ inches, but the value of w_r used in the calculations shall not exceed the clear width at the top of the deck.
- Minimum slab thickness above the top of the deck = 2 inches.
- Maximum stud diameter = ¾ inch. This requirement for formed steel deck is in addition to the usual maximum diameter of $2.5 t_f$.
- Minimum height of stud above the top of the deck = 1½ inches.
- Minimum cover above studs = ½ inch.
- The deck must be attached to the beam flange at intervals of no more than 18 inches, either by the studs or spot welds. This is for the purpose of resisting uplift.

Slab and Deck Weight

To simplify computation of the slab weight, we use the full depth of the slab, from bottom of deck to top of slab. Although this approach overestimates the volume of concrete, it is conservative. For the unit weight of reinforced concrete, we use the weight of plain concrete plus 5 pcf. Because slabs on formed metal deck are usually lightly reinforced (sometimes welded wire mesh, rather than reinforcing bars, is used), adding 5 pcf for reinforcement may seem excessive, but the deck itself can weigh between 2 and 3 psf.

An alternative approach is to use the thickness of the slab above the deck plus half the height of the rib as the thickness of concrete in computing the weight of the slab. In practice, the combined weight of the slab and deck can usually be found in tables furnished by the deck manufacturer.

Example 9.9   Floor beams are to be used with the formed steel deck shown in Figure 9.16 and a reinforced concrete slab whose total thickness is 4.75 inches. The deck ribs are perpendicular to the beams. The span length is 30 feet, and the beams are spaced at 10 feet center-to-center. The structural steel is A992, and the concrete strength is $f_c' = 4$ ksi. The slab and deck combination weighs 50 psf. The live load is 40 psf, and there is a partition load of 10 psf. No shoring is used, and there is a construction load of 20 psf.

    a. Select a W-shape.
    b. Design the shear connectors.
    c. Check deflections. The maximum permissible total deflection is $\frac{1}{240}$ of the span length.

Solution   Compute the loads (other than the weight of the steel shape). Before the concrete cures,

    Slab wt. = 50(10) = 500 lb/ft
    Construction load = 20(10) = 200 lb/ft

After the concrete cures,

    Partition load = 10(10) = 100 lb/ft
    Live load = 40(10) = 400 lb/ft

FIGURE 9.16

LRFD Solution   a. **Beam design**: Select a trial shape based on full composite behavior.

$$w_D = \text{slab wt.} = 500 \text{ lb/ft}$$

$$w_L = \text{live load} + \text{partition load} = 400 + 100 = 500 \text{ lb/ft}$$

$$w_u = 1.2w_D + 1.6w_L = 1.2(0.500) + 1.6(0.500) = 1.400 \text{ kips/ft}$$

$$M_u = \frac{1}{8}w_u L^2 = \frac{1}{8}(1.400)(30)^2 = 157.5 \text{ ft-kips}$$

Assume that $d = 16$ in., $a/2 = 0.5$ in., and estimate the beam weight from Equation 9.4:

$$w = \frac{3.4M_u}{\phi_b F_y(d/2 + t - a/2)} = \frac{3.4(157.5 \times 12)}{0.90(50)(16/2 + 4.75 - 0.5)} = 11.7 \text{ lb/ft}$$

Try a W16 × 26. Check the flexural strength before the concrete has cured.

$$w_u = 1.2w_D + 1.6w_L = 1.2(0.500 + 0.026) + 1.6(0.200) = 0.9512 \text{ kips/ft}$$

$$M_u = (1/8)(0.9512)(30)^2 = 107 \text{ ft-kips}$$

A W16 × 26 is compact for $F_y = 50$ ksi, and since the steel deck will provide adequate lateral support, the nominal strength, M_n, is equal to the plastic moment strength, M_p. From the Z_x table,

$$\phi_b M_p = 166 \text{ ft-kips} > 107 \text{ ft-kips} \qquad \text{(OK)}$$

After the concrete has cured, the total factored load to be resisted by the composite beam, adjusted for the weight of the steel shape, is

$$w_u = 1.2w_D + 1.6w_L = 1.2(0.500 + 0.026) + 1.6(0.500) = 1.431 \text{ kips/ft}$$

$$M_u = \frac{1}{8}w_u L^2 = \frac{1}{8}(1.431)(30)^2 = 161 \text{ ft-kips}$$

The effective slab width of the composite section will be the smaller of

$$\frac{\text{Span}}{4} = \frac{30(12)}{4} = 90 \text{ in. or Beam spacing} = 10(12) = 120 \text{ in.}$$

Use $b = 90$ in. For full composite action, the compressive force, C, in the concrete is the smaller of

$$A_s F_y = 7.68(50) = 384.0 \text{ kips}$$

or

$$0.85f_c'A_c = 0.85(4)[90(4.75 - 1.5)] = 994.5 \text{ kips}$$

where only the concrete above the top of the deck has been accounted for in the second equation, as illustrated in Figure 9.17. With $C = 384.0$ kips, the depth of the compressive stress distribution in the concrete is

$$a = \frac{C}{0.85f_c'b} = \frac{384.0}{0.85(4)(90)} = 1.255 \text{ in.}$$

FIGURE 9.17

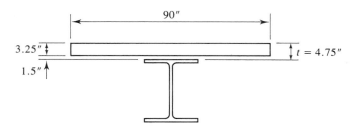

The moment arm of the internal resisting couple is

$$y = \frac{d}{2} + t - \frac{a}{2} = \frac{15.7}{2} + 4.75 - \frac{1.255}{2} = 11.97 \text{ in.}$$

and the design strength is

$$\phi_b M_n = \frac{0.90(384.0)(11.97)}{12}$$
$$= 345 \text{ ft-kips} > 161 \text{ ft-kips} \qquad \text{(OK)}$$

Check the shear:

$$V_u = \frac{w_u L}{2} = \frac{1.431(30)}{2} = 21.5 \text{ kips}$$

From the Z_x table,

$$\phi_v V_n = 106 \text{ kips} > 21.5 \text{ kips} \qquad \text{(OK)}$$

Answer   Use a W16 × 26.

b. **Shear connectors**: Because this beam has a substantial excess of moment strength, it will benefit from partial composite behavior. We must first find the shear connector requirements for full composite behavior and then reduce the number of connectors. For the fully composite beam, $C = V' = 384.0$ kips.

Maximum stud diameter $= 2.5t_f = 2.5(0.345) = 0.8625$ in.

or $\dfrac{3}{4}$ in.   (controls)

Try ¾-in. × 3-in. studs ($A_{sc} = 0.4418$ in.²), one at each section.

For $f_c' = 4$ ksi, the modulus of elasticity of the concrete is

$$E_c = w_c^{1.5}\sqrt{f_c'} = 145^{1.5}\sqrt{4} = 3492 \text{ ksi}$$

From AISC Equation I3-3, the shear strength of one connector is

$$Q_n = 0.5 A_{sc}\sqrt{f_c' E_c} \le R_g R_p A_{sc} F_u$$
$$= 0.5(0.4418)\sqrt{4(3492)} = 26.11 \text{ kips}$$
$$R_g R_p A_{sc} F_u = 1.0(0.6)(0.4418)(65)$$
$$= 17.23 \text{ kips} < 26.11 \text{ kips} \qquad \therefore \text{ use } Q_n = 17.23 \text{ kips}$$

The number of studs required between the end of the beam and midspan is

$$N_1 = \frac{V'}{Q_n} = \frac{384.0}{17.23} = 22.3$$

Use 23 for half the beam, or 46 total.

With one stud in each rib, the spacing is 6 inches, and the maximum number that can be accommodated is

$$\frac{30(12)}{6} = 60 > 46 \text{ total}$$

With one stud in every other rib, 30 will be furnished, which is fewer than what is required for full composite action. However, there is an excess of flexural strength, so partial composite action will probably be adequate.

Try 30 studs per beam, so that N_1 provided $= 30/2 = 15$.

$$\Sigma Q_n = 15(17.23) = 258.5 \text{ kips} < 384.0 \text{ kips} \qquad \therefore C = V' = 258.5 \text{ kips}$$

Because C is smaller than $A_s F_y$, part of the steel section must be in compression, and the plastic neutral axis is in the steel section.

To analyze this case, we must first determine whether the PNA is in the top flange or in the web. This can be done as follows. If the PNA were at the bottom of the flange, the entire flange would be in compression, and the resultant compressive force, as illustrated in Figure 9.18, would be

$$P_{yf} = b_f t_f F_y = 5.50(0.345)(50) = 94.88 \text{ kips}$$

The net force to be transferred at the interface between the steel and the concrete would be

$$T - C_s = T - P_{yf} = (A_s F_y - P_{yf}) - P_{yf} = 384.0 - 2(94.88)$$
$$= 194.2 \text{ kips}$$

This is less than the actual net tension force of 258.5 kips, so the top flange does not need to be in compression for its full thickness. This means that the PNA is in the flange.

FIGURE 9.18

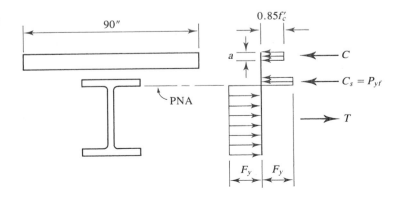

FIGURE 9.19

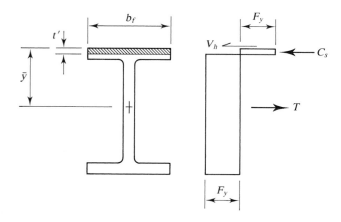

From Figure 9.19, the horizontal shear force to be transferred is

$$T - C_s = (A_s F_y - b_f t' F_y) - b_f t' F_y = V'$$
$$384.0 - 2[5.50t'(50)] = 258.5$$

Solving for the depth of compression in the flange, we obtain

$$t' = 0.2282 \text{ in.}$$

The tensile resultant force will act at the centroid of the area below the PNA. Before the moment strength can be computed, the location of this centroid must be determined. The calculations for $\bar{y}$, the distance from the top of the steel shape, are summarized in Table 9.5.

The depth of the compressive stress block in the concrete is

$$a = \frac{C}{0.85 f_c' b} = \frac{258.5}{0.85(4)(90)} = 0.8448 \text{ in.}$$

Moment arm for the concrete compressive force is

$$\bar{y} + t - \frac{a}{2} = 9.362 + 4.75 - \frac{0.8448}{2} = 13.69 \text{ in.}$$

Moment arm for the compressive force in the steel is

$$\bar{y} - \frac{t'}{2} = 9.362 - \frac{0.2282}{2} = 9.248 \text{ in.}$$

TABLE 9.5

Component	A	y	Ay
W16 × 26	7.68	15.7/2 = 7.85	60.29
Flange segment	−0.2282(5.50) = −1.255	0.2282/2 = 0.1141	−0.14
Sum	6.425		60.15

$$\bar{y} = \frac{\Sigma Ay}{\Sigma A} = \frac{60.15}{6.425} = 9.362 \text{ in.}$$

FIGURE 9.20

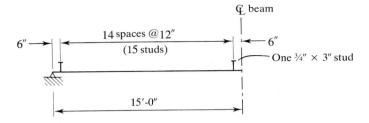

Taking moments about the tensile force and using the notation of Figure 9.18, we obtain the nominal strength:

$$M_n = C(13.69) + C_s(9.248)$$
$$= 258.5(13.69) + [0.2282(5.50)(50)](9.248) = 4119 \text{ in.-kips} = 343.3 \text{ ft-kips}$$

The design strength is

$$\phi_b M_n = 0.90(343.3) = 309 \text{ ft-kips} > 161 \text{ ft-kips} \qquad \text{(OK)}$$

The deck will be attached to the beam flange at intervals of 12 inches, so no spot welds will be needed to resist uplift.

Answer   Use the shear connectors shown in Figure 9.20.

c. **Deflections**: Before the concrete has cured,

$$w_D = w_{slab} + w_{beam} = 0.500 + 0.026 = 0.526 \text{ kips/ft}$$
$$\Delta_1 = \frac{5w_D L^4}{384EI_s} = \frac{5(0.526/12)(30 \times 12)^4}{384(29,000)(301)} = 1.098 \text{ in.}$$

The deflection caused by the construction load is

$$\Delta_2 = \frac{5w_{const} L^4}{384EI_s} = \frac{5(0.200/12)(30 \times 12)^4}{384(29,000)(301)} = 0.418 \text{ in.}$$

The total deflection before the concrete has cured is

$$\Delta_1 + \Delta_2 = 1.098 + 0.418 = 1.52 \text{ in.}$$

For deflections that occur after the concrete has cured, the moments of inertia of the transformed section will be needed. The modular ratio is

$$n = \frac{E_s}{E_c} = \frac{29,000}{3492} = 8.3 \quad \therefore \text{ use } n = 8$$

The effective width is

$$\frac{b}{n} = \frac{90}{8} = 11.25 \text{ in.}$$

Figure 9.21 shows the corresponding transformed section. The computations for the neutral axis location and the moment of inertia are summarized in Table 9.6.

FIGURE 9.21

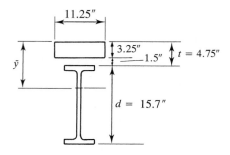

TABLE 9.6

Component	A	y	Ay	$\bar{I}$	d	$\bar{I} + Ad^2$
Concrete	36.56	1.625	59.41	32.18	1.906	165
W16 × 26	7.68	12.6	96.77	301	9.069	933
Sum	44.24		156.18			1098 in.⁴

$$\bar{y} = \frac{\Sigma Ay}{\Sigma A} = \frac{156.2}{44.24} = 3.531 \text{ in.}$$

Since partial composite action is being used, a reduced transformed moment of inertia must be used. From AISC Equation C-I3-3, this effective moment of inertia is

$$I_{\text{eff}} = I_s + \sqrt{\Sigma Q_n / C_f} (I_{tr} - I_s)$$
$$= 301 + \sqrt{258.5/384.0} (1098 - 301) = 954.9 \text{ in.}^4$$

The deflection caused by the live load is

$$\Delta_3 = \frac{5w_L L^4}{384 EI_{\text{eff}}} = \frac{5(0.400/12)(30 \times 12)^4}{384(29,000)(954.9)} = 0.2633 \text{ in.}$$

The deflection caused by the partition load is

$$\Delta_4 = \frac{5w_{\text{part}} L^4}{384 EI_{\text{eff}}} = \frac{5(0.100/12)(30 \times 12)^4}{384(29,000)(954.9)} = 0.0658 \text{ in.}$$

The total deflection is

$$\Delta_1 + \Delta_3 + \Delta_4 = 1.098 + 0.2633 + 0.0658 = 1.43 \text{ in.}$$

and

$$\frac{L}{240} = \frac{30(12)}{240} = 1.50 \text{ in.} > 1.43 \text{ in.} \qquad \text{(OK)}$$

Answer The deflection is satisfactory.

Note that although the strength of this composite beam is far larger than what is needed, the deflection is very close to the limit.

ASD Solution   a. **Beam design**: Select a trial shape based on full composite behavior.

$$w_D = \text{slab wt.} = 500 \text{ lb/ft}$$

$$w_L = \text{live load} + \text{partition load} = 400 + 100 = 500 \text{ lb/ft}$$

$$w_a = w_D + w_L = 0.500 + 0.500 = 1.000 \text{ kips/ft}$$

$$M_a = \frac{1}{8} w_a L^2 = \frac{1}{8}(1.000)(30)^2 = 112.5 \text{ ft-kips}$$

Assume that $d = 16$ in., $a/2 = 0.5$ in., and estimate the beam weight from Equation 9.6:

$$w = \frac{3.4\Omega_b M_a}{F_y\left(\frac{d}{2} + t - \frac{a}{2}\right)} = \frac{3.4(1.67)(112.5 \times 12)}{50\left(\frac{16}{2} + 4.75 - 0.5\right)} = 12.5 \text{ lb/ft}$$

Try a W16 × 26. Check the flexural strength before the concrete has cured.

$$w_a = w_{\text{slab}} + w_{\text{beam}} + w_{\text{const}} = 0.5000 + 0.026 + 0.200 = 0.7260 \text{ lb/ft}$$

$$M_a = \frac{1}{8} w_a L^2 = \frac{1}{8}(0.7260)(30)^2 = 81.7 \text{ ft-kips}$$

From the Z_x table,

$$\frac{M_n}{\Omega_b} = \frac{M_p}{\Omega_b} = 110 \text{ ft-kips} > 81.7 \text{ ft-kips} \qquad \text{(OK)}$$

After the concrete cures and composite behavior has been achieved, the load and moment are

$$w_a = w_{\text{slab}} + w_{\text{beam}} + w_L = 0.500 + 0.026 + 0.500 = 1.026 \text{ kips/ft}$$

$$M_a = \frac{1}{8}(1.026)(30)^2 = 115 \text{ ft-kips}$$

The effective slab width of the composite section is the smaller of

$$\frac{\text{Span}}{4} = \frac{30 \times 12}{4} = 90 \text{ in.} \qquad \text{or} \qquad \text{Beam spacing} = 10 \times 12 = 120 \text{ in.}$$

Use $b = 90$ in. For full composite behavior, the compressive force in the concrete at ultimate (equal to the horizontal shear at the interface between the concrete and steel) will be the smaller of

$$A_s F_y = 7.68(50) = 384.0 \text{ kips}$$

or

$$0.85 f'_c A_c = 0.85(4)[90(4.75 - 1.5)] = 994.5 \text{ kips}$$

Only the concrete above the top of the deck has been accounted for. With $C = 384.0$ kips, the depth of the compressive stress distribution in the concrete is

$$a = \frac{C}{0.85 f_c' b} = \frac{384.0}{0.85(4)(90)} = 1.255 \text{ in.}$$

$$y = \frac{d}{2} + t - \frac{a}{2} = \frac{15.7}{2} + 4.75 - \frac{1.255}{2} = 11.97 \text{ in.}$$

and the allowable flexural strength is

$$\frac{M_n}{\Omega_b} = \frac{Cy}{\Omega_b} = \frac{384.0(11.97)}{1.67} = 2752 \text{ in.-kips} = 229 \text{ ft-kips} > 115 \text{ ft-kips} \quad \text{(OK)}$$

Check shear:

$$V_a = \frac{w_a L}{2} = \frac{1.026(30)}{2} = 15.4 \text{ kips}$$

From the Z_x table,

$$\frac{V_n}{\Omega_v} = 70.5 \text{ kips} > 15.4 \text{ kips} \quad \text{(OK)}$$

Answer   Use a W16 × 26.

b. **Shear connectors.** The design of shear connectors is the same for both LRFD and ASD. From the LRFD solution, for one ¾-inch × 3-inch stud every other rib,

$$M_n = 343.3 \text{ in.-kips}$$

The allowable moment strength is therefore

$$\frac{M_n}{\Omega_b} = \frac{343.3}{1.67} = 206 \text{ ft-kips} > 115 \text{ ft-kips} \quad \text{(OK)}$$

Answer   Use the shear connectors shown in Figure 9.20.

c. **Deflections**: The computation of deflections is the same for LRFD and ASD. See the LRFD solution.

9.8   TABLES FOR COMPOSITE BEAM ANALYSIS AND DESIGN

When the plastic neutral axis is within the steel section, computation of the flexural strength can be laborious. Formulas to expedite this computation have been developed (Hansell et al., 1978), but the tables presented in Part 3 of the *Manual* are more convenient. Three tables are presented: strengths of various combinations of shapes and slabs; tables of "lower-bound" moments of inertia; and a table of shear stud strength Q_n for various combinations of stud size, concrete strength, and deck geometry.

FIGURE 9.22

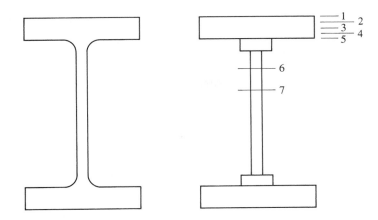

The available flexural strength is given in Table 3-19 of the *Manual* for seven specific locations of the plastic neutral axis, as shown in Figure 9.22: top of the flange (TFL), bottom of the top flange (BFL), three equally spaced levels within the top flange, and two locations in the web. Strengths given for PNA location 1 (TFL) are also valid for PNA locations within the slab.

The lowest PNA location, level 7, corresponds to the recommended lower limit of $\Sigma Q_n = 0.25 A_s F_y$ (Hansell et al., 1978). PNA location 6 corresponds to a ΣQ_n midway between ΣQ_n for level 5 and ΣQ_n for level 7. For each combination of shape and slab, two strengths are given in Table 3-19. For LRFD, the design strength $\phi_b M_n$ is tabulated. For ASD, the allowable strength M_n/Ω_b is tabulated.

To use the tables for analysis of a composite beam, first find the portion of the table corresponding to the steel shape and proceed as follows:

1. **Select ΣQ_n.** This is the *Manual*'s notation for the compressive force C, which is the smallest of $A_s F_y$, $0.85 f_c' A_c$, and the total shear connector strength (which we have been calling ΣQ_n).
2. **Select $Y2$.** The distance from the top of the steel shape to the resultant compressive force in the concrete is computed as

$$Y2 = t - \frac{a}{2}$$

This dimension is illustrated in Figure 9.23.

3. Read the available strength, interpolating if necessary.

FIGURE 9.23

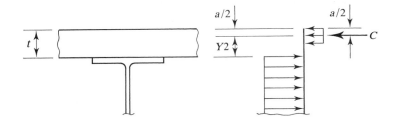

For design, the tables can be entered with the required strength, and a combination of steel shape and ΣQ_n can be selected. A value of $Y2$ will be needed, so the depth of the concrete compressive stress distribution will need to be assumed and then revised after an iteration.

The tables also give values of $\phi_b M_p$ and M_p/Ω_b for the steel shape, which may be needed for checking unshored beams during the curing of the concrete; and $Y1$, the distance from the top of the steel to the plastic neutral axis.

Example 9.10

Compute the available strength of the composite beam in Examples 9.1 and 9.2. Use the tables in Part 3 of the *Manual*.

Solution

From Example 9.1, the composite beam consists of a W16 × 36 of A992 steel, a slab with a thickness $t = 5$ inches, and an effective width $b = 87$ inches. The 28-day compressive strength of the concrete is $f_c' = 4$ ksi.

The compressive force in the concrete is the smaller of

$$A_s F_y = 10.6(50) = 530 \text{ kips}$$

or

$$0.85 f_c' A_c = 0.85(4)(5 \times 87) = 1487 \text{ kips}$$

Use $C = 530$ kips. The depth of the compressive stress block is

$$a = \frac{C}{0.85 f_c' b} = \frac{530}{0.85(4)(87)} = 1.792 \text{ in.}$$

The distance from the top of the steel to the compressive force C is

$$Y2 = t - \frac{a}{2} = 5 - \frac{1.792}{2} = 4.104 \text{ in.}$$

LRFD Solution

Enter the tables with $\Sigma Q_n = 530$ kips and $Y2 = 4.104$ inches. Since 530 kips is larger than the value of ΣQ_n for PNA location TFL, the plastic neutral axis is in the slab, and PNA location TFL can be used. By interpolation,

$$\phi_b M_n = 477 \text{ ft-kips}$$

which compares favorably with the results of Example 9.2 but involves about the same amount of effort. The value of the tables becomes obvious when the plastic neutral axis is within the steel shape.

Answer

Design strength = 477 ft-kips

ASD Solution

Enter the tables with $\Sigma Q_n = 530$ kips and $Y2 = 4.104$. By interpolation,

$$\frac{M_n}{\Omega_b} = 318 \text{ ft-kips}$$

which compares favorably with Example 9.2.

Answer

Allowable strength = 318 ft-kips

Table 3-20 gives values of a lower-bound moment of inertia, denoted by I_{LB}, for the same beams that are in the design strength tables. This moment of inertia is a conservative estimate of the moment of inertia of the transformed section. The chief simplifying assumption made in the construction of these tables is that only the area of concrete used in resisting the moment is effective in computing the moment of inertia. The force in the concrete is $C = \Sigma Q_n$, and the corresponding area in the transformed section is

$$A_c = \frac{\Sigma Q_n}{\text{stress in transformed area}} = \frac{\Sigma Q_n}{F_y}$$

As a further simplification, the moment of inertia of the concrete about its own centroidal axis is neglected. To illustrate this concept, one of the tabulated values will be derived in Example 9.11.

Example 9.11

A design has resulted in a W16 × 31 with $F_y = 50$ ksi, $\Sigma Q_n = 335$ kips (PNA location 3), and $Y2 = 4$ inches. Compute the lower-bound moment of inertia.

Solution

The area of concrete to be used is

$$A_c = \frac{\Sigma Q_n}{F_y} = \frac{335}{50} = 6.700 \text{ in.}^2$$

The corresponding transformed section is shown in Figure 9.24, and the computations are summarized in Table 9.7. To locate the centroid, take moments about an axis at the bottom of the steel shape.

FIGURE 9.24

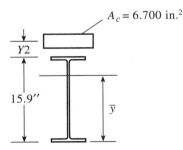

$A_c = 6.700$ in.^{2}

Y2

15.9''

$\overline{y}$

TABLE 9.7

Component	A	y	Ay	$\overline{I}$	d	$\overline{I} + Ad^2$
Concrete	6.700	19.9	133.3	—	6.89	318.1
W16 × 31	9.13	7.95	72.58	375	5.06	608.8
Sum	15.83		205.9			926.9 in.^{4}

$$\overline{y} = \frac{\Sigma Ay}{\Sigma A} = \frac{205.9}{15.83} = 13.01 \text{ in.}$$

The moment of inertia from the lower-bound moment of inertia tables is $I_{LB} = 926$ in.[4] Accounting for a slight difference due to rounding, this verifies the calculated result of 927 in.[4]

Answer $I_{LB} = 927$ in.[4]   ■

Despite the conservative assumptions used in deriving the lower-bound moment of inertia, deflections calculated with a lower-bound moment of inertia usually match experimental results better than deflections calculated with an elastic moment of inertia. This is because use of the elastic moment of inertia underestimates the deflection, and the lower-bound moment of inertia is 20 to 30% lower than the elastic one (Viest et al., 1997). The error in the elastic moment of inertia is due primarily to the Specification rules for effective slab width, which result in an overestimation of the width and a correspondingly larger moment of inertia (Viest et al., 1997).

Example 9.12   Use the Tables in Part 3 of the *Manual* and select a W-shape of A992 steel and shear connectors for the following conditions: The beam spacing is 5 feet 6 inches, and the span length is 30 feet. The slab has a total thickness of $4\frac{1}{2}$ inches and is supported by a formed steel deck, the cross section of which is shown in Figure 9.25. The 28-day compressive strength of the concrete is $f_c' = 4$ ksi. The loads consist of a 20-psf construction load, a 20-psf partition load, a ceiling weighing 5 psf, and a live load of 150 psf. The maximum live-load deflection cannot exceed $L/240$.

Solution   Loads to be supported before the concrete cures:

$$\text{Slab wt.} = \frac{4.5}{12}(150) = 56.25 \text{ psf (conservatively)}$$

$$w_{\text{slab}} = 56.25(5.5) = 309.4 \text{ lb/ft}$$

$$\text{Construction load} = 20(5.5) = 110.0 \text{ lb/ft}$$

We will account for the beam weight later.

After the concrete cures,

$$\text{Partition load} = 20(5.5) = 110.0 \text{ lb/ft}$$

$$\text{Live load: } 150(5.5) = 825.0 \text{ lb/ft}$$

$$\text{Ceiling: } 5(5.5) = 27.5 \text{ lb/ft}$$

FIGURE 9.25

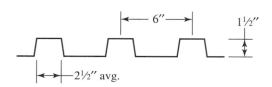

LRFD Solution   Loads to be carried by the composite section:

$$w_D = w_{slab} + w_{ceil} = 309.4 + 27.5 = 336.9 \text{ lb/ft}$$

$$w_L = 110 + 825 = 935 \text{ lb/ft}$$

$$w_u = 1.2w_D + 1.6w_L = 1.2(0.3369) + 1.6(0.935) = 1.900 \text{ kips/ft}$$

$$M_u = \frac{1}{8}w_u L^2 = \frac{1}{8}(1.900)(30)^2 = 214 \text{ ft-kips}$$

Assume $a = 2$ in.:

$$Y2 = t - \frac{a}{2} = 4.5 - \frac{2}{2} = 3.5 \text{ in.}$$

From Table 3-19 in the *Manual*, any combination of steel shape, ΣQ_n, and $Y2$ that furnishes a design strength of more than 214 ft-kips will be an acceptable trial beam. Two possibilities are summarized in Table 9.8.

The W14 × 22 is the lighter shape, but because ΣQ_n is larger, it will require more shear connectors. For this reason, **try the W14 × 26.**

Compute the design strength:

b = beam spacing or span length ÷ 4
  = 5.5(12) = 66 in.   or   30(12)/4 = 90 in.

Since 66 in. < 90 in., $b = 66$ in. Next, refine the value for $Y2$:

From $C = \Sigma Q_n$,

$$a = \frac{\Sigma Q_n}{0.85 f'_c b} = \frac{135}{0.85(4)(66)} = 0.6016 \text{ in.}$$

$$Y2 = t - \frac{a}{2} = 4.5 - \frac{0.6016}{2} = 4.199 \text{ in.}$$

From Table 3-19 in the *Manual*, by interpolation,

$$\phi_b M_n = 237 \text{ ft-kips} > 214 \text{ ft-kips} \qquad \text{(OK)}$$

Adjust M_u for the beam weight:

$$w_u = 1.900 + 1.2(0.026) = 1.931 \text{ kips/ft}$$

$$M_u = \frac{1}{8}w_u L^2 = \frac{1}{8}(1.931)(30)^2 = 217 \text{ ft-kips} < 237 \text{ ft-kips} \qquad \text{(OK)}$$

Check shear: From the Z_x table, $\phi_v V_n = 106$ kips.

$$V_u = \frac{w_u L}{2} = \frac{1.931(30)}{2} = 29.0 \text{ kips} < 106 \text{ kips} \qquad \text{(OK)}$$

TABLE 9.8

Shape	PNA Location	ΣQ_n	$\phi_b M_n$
W14 × 26	6	135	230
W14 × 22	3	241	230

Before the concrete cures,

$$w_D = 309.4 + 26 = 335.4 \text{ lb/ft}$$
$$w_L = 110.0 \text{ lb/ft}$$
$$w_u = 1.2w_D + 1.6w_L = 1.2(0.3354) + 1.6(0.110) = 0.5785 \text{ kips/ft}$$
$$M_u = \frac{1}{8}w_u L^2 = \frac{1}{8}(0.5785)(30)^2 = 65.1 \text{ ft-kips}$$
$$\phi_b M_n = \phi_b M_p = 151 \text{ ft-kips} > 65.1 \text{ ft-kips} \qquad \text{(OK)}$$

The live-load deflection will be checked after the shear connectors have been selected, because the plastic neutral axis location may change, and this would affect the lower-bound moment of inertia.

The maximum stud diameter is $2.5t_f = 2.5(0.420) = 1.05$ in., but the maximum diameter with formed steel deck $= \frac{3}{4}$ in.

The minimum length is $4d = 4(\frac{3}{4}) = 3$ in. or $h_r + 1.5 = 1.5 + 1.5 = 3$ in.

For a length of 3 in., the cover above the top of the stud is

$$4.5 - 3 = 1.5 \text{ in.} > 0.5 \text{ in.} \qquad \text{(OK)}$$

Try $\frac{3}{4} \times 3$ studs.

$$A_{sc} = \pi(0.75)^2 / 4 = 0.4418 \text{ in.}^2$$
$$Q_n = 0.5A_{sc}\sqrt{f_c' E_c} \le R_g R_p A_{sc} F_u$$

For one stud per rib, $R_g = 1.0$. In this book, we always conservatively use $R_p = 0.6$ with steel deck.

$$Q_n = 0.5(0.4418)\sqrt{4(3492)} = 26.11 \text{ kips}$$
$$R_g R_p A_{sc} F_u = 1.0(0.6)(0.4418)(65) = 17.23 \text{ kips} < 26.11 \text{ kips}$$
$$\therefore \text{ use } Q_n = 17.23 \text{ kips}$$

(Alternatively, the stud shear strength can be found from Table 3-21 in the *Manual*. We consider the stud to be a "weak" stud, in keeping with our practice of always using $R_p = 0.6$ with formed steel deck. The value of Q_n is found to be 17.2 kips.)

Number of studs:

$$N_1 = \frac{V'}{Q_n} = \frac{135}{17.23} = 7.84 \ \therefore \text{ use } 8$$

Total no. required $= 2(8) = 16$

The approximate spacing is $\dfrac{30(12)}{16} = 22.5$ in.

Minimum longitudinal spacing $= 6d = 6\left(\dfrac{3}{4}\right) = 4.5$ in.

Maximum longitudinal spacing $= 8t \le 36$ in.

$8t = 8(4.5) = 36$ in.

Try one stud every 3rd rib: Spacing $= 3(6) = 18$ in. < 36 in.   (OK)

$$\text{Total number} = \frac{30(12)}{18} = 20$$

For $N_1 = 20/2 = 10$, $\Sigma Q_n = 10(17.23) = 172.3$ kips. From $C = \Sigma Q_n$,

$$a = \frac{\Sigma Q_n}{0.85 f_c' b} = \frac{172.3}{0.85(4)(66)} = 0.7678$$

$$Y2 = t - \frac{a}{2} = 4.5 - \frac{0.7678}{2} = 4.116 \text{ in.}$$

From Table 3-19 in the *Manual*, for $\Sigma Q_n = 172.3$ kips and $Y2 = 4.116$ in.,

$$\phi_b M_n = 250 \text{ ft-kips (by interpolation)} > 217 \text{ ft-kips}$$ (OK)

Check the live load deflection: For $\Sigma Q_n = 172.3$ kips and $Y2 = 4.116$ in., the lower-bound moment of inertia from Table 3-20 in the *Manual* is

$$I_{LB} = 538 \text{ in.}^4 \quad \text{(by interpolation)}$$

and the live load deflection is

$$\Delta_L = \frac{5 w_L L^4}{384 E I_{LB}} = \frac{5(0.935/12)(30 \times 12)^4}{384(29,000)538} = 1.09 \text{ in.}$$

The maximum permissible live load deflection is

$$\frac{L}{240} = \frac{30(12)}{240} = 1.5 \text{ in.} > 1.09 \text{ in.}$$ (OK)

Answer   Use a W14 × 26 and 20 studs, ¾ × 3, one every third rib.

ASD Solution   Loads to be carried by the composite section:

$$w_D = w_{\text{slab}} + w_{\text{ceil}} = 309.4 + 27.5 = 336.9 \text{ lb/ft}$$

$$w_L = 110 + 825 = 935 \text{ lb/ft}$$

$$w_a = w_D + w_L = 0.3369 + 0.935 = 1.272 \text{ kips/ft}$$

$$M_a = \frac{1}{8} w_a L^2 = \frac{1}{8}(1.272)(30)^2 = 143 \text{ ft-kips}$$

Assume $a = 2$ in.:

$$Y2 = t - \frac{a}{2} = 4.5 - \frac{2}{2} = 3.5 \text{ in.}$$

From Table 3-19 in the *Manual*, any combination of steel shape, ΣQ_n, and $Y2$ that furnishes an allowable strength of more than 143 ft-kips will be an acceptable trial beam. Two possibilities are summarized in Table 9.9.

TABLE 9.9

Shape	PNA Location	ΣQ_n	M_n/Ω_b
W14 × 26	6	135	153
W14 × 22	3	241	153

The W14 × 22 is the lighter shape, but because ΣQ_n is larger, it will require more shear connectors. For this reason, **try the W14 × 26**. Compute the allowable strength:

b = beam spacing or span length ÷ 4

$\quad = 5.5(12) = 66$ in. $\qquad$ or $\qquad 30(12)/4 = 90$ in.

Since 66 in. <90 in., $b = 66$ in. Next, refine the value of $Y2$. From $C = \Sigma Q_n$,

$$a = \frac{\Sigma Q_n}{0.85 f'_c b} = \frac{135}{0.85(4)(66)} = 0.6016 \text{ in.}$$

$$Y2 = t - \frac{a}{2} = 4.5 - \frac{0.6016}{2} = 4.199 \text{ in.}$$

From Table 3-19 in the *Manual*, by interpolation,

$$\frac{M_n}{\Omega_b} = 158 \text{ ft-kips} > 143 \text{ ft-kips} \qquad \text{(OK)}$$

Adjust M_a for the beam weight:

$$w_a = 1.272 + 0.026 = 1.298 \text{ kips/ft}$$

$$M_a = \frac{1}{8} w_a L^2 = \frac{1}{8}(1.298)(30)^2 = 146 \text{ ft-kips} < 158 \text{ ft-kips} \qquad \text{(OK)}$$

Check shear: From the Z_x table, $V_n/\Omega_v = 70.9$ kips

$$V_a = \frac{w_a L}{2} = \frac{1.298(30)}{2} = 19.5 \text{ kips} < 70.9 \text{ kips} \qquad \text{(OK)}$$

Before the concrete cures,

$$w_D = 309.4 + 26 = 335.4 \text{ lb/ft}$$
$$w_L = 110.0 \text{ lb/ft}$$
$$w_a = w_D + w_L = 0.3354 + 0.1100 = 0.4454 \text{ kips/ft}$$
$$M_a = \frac{1}{8} w_a L^2 = \frac{1}{8}(0.4454)(30)^2 = 50.1 \text{ ft-kips}$$

From Table 3-19,

$$\frac{M_n}{\Omega_b} = \frac{M_p}{\Omega_b} = 100 \text{ ft-kips} > 50.1 \text{ ft-kips} \qquad \text{(OK)}$$

For the design of the shear connectors and the deflection check, see the LRFD solution. One ¾ × 3 stud will be used every third rib. From Table 3-19, the allowable strength for this composite beam is

$$\frac{M_n}{\Omega_b} = 166 \text{ ft-kips} > 146 \text{ ft-kips} \qquad \text{(OK)}$$

Answer   Use a W14 × 26 and 20 studs, ¾ × 3, one every third rib.

As Example 9.12 illustrates, the tables greatly simplify the design of a partially composite beam, in which the PNA lies within the steel shape.

9.9 CONTINUOUS BEAMS

In a simply supported beam, the point of zero moment is at the support. The number of connectors required between each support and the point of maximum positive moment will be half the total number required. In a continuous beam, the points of inflection are also points of zero moment and, in general, $2N_1$ connectors will be required for each span. Figure 9.26a shows a typical continuous beam and the regions

FIGURE 9.26

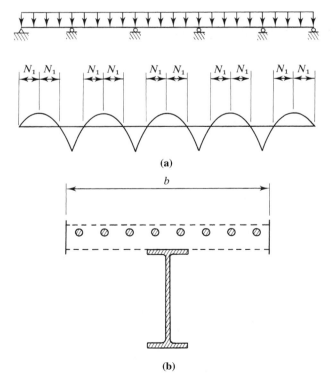

(a)

(b)

in which shear connectors would be required. In the negative moment zones, the concrete slab will be in tension and therefore ineffective. In these regions there will be no composite behavior in the sense that we have considered thus far. The only type of composite behavior possible is that between the structural steel beam and the longitudinal reinforcing steel in the slab. The corresponding composite cross section is shown in Figure 9.26b. If this concept is used, a sufficient number of shear connectors must be provided to achieve a degree of continuity between the steel shape and the reinforcing.

The AISC Specification in Section I3.2b offers two alternatives for negative moment.

1. Rely on the strength of the steel shape only.
2. Use a composite section that consists of the steel shape and the reinforcing steel, subject to the following conditions:
   a. The steel shape must be compact with adequate lateral support.
   b. Shear connectors must be present in the negative moment region (between the point of zero moment and the point of maximum negative moment).
   c. The reinforcement within the effective width must be adequately developed (anchored).
   The strength of this composite section should be based on a plastic stress distribution, in which the horizontal force to be transferred between the steel shape and the reinforcing steel should be taken as the smaller of $A_r F_{yr}$. and ΣQ_n, where
   A_r = area of reinforcing steel within the effective width of the slab
   F_{yr} = yield stress of the reinforcing steel

If this composite section is used, the resistance factor ϕ_b for LRFD is 0.90, and the safety factor Ω_b for ASD is 1.67.

The additional strength gained from considering composite action for negative moment is relatively small. If the steel shape alone is relied on to resist the negative moment, cover plates are sometimes added to the beam flanges to increase the moment strength.

9.10   COMPOSITE COLUMNS

Composite columns can take one of two forms: a pipe or HSS filled with plain concrete or a rolled steel shape encased in concrete with both vertical reinforcement and transverse reinforcement in the form of ties or spirals, as in a reinforced concrete column. Figure 9.27 illustrates these two types.

The current approach to the design and analysis of composite columns is based on the work of Leon and Aho (2002). Although the behaviors of encased and filled composite columns are based on the same general principles, there are enough differences, especially with regard to details, that the AISC Specification treats them separately. We do the same in this book.

FIGURE 9.27

Strength of Encased Composite Columns

The AISC Specification covers encased composite columns in Section I2.1. If buckling were not an issue, the member strength could be taken as the summation of the axial compressive strengths of the component materials:

$$P_o = A_s F_y + A_{sr} F_{yr} + 0.85 A_c f'_c$$ (AISC Equation I2-4)

where

A_s = cross-sectional area of the steel shape
F_y = yield stress of the steel shape
A_{sr} = cross-sectional area of the longitudinal steel reinforcing bars
F_{yr} = yield stress of the reinforcing steel
A_c = cross-sectional area of concrete

The strength P_o is termed the "squash" load.

Modern reinforcing bars are *deformed*—that is, the surface of the bar has protrusions that help the steel grip or bond with the concrete. The bar cross-sectional area to be used in computations is a nominal area equal to the area of a plain bar that has the same weight per foot of length as the deformed bar. Table 9.10 gives nominal diameters and areas for standard bar sizes as defined in ASTM (2005) and ACI (2005).

Because of slenderness effects, the strength predicted by AISC Equation I2-4 cannot be achieved. To account for slenderness, the relationship between P_o and P_e is used, where P_e is the Euler buckling load and is defined as

$$P_e = \frac{\pi^2 (EI)_{\text{eff}}}{(KL)^2}$$ (AISC Equation I2-5)

TABLE 9.10

Bar Number	Diameter (in.)	Area (in.²)
3	0.375	0.11
4	0.500	0.20
5	0.625	0.31
6	0.750	0.44
7	0.875	0.60
8	1.000	0.79
9	1.128	1.00
10	1.270	1.27
11	1.410	1.56
14	1.693	2.25
18	2.257	4.00

where $(EI)_{eff}$ is the effective flexural rigidity of the composite section and is given by

$$(EI)_{eff} = E_s I_s + 0.5 E_s I_{sr} + C_1 E_c I_c \qquad \text{(AISC Equation I2-6)}$$

where

I_s = moment of inertia of the steel shape about the axis of buckling

I_{sr} = moment of inertia of the longitudinal reinforcing bars about the axis of buckling

$$C_1 = 0.1 + 2\left(\frac{A_s}{A_c + A_s}\right) \le 0.3 \qquad \text{(AISC Equation I2-7)}$$

I_c = moment of inertia of the concrete section about the axis of buckling

The nominal strength equations are similar to those for noncomposite members. When $P_e \ge 0.44 P_o$,

$$P_n = P_o\left[0.658^{\left(\frac{P_o}{P_e}\right)}\right] \qquad \text{(AISC Equation I2-2)}$$

When $P_e < 0.44 P_o$,

$$P_n = 0.877 P_e \qquad \text{(AISC Equation I2-3)}$$

Note that if the reinforced concrete terms in AISC Equations I2-4 and I2-6 are omitted,

$$P_o = A_s F_y$$

$$(EI)_{eff} = E_s I_s$$

$$P_e = \frac{\pi^2 E_s I_s}{(KL)^2}$$

and

$$\frac{P_o}{P_e} = \frac{A_s F_y}{\pi^2 E_s I_s/(KL)^2} = \frac{F_y}{\pi^2 E_s A_s r^2/A_s (KL)^2} = \frac{F_y}{\pi^2 E_s/(KL/r)^2} = \frac{F_y}{F_e}$$

where F_e is defined in AISC Chapter E as the elastic buckling stress.

AISC Equation I2-2 then becomes

$$P_n = A_s\left[0.658^{\left(\frac{F_y}{F_e}\right)}\right] F_y \qquad (9.7)$$

and AISC Equation I2-3 becomes

$$P_n = A_s(0.877) F_e \qquad (9.8)$$

Equations 9.7 and 9.8 give the same strength as the equations in AISC Chapter E for noncomposite compression members.

For LRFD, the design strength is $\phi_c P_n$, where $\phi_c = 0.75$. For ASD, the allowable strength is P_n/Ω_c, where $\Omega_c = 2.00$.

Example 9.13
A composite compression member consists of a W12 × 136 encased in a 20-inch × 22-inch concrete column as shown in Figure 9.28. Four #10 bars are used for longitudinal reinforcement, and #3 ties spaced 13 inches center-to-center provide the lateral reinforcement. Assume a concrete cover of 2.5 inches to the center of the longitudinal reinforcement. The steel yield stress is $F_y = 50$ ksi, and Grade 60 reinforcing bars are used. The concrete strength is $f_c' = 5$ ksi. Compute the available strength for an effective length of 16 feet with respect to both axes.

FIGURE 9.28

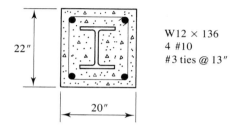

22″

20″

W12 × 136
4 #10
#3 ties @ 13″

Solution
Values needed for the AISC strength equations are as follows. For the W12 × 136, $A_s = 39.9$ in.^{2} and $I_s = I_y = 398$ in.^{4} For the longitudinal reinforcement,

$$A_{sr} = 4(1.27) = 5.08 \text{ in.}^2$$

$$I_{sr} = \sum Ad^2 = 4 \times 1.27 \left(\frac{20 - 2 \times 2.5}{2} \right)^2 = 285.8 \text{ in.}^2 \text{ (about the weak axis)}$$

For the concrete,

$$A_c = \text{net area of concrete} = 20(22) - A_s - A_{sr} = 440 - 39.9 - 5.08 = 395.0 \text{ in.}^2$$

$$E_c = w^{1.5} \sqrt{f_c'} = (145)^{1.5} \sqrt{5} = 3904 \text{ ksi}$$

$$I_c = \frac{22(20)^3}{12} = 14,670 \text{ in.}^4 \text{ (about the weak axis)}$$

From AISC Equation I2-4,

$$P_o = A_s F_y + A_{sr} F_{yr} + 0.85 A_c f_c' = 39.9(50) + 5.08(60) + 0.85(395.0)(5) = 3979 \text{ kips}$$

From AISC Equation I2-7,

$$C_1 = 0.1 + 2 \left(\frac{A_s}{A_c + A_s} \right) \le 0.3$$

$$= 0.1 + 2 \left(\frac{39.9}{395.0 + 39.9} \right) = 0.2835 < 0.3$$

From AISC Equation I2-6,

$$(EI)_{eff} = E_s I_s + 0.5 E_s I_{sr} + C_1 E_c I_c$$
$$= 29{,}000(398) + 0.5(29{,}000)(285.8) + 0.2835(3904)(14{,}670)$$
$$= 3.192 \times 10^7 \text{ kip-in.}^2$$

From AISC Equation I2.5,

$$P_e = \frac{\pi^2 (EI)_{eff}}{(KL)^2} = \frac{\pi^2 (3.192 \times 10^7)}{(16 \times 12)^2} = 8546 \text{ kips}$$

Then

$$0.44 P_o = 0.44(3979) = 1751 \text{ kips}$$
$$P_e = 8546 \text{ kips} > 1751 \text{ kips} \quad \therefore \text{ use AISC Equation I2-2}$$
$$P_n = P_o \left[0.658^{\left(\frac{P_o}{P_e} \right)} \right] = 3979 \left[0.658^{\left(\frac{3979}{8546} \right)} \right] = 3274 \text{ kips}$$

Answer For LRFD, the design strength is $\phi_c P_n = 0.75(3274) = 2456$ kips.
For ASD, the allowable strength is $P_n / \Omega_c = 3274/2.00 = 1637$ kips. ∎

In addition to strength requirements, the requirements of AISC I2.1a "Limitations" and I2.1f "Detailing Requirements" must be observed. These limitations and detailing requirements for encased composite columns can be summarized as follows:

1. The cross-sectional area of the steel shape must make up at least 1% of the total area.
2. At least four longitudinal bars must be used.
3. The longitudinal bars must have an area of at least 0.4% of the total area.
4. The transverse reinforcement (ties or spirals) must have an area of at least 0.009 in.^{2} per inch of tie spacing.
5. Spacing of the transverse reinforcement must be the smallest of 16 longitudinal bar diameters, 48 tie diameters, or 0.5 times the smaller dimension of the composite column.
6. There must be at least 1.5 inches of clear concrete cover to the reinforcing steel.

To ensure that the steel and concrete act as a unit, there must be a mechanism for transferring the load from one component to the other. Although this can be done to a certain extent by chemical bond between the concrete and steel, shear connectors are usually required. The design of the shear connectors depends on whether the axial load is applied directly to the encased steel shape or to the concrete. If the load is applied directly to the concrete in bearing, the bearing strength of the concrete must also be investigated. Load transfer requirements are covered in AISC I2.1e "Load Transfer" I2.1f "Detailing Requirements", and I2.1g "Strength of Stud Shear Connectors".

Strength of Filled Composite Columns

The AISC provisions for this type of composite compression member (AISC I2.2) are similar to those for encased shapes. The equations for P_o and $(EI)_{eff}$ are slightly different, but the equations for P_e and P_n are the same.

$$P_o = A_s F_y + A_{sr} F_{yr} + C_2 A_c f'_c \qquad \text{(AISC Equation I2-13)}$$

where

$C_2 = 0.85$ for rectangular sections

$\quad = 0.95$ for circular sections (this increase is to account for the beneficial effect of a circular section on the confinement of the concrete)

$$(EI)_{eff} = E_s I_s + E_s I_{sr} + C_3 E_c I_c \qquad \text{(AISC Equation I2-14)}$$

where

$$C_3 = 0.6 + 2\left(\frac{A_s}{A_c + A_s}\right) \le 0.9 \qquad \text{(AISC Equation I2-15)}$$

Example 9.14

A Pipe 8 Std with an effective length of 13 feet is filled with normal-weight concrete and used as a column. No longitudinal reinforcement is used. The concrete strength is $f'_c = 4$ ksi. Compute the available compressive strength.

Solution

The following dimensions and properties from Part 1 of the *Manual* will be needed: $A_s = 7.85$ in.^{2}, outside diameter = 8.63 in., inside diameter = 7.98 in., design wall thickness = 0.300 in., and $I_s = 68.1$ in.^{4} The following values will also be required.

$$A_c = \frac{\pi d_{inside}^4}{4} = \frac{\pi (7.98)^2}{4} = 50.01 \text{ in.}^2$$

$$E_c = w^{1.5}\sqrt{f'_c} = (145)^{1.5}\sqrt{4} = 3492 \text{ ksi}$$

$$I_c = \frac{\pi d_{inside}^4}{64} = \frac{\pi (7.98)^4}{64} = 199.1 \text{ in.}^4$$

From AISC Equation I2-13,

$$P_o = A_s F_y + A_{sr} F_{yr} + C_2 A_c f'_c = 7.85(35) + 0 + 0.95(50.01)(4) = 464.8 \text{ kips}$$

From AISC Equation I2-15,

$$C_3 = 0.6 + 2\left(\frac{A_s}{A_c + A_s}\right) \le 0.9$$

$$= 0.6 + 2\left(\frac{7.85}{50.01 + 7.85}\right) = 0.8713 < 0.9$$

From AISC Equation I2-14,

$$(EI)_{eff} = E_s I_s + E_s I_{sr} + C_3 E_c I_c$$
$$= 29,000(68.1) + 0 + 0.8713(3492)(199.1) = 2.581 \times 10^6 \text{ kip-in.}^2$$

From AISC Equation I2-5,

$$P_e = \frac{\pi^2 (EI)_{\text{eff}}}{(KL)^2} = \frac{\pi^2 (2.581 \times 10^6)}{(13 \times 12)^2} = 1047 \text{ kips}$$

Then

$$0.44 P_o = 0.44 (464.8) = 204.5 \text{ kips}$$

$$P_e = 1047 \text{ kips} > 205.6 \text{ kips} \quad \therefore \text{ use AISC Equation I2-2}$$

$$P_n = P_o \left[0.658^{\left(\frac{P_o}{P_e} \right)} \right] = 464.8 \left[0.658^{\left(\frac{464.8}{1047} \right)} \right] = 386.0 \text{ kips}$$

Answer   For LRFD, the design strength is $\phi_c P_n = 0.75(386.0) = 290$ kips.

For ASD, the allowable strength is $P_n / \Omega_c = 386.0 / 2.00 = 193$ kips ∎

Transfer of load between the hollow steel shape and the concrete core can usually be accomplished through bond or direct bearing. The Commentary to AISC Section I2.2e provides guidance on this issue. If the axial load is applied directly to the concrete, the bearing strength of the concrete must be investigated using the provisions of AISC I2.2e. If shear connectors are required, AISC I2.2f, "Detailing Requirements" gives the required spacing.

Additional requirements for filled composite columns are given in AISC I2.2a, "Limitations":

1. The cross-sectional area of the steel shape must make up at least 1% of the total area (the same as for encased shapes).
2. To prevent local buckling in HSS filled with concrete, width–thickness ratios of the HSS cross-sectional elements must be limited.
   For each rectangular section face of width b,

$$\text{Maximum } \frac{b}{t} = 2.26 \sqrt{E/F_y}$$

For circular sections of outer diameter D,

$$\text{Maximum } \frac{D}{t} = 0.15 E/F_y$$

Tables for Analysis and Design

Part 4 of the *Manual*, "Design of Compression Members," contains tables that give the compressive strength of concrete-filled HSS (rectangular, square, and round) and steel pipe. Tables 4-13 through 4-20 give capacities for both $f'_c = 4$ ksi and $f'_c = 5$ ksi concrete. For those cases where $K_x L \neq K_y L$ (rectangular HSS), Tables 4-13 and 4-14 give values of r_{mx}/r_{my}, the ratio of weak axis radius of gyration to strong axis radius of gyration. Both design strength (LRFD) and allowable strength (ASD) are given.

The *Manual* does not contain design aids for concrete-encased steel shapes.

Example 9.15

A 16-foot-long compression member must support a total service load of 500 kips, consisting of equal parts of dead load and live load. The member is pinned at both ends, with additional support at mid-height in the weak direction. Use the tables in Part 4 of the *Manual* to select a concrete-filled rectangular HSS. Use $f'_c = 5$ ksi.

Solution

Strong-axis buckling will control when

$$\frac{K_x L}{r_{mx}/r_{my}} > K_y L; \text{ that is, whenever } r_{mx}/r_{my} < \frac{K_x L}{K_y L} = \frac{16}{8} = 2$$

An inspection of Table 4-14 shows that r_{mx}/r_{my} is always < 2, therefore x-axis buckling always controls. Our strategy will be to select the lightest shape for various values of r_{mx}/r_{my}, then choose the lightest shape overall. The ratios given in Table 4-14 can be placed into six groups: those with approximate values of r_{mx}/r_{my} of 1.2, 1.3, 1.4, 1.5, 1.6, and 1.8.

LRFD Solution

The factored axial load is

$$P_u = 1.2(250) + 1.6(250) = 700 \text{ kips}$$

Starting with the smallest ratio of 1.2, we compute the approximate value

$$\frac{K_x L}{r_{mx}/r_{my}} = \frac{16}{1.2} = 13.3$$

An HSS12 × 10 × ⅜, weighing 52.9 lb/ft, is a possibility. For an effective length of

$$KL = \frac{K_x L}{r_{mx}/r_{my}} = \frac{16}{1.17} = 13.7 \text{ ft}$$

the design strength is 724 kips (for $KL = 14$ ft). The results of all searches are summarized in the following tabulation.

Approximate r_{mx}/r_{my}	$\dfrac{K_x L}{r_{mx}/r_{my}}$ (ft)	Shape	Actual r_{mx}/r_{my}	$\phi_c P_n$ (kips)	Weight (lb/ft)
1.2	13.3	HSS12 × 10 × ⅜	1.17	724	52.9
1.3	12.3	HSS14 × 10 × ¼	1.34	702	39.5
1.4	11.4	HSS12 × 8 × ½	1.41	> 719	62.3
1.5	10.7	HSS20 × 12 × ⅝	1.54	> 1750	127
1.6	10.0	HSS20 × 12 × ⅜	1.55	> 1350	78.4
1.8	8.9	HSS16 × 8 × 5⁄16	1.81	> 757	48.9

The HSS14 × 10 × ¼ is the lightest shape.

Answer   Use an HSS14 × 10 × ¼.

ASD Solution   The required strength is

$$P_a = 500 \text{ kips}$$

Starting with the smallest ratio of 1.2, we compute the approximate value

$$\frac{K_x L}{r_{mx}/r_{my}} = \frac{16}{1.2} = 13.3$$

An HSS12 × 10 × ½, weighing 69.1 lb/ft, is a possibility. For an effective length of

$$KL = \frac{K_x L}{r_{mx}/r_{my}} = \frac{16}{1.17} = 13.7 \text{ ft}$$

the allowable strength is 565 kips (for $KL = 14$ ft). The results of all searches are summarized in the following tabulation.

Approximate r_{mx}/r_{my}	$\dfrac{K_x L}{r_{mx}/r_{my}}$ (ft)	Shape	Actual r_{mx}/r_{my}	P_n/Ω_c (kips)	Weight (lb/ft)
1.2	13.3	HSS12 × 10 × ½	1.17	565	69.1
1.3	12.3	HSS14 × 10 × ⁵⁄₁₆	1.33	518	48.9
1.4	11.4	HSS12 × 8 × ⅝	1.40	546	76.1
1.5	10.7	HSS20 × 12 × ⅝	1.54	> 1170	127
1.6	10.0	HSS20 × 12 × ⅜	1.55	> 901	78.4
1.8	8.9	HSS16 × 8 × ⁵⁄₁₆	1.81	506	48.9

The lightest weight is 48.9 lb/ft, a property of both the HSS14 × 10 × ⁵⁄₁₆ and the HSS16 × 8 × ⁵⁄₁₆, but the HSS14 × 10 × ⁵⁄₁₆ has a larger allowable strength.

Answer   Use an HSS14 × 10 × ⁵⁄₁₆.   ■

Problems

Notes   Unless otherwise indicated, the following conditions apply to all problems.

1. No shoring is used.
2. The slab formwork or deck provides continuous lateral support of the beam during the construction phase.
3. Normal-weight concrete is used.

Introduction: Analysis of Composite Beams

9.1-1   A W18 × 40 floor beam supports a 4-inch-thick reinforced concrete slab with an effective width b of 81 inches. Sufficient shear connectors are provided to make the beam fully composite. The 28-day compressive strength of the concrete is $f'_c = 4$ ksi.

    a. Compute the moment of inertia of the transformed section.

    b. For a positive service load moment of 290 ft-kips, compute the stress at the top of the steel (indicate whether tension or compression), the stress at the bottom of the steel, and the stress at the top of the concrete.

9.1-2 A W21 × 57 floor beam supports a 5-inch-thick reinforced concrete slab with an effective width b of 75 inches. Sufficient shear connectors are provided to make the beam fully composite. The 28-day compressive strength of the concrete is $f'_c = 4$ ksi.

    a. Compute the moment of inertia of the transformed section.

    b. For a positive service load moment of 300 ft-kips, compute the stress at the top of the steel (indicate whether tension or compression), the stress at the bottom of the steel, and the stress at the top of the concrete.

9.1-3 A W24 × 55 floor beam supports a 4½-inch-thick reinforced concrete slab with an effective width b of 78 inches. Sufficient shear connectors are provided to make the beam fully composite. The 28-day compressive strength of the concrete is $f'_c = 4$ ksi.

    a. Compute the moment of inertia of the transformed section.

    b. For a positive service load moment of 450 ft-kips, compute the stress at the top of the steel (indicate whether tension or compression), the stress at the bottom of the steel, and the stress at the top of the concrete.

9.1-4 Compute the nominal flexural strength of the composite beam of Problem 9.1-1. Use $F_y = 50$ ksi.

9.1-5 Compute the nominal flexural strength of the composite beam in Problem 9.1-2. Use $F_y = 50$ ksi.

9.1-6 Compute the nominal flexural strength of the composite beam in Problem 9.1-3. Use $F_y = 50$ ksi.

Strength of Unshored Composite Beams

9.2-1 A W14 × 22 acts compositely with a 4-inch-thick floor slab whose effective width b is 90 inches. The beams are spaced at 7 feet 6 inches, and the span length is 30 feet. The superimposed loads are as follows: construction load = 20 psf, partition load = 10 psf, weight of ceiling and light fixtures = 5 psf, and live load = 60 psf. A992 steel is used, and $f'_c = 4$ ksi. Determine whether the flexural strength is adequate.

    a. Use LRFD.

    b. Use ASD.

9.2-2 A composite floor system consists of W18 × 97 beams and a 5-inch-thick reinforced concrete floor slab. The effective width of the slab is 84 inches. The span length is 30 feet, and the beams are spaced at 8 feet. In addition to the weight of the slab, there is a construction load of 20 psf and a uniform live load of 800 psf. Determine whether the flexural strength is adequate. Assume that there is no lateral support during the construction phase. A992 steel is used, and $f'_c = 4$ ksi.

    a. Use LRFD.

    b. Use ASD.

Effective Flange Width

9.3-1 A fully composite floor system consists of W12 × 16 floor beams supporting a 4-inch-thick reinforced concrete floor slab. The beams are spaced at 7 feet, and the span length is 25 feet. The superimposed loads consist of a construction load of 20 psf, a partition load of 15 psf, and a live load of 125 psf. A992 steel is used, and $f'_c = 4$ ksi. Determine whether this beam satisfies the provisions of the AISC Specification.

    a. Use LRFD.
    b. Use ASD.

9.3-2 A fully composite floor system consists of W16 × 50 floor beams of A992 steel supporting a 4½-inch-thick reinforced concrete floor slab, with $f'_c = 4$ ksi. The beams are spaced at 10 feet, and the span length is 35 feet. The superimposed loads consist of a construction load of 20 psf and a live load of 160 psf. Determine whether this beam satisfies the provisions of the AISC Specification.

    a. Use LRFD.
    b. Use ASD.

Shear Connectors

9.4-1 A fully composite floor system consists of 40-foot-long, W21 × 57 steel beams spaced at 9 feet center-to-center and supporting a 6-inch-thick reinforced concrete floor slab. The steel is A992 and the concrete strength is $f'_c = 4$ ksi. There is a construction load of 20 psf and a live load of 250 psf.

    a. Determine whether this beam is adequate for LRFD.
    b. Determine whether the beam is adequate for ASD.
    c. How many ¾-inch × 3-inch shear connectors are required for full composite behavior?

9.4-2 A fully composite floor system consists of 27-foot-long, W14 × 22 steel beams spaced at 8 feet and supporting a 4-inch-thick reinforced concrete floor slab. The steel is A572 Grade 50 and the concrete strength is $f'_c = 4$ ksi. There is a construction load of 20 psf, a partition load of 20 psf, and a live load of 120 psf.

    a. Determine whether this beam is adequate for LRFD.
    b. Determine whether the beam is adequate for ASD.
    c. How many ¾-inch × 3-inch shear connectors are required for full composite behavior?

9.4-3 How many ¾-inch × 3-inch shear studs are required for the beam in Problem 9.1-1 for full composite behavior? Use $F_y = 50$ ksi.

9.4-4 How many ⅞-inch × 3½-inch shear studs are required for the beam in Problem 9.1-2 for full composite behavior? Use $F_y = 50$ ksi.

9.4-5 Design shear connectors for the beam in Problem 9.3-1. Show the results on a sketch similar to Figure 9.11.

Design

9.5-1   A fully composite floor system consists of 36-foot-long steel beams spaced at 6'-6" center-to-center supporting a 4½-inch-thick reinforced concrete floor slab. The steel yield stress is 50 ksi and the concrete strength is $f'_c = 4$ ksi. There is a construction load of 20 psf and a live load of 175 psf. Select a W16 shape.

    a. Use LRFD.

    b. Use ASD.

    c. Select shear connectors and show the layout on a sketch similar to Figure 9.13.

9.5-2   A floor system has the following characteristics:

- span length = 40 ft
- beam spacing = 5 ft
- slab thickness = 4 in.

The superimposed loads consist of a 20-psf construction load, a partition load of 20 psf, and a live load of 125 psf. Use $F_y = 50$ ksi, $f'_c = 4$ ksi, and select a W-shape for a fully composite floor system.

    a. Use LRFD.

    b. Use ASD.

    c. Select shear connectors and show your recommended layout on a sketch similar to the one in Figure 9.13.

9.5-3   A floor system has the following characteristics:

- span length = 30 ft
- beam spacing = 7 ft
- slab thickness = 5 in.

The superimposed loads consist of a construction load of 20 psf and a live load of 800 psf. Use $F_y = 50$ ksi, $f'_c = 4$ ksi, and select a W-shape for a fully composite floor system. Assume that there is no lateral support during the construction phase.

    a. Use LRFD.

    b. Use ASD.

    c. Select shear connectors and show the layout on a sketch similar to Figure 9.13.

Deflections

9.6-1   Compute the following deflections for the beam in Problem 9.2-1.

    a. Maximum deflection before the concrete has cured.

    b. Maximum total deflection after composite behavior has been attained.

9.6-2   Compute the following deflections for the beam in Problem 9.2-2.

    a. Maximum deflection before the concrete has cured.

    b. Maximum total deflection after composite behavior has been attained.

9.6-3   For the beam of Problem 9.3-1,

    a. Compute the deflections that occur before and after the concrete has cured.

    b. If the live-load deflection exceeds $L/360$, select another steel shape using either LRFD or ASD.

9.6-4   For the beam of Problem 9.4-1;

a. Compute the deflections that occur before and after the concrete has cured.

b. If the total deflection after the concrete has cured exceeds $L/240$, select another steel shape using either LRFD or ASD.

9.6-5   For the beam of Problem 9.4-2,

a. Compute the deflections that occur before and after the concrete has cured.

b. If the live load deflection exceeds $L/360$, select another steel shape using either LRFD or ASD.

Composite Beams with Formed Steel Deck

9.7-1   A fully composite floor system consists of W18 × 35 beams spaced at 6 feet, spanning 30 feet, and supporting a formed steel deck with a concrete slab. The total depth of the slab is 4½ inches, and the deck is 2 inches high. The steel and concrete strengths are $F_y = 50$ ksi and $f'_c = 4$ ksi.

a. Compute the transformed moment of inertia and compute the deflection for a service load of 1 kip/ft.

b. Compute the nominal strength of the composite section.

9.7-2   A composite floor system uses formed steel deck of the type shown in Figure P9.7-2. The beams are W18 × 50, and the slab has a total thickness of 4½ inches from top of slab to bottom of deck. The effective slab width is 90 inches, and the span length is 30 feet. The structural steel is A992, and the concrete strength is $f'_c = 4$ ksi. Compute the nominal flexural strength with two ¾-inch × 3½-inch studs per rib.

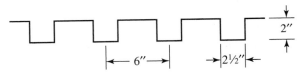

FIGURE P9.7-2

9.7-3   Determine the nominal strength of the following composite beam:

- W14 × 26, $F_y = 50$ ksi
- $f'_c = 4$ ksi
- effective slab width = 66 in.
- total slab thickness = 4½ in.
- span length = 30 ft
- formed steel deck is used (see Figure P9.7-3)
- studs are ¾-inch × 3-inch, one in every other rib

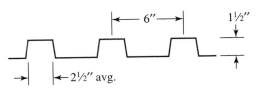

FIGURE P9.7-3

9.7-4   A composite floor system consists of $W18 \times 40$ steel beams supporting a formed steel deck and concrete slab. The deck is $1\frac{1}{2}$ inches deep, and the total depth from bottom of deck to top of slab is $4\frac{1}{2}$ inches. Lightweight concrete with a 28-day compressive strength of 4 ksi is used (unit weight = 115 psf). The beams are spaced at 10 feet, and the span length is 40 feet. There is a 20-psf construction load, a partition load of 20 psf, a ceiling weight of 5 psf, a mechanical load of 5 psf, and a live load of 120 psf. Thirty-four $\frac{3}{4}$-inch $\times$ 3-inch shear studs are used for each beam. Is the strength of this system adequate?

a. Use LRFD.
b. Use ASD.

Tables for Composite Beam Analysis and Design

9.8-1   Use the composite beam tables to solve Problem 9.7-3.

9.8-2   A simply supported, uniformly loaded composite beam consists of a $W16 \times 36$ of A992 steel and a reinforced concrete slab with $f_c' = 4$ ksi. The slab is supported by a 2-inch deep steel deck, and the total slab thickness is 5 inches, from bottom of deck to top of concrete. The effective slab width is 90 inches. The shear studs are $\frac{3}{4}$ inch in diameter. Determine the available strength (for both LRFD and ASD) for each of the following cases:

a. 44 shear connectors per beam.
b. 24 shear connectors per beam.

9.8-3   A beam must be designed to the following specifications:

- span length = 35 ft
- beam spacing = 10 ft
- 2-in. deck with 3 in. of lightweight concrete fill ($w_c = 115$ pcf) for a total depth of $t = 5$ in. Total weight of deck and slab = 51 psf
- construction load = 20 psf
- partition load = 20 psf
- miscellaneous dead load = 10 psf
- live load = 80 psf
- $F_y = 50$ ksi, $f_c' = 4$ ksi

Assume continuous lateral support and use LRFD.

a. Design a noncomposite beam. Compute the total deflection (there is no limit to be checked).
b. Design a composite beam and specify the size and number of shear studs required. Compute the maximum total deflection as follows:
1. Use the transformed section.
2. Use the lower-bound moment of inertia.

9.8-4   Same as Problem 9.8-3, but use ASD.

9.8-5   The beams in a composite floor system support a formed steel deck and concrete slab. The concrete has a 28-day compressive strength of 4 ksi and the total slab thickness is 4 inches. The deck rib height is 2 inches. The beams are spaced at 8 feet, and the span length is 36 feet. There is a 20 psf construction load, a 20 psf partition load, a ceiling load of 8 psf, and a live load of 100 psf. Use LRFD.

a. Select a W16-shape with $F_y = 50$ ksi. Use partial composite action and do not interpolate in the composite beam tables.

b. Determine the **total** number of ¾-inch × 3½-inch shear studs required. Do not consider the spacing of the studs.

9.8-6   Same as Problem 9.8-5, but use ASD.

9.8-7   Use the composite beam tables to solve Problem 9.5-3 with the following modifications:
   - Continuous lateral support is provided during the construction phase.
   - Use partial composite action.
   - The live-load deflection cannot exceed $L/360$. Use the lower-bound moment of inertia.

9.8-8   A composite floor system consists of steel beams supporting a formed steel deck and concrete slab. The deck is shown in Figure P9.8-8, and the total depth from bottom of deck to top of slab is 6½ inches. Lightweight concrete is used (unit weight = 115 pcf), and the 28-day compressive strength is 4 ksi. The deck and slab combination weighs 53 psf. The beams are spaced at 12 feet, and the span length is 40 feet. There is a 20-psf construction load, a partition load of 20 psf, other dead load of 10 psf, and a live load of 160 psf. The maximum permissible live-load deflection is $L/360$. Use the composite beam tables and select a W-shape with $F_y = 50$ ksi. Design the shear connectors. Use partial composite action and a lower-bound moment of inertia.

a. Use LRFD.
b. Use ASD.

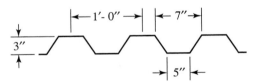

FIGURE P9.8-8

9.8-9   Use the composite beam tables and select a W-shape and shear connectors for the following conditions:
   - span length = 18'-6"
   - beam spacing = 9 ft
   - total slab thickness = 5½ in. (the slab and deck combination weighs 57 psf). Lightweight concrete with a unit weight of 115 pcf is used
   - construction load = 20 psf
   - partition load = 20 psf
   - live load = 225 psf
   - $F_y = 50$ ksi and $f'_c = 4$ ksi

A cross section of the formed steel deck is shown in Figure P9.8-9. The maximum live-load deflection cannot exceed $L/360$ (use a lower-bound moment of inertia).

a. Use LRFD.
b. Use ASD.

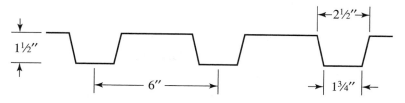

FIGURE P9.8-9

9.8-10 Use the composite beam tables and select a W-shape and shear connectors for the following conditions:

- span length = 35 ft
- beam spacing = 12 ft
- formed steel deck with rib spacing = 6 in., h_r = 2 in., w_r = 2¼ in.
- total slab thickness = 5½ in.
- partition load = 15 psf, live load = 100 psf
- maximum permissible live-load deflection = $L/360$ (use a lower-bound moment of inertia)

Use steel with a yield stress of F_y = 50 ksi and lightweight concrete weighing 115 pcf with f_c' = 4 ksi.

a. Use LRFD.
b. Use ASD.

Composite Columns

9.10-1 An HSS9 × 7 × ⅜ filled with concrete is used as a composite column, as shown in Figure P9.10-1. The steel has a yield stress of F_y = 46 ksi, and the concrete has a compressive strength of f_c' = 4 ksi. Compute the nominal strength of the column and check your answer with the tables in Part 4 of the *Manual*.

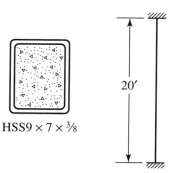

HSS9 × 7 × ⅜

20′

FIGURE P9.10-1

9.10-2 The composite compression member shown in Figure P9.10-2 has lateral support at midheight in the weak direction only. Compute the nominal strength. The steel yield stress is 50 ksi, and f_c' = 8 ksi. Grade 60 reinforcing bars are used.

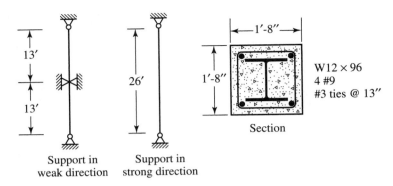

FIGURE P9.10-2

9.10-3 Design a rectangular (not square) HSS filled with concrete to resist an axial dead load of 165 kips and an axial live load of 500 kips. The effective lengths with respect to the two principle axes are $K_xL = 36$ feet and $K_yL = 12$ feet. Use $F_y = 46$ ksi and $f_c' = 5$ ksi. The tables in Part 4 of the *Manual* may be used.

a. Use LRFD.
b. Use ASD.

9.10-4 Use the tables in Part 4 of the *Manual* and select a concrete-filled pipe for the conditions shown in Figure P9.10-4. Select the smallest-diameter pipe that will work. Use $f_c' = 4$ ksi.

a. Use LRFD.
b. Use ASD.

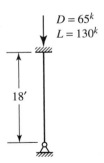

FIGURE P9.10-4

10

Plate Girders

10.1 INTRODUCTION

In this chapter, we consider large flexural members (girders) that are composed of plate elements—in particular, those with noncompact or slender webs. In Chapter 5, "Beams," we covered hot-rolled shapes, and for all the standard sections in the *Manual*, the webs are compact. Some have noncompact flanges, but none have slender flanges. With shapes built up from plates, however, both flanges and webs can be compact, noncompact, or slender. These built-up shapes usually are used when the bending moments are larger than standard hot-rolled shapes can resist, usually because of a large span. These girders are invariably very deep, resulting in noncompact or slender webs.

The AISC Specification covers flexural members with slender webs in Section F5, "Doubly Symmetric and Singly Symmetric I-Shaped Members with Slender Webs Bent About Their Major Axis." This is the category usually thought of as *plate girders*. Flexural members with noncompact webs are covered in Section F4, "Other I-shaped Members with Compact or Noncompact Webs Bent About Their Major Axis." This section deals with both doubly and singly symmetric sections. Interestingly, noncompact webs are more difficult to deal with than slender webs. In a User Note in Section F4, the Specification permits members covered by Section F4 to be designed by the provisions of Section F5. In this book, we do this and use Section F5 for girders with either noncompact or slender webs. We refer to both types as plate girders. Shear provisions for all flexural members are covered in AISC Chapter G, "Design of Members for Shear." Other requirements are given in AISC F13, "Proportions of Beams and Girders."

A plate girder cross section can take several forms. Figure 10.1 shows some of the possibilities. The usual configuration is a single web with two equal flanges, with all parts connected by welding. The box section, which has two webs as well as two flanges, is a torsionally superior shape and can be used when large unbraced lengths are necessary. Hybrid girders, in which the steel in the flanges is of a higher strength than that in the web or webs, are sometimes used.

Before the widespread use of welding, connecting the components of the cross section was a major consideration in the design of plate girders. All of the connections were made by riveting, so there was no way to attach the flange directly to the web, and additional cross-sectional elements were introduced for the specific purpose

FIGURE 10.1

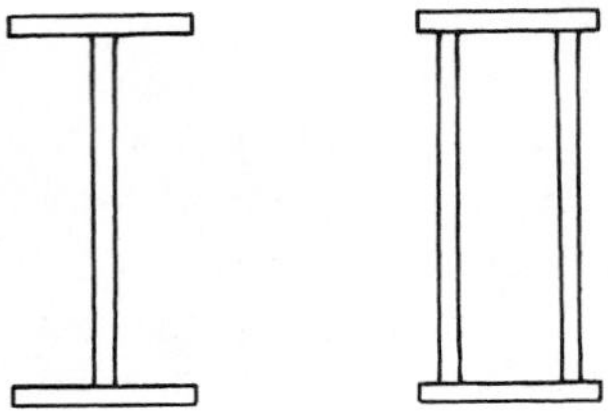

(a) Welded    (b) **Riveted without Stiffeners**

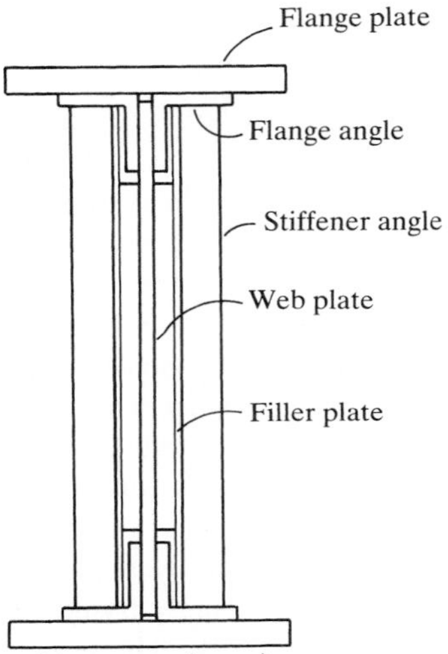

Flange plate

Flange angle

Stiffener angle

Web plate

Filler plate

(rivets not shown)

(c) **Riveted with Stiffeners**

of transmitting the load from one component to the other. The usual technique was to use a pair of angles, placed back-to-back, to attach the flange to the web. One pair of legs was attached to the web, and the other pair attached to the flange, as shown in Figure 10.1b. If web stiffeners were needed, pairs of angles were used for that purpose also. To avoid a conflict between the stiffener angles and the flange angles, filler plates were added to the web so that the stiffeners could clear the flange angles, as shown in Figure 10.1c. If a variable cross section was desired, one or more cover plates of different lengths were riveted to the flanges. (Although cover plates can also be used with welded plate girders, a simpler approach is to use different thicknesses of flange plate, welded end-to-end, at different locations along the length of the girder.) It should be evident that the welded plate girder is far superior to the riveted or bolted girder in terms of simplicity and efficiency. We consider only I-shaped welded plate girders in this chapter.

Before considering specific AISC Specification requirements, we need to examine, in a very general way, the peculiarities of plate girders as opposed to ordinary rolled beams.

10.2 GENERAL CONSIDERATIONS

Structural steel design is largely a matter of providing for stability, either locally or in an overall sense. Many standard hot-rolled structural shapes are proportioned so that local stability problems have been eliminated or minimized. When a plate girder is used, however, the designer must account for factors that in many cases would not be a problem with a hot-rolled shape. Deep, thin webs account for many of the special problems associated with plate girders, including local instability. A thorough understanding of the basis of the AISC provisions for plate girders requires a background in stability theory, particularly plate stability. Such a treatment is beyond the scope of this book, however, and the emphasis in this chapter is on the qualitative basis of the Specification requirements and their application. For those interested in delving deeper, the *Guide to Stability Criteria for Metal Structures* (Galambos, 1998) is a good starting point, and *Buckling Strength of Metal Structures* (Bleich, 1952) and *Theory of Elastic Stability* (Timoshenko and Gere, 1961) will provide the fundamentals of stability theory.

In some cases, plate girders rely on the strength available after the web has buckled, so most of the flexural strength will come from the flanges. The limit states considered are yielding of the tension flange and buckling of the compression flange. Compression flange buckling can take the form of vertical buckling into the web or flange local buckling (FLB), or it can be caused by lateral-torsional buckling (LTB).

At a location of high shear in a girder web, usually near the support and at or near the neutral axis, the principal planes will be inclined with respect to the longitudinal axis of the member, and the principal stresses will be diagonal tension and diagonal compression. The diagonal tension poses no particular problem, but the diagonal compression can cause the web to buckle. This problem can be addressed in one of three ways: (1) The depth-to-thickness ratio of the web can be made small enough that the problem is eliminated, (2) web stiffeners can be used to form panels with increased shear strength, or (3) web stiffeners can be used to form panels that resist the diagonal compression through *tension-field action*. Figure 10.2 illustrates the concept of tension-field action. At the point of impending buckling, the web loses its ability to support the diagonal compression, and this stress is shifted to the transverse stiffeners and the flanges. The stiffeners resist the vertical component of the diagonal compression, and the flanges resist the horizontal component. The web will need to resist only the diagonal tension, hence the term *tension-field action*. This behavior can be likened to that of a Pratt truss, in which the vertical web members carry compression and the diagonals carry tension, as shown in Figure 10.2b. Since the tension field does not actually exist until the web begins to buckle, its contribution to the web shear strength will not exist until the web buckles. The total strength will consist of the strength prior to buckling plus the postbuckling strength deriving from tension-field action.

If an unstiffened web is incapable of resisting the applied shear, appropriately spaced stiffeners are used to develop tension-field action. Cross-sectional requirements for these stiffeners, called *intermediate stiffeners,* are minimal because their primary purpose is to provide stiffness rather than resist directly applied loads.

Additional stiffeners may be required at points of concentrated loads for the purpose of protecting the web from the direct compressive load. These members are

FIGURE 10.2

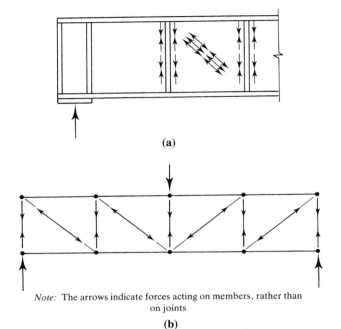

(a)

Note: The arrows indicate forces acting on members, rather than on joints

(b)

FIGURE 10.2

called *bearing stiffeners,* and they must be proportioned to resist the applied loads. They can also simultaneously serve as intermediate stiffeners. Figure 10.3 shows a bearing stiffener consisting of two rectangular plates, one on each side of the girder web. The plates are notched, or clipped, at the inside top and bottom corners so as to avoid the flange-to-web welds. If the stiffeners are conservatively assumed to resist the total applied load P (this assumption neglects any contribution by the web), the bearing stress on the contact surfaces may be written as

$$f_p = \frac{P}{A_{pb}}$$

FIGURE 10.3

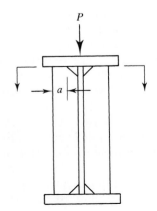
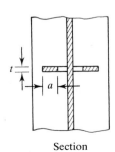

Section

FIGURE 10.4

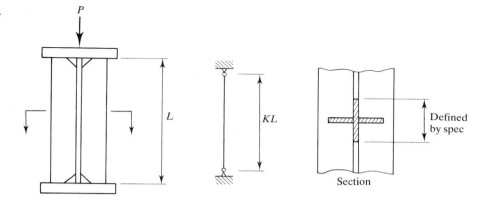

where

$$A_{pb} = \text{projected bearing area}$$
$$= 2at \qquad \text{(see Figure 10.3)}$$

or, expressing the bearing load in terms of the stress,

$$P = f_p A_{pb} \tag{10.1}$$

In addition, the pair of stiffeners, together with a short length of web, is treated as a column with an effective length less than the web depth and is investigated for compliance with the same Specification provisions as any other compression member. This cross section is illustrated in Figure 10.4. The compressive strength should always be based on the radius of gyration about an axis in the plane of the web, as instability about the other principal axis is prevented by the web itself.

Other limit states resulting from the application of concentrated loads to the top flange are web yielding, web crippling (buckling), and sidesway web buckling. Sidesway web buckling occurs when the compression in the web causes the *tension* flange to buckle laterally. This phenomenon can occur if the flanges are not adequately restrained against movement relative to one another by stiffeners or lateral bracing.

The welds for connecting the components of a plate girder are designed in much the same way as for other welded connections. The flange-to-web welds must resist the horizontal shear at the interface between the two components. This applied shear, called the *shear flow*, is usually expressed as a force per unit length of girder to be resisted by the weld. From Chapter 5, the shear flow, based on elastic behavior, is given by

$$f = \frac{VQ}{I_x}$$

where Q is the moment, about the neutral axis, of the area between the horizontal shear plane and the outside face of the section. This expression is Equation 5.7 for shearing stress multiplied by the width of the shear plane. Because the applied shear force V will usually be variable, the spacing of intermittent welds, if used, can also vary.

10.3   AISC REQUIREMENTS FOR PROPORTIONS OF PLATE GIRDERS

Whether a girder web is noncompact or slender depends on h/t_w, the width–thickness ratio of the web, where h is the depth of the web from inside face of flange to inside face of flange and t_w is the web thickness. From AISC B4, Table B4.1, the web of a doubly symmetric I-shaped section is noncompact if

$$3.76\sqrt{\frac{E}{F_y}} < \frac{h}{t_w} \le 5.70\sqrt{\frac{E}{F_y}}$$

and the web is slender if

$$\frac{h}{t_w} > 5.70\sqrt{\frac{E}{F_y}}$$

For singly symmetric I-shaped sections, the web is noncompact if

$$\frac{\dfrac{h_c}{h_p}\sqrt{\dfrac{E}{F_y}}}{\left(0.54\dfrac{M_p}{M_y} - 0.09\right)^2} < \frac{h_c}{t_w} \le 5.70\sqrt{\frac{E}{F_y}}$$

and it is slender if

$$\frac{h_c}{t_w} > 5.70\sqrt{\frac{E}{F_y}}$$

where

h_c = twice the distance from the elastic neutral axis (the centroidal axis) to the inside face of the compression flange. ($h_c/2$ defines the part of the web that is in compression for elastic bending. $h_c = h$ for girders with equal flanges). See Figure 10.5.

h_p = twice the distance from the plastic neutral axis to the inside face of the compression flange. ($h_p/2$ defines the part of the web in compression for the plastic moment. $h_p = h$ for girders with equal flanges). See Figure 10.5.

M_p = plastic moment = $F_y Z_x$

M_y = yield moment = $F_y S_x$

To prevent vertical buckling of the compression flange into the web, AISC F13.2 imposes an upper limit on the web slenderness. The limiting value of h/t_w is a function of the aspect ratio, a/h, of the girder panels, which is the ratio of intermediate stiffener spacing to web depth (see Figure 10.6).

FIGURE 10.5

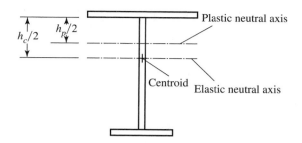

For $\dfrac{a}{h} \le 1.5,$

$$\dfrac{h}{t_w} \le 11.7 \sqrt{\dfrac{E}{F_y}}$$ (AISC Equation F13-3)

For $\dfrac{a}{h} > 1.5,$

$$\dfrac{h}{t_w} \le \dfrac{0.42E}{F_y}$$ (AISC Equation F13-4)

where a is the clear distance between stiffeners.

In all girders without web stiffeners, AISC F13.2 requires that h/t_w be no greater than 260 and that the ratio of the web area to the compression flange area be no greater than 10.

For singly symmetric sections, the proportions of the cross section must be such that

$$0.1 \le \dfrac{I_{yc}}{I_y} \le 0.9$$ (AISC Equation F13-2)

where

I_{yc} = moment of inertia of the compression flange about the y axis
I_y = moment of inertia of the entire cross section about the y axis

FIGURE 10.6

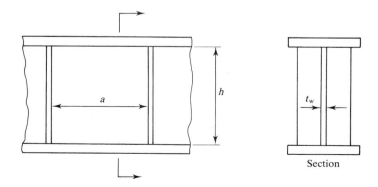

Section

10.4   FLEXURAL STRENGTH

The nominal flexural strength M_n of a plate girder is based on one of the limit states of tension flange yielding, compression flange yielding or local buckling (FLB), or lateral-torsional buckling (LTB).

Tension Flange Yielding

From Chapter 5, the maximum bending stress in a flexural member bent about its strong axis is

$$f_b = \frac{M}{S_x}$$

where S_x is the elastic section modulus about the strong axis. Expressing the bending moment as a function of the section modulus and stress gives

$$M = f_b S_x$$

AISC F5 gives the nominal flexural strength based on tension flange yielding as

$$M_n = F_y S_{xt} \qquad \text{(AISC Equation F5-10)}$$

where S_{xt} = elastic section modulus referred to the tension side.

Compression Flange Strength

The compression flange nominal strength is given by

$$M_n = R_{pg} F_{cr} S_{xc} \qquad \text{(AISC Equation F5-7)}$$

where
R_{pg} = bending strength reduction factor
F_{cr} = critical compressive flange stress, based on either yielding or local buckling
S_{xc} = elastic section modulus referred to the compression side

The bending strength reduction factor is given by

$$R_{pg} = 1 - \frac{a_w}{1200 + 300 a_w} \left(\frac{h_c}{t_w} - 5.7 \sqrt{\frac{E}{F_y}} \right) \le 1.0 \qquad \text{(AISC Equation F5-6)}$$

where

$$a_w = \frac{h_c t_w}{b_{fc} t_{fc}} \le 10 \qquad \text{(AISC Equation F4-11)}$$

b_{fc} = width of the compression flange
t_{fc} = thickness of the compression flange

(The upper limit of 10 in Equation F4-11 is not actually part of the AISC Equation, but AISC F5.2 stipulates it.)

The critical compression flange stress F_{cr} depends on whether the flange is compact, noncompact, or slender. The AISC Specification uses the generic notation λ, λ_p, and λ_r to define the flange width–thickness ratio and its limits. From AISC Table B4.1,

$$\lambda = \frac{b_f}{2t_f}$$

$$\lambda_p = 0.38\sqrt{\frac{E}{F_y}}$$

$$\lambda_r = 0.95\sqrt{\frac{k_c E}{F_L}}$$

$$k_c = \frac{4}{\sqrt{h/t_w}} \text{ and } (0.35 \le k_c \le 0.76)$$

$F_L = 0.7F_y$ for girders with slender webs. (See AISC Table B4.1 for compact and noncompact webs.)

If $\lambda \le \lambda_p$, the flange is compact. The limit state of yielding will control, and $F_{cr} = F_y$, resulting in

$$M_n = R_{pg} F_y S_{xc} \qquad \text{(AISC Equation F5-1)}$$

If $\lambda_p < \lambda \le \lambda_r$, the flange is noncompact. Inelastic FLB will control, and

$$F_{cr} = \left[F_y - 0.3F_y\left(\frac{\lambda - \lambda_p}{\lambda_r - \lambda_p} \right) \right] \qquad \text{(AISC Equation F5-8)}$$

If $\lambda > \lambda_r$, the flange is slender, elastic FLB will control, and

$$F_{cr} = \frac{0.9Ek_c}{\left(\dfrac{b_f}{2t_f} \right)^2} \qquad \text{(AISC Equation F5-9)}$$

Lateral-Torsional Buckling

The nominal flexural torsional bucking strength is given by

$$M_n = R_{pg} F_{cr} S_{xc} \qquad \text{(AISC Equation F5-2)}$$

Whether lateral-torsional buckling will occur depends on the amount of lateral support—that is, the unbraced length L_b. If the unbraced length is small enough, yielding or flange local buckling will occur before lateral-torsional buckling. The length parameters are L_p and L_r, where

$$L_p = 1.1r_t\sqrt{\frac{E}{F_y}} \qquad \text{(AISC Equation F4-7)}$$

FIGURE 10.7

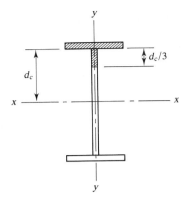

$$L_r = \pi r_t \sqrt{\frac{E}{0.7F_y}}$$ (AISC Equation F5-5)

r_t = radius of gyration about the weak axis for a portion of the cross section consisting of the compression flange and one-third of the compressed part of the web. For a doubly symmetric girder, this dimension will be one-sixth of the web depth. (See Figure 10.7.) This definition is a conservative approximation of r_t (see the user note in AISC F4.2). The exact definition is given by AISC Equation F4-10.

If $L_b \le L_p$, there is no lateral torsional buckling.

If $L_p < L_b \le L_r$, Failure will be by inelastic LTB, and

$$F_{cr} = C_b \left[F_y - 0.3F_y \left(\frac{L_b - L_p}{L_r - L_p} \right) \right] \le F_y$$ (AISC Equation F5-3)

If $L_b > L_r$, failure will be by elastic LTB, and

$$F_{cr} = \frac{C_b \pi^2 E}{\left(\frac{L_b}{r_t} \right)^2} \le F_y$$ (AISC Equation F5-4)

C_b is defined by AISC Equation F1-1 and is covered in Chapter 5 of this book.

As with all other flexural members covered in Chapter F of the Specification, the resistance factor for LRFD is $\phi_b = 0.90$, and the safety factor for ASD is $\Omega_b = 1.67$. The computation of flexural strength is illustrated in Example 10.1, part a.

10.5   SHEAR STRENGTH

The shear strength of a plate girder is a function of the depth-to-thickness ratio of the web and the spacing of any intermediate stiffeners that may be present. The shear capacity has two components: the strength before buckling and the postbuckling strength.

The postbuckling strength relies on tension-field action, which is made possible by the presence of intermediate stiffeners. If stiffeners are not present or are spaced too far apart, tension-field action will not be possible, and the shear capacity will consist only of the strength before buckling. The AISC Specification covers shear strength in Chapter G, "Design of Members for Shear." In that coverage, the constants k_v and C_v are used. AISC defines k_v, which is a plate-buckling coefficient, in Section G2 as follows:

$$k_v = 5 + \frac{5}{(a/h)^2}$$

$$= 5 \text{ if } \frac{a}{h} > 3$$

$$= 5 \text{ if } \frac{a}{h} > \left[\frac{260}{(h/t_w)} \right]^2$$

$$= 5 \text{ in unstiffened webs with } \frac{h}{t_w} < 260$$

For C_v, which can be defined as the ratio of the critical web shear stress to the web shear yield stress,

If $\dfrac{h}{t_w} \leq 1.10 \sqrt{\dfrac{k_v E}{F_y}}$,

$$C_v = 1.0 \qquad \text{(AISC Equation G2-3)}$$

If $1.10 \sqrt{\dfrac{k_v E}{F_y}} < \dfrac{h}{t_w} \leq 1.37 \sqrt{\dfrac{k_v E}{F_y}}$,

$$C_v = \frac{1.10\sqrt{k_v E/F_y}}{h/t_w} \qquad \text{(AISC Equation G2-4)}$$

If $\dfrac{h}{t_w} > 1.37 \sqrt{\dfrac{k_v E}{F_y}}$,

$$C_v = \frac{1.51 E k_v}{(h/t_w)^2 F_y} \qquad \text{(AISC Equation G2-5)}$$

Whether the shear strength is based on web shear yielding or web shear buckling depends on the web width–thickness ratio h/t_w. If

$$\frac{h}{t_w} \leq 1.10 \sqrt{\frac{k_v E}{F_y}}$$

the strength is based on shear yielding, and

$$V_n = 0.6 F_y A_w \qquad \text{(AISC Equation G3-1)}$$

where A_w = cross-sectional area of the web. If

$$\frac{h}{t_w} > 1.10\sqrt{\frac{k_v E}{F_y}},$$

the strength will be based on shear buckling or shear buckling plus tension-field action. If tension-field behavior exists,

$$V_n = 0.6F_y A_w \left(C_v + \frac{1 - C_v}{1.15\sqrt{1 + (a/h)^2}} \right) \qquad \text{(AISC Equation G3-2)}$$

AISC Equation G3-2 can also be written as

$$V_n = 0.6F_y A_w C_v + 0.6F_y A_w \frac{1 - C_v}{1.15\sqrt{1 + (a/h)^2}} \qquad (10.2)$$

The first term in Equation 10.2 gives the web shear buckling strength and the second term gives the post-buckling strength. If there is no tension-field action, the second term in Equation 10.2 is omitted, resulting in

$$V_n = 0.6\, F_y\, A_w C_v \qquad \text{(AISC Equation G2-1)}$$

Solution of AISC Equations G2-1 (without tension field) and G3-2 (with tension field) is facilitated by curves given in Part 3 of the *Manual*. Tables 3-16a and 3-16b present curves that relate the variables of these two equations for steel with a yield stress of 36 ksi, and Tables 3-17a and 3-17b do the same for steels with a yield stress of 50 ksi.

What are the conditions under which a tension field cannot be developed? A tension field ordinarily cannot be fully developed in an end panel. This can be understood by considering the horizontal components of the tension fields shown in Figure 10.8. (The vertical components are resisted by the stiffeners.) The tension field in panel *CD* is balanced on the left side in part by the tension field in panel *BC*. Thus interior panels are anchored by adjacent panels. Panel *AB*, however, has no such anchorage on the left side. Although anchorage could be provided by an end stiffener specially designed to resist the bending induced by a tension field, that usually is not done. (Because the tension field does not cover the full depth of the web, the interior stiffeners are also subjected to a certain amount of bending caused by the offset of the tension fields in the adjacent panels, but this bending is not significant.) Hence the anchorage for panel *BC* must be provided on the left side by a *beam-shear panel* rather than the tension-field panel shown.

FIGURE 10.8

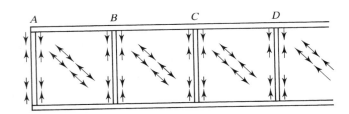

AISC G3.1 lists all of the conditions under which a tension field cannot be used:

a. In end panels

b. When $\dfrac{a}{h} > 3$ or $\dfrac{a}{h} > \left[\dfrac{260}{(h/t_w)}\right]^2$ (Each of these cases corresponds to $k_v = 5$.)

c. When $\dfrac{2A_w}{(A_{fc} + A_{ft})} > 2.5$

d. When $\dfrac{h}{b_{fc}}$ or $\dfrac{h}{b_{ft}} > 6$

where

A_w = area of the web
A_{fc} = area of the compression flange
A_{ft} = area of the tension flange
b_{fc} = width of the compression flange
b_{ft} = width of the tension flange

Summary

Computation of the nominal shear strength can be condensed to the following procedure:

1. Compute the aspect ratio a/h and the web slenderness ratio h/t_w.
2. Compute k_v and C_v.
3. Determine whether tension-field action can be used.
   a. If tension-field action is permitted,

$$V_n = 0.6F_y A_w \left(C_v + \dfrac{1 - C_v}{1.15\sqrt{1 + (a/h)^2}} \right)$$

   b. If tension-field action is not permitted,

$$V_n = 0.6F_y A_w C_v$$

Note that there is no requirement that tension-field behavior must be used, although its use will result in a more economical design.

This same procedure also is used for determining the shear strength of hot-rolled shapes with unstiffened webs (see Chapter 5). For those shapes, a/h does not apply, $k_v = 5$, and there is no tension field.

For LRFD, the resistance factor is $\phi_v = 0.90$. For ASD, the safety factor is $\Omega = 1.67$. Recall that these factors are different for some *hot-rolled* shapes.

Shear strength computation is illustrated in Example 10.1, part b.

Intermediate Stiffeners

If intermediate stiffeners are necessary to obtain enough shear strength when tension-field action is being used, the minimum cross-sectional area of either a single stiffener or a pair is given by AISC G3.3 as

$$A_{st} > \dfrac{F_y}{F_{yst}} \left[0.15 D_s h t_w (1 - C_v) \dfrac{V_r}{V_c} - 18t_w^2 \right] \geq 0 \qquad \text{(AISC Equation G3-3)}$$

where

A_{st} = total cross-sectional area of the stiffener required when tension-field
     action is used
F_{yst} = yield stress of the stiffener
D_s = a function of the configuration of the stiffener
     = 1.0 for stiffeners in pairs (angles or plates)
     = 1.8 for single-angle stiffeners
     = 2.4 for single-plate stiffeners
V_r = required shear strength at the location of the stiffener (= V_u for LFRD
     and = V_a for ASD)
V_c = available shear strength (= $\phi_v V_n$ for LRFD and = V_n/Ω_v for ASD)

The area specified by AISC Equation G3-3 is needed to resist the vertical component of diagonal compression in the panel. In addition, when there is a tension field, the stiffener width–thickness ratio is limited to

$$\frac{b}{t} \le 0.56\sqrt{\frac{E}{F_{yst}}} \tag{10.3}$$

If there is no tension field, AISC G2.2 specifies a minimum moment of inertia of the stiffener, taken about an axis in the plane of the web (or for a single stiffener, about the face of the stiffener in contact with the web):

$$I_{st} = a t_w^3 j$$

where

$$j = \frac{2.5}{(a/h)^2} - 2 \ge 0.5 \tag{AISC Equation G2-6}$$

Unless they also serve as bearing stiffeners, intermediate stiffeners are not required to bear against the tension flange, so their length can be somewhat less than the web depth h, and fabrication problems associated with close fit can be avoided. According to Section G2.2 of the Specification, the length should be within limits established by the distance between the weld connecting the stiffener to the web and the weld connecting the web to the tension flange. This distance, labeled c in Figure 10.9, should be between four and six times the web thickness.

Proportioning the intermediate stiffeners by the AISC rules does not require the computation of any forces, but a force must be transmitted from the stiffener to the

FIGURE 10.9

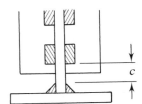

web, and the connection should be designed for this force. Basler (1961) recommends the use of a shear flow of

$$f = 0.045h\sqrt{\frac{F_y^3}{E}} \text{ kips/in.}$$

(10.4)

The minimum intermittent fillet weld will likely be adequate (Salmon and Johnson, 1996). AISC G2.2 requires the clear distance between intermittent fillet welds be no more than $16t_w$ or 10 inches.

10.6   BEARING STIFFENERS

Bearing stiffeners are required when the web has insufficient strength for any of the limit states of web yielding, web crippling, or sidesway web buckling. These limit states are covered in Chapter J of the Specification "Design of Connections". For web yielding, when the load is at a distance greater than the girder depth from the end, the nominal strength is

$$R_n = (5k + N)F_{yw}t_w$$

(AISC Equation J10-2)

When the load is less than this distance from the end,

$$R_n = (2.5k + N)F_{yw}t_w$$

(AISC Equation J10-3)

where

k = distance from the outer face of the flange to the toe of the fillet in the web (for rolled beams) or to the toe of the weld (for welded girders)
N = length of bearing of the concentrated load, measured in the direction of the girder longitudinal axis (not less than k for an end reaction)
F_{yw} = yield stress of the web

For LRFD, the resistance factor is $\phi = 1.00$. For ASD, the safety factor is $\Omega = 1.50$. (We covered this limit state in Chapter 5.)

For web crippling, when the load is at least *half* the girder depth from the end,

$$R_n = 0.80t_w^2\left[1 + 3\left(\frac{N}{d}\right)\left(\frac{t_w}{t_f}\right)^{1.5}\right]\sqrt{\frac{EF_{yw}t_f}{t_w}}$$

(AISC Equation J10-4)

When the load is less than this distance from the end of the girder,

$$R_n = 0.40t_w^2\left[1 + 3\left(\frac{N}{d}\right)\left(\frac{t_w}{t_f}\right)^{1.5}\right]\sqrt{\frac{EF_{yw}t_f}{t_w}}, \quad \text{for } \frac{N}{d} \leq 0.2$$

(AISC Equation J10-5a)

and

$$R_n = 0.40t_w^2 \left[1 + \left(\frac{4N}{d} - 0.2 \right) \left(\frac{t_w}{t_f} \right)^{1.5} \right] \sqrt{\frac{EF_{yw}t_f}{t_w}}, \quad \text{for} \quad \frac{N}{d} > 0.2$$

$$\text{(AISC Equation J10-5b)}$$

where

d = overall depth of girder
t_f = thickness of girder flange

For LRFD, the resistance factor is $\phi = 0.75$. For ASD, the safety factor is $\Omega = 2.00$. (We also covered this limit state in Chapter 5.)

Bearing stiffeners are required to prevent sidesway web buckling only under a limited number of circumstances. Sidesway web buckling should be checked when the compression flange is not restrained against movement relative to the tension flange. If the compression flange is restrained against rotation, the nominal strength is

$$R_n = \frac{C_r t_w^3 t_f}{h^2} \left[1 + 0.4 \left(\frac{h/t_w}{\ell/b_f} \right)^3 \right] \qquad \text{(AISC Equation J10-6)}$$

[This equation need not be checked if $(h/t_w)/(\ell/b_f) > 2.3$.]

If the flange is not restrained against rotation,

$$R_n = \frac{C_r t_w^3 t_f}{h^2} \left[0.4 \left(\frac{h/t_w}{\ell/b_f} \right)^3 \right] \qquad \text{(AISC Equation J10-7)}$$

[This equation need not be checked if $(h/t_w)/(\ell/b_f) > 1.7$.] where

$C_r = 960{,}000$ ksi when $M_u < M_y$ (for LRFD) or $1.5\,M_a < M_y$ (for ASD)
$\quad = 480{,}000$ ksi when $M_u \geq M_y$ (for LRFD) or $1.5\,M_a \geq M_y$ (for ASD)
(All moments are at the location of the load.)
ℓ = largest unbraced length of either flange (at the load point)

For LRFD, $\phi = 0.85$. For ASD, $\Omega = 1.76$.

Although the web can be proportioned to directly resist any applied concentrated loads, bearing stiffeners are usually provided. If stiffeners are used to resist the full concentrated load, the limit states of web yielding, web crippling, and sidesway web buckling do not need to be checked.[*]

The nominal bearing strength of a stiffener is given in AISC J7 as

$$R_n = 1.8 F_y A_{pb} \qquad \text{(AISC Equation J7-1)}$$

(This equation is the same as Equation 10.1 with a bearing stress of $f_p = 1.8 F_y$.)

---

[*] If the compression flange is not restrained at the load point and AISC Equation J10-7 is not satisfied, lateral bracing must be provided at both flanges at the load point (AISC J10-4).

For LRFD, the resistance factor is $\phi = 0.75$. For ASD, the safety factor is $\Omega = 2.00$.

AISC J10.8 requires that full-depth stiffeners be used in pairs and analyzed as axially loaded columns subject to the following guidelines:

- The cross section of the axially loaded member consists of the stiffener plates and a length of the web (see Figure 10.4). This length can be no greater than 12 times the web thickness for an end stiffener or 25 times the web thickness for an interior stiffener.
- The effective length should be taken as 0.75 times the actual length—that is, $KL = 0.75h$.
- The nominal axial strength is based on the provisions of AISC J4.4, "Strength of Elements in Compression," which are as follows:

For $\dfrac{KL}{r} \leq 25$,

$$P_n = F_y A_g \qquad\qquad\qquad\qquad \text{(AISC Equation J4-6)}$$

This is the "squash load" for the stiffener—that is, the load that causes compression yielding, with no buckling. For LRFD, the resistance factor for this limit state is $\phi = 0.90$; for ASD, the safety factor is $\Omega = 1.67$.

For $\dfrac{KL}{r} > 25$, the usual requirements for compression members in AISC E apply.

- The weld connecting the stiffener to the web should have the capacity to transfer the unbalanced force. Conservatively, the weld can be designed to carry the entire concentrated load. If the stiffener bears on the compression flange, it need not be welded to the flange.

Although no width–thickness ratio limit is given in the Specification for bearing stiffeners, the requirement of Equation 10.3 for intermediate stiffeners can be used as guide in proportioning bearing stiffeners:

$$\frac{b}{t} \leq 0.56 \sqrt{\frac{E}{F_{yst}}} \qquad\qquad\qquad\qquad (10.5)$$

Bearing-stiffener analysis is illustrated in Example 10.1, part c.

Example 10.1   The plate girder shown in Figure 10.10 must be investigated for compliance with the AISC Specification. The loads are service loads with a live-load–to–dead-load ratio of 3.0. The uniform load of 4 kips/ft includes the weight of the girder. The compression flange has lateral support at the ends and at the points of application of the concentrated loads. The compression flange is restrained against rotation at these same points. Bearing stiffeners are provided as shown at the ends and at the concentrated loads. They are clipped 1 inch at the inside edge, both top and bottom, to clear the

FIGURE 10.10

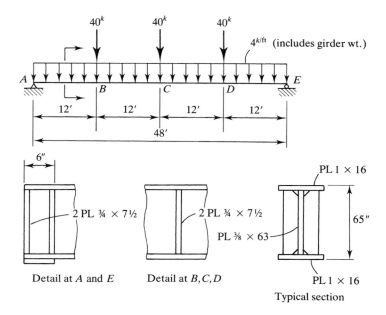

Detail at *A* and *E*   Detail at *B,C,D*   Typical section

flange-to-web welds. There are no intermediate stiffeners, and A36 steel is used throughout. Assume that all welds are adequate and check

   a. Flexural strength
   b. Shear strength
   c. Bearing stiffeners

LRFD Solution   Check the web width–thickness ratio.

$$\frac{h}{t_w} = \frac{63}{3/8} = 168$$

$$5.70\sqrt{\frac{E}{F_y}} = 5.70\sqrt{\frac{29,000}{36}} = 161.8$$

Since $h/t_w > 5.70\sqrt{E/F_y}$, the web is slender and the provisions of AISC F5 apply.

The web must satisfy the slenderness limitation of AISC F13.2. The limiting value of h/t_w will depend on the aspect ratio a/h. For this plate girder, the bearing stiffeners will serve as intermediate stiffeners, and

$$\frac{a}{h} \approx \frac{12(12)}{63} = 2.286$$

(This ratio is approximate because *a* is not exactly 12 feet. In the interior panels, 12 feet is the center-to-center stiffener spacing rather than the clear spacing. In the end panels, *a* is less than 12 feet because of the double stiffeners at the supports.)

TABLE 10.1

Component	A	$\bar{I}$	d	$\bar{I} + Ad^2$
Web	—	7814	—	7,814
Flange	16	—	32	16,380
Flange	16	—	32	16,380
				40,574

Because a/h is greater than 1.5, AISC Equation F13-4 applies:

$$\frac{0.42E}{F_y} = \frac{0.42(29,000)}{36} = 338 > \frac{h}{t_w} \quad \text{(OK)}$$

a. **Flexural strength**: To determine the flexural strength, the elastic section modulus will be needed. Because of symmetry.

$$S_{xt} = S_{xc} = S_x$$

The computations for I_x, the moment of inertia about the strong axis, are summarized in Table 10.1. The moment of inertia of each flange about its centroidal axis was neglected because it is small relative to the other terms. The elastic section modulus is

$$S_x = \frac{I_x}{c} = \frac{40,570}{32.5} = 1248 \text{ in.}^3$$

From AISC Equation F5-10, the tension flange strength based on yielding is

$$M_n = F_y S_{xt} = 36(1248) = 44,930 \text{ in.-kips}$$
$$= 3744 \text{ ft-kips}$$

The compression flange strength is given by AISC Equation F5-7:

$$M_n = R_{pg} F_{cr} S_{xc}$$

where the critical stress F_{cr} is based on either flange local buckling or yielding. For flange local buckling, the relevant slenderness parameters are

$$\lambda = \frac{b_f}{2t_f} = \frac{16}{2(1.0)} = 8$$

$$\lambda_p = 0.38\sqrt{\frac{E}{F_y}} = 0.38\sqrt{\frac{29,000}{36}} = 10.79$$

Since $\lambda < \lambda_p$, there is no flange local buckling. The compression flange strength is therefore based on yielding, and $F_{cr} = F_y = 36$ ksi.

To compute the bending strength reduction factor R_{pg}, the value of a_w will be needed:

$$a_w = \frac{h_c t_w}{b_{fc} t_{fc}} = \frac{63(3/8)}{16(1.0)} = 1.477 < 10$$

From AISC Equation F5-6,

$$R_{pg} = 1 - \frac{a_w}{1200 + 300a_w}\left(\frac{h_c}{t_w} - 5.7\sqrt{\frac{E}{F_y}}\right) \le 1.0$$

$$= 1 - \frac{1.447}{1200 + 300(1.447)}\left(168 - 5.7\sqrt{\frac{29,000}{36}}\right) = 0.9945$$

From AISC Equation F5-7, the nominal flexural strength for the compression flange is

$$M_n = R_{pg}F_{cr}S_{xc} = 0.9945(36)(1248)$$
$$= 44,680 \text{ in.-kips} = 3723 \text{ ft-kips}$$

To check lateral-torsional buckling we need the slenderness ratio r_t. From Figure 10.11,

$$I_y = \frac{1}{12}(1)(16)^3 + \frac{1}{12}(10.5)\left(\frac{3}{8}\right)^3 = 341.4 \text{ in.}^4$$

$$A = 16(1) + 10.5\left(\frac{3}{8}\right) = 19.94 \text{ in.}^2$$

$$r_t = \sqrt{\frac{I_y}{A}} = \sqrt{\frac{341.4}{19.94}} = 4.138 \text{ in.}$$

Check the unbraced length.

$$L_b = 12 \text{ ft}$$

$$L_p = 1.1r_t\sqrt{\frac{E}{F_y}} = 1.1(4.138)\sqrt{\frac{29,000}{36}} = 129.2 \text{ in.} = 10.77 \text{ ft}$$

$$L_r = \pi r_t\sqrt{\frac{E}{0.7F_y}} = \pi(4.138)\sqrt{\frac{29,000}{0.7(36)}} = 441.0 \text{ in.} = 36.75 \text{ ft}$$

FIGURE 10.11

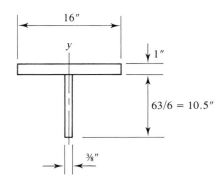

Since $L_p < L_b < L_r$, the girder is subject to inelastic lateral-torsional buckling. From AISC Equation F5-3,

$$F_{cr} = C_b \left[F_y - 0.3F_y \left(\frac{L_b - L_p}{L_r - L_p} \right) \right] \leq F_y$$

For the computation of C_b, refer to Figure 10.12, which shows the loading, shear, and bending moment diagrams based on factored loads. (Verification of the values shown is left as an exercise for the reader.) Compute C_b for the 12-foot unbraced segment adjacent to the midspan. Dividing this segment into four equal spaces, we get points *A, B,* and *C* located at 15 ft, 18 ft and 21 ft from the left end of the girder. The corresponding bending moments are

$$M_A = 234(15) - 60(3) - 6(15)^2/2 = 2655 \text{ ft-kips}$$
$$M_B = 234(18) - 60(6) - 6(18)^2/2 = 2880 \text{ ft-kips}$$
$$M_C = 234(21) - 60(9) - 6(21)^2/2 = 3051 \text{ ft-kips}$$

From AISC Equation F1-1,

$$\begin{aligned} C_b &= \frac{12.5M_{\text{max}}}{2.5M_{\text{max}} + 3M_A + 4M_B + 3M_C} R_m \leq 3.0 \\ &= \frac{12.5(3168)}{2.5(3168) + 3(2655) + 4(2880) + 3(3051)}(1.0) = 1.083 < 3.0 \end{aligned}$$

FIGURE 10.12

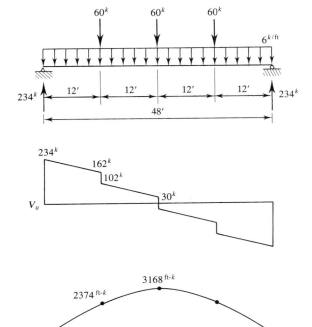

where R_m has a value of 1.0 for all doubly symmetric sections.

$$F_{cr} = C_b \left[F_y - 0.3 F_y \left(\frac{L_b - L_p}{L_r - L_p} \right) \right] \le F_y$$

$$= 1.083 \left[36 - (0.3 \times 36) \left(\frac{12 - 10.77}{36.75 - 10.77} \right) \right] = 38.43 \text{ ksi}$$

Since 38.43 ksi > F_y = 36 ksi, use F_{cr} = 36 ksi (same as for the other limit states). The nominal flexural strength is therefore based on yielding of the compression flange, and

$$M_n = 3723 \text{ ft-kips}$$

The design strength is $\phi_b M_n = 0.90(3723) = 3350$ ft-kips, which must be compared to M_u, the maximum factored load moment. From Figure 10.12, the maximum factored load moment is

$$M_u = 3168 \text{ ft-kips} < 3350 \text{ ft-kips} \qquad \text{(OK)}$$

Answer   The flexural strength is adequate.

b. **Shear strength**: The shear strength is a function of the web–slenderness ratio h/t_w and the aspect ratio a/h. We will first determine whether tension-field action can be used in regions other than the end panels. (It cannot be used in the end panels.) Referring to the conditions of AISC G3.1,

a. $\dfrac{a}{h} \approx 2.286 < 3 \qquad \text{(OK)}$

b. $\dfrac{a}{h} < \left[\dfrac{260}{(h/t_w)} \right]^2 = \left[\dfrac{260}{(168)} \right]^2 = 2.40 \qquad \text{(OK)}$

c. $A_w = h t_w = 63(3/8) = 23.63 \text{ in.}^2$

$A_{fc} = A_{ft} = b_f t_f = 16(1.0) = 16 \text{ in.}^2$

$\dfrac{2 A_w}{(A_{fc} + A_{ft})} = \dfrac{2(23.63)}{(16+16)} = 1.48 < 2.5 \qquad \text{(OK)}$

d. $\dfrac{h}{b_{fc}} = \dfrac{h}{b_{ft}} = \dfrac{63}{16} = 3.94 < 6 \qquad \text{(OK)}$

Therefore tension-field action can be used. Determine k_v and C_v.

$$k_v = 5 + \frac{5}{(a/h)^2} = 5 + \frac{5}{(2.286)^2} = 5.957$$

Establish the range of h/t_w:

$$1.10 \sqrt{\frac{k_v E}{F_y}} = 1.10 \sqrt{\frac{5.957(29,000)}{36}} = 76.20$$

$$1.37 \sqrt{\frac{k_v E}{F_y}} = 1.37 \sqrt{\frac{5.957(29,000)}{36}} = 94.90$$

Since $h/t_w = 168 > 1.37\sqrt{k_v E/F_y}$, C_v is found from AISC Equation G2-5:

$$C_v = \frac{1.51 E k_v}{(h/t_w)^2 F_y} = \frac{1.51(29{,}000)(5.957)}{(168)^2(36)} = 0.2567$$

AISC Equation G3-2, which accounts for tension-field action, will be used to determine the nominal shear strength (except for the end panels):

$$V_n = 0.6 F_y A_w \left(C_v + \frac{1 - C_v}{1.15\sqrt{1 + (a/h)^2}} \right)$$

$$= 0.6(36)(23.63)\left[0.2567 + \frac{1 - 0.2567}{1.15\sqrt{1 + (2.286)^2}} \right] = 263.2 \text{ kips}$$

The design shear strength is

$$\phi_v V_n = 0.90(263.2) = 237 \text{ kips}$$

From Figure 10.12, the maximum factored load shear in the middle half of the girder is 102 kips, so the shear strength is adequate where tension-field action is permitted.

For the end panels, tension-field action is not permitted, and the shear strength must be computed from AISC Equation G2-1:

$$V_n = 0.6 F_y A_w C_v = 0.6(36)(23.62)(0.2567) = 131.0 \text{ kips}$$

The design strength is

$$\phi_v V_n = 0.90(131.0) = 118 \text{ kips}$$

The maximum factored load shear in the end panel is

$$V_u = 234 \text{ kips} > 118 \text{ kips} \qquad \text{(N.G.)}$$

Two options are available for increasing the shear strength: either decrease the web slenderness (probably by increasing its thickness) or decrease the aspect ratio of each end panel by adding an intermediate stiffener. Stiffeners are added in this example.

The location of the first intermediate stiffener will be determined by the following strategy: First, equate the shear strength from AISC Equation G2-1 to the required shear strength and solve for the required value of C_v. Next, solve for k_v from Equation G2-5, then solve for a/h.

The required shear strength is

$$\text{Required } V_n = \frac{V_u}{\phi_v} = \frac{234}{0.90} = 260.0 \text{ kips}$$

From AISC Equation G2-1,

$$V_n = 0.6F_y A_w C_v$$

$$C_v = \frac{V_n}{0.6F_y A_w} = \frac{260.0}{0.6(36)(23.63)} = 0.5094$$

$$C_v = \frac{1.51Ek_v}{(h/t_w)^2 F_y} \qquad \text{(AISC Equation G2-5)}$$

$$k_v = \frac{C_v(h/t_w)^2 F_y}{1.51E} = \frac{0.5094(168)^2(36)}{1.51(29,000)} = 11.82$$

$$k_v = 5 + \frac{5}{(a/h)^2}$$

$$\frac{a}{h} = \sqrt{\frac{5}{k_v - 5}} = \sqrt{\frac{5}{11.82 - 5}} = 0.8562$$

The required stiffener spacing is

$$a = 0.8562h = 0.8562(63) = 53.9 \text{ in.}$$

Although a is defined as the clear spacing, we will treat it conservatively as a center-to-center spacing and place the first intermediate stiffener at 54 inches from the end of the girder. This placement will give a design strength that approximately equals the maximum factored load shear of 234 kips. No additional stiffeners will be needed, since the maximum factored load shear outside of the end panels is less than the design strength of 237 kips.

The determination of the stiffener spacing can be facilitated by the use of the design curves in Part 3 of the *Manual;* we illustrate this technique in Example 10.2.

Answer The shear strength is inadequate. Add one intermediate stiffener 54 inches from each end of the girder.

c. **Bearing stiffeners**: Bearing stiffeners are provided at each concentrated load, so there is no need to check the provisions of AISC Chapter J for web yielding, web crippling, or sidesway web buckling.

For the interior bearing stiffeners, first compute the bearing strength. From Figure 10.13,

$$A_{pb} = 2at = 2(7.5 - 1)(0.75) = 9.750 \text{ in.}^2$$

From AISC Equation J7-1,

$$R_n = 1.8F_y A_{pb} = 1.8(36)(9.750) = 631.8 \text{ kips}$$
$$\phi R_n = 0.75(631.8) = 474 \text{ kips} > 60 \text{ kips} \qquad \text{(OK)}$$

Check the strength of the stiffener as a compression member. Referring to Figure 10.13, we can use a web length of 9.375 inches, resulting in a cross-sectional area for the "column" of

$$A = 2(0.75)(7.5) + \left(\frac{3}{8}\right)(9.375) = 14.77 \text{ in.}^2$$

FIGURE 10.13

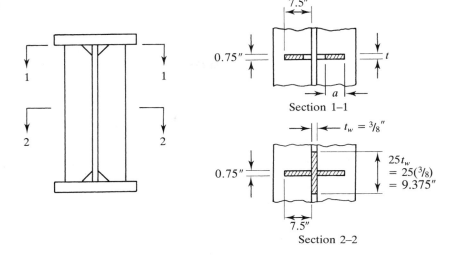

The moment of inertia of this area about an axis in the web is

$$I = \Sigma(\bar{I} + Ad^2)$$

$$= \frac{9.375(3/8)^3}{12} + 2\left[\frac{0.75(7.5)^3}{12} + 7.5(0.75)\left(\frac{7.5}{2} + \frac{3/8}{2}\right)^2\right]$$

and the radius of gyration is

$$r = \sqrt{\frac{I}{A}} = \sqrt{\frac{227.2}{14.77}} = 3.922 \text{ in.}$$

The slenderness ratio is

$$\frac{KL}{r} = \frac{Kh}{r} = \frac{0.75(63)}{3.922} = 12.05 < 25$$

$$\therefore P_n = F_y A_g = 36(14.77) = 531.7 \text{ kips}$$

and

$$\phi P_n = 0.90(531.7) = 479 \text{ kips} > 60 \text{ kips} \qquad \text{(OK)}$$

For the bearing stiffeners at the supports, from Figure 10.14, the nominal bearing strength is

$$R_n = 1.8F_y A_{pb} = 1.8(36)[4(6.5)(0.75)] = 1264 \text{ kips}$$

and

$$\phi R_n = 0.75(1264) = 948 \text{ kips} > 234 \text{ kips} \qquad \text{(OK)}$$

FIGURE 10.14

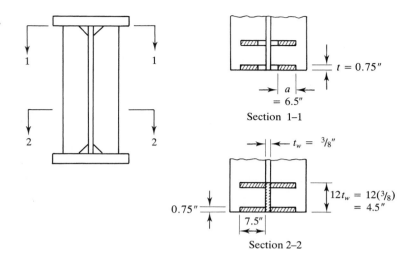

Section 1–1

Section 2–2

Check the stiffener–web assembly as a compression member. Referring to Figure 10.14, the moment of inertia about an axis in the plane of the web is

$$I = \Sigma(\bar{I} + Ad^2)$$

$$= \frac{4.5(3/8)^3}{12} + 4\left[\frac{0.75(7.5)^3}{12} + 7.5(0.75)\left(\frac{7.5}{2} + \frac{3/8}{2}\right)^2\right] = 454.3 \text{ in.}^4$$

and the area and radius of gyration are

$$A = 4.5\left(\frac{3}{8}\right) + 4(0.75)(7.5) = 24.19 \text{ in.}^2$$

and

$$r = \sqrt{\frac{I}{A}} = \sqrt{\frac{454.3}{24.19}} = 4.334 \text{ in.}$$

The slenderness ratio is

$$\frac{Kh}{r} = \frac{0.75(63)}{4.334} = 10.90 < 25$$

$$\therefore P_n = F_y A_g = 36(24.19) = 870.8 \text{ kips}$$

$$\phi P_n = 0.90(870.8) = 784 \text{ kips} > 234 \text{ kips} \qquad \text{(OK)}$$

Answer The bearing stiffeners are adequate.

ASD Solution The allowable strength solution of this problem is practically identical to the load and resistance factor design solution. With one exception, the only differences are in the loads and whether a resistance factor or a safety factor is used. Because of this, the ASD solution presented here will make use of most of the LRFD calculations, which will not be repeated. From the LRFD solution, the web–slenderness limitation of AISC F13.2 is satisfied.

a. **Flexural strength**: This is the same as the LRFD solution up to the computation of C_b. For this, we need the service load moments. Figure 10.15 shows the loading, shear, and bending moment diagrams for the service loads (verification of the values is left as an exercise for the reader). Compute C_b for the 12-foot unbraced segment adjacent to midspan. Dividing this segment into four equal spaces, we get points A, B, and C located at 15 ft, 18 ft, and 21 ft from the left end of the girder. The corresponding bending moments are

$$M_A = 156(15) - 40(3) - 4(15)^2/2 = 1770 \text{ ft-kips}$$
$$M_B = 156(18) - 40(6) - 4(18)^2/2 = 1920 \text{ ft-kips}$$
$$M_C = 156(21) - 40(9) - 4(21)^2/2 = 2034 \text{ ft-kips}$$

From AISC Equation F1-1,

$$C_b = \frac{12.5 M_{max}}{2.5 M_{max} + 3M_A + 4M_B + 3M_C} R_m \leq 3.0$$

$$= \frac{12.5(2112)}{2.5(2112) + 3(1770) + 4(1920) + 3(2034)} (1.0) = 1.083 < 3.0$$

FIGURE 10.15

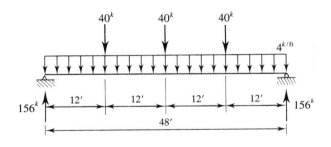

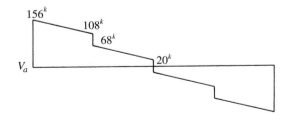

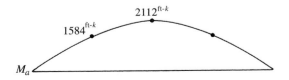

where R_m has a value of 1.0 for all doubly symmetric sections. This value of C_b is the same as that obtained in the LRFD solution (this is typically the case), so the critical stress for lateral-torsional buckling is the same as for the LRFD solution. The nominal flexural strength is therefore based on yielding of the compression flange, and

$$M_n = 3723 \text{ ft-kips}$$

The allowable strength is

$$\frac{M_n}{\Omega_b} = \frac{3723}{1.67} = 2229 \text{ ft-kips}$$

From Figure 10.15, the maximum moment is

$$M_a = 2112 \text{ ft-kips} < 2229 \text{ ft-kips} \qquad \text{(OK)}$$

Answer The flexural strength is adequate.

b. **Shear strength**: From the LRFD solution, the nominal shear strength (except for the end panels), is

$$V_n = 263.2 \text{ kips}$$

The allowable strength is

$$\frac{V_n}{\Omega_v} = \frac{263.2}{1.67} = 157.6 \text{ kips}$$

From Figure 10.15, the maximum shear in the middle half of the girder is 68 kips, so the shear strength is adequate where tension-field action is permitted.

From the LRFD solution, the nominal shear strength in the end panels is

$$V_n = 131.0 \text{ kips}$$

and the allowable strength is

$$\frac{V_n}{\Omega_v} = \frac{131.0}{1.67} = 78.4 \text{ kips}$$

The maximum shear in the end panel is

$$V_a = 156 \text{ kips} > 78.4 \text{ kips} \qquad \text{(N.G.)}$$

Two options are available for increasing the shear strength: either decrease the web slenderness or decrease the aspect ratio of each end panel by adding an intermediate stiffener. Stiffeners will be added in this example.

The location of the first intermediate stiffener will be determined by the following strategy: First, equate the shear strength from AISC Equation G2-1 to the required shear strength and solve for the required value of C_v. Next, solve for k_v from Equation G2-5, then solve for a/h. The required shear strength is

$$\text{Required } V_n = \Omega_v V_a = 1.67(156) = 260.\,5 \text{ kips}$$

This is the same as the required V_n from the LRFD solution. The slight difference is caused by the rounding of the safety factor Ω from $\frac{5}{3}$ to 1.67 (this rounding was noted in Chapter 2). Therefore, we will use the results of the LRFD solution:

Required $a = 53.9$ in.

Although a is defined as the clear spacing, we will treat it conservatively as a center-to-center spacing and place the first intermediate stiffener at 54 inches from the end of the girder. This placement will give a strength that approximately equals the maximum shear of 156 kips. Since the allowable shear strength outside of the end panels is 157.6 kips, which is more than the maximum shear in the girder, no additional stiffeners will be needed.

The determination of the stiffener spacing can be facilitated by the use of the design curves in Part 3 of the *Manual*.

Answer   The shear strength is inadequate. Add one intermediate stiffener 54 inches from each end of the girder.

c. **Bearing stiffeners**: Bearing stiffeners are provided at each concentrated load, so there is no need to check for web yielding, web crippling, or sidesway web buckling.

For the interior bearing stiffeners, the nominal bearing strength (see the LRFD solution) is

$R_n = 631.8$ kips

and the allowable strength is

$$\frac{R_n}{\Omega} = \frac{631.8}{2.00} = 316 \text{ kips} > 40 \text{ kips} \qquad \text{(OK)}$$

From the LRFD solution, the nominal strength of the stiffener assembly as a compression member is

$P_n = 531.7$ kips

The allowable strength is

$$\frac{P_n}{\Omega} = \frac{531.7}{1.67} = 318 \text{ kips} > 40 \text{ kips} \qquad \text{(OK)}$$

For the bearing stiffeners at the supports, the nominal bearing strength from the LRFD solution is

$R_n = 1264$ kips

and the allowable strength is

$$\frac{R_n}{\Omega} = \frac{1264}{2.00} = 632 \text{ kips} > 156 \text{ kips} \qquad \text{(OK)}$$

From the LRFD solution, the nominal strength of the stiffener assembly as a compression member is

$$P_n = 870.8 \text{ kips}$$

and the allowable strength is

$$\frac{P_n}{\Omega} = \frac{870.8}{1.67} = 521 \text{ kips} > 156 \text{ kips} \qquad \text{(OK)}$$

Answer   The bearing stiffeners are adequate.

10.7 DESIGN

The primary task in plate girder design is to determine the size of the web and the flanges. If a variable moment of inertia is desired, decisions must be made regarding the method of varying the flange size—that is, whether to use cover plates or different thicknesses of flange plate at different points along the length of the girder. A decision about whether to use intermediate stiffeners must be made early in the process because it will affect the web thickness. If bearing stiffeners are needed, they must be designed. Finally, the various components must be connected by properly designed welds. The following step-by-step procedure is recommended.

1. **Select the overall depth**. As a rule of thumb, a well-proportioned girder will have a depth of one tenth to one twelfth the span length. As with any beam design, constraints on the maximum depth could establish the depth by default. Building code limitations on the depth-to-span ratio or the deflection would also influence the selection.

2. **Select a trial web size**. The web depth can be estimated by subtracting twice the flange thickness from the overall depth selected. Of course, at this stage of the design, the flange thickness must also be estimated, but the consequences of a poor estimate are minor. The web thickness t_w can then be found by using the following limitations as a guide.
   For $a/h \leq 1.5$,

$$\frac{h}{t_w} \leq 11.7 \sqrt{\frac{E}{F_y}} \qquad \text{(AISC Equation F13-3)}$$

   For $a/h > 1.5$,

$$\frac{h}{t_w} \leq \frac{0.42E}{F_y} \qquad \text{(AISC Equation F13-4)}$$

Once h and t_w have been selected, determine whether the web width–thickness ratio qualifies this member as a slender-web flexural member. If so, the provisions of AISC F5 can be used. (If the web is noncompact, AISC F5 can still be used, but it will be conservative.)

3. **Estimate the flange size**. The required flange area can be estimated from a simple formula derived as follows. Let

$$I_x = I_{\text{web}} + I_{\text{flanges}}$$

$$\approx \frac{1}{12}t_w h^3 + 2A_f y^2 \approx \frac{1}{12}t_w h^3 + 2A_f (h/2)^2 \qquad (10.6)$$

where

A_f = cross-sectional area of one flange

y = distance from the elastic neutral axis to the centroid of the flange

The contribution of the moment of inertia of each flange about its own centroidal axis was neglected in Equation 10.6. The section modulus can be estimated as

$$S_x = \frac{I_x}{c} \approx \frac{t_w h^3/12}{h/2} + \frac{2A_f (h/2)^2}{h/2} = \frac{t_w h^2}{6} + A_f h$$

If we assume that compression flange buckling will control the design, we can find the required section modulus from AISC Equation F5-7:

$$M_n = R_{pg} F_{cr} S_{xc}$$

$$\text{Required } S_{xc} = \frac{M_{n\,req}}{R_{pg} F_{cr}}$$

where $M_{n\,req}$ is the required nominal moment strength. Equating the required section modulus to the approximate value, we have

$$\frac{M_{n\,req}}{R_{pg} F_{cr}} = \frac{t_w h^2}{6} + A_f h$$

$$A_f = \frac{M_{n\,req}}{h R_{pg} F_{cr}} - \frac{t_w h}{6}$$

If we assume that $R_{pg} = 1.0$ and $F_{cr} = F_y$, the required area of one flange is

$$A_f = \frac{M_{n\,req}}{h F_y} - \frac{A_w}{6}$$

where

$M_{n\,req}$ = Required nominal flexural strength

   $= M_u / \phi_b$ for LRFD

   $= \Omega_b M_a$ for ASD

A_w = web area

Once the required flange area has been determined, select the width and thickness. If the thickness originally used in the estimate of the web depth is retained, no adjustment in the web depth will be needed. At this point, an estimated girder weight can be computed, then $M_{n\,req}$ and A_f should be recomputed.

4. **Check the bending strength of the trial section**.
5. **Check shear**. If an end panel is being considered, or if intermediate stiffeners are not used, the shear strength can be found from AISC Equation G2-1, which gives the strength in the absence of a tension field. The design curves in Part 3 of the *Manual* can be used for this purpose. If the curves are not used, the intermediate stiffener spacing can be determined as follows:
   a. Equate the required shear strength to the shear strength given by AISC Equation G2-1 and solve for the required value of C_v.
   b. Solve for k_v from AISC Equation G2-4 or G2-5.
   c. Solve for the required value of a/h from

$$k_v = 5 + \frac{5}{(a/h)^2}$$

   If tension-field action is used, either a trial-and-error approach or the design curves in Part 3 of the *Manual* can be used to obtain the required a/h.
   If there is a tension field, the intermediate stiffeners must be proportioned to satisfy the area requirement of AISC Equation G3-3 and the width–thickness requirement of Equation 10.3. If tension-field action is not used, the moment of inertia requirement of AISC G2.2 must be satisfied.

6. **Design bearing stiffeners**. To determine whether bearing stiffeners are needed, check the web resistance to concentrated loads (web yielding, web crippling, and web sidesway buckling). Alternatively, bearing stiffeners can be provided to fully resist the concentrated loads, and the web need not be checked. If bearing stiffeners are used, the following design procedure can be used.
   a. Try a width that brings the edge of the stiffener near the edge of the flange and a thickness that satisfies Equation 10.5:

$$\frac{b}{t} \le 0.56\sqrt{\frac{E}{F_{yst}}}$$

   b. Compute the cross-sectional area needed for bearing strength. Compare this area with the trial area and revise if necessary.
   c. Check the stiffener–web assembly as a compression member.

7. **Design the flange-to-web welds, stiffener-to-web welds, and any other connections** (flange segments, web splices, etc.).

Example 10.2   Design a simply supported plate girder to span 60 feet and support the service loads shown in Figure 10.16a. The maximum permissible depth is 65 inches. Use A36 steel and E70XX electrodes and assume that the girder has continuous lateral support. The ends have bearing-type supports and are not framed. Use LRFD.

Solution   The factored loads, excluding the girder weight, are shown in Figure 10.16b. Determine the overall depth.

FIGURE 10.16

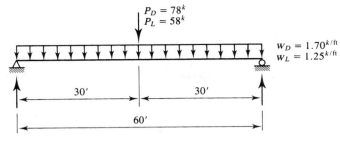

(a) Service Loads

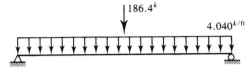

(b) Factored Loads
(Girder weight not included)

$$\frac{\text{Span length}}{10} = \frac{60(12)}{10} = 72 \text{ in.}$$

$$\frac{\text{Span length}}{12} = \frac{60(12)}{12} = 60 \text{ in.}$$

Use the maximum permissible depth of 65 inches.

Try a flange thickness of $t_f = 1.5$ inches and a web depth of

$$h = 65 - 2(1.5) = 62 \text{ in.}$$

To determine the web thickness, first examine the limiting values of h/t_w.

From AISC Equations F13-3 and F13-4:

For $\dfrac{a}{h} \leq 1.5$,

$$\frac{h}{t_w} \leq 11.7\sqrt{\frac{E}{F_y}} = 11.7\sqrt{\frac{29,000}{36}} = 332.1$$

$$t_w \geq \frac{62}{332.1} = 0.187 \text{ in.}$$

For $\dfrac{a}{h} > 1.5$,

$$\frac{h}{t_w} \leq \frac{0.42E}{F_y} = \frac{0.42(29,000)}{36} = 338.3$$

$$t_w \geq \frac{62}{338.3} = 0.183 \text{ in.}$$

Try a ⁵⁄₁₆ × 62 web plate. Determine whether the web is slender.

$$\frac{h}{t_w} = \frac{62}{5/16} = 198.4$$

$$5.70\sqrt{\frac{E}{F_y}} = 5.70\sqrt{\frac{29,000}{36}} = 161.8 < 198.4 \quad \therefore \quad \text{the web is slender}$$

Determine the required flange size. From Figure 10.16b, the maximum factored-load bending moment is

$$M_u = \frac{186.4(60)}{4} + \frac{4.040(60)^2}{8} = 4614 \text{ ft-kips}$$

From Equation 10.7, the required flange area is

$$A_f = \frac{M_{n\,req}}{hF_y} - \frac{A_w}{6}$$

$$= \frac{M_u/\phi_b}{hF_y} - \frac{A_w}{6} = \frac{(4614 \times 12)/0.90}{62(36)} - \frac{62(5/16)}{6} = 24.33 \text{ in.}^2$$

If the original estimate of the flange thickness is retained, the required width is

$$b_f = \frac{A_f}{t_f} = \frac{24.33}{1.5} = 16.2 \text{ in.}$$

Try a 1½ × 18 flange plate. The girder weight can now be computed.

Web area:	$62(5/16) = 19.38 \text{ in.}^2$
Flange area:	$2(1.5 \times 18) = 54.00 \text{ in.}^2$
Total:	73.38 in.^2

$$\text{Weight} = \frac{73.38}{144}(490) = 250 \text{ lb/ft}$$

The adjusted bending moment is

$$M_u = 4614 + \frac{(1.2 \times 0.250)(60)^2}{8} = 4749 \text{ ft-kips}$$

Figure 10.17 shows the trial section, and Figure 10.18 shows the shear and bending moment diagrams for the factored loads, which include the girder weight of 250 lb/ft.

Check the flexural strength of the trial section. From Figure 10.17, the moment of inertia about the axis of bending is

$$I_x = \frac{(5/16)(62)^3}{12} + 2(1.5)(18)(31.75)^2 = 60,640 \text{ in.}^4$$

and the elastic section modulus is

$$S_x = \frac{I_x}{c} = \frac{60,640}{32.5} = 1866 \text{ in.}^3$$

FIGURE 10.17

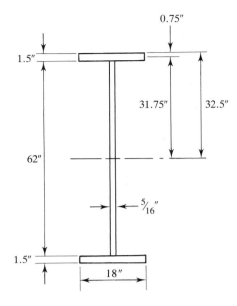

An examination of AISC Equations F5-7 and F5-10 shows that for a girder with a symmetrical cross section, the flexural strength will never be controlled by tension-flange yielding; therefore only compression-flange buckling will be investigated. Furthermore, as this girder has continuous lateral support, lateral-torsional buckling need not be considered.

FIGURE 10.18

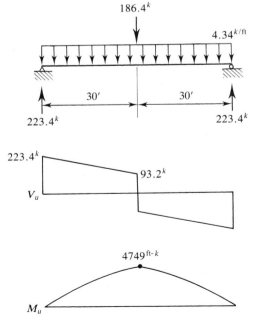

Determine whether the compression flange is compact, noncompact, or slender.

$$\lambda = \frac{b_f}{2t_f} = \frac{18}{2(1.5)} = 6.0$$

$$\lambda_p = 0.38\sqrt{\frac{E}{F_y}} = 0.38\sqrt{\frac{29,000}{36}} = 10.79$$

Since $\lambda < \lambda_p$, there is no flange local buckling. The compression-flange strength is therefore based on yielding, and $F_{cr} = F_y = 36$ ksi.

To compute the plate girder strength reduction factor R_{pg}, the value of a_w will be needed:

$$a_w = \frac{h_c t_w}{b_{fc} t_{fc}} = \frac{62(5/16)}{18(1.5)} = 0.7176 < 10$$

From AISC Equation F5-6,

$$R_{pg} = 1 - \frac{a_w}{1200 + 300a_w}\left(\frac{h_c}{t_w} - 5.7\sqrt{\frac{E}{F_y}}\right) \le 1.0$$

$$= 1 - \frac{0.7176}{1200 + 300(0.7176)}\left(198.4 - 5.7\sqrt{\frac{29,000}{36}}\right) = 0.9814$$

From AISC Equation F5-7, the nominal flexural strength for the compression flange is

$$M_n = R_{pg}F_{cr}S_{xc} = 0.9814(36)(1866) = 65,930 \text{ in.-kips} = 5494 \text{ ft-kips}$$

and the design strength is

$$\phi_b M_n = 0.90(5494) = 4945 \text{ ft-kips} > 4749 \text{ ft-kips} \qquad \text{(OK)}$$

Although this capacity is somewhat more than needed, the excess will compensate for the weight of stiffeners and other incidentals that we have not accounted for.

Answer   Use a $\frac{5}{16} \times 62$ web and $1\frac{1}{2} \times 18$ flanges, as shown in Figure 10.17.

Check the shear strength. The shear is maximum at the support, but tension-field action cannot be used in an end panel. Table 3-16a in Part 3 of the *Manual* will be used to obtain the required size of the end panel. The curves will be entered with values of h/t_w and $\phi_v V_n/A_w$, where

$$\frac{h}{t_w} = 198.4$$

$$A_w = 62\left(\frac{5}{16}\right) = 19.38 \text{ in.}^2$$

$$\frac{\phi_v V_n}{A_w} = \frac{223.4}{19.38} = 11.5 \text{ ksi}$$

Using $h/t_w = 198$ and $\phi_v V_n/A_w = 12$ ksi, we get a value of a/h of approximately 0.60. The corresponding intermediate stiffener spacing is

$$a = 0.60h = 0.60(62) = 37.2 \text{ in.}$$

Although the required stiffener spacing is a clear distance, the use of center-to-center distances is somewhat simpler and will be slightly conservative. In addition, because of the approximations involved in using the curves, we will be conservative in rounding the value of a. Use a distance of 36 inches from the center of the end bearing stiffener to the center of the first intermediate stiffener. (The calculations are not shown here, but for $a = 36$ in., the computed value of $\phi_v V_n$ is 231 kips, which is greater than the required value of 223.4 kips.)

Determine the intermediate stiffener spacings needed for shear strength outside the end panels. At a distance of 36 inches from the left end, the shear is

$$V_u = 223.4 - 4.34 \left(\frac{36}{12} \right) = 210.4 \text{ kips}$$

$$\text{Required } \frac{\phi_v V_n}{A_w} = \frac{210.4}{19.38} = 10.86 \text{ ksi}$$

Tension-field action can be used outside the end panels, so the curves of AISC Table 3-16b will be used. For $h/t_w = 198$ and $\phi_v V_n/A_w = 11$ ksi,

$$\frac{a}{h} \approx 1.60$$

The required stiffener spacing is

$$a = 1.60h = 1.60(62) = 99.2 \text{ in.}$$

This maximum spacing will apply for the remaining distance to the centerline of the girder. This distance is

$$30(12) - 36 = 324 \text{ in.}$$

The number of spaces required is

$$\frac{324}{99.2} = 3.27$$

Use four spaces. This results in a spacing of

$$\frac{324}{4} = 81 \text{ in.}$$

Before proceeding, check the conditions of AISC G3.1 to be sure that tension-field action can be used for this girder and this stiffener spacing.

a. $\dfrac{a}{h} = \dfrac{81}{62} = 1.306 < 3$

(This condition was automatically satisfied by use of the curves.)

b. $\dfrac{a}{h} < \left[\dfrac{260}{(h/t_w)} \right]^2 = \left[\dfrac{260}{(198.4)} \right]^2 = 1.717$

(This condition was automatically satisfied by use of the curves.)

c. $A_w = ht_w = 62(5/16) = 19.38$ in.^{2}

$A_{fc} = A_{ft} = b_f t_f = 18(1.5) = 27.0$ in.^{2}

$$\frac{2A_w}{(A_{fc} + A_{ft})} = \frac{2(19.38)}{(27 + 27)} = 0.7178 < 2.5 \qquad \text{(OK)}$$

d. $\dfrac{h}{b_{fc}} = \dfrac{h}{b_{ft}} = \dfrac{62}{18} = 3.444 < 6 \qquad$ (OK)

The following spacings will be used from the each end of the girder: one at 36 inches and four at 81 inches, as shown in Figure 10.19.

The cross section of the intermediate stiffeners is based on three criteria:

1. A minimum area (when there is a tension field).
2. A minimum width-thickness ratio (when there is a tension field).
3. A minimum moment of inertia (in the absence of a tension field).

For convenience, we will apply all three criteria to all intermediate stiffeners and use one size. From AISC Equation G3-3,

$$A_{st} > \frac{F_y}{F_{yst}} \left[0.15 D_s h t_w (1 - C_v) \frac{V_r}{V_c} - 18 t_w^2 \right] \geq 0$$

For stiffeners in pairs, $D_s = 1.0$. Determine C_v. For $a = 81$ in.,

$$\frac{a}{h} = 1.306$$

$$k_v = 5 + \frac{5}{(a/h)^2} = 5 + \frac{5}{(1.306)^2} = 7.931$$

$$1.10\sqrt{\frac{k_v E}{F_y}} = 1.10\sqrt{\frac{7.931(29,000)}{36}} = 87.92$$

$$1.37\sqrt{\frac{k_v E}{F_y}} = 1.37\sqrt{\frac{7.931(29,000)}{36}} = 109.5$$

$$\frac{h}{t_w} = 198.4 > 109.5 \qquad \therefore C_v = \frac{1.51 E k_v}{(h/t_w)^2 F_y} = \frac{1.51(29,000)(7.931)}{(198.4)^2(36)} = 0.2451$$

FIGURE 10.19

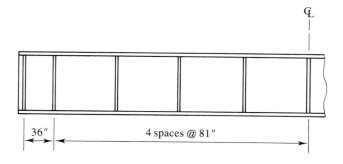

36"   4 spaces @ 81"

Conservatively, use $V_r/V_c = 1$. Then

$$A_{st} > \frac{36}{36}[0.15(1.0)(62)(5/16)(1 - 0.2451)(1.0) - 18(5/16)^2] = 0.436 \text{ in.}^2$$

From Equation 10.3,

$$\frac{b}{t} \leq 0.56\sqrt{\frac{E}{F_{yst}}} = 0.56\sqrt{\frac{29,000}{36}} = 15.9$$

For the moment of inertia requirement, from AISC Equation G2-6,

$$j = \frac{2.5}{(a/h)^2} - 2 \geq 0.5$$

$$j = \frac{2.5}{(1.306)^2} - 2 = -0.5343 \quad \therefore \text{ use } j = 0.5$$

$$I_{st} = at_w^3 j = 81(5/16)^3(0.5) = 1.24 \text{ in.}^4$$

Try two ¼ × 4 plates:

$$\frac{b}{t} = \frac{4}{1/4} = 16 \approx 15.9 \qquad (\text{say OK})$$

$$A_{st} \text{ provided} = 2(4)\left(\frac{1}{4}\right) = 2.0 \text{ in.}^2 > 0.436 \text{ in.}^2 \quad (\text{OK})$$

From Figure 10.20 and the parallel-axis theorem,

$$I_{st} = \Sigma(\bar{I} \times Ad^2)$$
$$= \left[\frac{0.25(4)^3}{12} + 0.25(4)(2 + 5/32)^2\right] \times 2 \text{ stiffeners}$$
$$= 11.97 \text{ in.}^4 > 1.24 \text{ in.}^4 \quad (\text{OK})$$

FIGURE 10.20

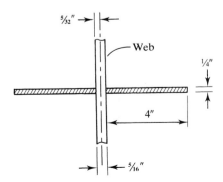

To determine the length of the stiffeners, first compute the distance between the stiffener-to-web weld and the web-to-flange weld (see Figure 10.9):

$$\text{Minimum distance} = 4t_w = 4\left(\frac{5}{16}\right) = 1.25 \text{ in.}$$

$$\text{Maximum distance} = 6t_w = 6\left(\frac{5}{16}\right) = 1.875 \text{ in.}$$

If we assume a flange-to-web weld size of $\frac{5}{16}$ inch and 1.25 inches between welds, the approximate length of the stiffener is

$$h - \text{weld size} - 1.25 = 62 - 0.3125 - 1.25$$
$$= 60.44 \text{ in.} \quad \therefore \text{use 60 in.}$$

Answer Use two plates $\frac{1}{4} \times 4 \times 5'$-$0''$ for the intermediate stiffeners.

Bearing stiffeners will be provided at the supports and at midspan. Since there will be a stiffener at each concentrated load, there is no need to investigate the resistance of the web to these loads. If the stiffeners were not provided, the web would need to be protected from yielding and crippling. To do so, enough bearing length, N, as required by AISC Equations J10-2 through J10-5, must be provided. Sidesway web buckling would not be an applicable limit state because the girder has continuous lateral support (making the unbraced length $\ell = 0$ and $(h/t_w)/(\ell/b_f) > 2.3$).

Try a stiffener width, b, of 8 inches. The total combined width will be $2(8) + \frac{5}{16}$ (the web thickness) = 16.31 inches, or slightly less than the flange width of 18 inches. From Equation 10.5,

$$\frac{b}{t} \le 0.56\sqrt{\frac{E}{F_{yst}}}$$

$$t > \frac{b}{0.56}\sqrt{\frac{F_{yst}}{E}} = \frac{8}{0.56}\sqrt{\frac{36}{29,000}} = 0.503 \text{ in.}$$

Try two $\frac{3}{4} \times 8$ stiffeners. Assume a $\frac{3}{16}$-inch web-to-flange weld and a $\frac{1}{2}$-inch cut-out in the stiffener. Check the stiffener at the support. The bearing strength is

$$R_n = 1.8F_y A_{pb} = 1.8(36)(0.75)(8 - 0.5) \times 2 = 729.0 \text{ kips}$$
$$\phi R_n = 0.75(729.0) = 547 \text{ kips} > 223.4 \text{ kips}$$

Check the stiffener as a column. The length of web acting with the stiffener plates to form a compression member is 12 times the web thickness for an end stiffener (AISC J10.8). As shown in Figure 10.21, this length is $12(\frac{5}{16}) = 3.75$ in. If we locate

FIGURE 10.21

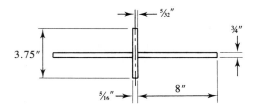

FIGURE 10.22

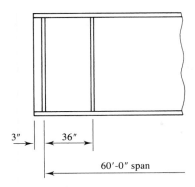

3" |← 36" →|

60'-0" span

the stiffener centrally within this length, the point of support (location of the girder reaction) will be approximately $3.75/2 = 1.875$ inches from the end of the girder. Use 3 inches, as shown in Figure 10.22, but base the computations on a total length of web of 3.75 inches, which gives

$$A = 2(8)\left(\frac{3}{4}\right) + \left(\frac{5}{16}\right)(3.75) = 13.17 \text{ in.}^2$$

$$I = \frac{3.75(5/16)^3}{12} + 2\left[\frac{0.75(8)^3}{12} + 8\left(\frac{3}{4}\right)\left(4 + \frac{5}{32}\right)^2\right] = 271.3 \text{ in.}^4$$

$$r = \sqrt{\frac{I}{A}} = \sqrt{\frac{271.3}{13.17}} = 4.539 \text{ in.}$$

$$\frac{KL}{r} = \frac{Kh}{r} = \frac{0.75(62)}{4.539} = 10.24 < 25$$

$$\therefore P_n = F_y A_g = 36(13.17) = 474.1 \text{ kips}$$

$$\phi P_n = 0.90(474.1) = 427 \text{ kips} > 223.4 \text{ kips} \qquad \text{(OK)}$$

Since the load at midspan is smaller than the reaction, use the same stiffener at midspan.

Answer Use two plates $\frac{3}{4} \times 8 \times 5'\text{-}2''$ for the bearing stiffeners.

At this point, all components of the girder have been sized. The connections of these elements will now be designed. E70 electrodes, with a design strength of 1.392 kips/inch per sixteenth of an inch in size, will be used for all welds.

For the flange-to-web welds, compute the horizontal shear flow at the flange-to-web junction:

$$\text{Maximum } V_u = 223.4 \text{ kips}$$
$$Q = \text{flange area} \times 31.75 \text{ (see Figure 10.17)}$$
$$= 1.5(18)(31.75) = 857.2 \text{ in.}^3$$
$$I_x = 60,640 \text{ in.}^4$$

$$\text{Maximum } \frac{V_u Q}{I_x} = \frac{223.4(857.2)}{60,640} = 3.158 \text{ kips/in.}$$

For the plate thicknesses being welded, the minimum weld size, w, is $\frac{3}{16}$ inch. If intermittent welds are used, their minimum length is

$$L_{min} = 4 \times w \ge 1.5 \text{ in.}$$

$$= 4\left(\frac{3}{16}\right) = 0.75 \text{ in.} \quad \therefore \text{ use } 1.5 \text{ in.}$$

Try $\frac{3}{16}$-in. × $1\frac{1}{2}$-in. fillet welds:

Capacity per inch $= 1.392 \times 3$ sixteenths $\times 2$ welds $= 8.352$ kips/in.

Check the capacity of the base metal. The web is the thinner of the connected parts and controls. From Equation 7.35, the base metal shear *yield* strength per unit length is

$$\phi R_n = 0.6 F_y t = 0.6(36)\left(\frac{5}{16}\right) = 6.750 \text{ kips/in.}$$

From Equation 7.36, the base metal shear *rupture* strength per unit length is

$$\phi R_n = 0.45 F_u t = 0.45(58)\left(\frac{5}{16}\right) = 8.156 \text{ kips/in.}$$

The base metal shear strength is therefore 6.750 kips/in. < 8.352 kips/in.

Use a total weld capacity of 6.750 kips/in. The capacity of a 1.5-inch length of a pair of welds is

$$6.750(1.5) = 10.13 \text{ kips}$$

To determine the spacing, let

$$\frac{10.13}{s} = \frac{V_u Q}{I_x}$$

where s is the center-to-center spacing of the welds in inches and

$$s = \frac{10.13}{V_u Q / I_x} = \frac{10.13}{3.158} = 3.21 \text{ in.}$$

Using a center-to-center spacing of 3 inches will give a clear spacing of $3 - 1.5 = 1.5$ inches. The AISC Specification refers to spacing of intermittent fillet welds in Section F13, "Proportions of Beams and Girders," under "Cover Plates." The provisions of AISC D4 and E6 are to be used, although only Section E6 is relevant. This calls for a maximum clear spacing of

$$d \le 0.75 \sqrt{\frac{E}{F_y}} t, \text{ but no greater than 12 in.}$$

Adapting these limits to the present case yields

$$0.75 \sqrt{\frac{E}{F_y}} t = 0.75 \sqrt{\frac{29,000}{36}} (1.5) = 31.9 \text{ in.} > 12 \text{ in.}$$

The maximum permissible clear spacing is therefore 12 inches, and the required clear spacing of 1.5 inches is satisfactory.

There is no minimum spacing given in the Specification, but the AISC publication, "Detailing for Steel Construction," (AISC, 2002) states that intermittent welds are more economical than continuous welds only if the center-to-center spacing is more than the length of the weld. In this example, the spacing is the same as the length, so either type could be used.

Although the 3-inch center-to-center spacing can be used for the entire length of the girder, an increased spacing can be used where the shear is less than the maximum of 223.4 kips. Three different spacings will be investigated:

1. The closest required spacing of 3 inches.
2. The maximum permissible center-to-center spacing of $12 + 1.5 = 13.5$ inches.
3. An intermediate spacing of 5 inches.

The 5-inch spacing can be used when

$$\frac{V_u Q}{I_x} = \frac{10.13}{s} \quad \text{or} \quad V_u = \frac{10.13 I_x}{Q s} = \frac{10.13\,(60{,}640)}{857.2(5)} = 143.3 \text{ kips}$$

Refer to Figure 10.18 and let x be the distance from the left support, giving

$$V_u = 223.4 - 4.34x = 143.3 \text{ kips}$$
$$x = 18.46 \text{ ft}$$

The 13.5-inch spacing can be used when

$$V_u = \frac{10.13 I_x}{Q s} = \frac{10.13\,(60{,}640)}{857.2(13.5)} = 53.08 \text{ kips}$$

Figure 10.18 shows that the shear never gets this small, so the maximum spacing never controls.

Answer   Use $\frac{3}{16}$-inch $\times$ $1\frac{1}{2}$-inch fillet welds for the flange-to-web welds, spaced as shown in Figure 10.23.

For the intermediate stiffener welds:

Minimum weld size $= \dfrac{1}{8}$ in.   (based on the stiffener thickness of $\dfrac{1}{4}$ in.)

Minimum length $= 4\left(\dfrac{1}{8}\right) = 0.5$ in. < 1.5 in.   use 1.5 in.

FIGURE 10.23

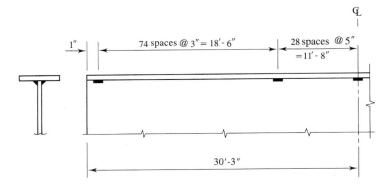

Use two welds per stiffener for a total of four. The capacity per inch for two ⅛-inch fillet welds per stiffener plate is

$$1.392 \times 2 \times 2 = 5.568 \text{ kips/in.}$$

Check the shear strength of the stiffener (the thinner of the two connected parts). From Equation 7.35, the shear *yield* strength per unit length is

$$\phi R_n = 0.6 F_y t = 0.6(36)\left(\frac{1}{4}\right) = 5.400 \text{ kips/in.}$$

From Equation 7.36, the base metal shear *rupture* strength per unit length is

$$\phi R_n = 0.45 F_u t = 0.45(58)\left(\frac{1}{4}\right) = 6.525 \text{ kips/in.}$$

The base metal shear strength is therefore 5.400 kips/in. per stiffener. This is less than the shear strength of two welds (using two welds for each plate), so use a weld strength of 5.400 kips/in. For the two plates (four welds), use

$$5.400 \times 2 = 10.80 \text{ kips/in.}$$

From Equation 10.4, the shear to be transferred is

$$f = 0.045h\sqrt{\frac{F_y^3}{E}} = 0.045(62)\sqrt{\frac{(36)^3}{29{,}000}} = 3.539 \text{ kips/in.}$$

Use intermittent welds. The capacity of a 1.5-inch length of the 4 welds is

$$1.5(10.80) = 16.20 \text{ kips}$$

Equating the shear strength per inch and the required strength gives

$$\frac{16.20}{s} = 3.539 \text{ kips/in. or } s = 4.58 \text{ in.}$$

From AISC G2.2, the maximum clear spacing is 16 times the web thickness but no greater than 10 inches, or

$$16t_w = 16\left(\frac{5}{16}\right) = 5 \text{ in.}$$

Use a center-to-center spacing of 4½ inches, resulting in a clear spacing of

$$4.5 - 1.5 = 3 \text{ in.} < 5 \text{ in.} \quad \text{(OK)}$$

Answer Use ⅛-inch × 1½-inch fillet welds for intermediate stiffeners, spaced as shown in Figure 10.24.

For the bearing stiffener welds:

Minimum weld size $= \dfrac{3}{16}$ in.   (based on the thickness $t_w = \dfrac{5}{16}$ in.)

Minimum length $= 4\left(\dfrac{3}{16}\right) = 0.75$ in. < 1.5 in.   ∴ use 1.5 in.

FIGURE 10.24

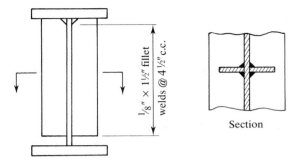

Section

Use two welds per stiffener for a total of four. The capacity per inch for two $\frac{3}{16}$-inch fillet welds per stiffener plate is

$$1.392 \times 3 \times 2 = 8.352 \text{ kips/in.}$$

Check the shear strength of the web. From the design of the flange-to-web welds, the base metal shear strength is 6.750 kips/in. per stiffener. This is less than the shear strength of two welds (using two welds for each plate), so use a weld strength of 6.750 kips/in. For the two plates (four welds), use

$$6.750 \times 2 = 13.\,50 \text{ kips/in}$$

The capacity of a 1.5-inch length of four welds is

$$1.5(13.50) = 20.25 \text{ kips}$$

For the end bearing stiffener, the applied load per inch is

$$\frac{\text{Reaction}}{\text{Length available for weld}} = \frac{223.4}{62 - 2(0.5)} = 3.662 \text{ kips/in.}$$

From $\dfrac{20.25}{s} = 3.662$, $s = 5.53$ inches. (Note that a smaller weld spacing is required for the internediate stiffeners.)

FIGURE 10.25

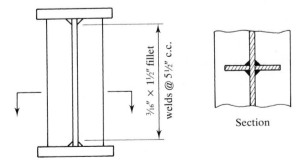

Section

Answer   Use $\frac{3}{16}$-inch $\times$ $1\frac{1}{2}$-inch fillet welds for all bearing stiffeners, spaced as shown in Figure 10.25.

The girder designed in this example is not necessarily the most economical one. Other possibilities include a girder with a thinner web and more intermediate stiffeners and a girder with a thicker web and no intermediate stiffeners. Variables that affect economy include weight (volume of steel required) and fabricating costs. Although girders with intermediate stiffeners will usually require less steel, the savings can be offset by the additional fabrication cost. Variable flange thicknesses can also be considered. This alternative will definitely save weight, but here also, fabrication costs must be considered. A practical approach to achieving an economical design is to prepare several alternatives and compare their costs, using estimates of material and fabricating costs. *Design of Welded Structures* (Blodgett, 1966) contains many useful suggestions for the design of economical welded plate girders.

Problems

Flexural Strength

10.4-1 Determine the nominal flexural strength of the following welded shape: The flanges are 1 inch × 10 inches, the web is $\frac{3}{8}$ inch × 45 inches, and the member is simply supported, uniformly loaded, and has continuous lateral support. A572 Grade 50 steel is used.

10.4-2 Determine the nominal flexural strength of the following welded shape: The flanges are 3 inches × 22 inches, the web is $\frac{1}{2}$ inch × 70 inches, and the member is simply supported, uniformly loaded, and has continuous lateral support. A572 Grade 50 steel is used.

10.4-3 Determine the nominal flexural strength of the following welded shape: The flanges are $\frac{7}{8}$ inch × 12 inches, the web is $\frac{3}{8}$ inch × 60 inches, and the member is simply supported, uniformly loaded, and has a span length of 40 feet. Lateral support is provided at the ends and at midspan. A572 Grade 50 steel is used.

10.4-4 Determine the nominal flexural strength of the following welded shape: The flanges are $\frac{3}{4}$ inch × 18 inches, the web is $\frac{1}{4}$ inch × 52 inches, and the member is simply supported, uniformly loaded, and has a span length of 50 feet. Lateral support is provided at the ends and at midspan. A572 Grade 50 steel is used.

10.4-5 An 80-foot-long plate girder is fabricated from a $\frac{1}{2}$-inch × 78-inch web and two $2\frac{1}{2}$-inch × 22-inch flanges. Continuous lateral support is provided. The steel is A572 Grade 50. The loading consists of a uniform service dead load of 1.0 kip/ft (including the weight of the girder), a uniform service live load of 2.0 kips/ft, and a concentrated service live load of 475 kips at midspan. Stiffeners are placed at each end and at 4 feet, 16 feet, and 28 feet from each end. One stiffener is placed at midspan. Determine whether the flexural strength is adequate.

a. Use LRFD.
b. Use ASD.

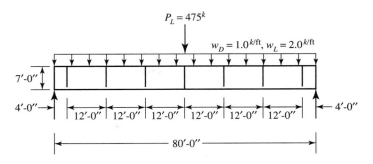

FIGURE P10.4-5

Shear Strength

10.5-1 For the girder of Problem 10.4-2:

a. Compute the nominal shear strength of the end panel if the first intermediate stiffener is placed 70 inches from the support.

b. Compute the nominal shear strength of an interior panel with a stiffener spacing of 200 inches.

c. What is the nominal shear strength if no intermediate stiffeners are used?

10.5-2 A welded plate girder has 1-inch × 30-inch flanges and a $\frac{9}{16}$-inch × 90-inch web. It must support a service dead load of 4 kips/ft (including the self-weight of the girder) and a service live load of 5 kips/ft on a simple span of 75 feet. The steel has a yield stress of 36 ksi. If a stiffener is placed at the end of the girder, what is the required distance to the next stiffener?

a. Use LRFD.

b. Use ASD.

10.5-3 A plate girder cross section consists of two flanges, $1\frac{1}{2}$ inches × 15 inches, and a $\frac{5}{16}$-inch × 66-inch web. A572 Grade 50 steel is used. The span length is 55 feet, the service live load is 2.0 kips/ft, and the dead load is 0.225 kips/ft, including the weight of the girder. Bearing stiffeners are placed at the ends, and intermediate stiffeners are placed at 6'-2" and 12'-9" from each end. Does this girder have enough shear strength?

a. Use LRFD.

b. Use ASD.

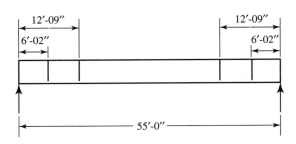

FIGURE P10.5-3

10.5-4   Determine whether the girder of Problem 10.4-5 has enough shear strength.

    a. Use LRFD.

    b. Use ASD.

Bearing Stiffeners

10.6-1   An interior bearing stiffener consists of two plates, $\frac{1}{2}$ inch × 6 inches, one on each side of a $\frac{5}{16}$-inch × 56-inch plate girder web. The stiffeners are clipped $\frac{1}{2}$ inch to clear the flange-to-web welds. All steel is A36.

    a. Use LRFD and determine the maximum factored concentrated load that can be supported.

    b   Use ASD and determine the maximum service concentrated load that can be supported.

10.6-2   The details of an end bearing stiffener are shown in Figure P10.6-2. The stiffener plates are $\frac{9}{16}$-inch thick, and the web is $\frac{3}{16}$-inch thick. The stiffeners are clipped $\frac{1}{2}$ inch to provide clearance for the flange-to-web welds. All steel is A572 Grade 50.

    a. Use LRFD and determine the maximum factored concentrated load that can be supported.

    b. Use ASD and determine the maximum service concentrated load that can be supported.

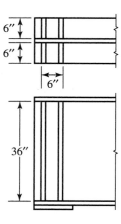

FIGURE P10.6-2

Design

10.7-1   A plate girder must be designed to resist a service-load bending moment of 12,000 ft-kips consisting of 25% dead load and 75% live load. The total depth of the girder will be 101 inches, and continuous lateral support will be provided. Use A36 steel and select a trial cross section. Assume that the moment includes an accurate estimate of the girder weight.

    a. Use LRFD.

    b. Use ASD.

10.7-2   A plate girder must be designed to resist a dead-load bending moment of 3800 ft-kips and a live-load bending moment of 7800 ft-kips. The total depth of the girder will be 78 inches, and the unbraced length is 25 feet. Use A572 Grade 50 steel and select a trial cross section. Assume that the moment includes an accurate estimate of the girder weight. Use $C_b = 1.67$.

a. Use LRFD.
b. Use ASD.

10.7-3 Use LRFD and determine the flange and web dimensions for a plate girder of the type shown in Figure P10.7-3 for the following conditions:

- Span length = 50 ft.
- Girder is simply supported with lateral bracing at 12 ft-6 in. intervals.
- *Superimposed* service dead load = 0.5 kips/ft (does not include the girder weight).
- Concentrated service live load = 130 kips applied at midspan.
- Steel is A572 Grade 50.

Select the flange and web dimensions so that intermediate stiffeners are not required. Assume that a bearing stiffener will be used at the concentrated load.

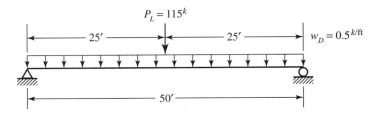

FIGURE P10.7-3

10.7-4 A plate girder must be designed for the conditions shown in Figure P10.7-4. The given loads are factored, and the uniformly distributed load includes a conservative estimate of the girder weight. Lateral support is provided at the ends and at the load points. Use LRFD for the following:

a. Select the flange and web dimensions so that intermediate stiffeners will be required. Use $F_y = 50$ ksi and a total depth of 50 inches. Bearing stiffeners will be used at the ends and at the load points, but do not proportion them.
b. Determine the locations of the intermediate stiffeners, but do not proportion them.

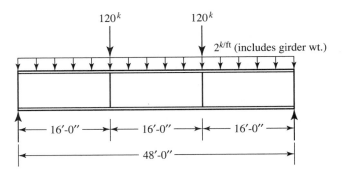

FIGURE P10.7-4

10.7-5   Use LRFD and design the bearing stiffeners for the girder of Problem 10.7-4. Use $F_y = 50$ ksi.

10.7-6   Design flange-to-web welds for the girder of Problem 10.4-5.
a. Use LRFD.
b. Use ASD.

10.7-7   A 66-foot long plate girder must support a uniformly distributed load and concentrated loads at the one-third points. The uniform load consists of a 1.3-kip/ft dead load and a 2.3 kip/ft live load. Each concentrated load consists of a 28-kip dead load and a 49-kip live load. There is lateral support at the ends and at the one-third points. Use A572 Grade 50 steel, a total depth of 80 inches, and LRFD.
a. Select the girder cross section and the required spacing of intermediate stiffeners.
b. Determine the size of intermediate and bearing stiffeners.
c. Design all welds.

10.7-8   Design a plate girder for the following conditions:
- Span length = 100 ft.
- There is a uniformly distributed service load consisting of a live load of 0.7 kips/ft and a superimposed dead load of 0.3 kips per foot.
- There are concentrated service loads at the quarter points consisting of 50 kips dead load and 150 kips live load each.
- The compression flange will be laterally supported at the ends and at the quarter points.

Use A572 Grade 50 steel.
a. Select the girder cross section and the required spacing of intermediate stiffeners.
b. Determine the size of intermediate and bearing stiffeners.
c. Design all welds.

10.7-9   A plate girder *ABCDE* will be used in a building to provide a large column-free area as shown in Figure P10.7-9. The uniform load on the girder consists of 1.9 kips/ft of dead load (not including the weight of the girder) and 2.8 kips/ft of live load. In addition, the girder must support column loads at *B*, *C*, and *D* consisting of 112 kips dead load and 168 kips live load at each location. Assume that the girder is simply supported and that the column loads act as concentrated loads; that is, there is no continuity with the girder. Assume lateral support of the compression flange at 10-foot intervals. Use LRFD and $F_y = 50$ ksi for all components.

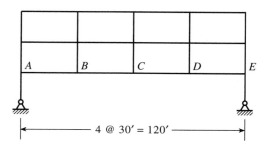

FIGURE P10.7-9

a. Design the girder cross section. Use a total depth of 10 feet.
b. Determine the location and size of intermediate stiffeners.
c. Determine the required size of bearing stiffeners at B, C, and D (assume that there will be a framed connection at each end of the girder).
d. Design all welds.

Plastic Analysis and Design

A.1 INTRODUCTION

We introduced the concept of plastic collapse in Section 5.2, "Bending Stress and the Plastic Moment." Failure of a structure will take place at a load that forms enough plastic hinges to create a mechanism that will undergo uncontained displacement without any increase in the load. In a statically determinate beam, only one plastic hinge is required. As shown in Figure A.1, the hinge will form where the moment is maximum—in this case, at midspan. When the bending moment is large enough to cause the entire cross section to yield, any further increase in moment cannot be countered, and the plastic hinge has formed. This hinge is similar to an ordinary hinge except that the plastic hinge will have some moment resistance, much like a "rusty" hinge.

The plastic moment capacity, denoted M_p, is the bending moment at which a plastic hinge forms. It is equal and opposite to the internal resisting moment corresponding to the stress distribution shown in Figure A.1c. The plastic moment can be computed for a given yield stress and cross-sectional shape, as indicated in Figure A.2. If the stress distribution in the fully plastic condition is replaced by two equal and opposite statically equivalent concentrated forces, a couple is formed. The magnitude of each of these forces equals the yield stress times one half of the total cross-sectional area. The moment produced by this internal couple is

$$M_p = F_y \frac{A}{2} a = F_y Z_x$$

where A is the total cross-sectional area, a is the distance between centroids of the two half areas, and Z_x is the plastic section modulus. The factor of safety between first yielding and the fully plastic state can be expressed in terms of the section moduli. From Figure A.1b, the moment causing first yield can be written as

$$M_y = F_y S_x \quad \text{and} \quad \frac{M_p}{M_y} = \frac{F_y Z_x}{F_y S_x} = \frac{Z_x}{S_x}$$

This ratio is a constant for a given cross-sectional shape and is called the *shape factor*. For a beam designed by allowable stress theory, it is a measure of the reserve capacity. For W-shapes, the shape factor is between 1.1. and 1.2.

FIGURE A.1

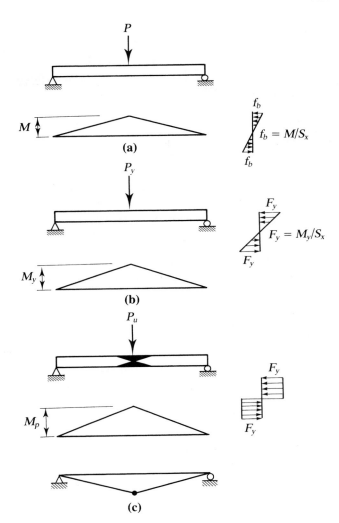

FIGURE A.2

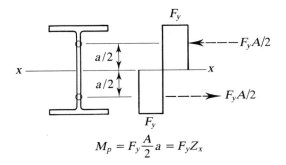

$$M_p = F_y \frac{A}{2} a = F_y Z_x$$

In a statically indeterminate beam or frame, more than one plastic hinge will be required for the formation of a collapse mechanism. These hinges will form sequentially, although it is not always necessary to know the sequence. Plastic analysis of statically indeterminate structures will be considered after a discussion of specification requirements.

A.2 AISC REQUIREMENTS

Plastic analysis and design is covered in Appendix 1 of the AISC Specification, "Inelastic Analysis and Design." Because plastic analysis is based on an ultimate load condition, load and resistance factor design (rather than allowable strength design) is appropriate. In this book, we limit our coverage to hot-rolled I shapes.

For a collapse mechanism to form, the structure must remain stable as the load increases, and the plastic hinges that form early must have enough rotation capacity (ductility) for the formation of the remaining plastic hinges. In hot-rolled I-shaped beams, these two conditions will be met if the steel yield stress is no greater than 65 ksi and the shape is compact—that is, if

$$\frac{b_f}{2t_f} \leq 0.38\sqrt{\frac{E}{F_y}} \quad \text{and} \quad \frac{h}{t_w} \leq 3.76\sqrt{\frac{E}{F_y}}$$

The web slenderness must also be small enough so that web buckling will not occur. This means that the nominal shear strength should be

$$V_n = 0.6F_y A_w C_v \qquad \text{(AISC Equation G2-1)}$$

where $C_v = 1.0$. For this to be true,

$$\frac{h}{t_w} \leq 2.24\sqrt{\frac{E}{F_y}}$$

To prevent lateral buckling, the value of L_b, the unbraced length of the compression flange, is limited to L_{pd} adjacent to plastic hinge locations. For I-shaped members,

$$L_{pd} = \left[0.12 + 0.076\left(\frac{M_1}{M_2}\right)\right]\left(\frac{E}{F_y}\right)r_y$$

$$\text{(AISC Equation A-1-7)}$$

In this equation, M_1 is the smaller moment at the end of the unbraced length and M_2 is the larger. The ratio M_1/M_2 is positive when M_1 and M_2 bend the segment in reverse curvature and negative when they cause single-curvature bending.

For compact shapes with adequate lateral bracing, the design strength ϕM_n may be taken as ϕM_p where

$$M_p = F_y Z < 1.6F_y S \qquad \text{(AISC Equation A-1-6)}$$

and $\phi = 0.90$.

The upper limit of $1.6F_yS$ is to prevent excessive deformations under service loads. This upper limit will never control for W-shapes bent about the strong axis.

Other specification requirements are as follows:

- Plastic analysis and design is permitted only for $F_y \leq 65$ ksi.
- For beam–columns, the compactness criterion for webs is more stringent. For I-shaped sections:

$$\text{For } \frac{P_u}{\phi_b P_y} \leq 0.125, \quad \frac{h}{t_w} \leq 3.76\sqrt{\frac{E}{F_y}}\left(1 - \frac{2.75P_u}{\phi_b P_y}\right) \quad \text{(AISC Equation A-1-1)}$$

$$\text{For } \frac{P_u}{\phi_b P_y} > 0.125, \quad \frac{h}{t_w} \leq 1.12\sqrt{\frac{E}{F_y}}\left(2.33 - \frac{P_u}{\phi_b P_y}\right) \geq 1.49\sqrt{\frac{E}{F_y}}$$
$$\text{(AISC Equation A-1-2)}$$

So, where $P_y = F_y A_g$, the compressive load causes yielding. This is called the "squash" load.
- For low-rise moment frames (unbraced frames) with small axial loads, a first-order plastic analysis (the type covered in this appendix) is acceptable, provided the moment amplification provisions and interaction equations of the Specification are used. For other moment frames, a second-order plastic analysis is required. Second-order analysis will not be covered in this appendix.
- For all moment frames, the required compressive design strength of columns cannot exceed $\phi_c(0.75F_y A_g)$, where $\phi_c = 0.90$.
- For braced frames, the compressive design strength of columns cannot exceed $\phi_c(0.85F_y A_g)$.
- The slenderness ratio L/r of columns (in braced or unbraced frames) must not exceed $4.71\sqrt{E/F_y}$.

A.3 ANALYSIS

If more than one collapse mechanism is possible, as with the continuous beam illustrated in Figure A.3, the correct one can be found and analyzed with the aid of three basic theorems of plastic analysis, given here without proof.

1. **Lower-bound theorem (static theorem)**: If a safe distribution of moment (one in which the moment is less than or equal to M_p everywhere) can be found, and it is statically admissible with the load (equilibrium is satisfied), then the corresponding load is less than or equal to the collapse load.
2. **Upper-bound theorem (kinematic theorem)**: The load that corresponds to an assumed mechanism must be greater than or equal to the collapse load. As a consequence, if all possible mechanisms are investigated, the one requiring the smallest load is the correct one.

FIGURE A.3

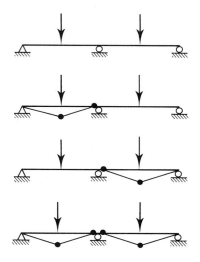

3. **Uniqueness theorem**: If there is a safe and statically admissible distribution of moment in which enough plastic hinges form to produce a collapse mechanism, the corresponding load is the collapse load; that is, if a mechanism satisfies both the upper- and lower-bound theorems, it is the correct one.

Analysis based on the lower-bound theorem is called the *equilibrium method* and is illustrated in Example A.1.

Example A.1   Find the ultimate load for the beam shown in Figure A.4a by the equilibrium method of plastic analysis. Assume continuous lateral support and use $F_y = 50$ ksi.

FIGURE A.4

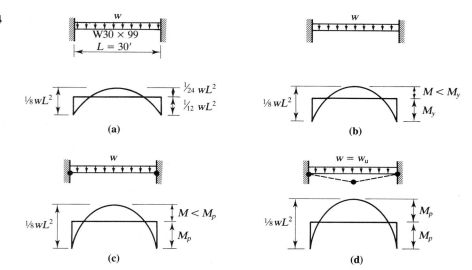

Solution A W30 × 99 with F_y = 50 ksi is compact and, with continuous lateral support, the lateral bracing requirement is satisfied; therefore, plastic analysis is acceptable.

The loading history of the beam, from working load to collapse load, is traced in Figure A.4a–d. At working loads, before yielding begins anywhere, the distribution of bending moments will be as shown in Figure A.4a, with the maximum moment occurring at the fixed ends. As the load is gradually increased, yielding begins at the supports when the bending moment reaches $M_y = F_y S_x$. Further increase in the load will cause the simultaneous formation of plastic hinges at each end at a moment of $M_p = F_y Z_x$. At this level of loading, the structure is still stable, the beam having been rendered statically determinate by the formation of the two plastic hinges. Only when a third hinge forms will a mechanism be created. This happens when the maximum positive moment attains a value of M_p. By virtue of the uniqueness theorem, the corresponding load is the collapse load because the distribution of moment is safe and statically admissible.

At all stages of loading, the sum of the absolute values of the maximum negative and positive moments is $wL^2/8$. At collapse, this sum becomes

$$M_p + M_p = \frac{1}{8} w_u L^2 \quad \text{or} \quad w_u = \frac{16M_p}{L^2}$$

Factored loads must be compared with factored strengths, so $\phi_b M_p$ rather than M_p should have been used in the preceding equations. To keep notation simple, however, we use M_p in all examples until the final step, when we substitute $\phi_b M_p$ for it. The correct result for this example is

$$w_u = \frac{16\phi_b M_p}{L^2}$$

For a W30 × 99,

$$M_p = F_y Z_x = \frac{50(312)}{12} = 1300 \text{ ft-kips}$$

and

$$\phi_b M_p = 0.9(1300) = 1170 \text{ ft-kips}$$

The value of $\phi_b M_p$ also can be obtained directly from the Z_x table in Part 3 of the *Manual*.

Answer $$w_u = \frac{16(1170)}{(30)^2} = 20.8 \text{ kips/ft}$$

■

Example A.2 If the beam in Example A.1 does not have continuous lateral support, determine where it must be braced.

Solution Try lateral bracing at midspan. Since the ends are fixed, they are also points of lateral support. Therefore, the unbraced length L_b is 15 feet, with $M_1 = M_2 = M_p$. These end moments are of opposite signs, so the beam is bent in reverse curvature and $M_1/M_2 = +1$. From AISC Equation A-1-7, the maximum permitted unbraced length is

$$L_{pd} = \left[0.12 + 0.076\left(\frac{M_1}{M_2}\right)\right]\left(\frac{E}{F_y}\right)r_y = \left[0.12 + 0.076(1.0)\right]\left(\frac{29,000}{50}\right)(2.10)$$

$$= 238.7 \text{ in.} = 19.9 \text{ ft}$$

$$L_b = 15 \text{ ft} < 19.9 \text{ ft} \qquad \text{(OK)}$$

Answer Use one lateral brace at midspan.

The *mechanism method* is based on the upper-bound theorem and requires that all possible collapse mechanisms be investigated. The one requiring the smallest load will control, and the corresponding load is the collapse load. The analysis of each mechanism is accomplished by application of the principle of virtual work. An assumed mechanism is subjected to virtual displacements consistent with the possible mechanism motion, and the external work is equated to the internal work. A relationship can then be found between the load and the plastic moment capacity, M_p. This technique will be illustrated in Examples A.3 and A.4.

Example A.3 The continuous beam shown in Figure A.5 has a compact cross section with a design strength $\phi_b M_p$ of 1040 ft-kips. Use the mechanism method to find the collapse load P_u. Assume continuous lateral support.

Solution In keeping with the notation adopted in Example A.1, we will use M_p in the solution and substitute $\phi_b M_p$ in the final step.

There are two possible failure mechanisms for this beam. As indicated in Figure A.5, they are similar, with each segment undergoing a rigid-body motion. To investigate the mechanism in span *AB,* impose a virtual rotation θ at *A*. The corresponding rotations at the plastic hinges will be as shown in Figure A.5b, and the vertical displacement of the load will be 10θ. From the principle of virtual work,

External work = internal work

$$P_u(10\theta) = M_p(2\theta) + M_p\theta$$

(No internal work is done at *A* because there is no plastic hinge.) Solving for the collapse load gives

$$P_u = \frac{3M_p}{10}$$

FIGURE A.5

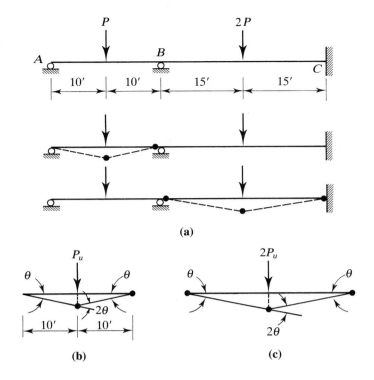

(a)

(b)   (c)

The mechanism for span BC is slightly different: All three hinges are plastic hinges. The external and internal virtual work in this case are

$$2P_u(15\theta) = M_p\theta + M_p(2\theta) + M_p\theta \quad \text{and} \quad P_u = \frac{2}{15}M_p$$

This second possibility requires the smaller load and is therefore the correct mechanism. The collapse load will be obtained using $\phi_b M_p$ in place of M_p.

Answer
$$P_u = \frac{2}{15}\phi_b M_p = \frac{2}{15}(1040) = 139 \text{ kips}$$

Example A.4 Determine the collapse load P_u for the rigid frame shown in Figure A.6. Each member of the frame is a W21 × 147 with $F_y = 50$ ksi. Assume continuous lateral support.

Solution A W21 × 147 is compact for $F_y = 50$ ksi, and with continuous lateral support provided, the conditions for the use of plastic analysis are satisfied.

As indicated in Figure A.6, there are three possible failure modes for this frame: a beam mechanism in member BC, a sway mechanism, and one that is a combination

FIGURE A.6

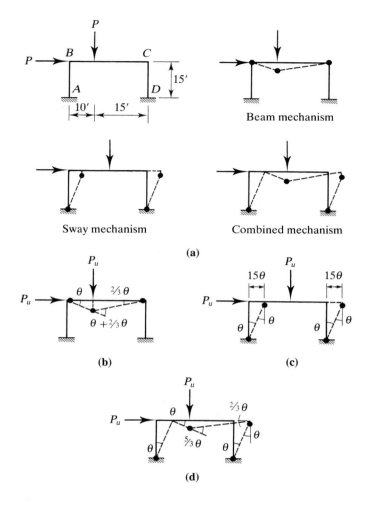

(a)

(b)   (c)

(d)

of the first two. We begin the analysis of each mechanism by imposing a virtual rotation of θ at one of the hinges and expressing the remaining notations and displacements as a function of this angle.

The virtual displacement of the beam mechanism is shown in Figure A.6b. From the equivalence of external and internal work,

$$P_u(10\theta) = M_p\theta + M_p\left(\frac{5}{3}\theta\right) + M_p\left(\frac{2}{3}\theta\right)$$

where M_p has been used in place of $\phi_b M_p$. Solving for P_u gives

$$P_u = 0.3333M_p$$

If axial strains in member BC are neglected, the sway mechanism will displace in the manner shown in Figure A.6c, with the same horizontal displacement at B and C.

As a consequence, the rotations of all the plastic hinges are the same:

$$P_u(15\theta) = M_p(4\theta) \quad \text{or} \quad P_u = 0.2667 M_p$$

From Figure A.6d, the principle of virtual work for the combined mechanism gives

$$P_u(15\theta) + P_u(10\theta) = M_p\theta + M_p\left(\frac{5}{3}\theta\right) + M_p\left(\frac{2}{3}\theta + \theta\right) + M_p\theta$$

$$P_u = 0.2133 M_p \qquad \text{(controls)}$$

Answer The collapse load for the frame is $P_u = 0.2133\phi_b M_p = 0.2133(1400) = 299$ kips. ■

Note that there is a certain similarity between the two methods of analysis. Although the equilibrium method does not require a consideration of all possible mechanisms, it does require that you recognize a mechanism when the assumed distribution of moment is consistent with one. Both methods require the assumption of failure mechanisms, but in the equilibrium method each assumption is checked for a safe and statically admissible distribution of moment, and it may not be necessary to investigate all possible mechanisms.

A.4 DESIGN

The design process is similar to analysis, except that the unknown being sought is the required plastic moment capacity, M_p. The collapse load is known in advance, having been obtained by multiplying the service loads by the load factors.

Example A.5 The three-span continuous beam shown in Figure A.7 must support the gravity service loads given. Each load consists of 25% dead load and 75% live load. Cover plates will be used in spans BC and CD to give the relative moment strengths indicated. Assume continuous lateral support and select a shape of A992 steel.

Solution The collapse loads are obtained by multiplying the service loads by the appropriate load factors. For the 45-kip service load,

$$P_u = 1.2(0.25 \times 45) + 1.6(0.75 \times 45) = 67.5 \text{ kips}$$

For the 57-kip service load,

$$P_u = 1.2(0.25 \times 57) + 1.6(0.75 \times 57) = 85.5 \text{ kips}$$

Three mechanisms must be investigated, one in each span. Figure A.7c–e shows each mechanism after being subjected to a virtual displacement. When a plastic hinge forms at a support where members of unequal strength meet, it will form when the bending moment equals the plastic moment capacity of the weaker member.

FIGURE A.7

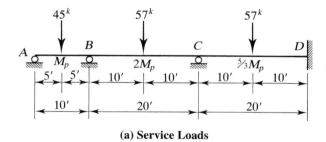

(a) Service Loads

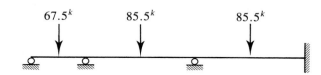

(b) Collapse Loads

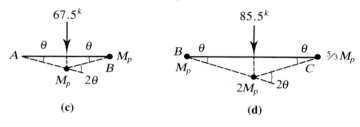

(c)   **(d)**

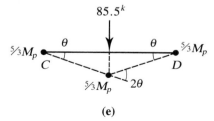

(e)

For span AB,

External work = internal work

$$67.5(5\theta) = M_p(2\theta + \theta) \quad \text{or} \quad M_p = 112.5 \text{ ft-kips}$$

For span BC,

$$85.5(10\theta) = M_p\theta + 2M_p(2\theta) + \frac{5}{3}M_p\theta \quad \text{or} \quad M_p = 128.2 \text{ ft-kips}$$

For span CD,

$$85.5(10\theta) = \frac{5}{3}M_p(\theta + 2\theta + \theta) \quad \text{or} \quad M_p = 128.2 \text{ ft-kips}$$

The upper-bound theorem may be interpreted as follows: The value of the plastic moment corresponding to an assumed mechanism is less than or equal to the plastic moment for the collapse load. Thus, the mechanism requiring the largest moment capacity is the correct one. Both of the last two mechanisms evaluated in this design problem require the same controlling value of M_p and will therefore occur simultaneously. The required strength is actually the required *design* strength, so

Required $\phi_b M_p = 128.2$ ft-kips

Before making a selection, remember that the shape must be compact. To prevent shear buckling of the web, the web width–thickness ratio must satisfy the requirement

$$\frac{h}{t_w} \leq 2.24\sqrt{\frac{E}{F_y}}$$

From the Z_x table, three shapes satisfy the moment requirement with the least weight: a W12 × 26, a W14 × 26, and a W16 × 26. All are compact. The W16 × 26, however, does not satisfy the h/t_w limit. (There is a "v" footnote in both the Dimensions and Properties table in Part 1 of the *Manual* and in the Z_x table.) Since there is no restriction on depth, the W14 × 26, with a flexural design strength of 151 ft-kips, will be tried, because it has more strength than the W12 × 26. In addition, because it has the larger moment of inertia, it will have a smaller deflection.

Try a W14 × 26 and check the shear (refer to Figure A.8).

FIGURE A.8

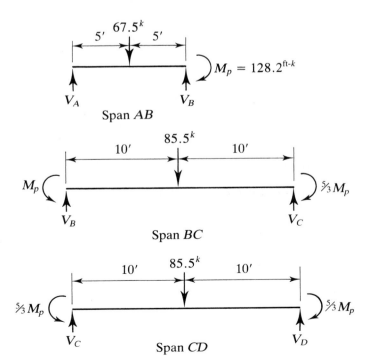

Span *AB*

Span *BC*

Span *CD*

For span AB,

$$\Sigma M_B = V_A(10) - 67.5(5) + 128.2 = 0$$
$$V_A = 20.93 \text{ kips}$$
$$V_B = 20.93 - 67.5 = -46.57 \text{ kips}$$

For span BC,

$$\Sigma M_B = -M_p + 85.5(10) + \left(\frac{5}{3}\right)M_p - V_C(20) = 0$$
$$V_C = \frac{85.5(10) + (2/3)M_p}{20} = \frac{855 + (2/3)(128.2)}{20} = 47.02 \text{ kips}$$
$$V_B = 85.5 - 47.02 = 38.48 \text{ kips}$$

For span CD,

$$\Sigma M_C = -\frac{5}{3}M_p + \frac{5}{3}M_p + 85.5(10) - V_D(20) = 0$$
$$V_D = 42.75 \text{ kips} = V_C$$

The maximum shear is therefore V_C from span BC, or 47.02 kips.

From the Z_x table in Part 5 of the *Manual,* the shear design strength of a W14 × 26 is

$$\phi_v V_n = 106 \text{ kips} > 47.02 \text{ kips} \qquad (\text{OK})$$

Answer   Use a W14 × 26.

A.5   CONCLUDING REMARKS

Analysis of a mechanism subjected to distributed loads presents an additional complication that has not been covered here. Any realistic problem in plastic analysis or design will certainly include this type of loading. In addition, members subjected to both axial compression and bending, as in the rigid frame in Example A.4, should be investigated for interaction of these effects.

With regard to analysis methods in general, a more formal and systematic approach to the equilibrium method exists and is discussed in considerable detail in *The Plastic Methods of Structural Analysis* (Neal, 1977). A more rigorous formulation of the mechanism method is also possible. With this approach, known as the *method of inequalities,* the correct mechanism can be determined directly by linear programming techniques. For most of the structures for which plastic design would ordinarily be used, however, the mechanism method as presented in this appendix usually will be adequate.

References

ABBREVIATIONS

AASHTO	American Association of State Highway and Transportation Officials
ACI	American Concrete Institute
AISC	American Institute of Steel Construction
AISI	American Iron and Steel Institute
AREMA	American Railway Engineering and Maintenance-of-Way Association
ASCE	American Society of Civil Engineers
ASTM	American Society for Testing and Materials
AWS	American Welding Society
BOCA	Building Officials and Code Administrators International
FEMA	Federal Emergency Management Agency
ICBO	International Conference of Building Officials
ICC	International Code Council
RCSC	Research Council on Structural Connections
SBCCI	Southern Building Code Congress International
SJI	Steel Joist Institute
SSRC	Structural Stability Research Council

Ad Hoc Committee on Serviceability. 1986. "Structural Serviceability: A Critical Appraisal and Research Needs." *Journal of Structural Engineering*, ASCE 112 (no. 12): 2646–2664.

Aminmansour, Abbas. 2000. "A New Approach for Design of Steel Beam-Columns." *Engineering Journal*, AISC 37 (no. 2): 41–72.

American Association of State Highway and Transportation Officials. 2002. *Standard Specifications for Highway Bridges*. Washington.

American Association of State Highway and Transportation Officials. 2004. *AASHTO LRFD Bridge Design Specifications*. Washington.

American Concrete Institute. 2005. *Building Code Requirements for Structural Concrete and Commentary* (ACI 318–05). Detroit.

American Institute of Steel Construction. 1989a. *Manual of Steel Construction: Allowable Stress Design*. 9th ed. Chicago.

American Institute of Steel Construction. 1989b. *Specification for Structural Steel Buildings: Allowable Stress Design and Plastic Design*. Chicago.

American Institute of Steel Construction. 1994. *LRFD Manual of Steel Construction, Metric Conversion of the 2nd edition*. Chicago.

American Institute of Steel Construction. 1997a. *Designing Steel for Serviceability*. Seminar Series. Chicago.

American Institute of Steel Construction. 1997b. *Torsional Analysis of Steel Members*. Chicago.

American Institute of Steel Construction. 2000a. *LRFD Specification for Steel Hollow Structural Sections*, Chicago.

American Institute of Steel Construction. 2000b. *LRFD Specification for Single-Angle Members*, Chicago.

American Institute of Steel Construction. 2002. *Detailing for Steel Construction*. 2nd ed. Chicago.

American Institute of Steel Construction. 2005a. *Specification for Structural Steel Buildings*. Chicago.

American Institute of Steel Construction. 2005b. *Steel Construction Manual*. 13th ed. Chicago.

American Institute of Steel Construction. 2005c. *Code of Standard Practice for Steel Buildings and Bridges,* Chicago.

American Iron and Steel Institute. 2001. *North American Specification for the Design of Cold-Formed Steel Structural Members.* Washington.

American Railway Engineering and Maintenance-of-Way Association. 2005. *Manual of Railway Engineering.* Landover, MD.

American Society for Testing and Materials. 2005a. *Annual Book of ASTM Standards.* Philadelphia.

American Society for Testing and Materials. 2005b. "Steel—Structural, Reinforcing, Pressure Vessel, Railway." *Annual Book of ASTM Standards.* Vol. 1.04. Philadelphia.

American Society for Testing and Materials. 2005c. "Fasteners; Rolling Element Bearings." *Annual Book of ASTM Standards.* Vol. 1.08. Philadelphia.

American Society of Civil Engineers. 2002. *Minimum Design Loads for Buildings and Other Structures.* SEI/ASCE 7–02. Reston, VA.

American Welding Society. 2004. *Structural Welding Code—Steel (AWS D1.1:2004).* Miami.

Amrine, J. J. and Swanson, J. A. 2004. "Effects of Variable Pretension on the Behavior of Bolted Connections with Prying." Engineering Journal, AISC 41 (no. 3): 107–116.

Basler, K. 1961. "Strength of Plate Girders in Shear." *Journal of the Structural Division,* ASCE 87 (no. ST7): 151–197.

Bethlehem Steel. 1969. *High-Strength Bolting for Structural Joints.*

Bickford, John H. 1981. *An Introduction to the Design and Behavior of Bolted Joints.* New York: Marcel Dekker.

Birkemore, Peter C. and Gilmor, Michael I. 1978. "Behavior of Bearing Critical Double-Angle Beam Connections." *Engineering Journal,* AISC 15 (no. 4): 109–115.

Bjorhovde, R.; Galambos, T. V.; and Ravindra, M. K. 1978. "LRFD Criteria for Steel Beam-Columns." *Journal of the Structural Division,* ASCE 104 (no. ST9): 1371–1387.

Bleich, Friedrich. 1952. *Buckling Strength of Metal Structures.* New York: McGraw–Hill.

Blodgett, O. W. 1966. *Design of Welded Structures.* Cleveland, OH: The James F. Lincoln Arc Welding Foundation.

Building Officials and Code Administrators International, Inc. 1999. *The BOCA National Building Code.* Chicago.

Burgett, Lewis B. 1973. "Selection of a 'Trial' Column Section." *Engineering Journal,* AISC 10 (no. 2): 54–59.

Butler, L. J.; Pal, S.; and Kulak, G. L. 1972. "Eccentrically Loaded Welded Connections." *Journal of the Structural Division,* ASCE 98 (no. ST5): 989–1005.

Carter, Charles J. 1999. *Stiffening of Wide-Flange Columns at Moment Connections: Wind and Seismic Applications,* Steel Design Guide Series No. 13. Chicago: AISC.

Carter, Charles J. 2004. "Are You Properly Specifying Materials? *Modern Steel Construction,* AISC 44 (no. 1): 41–47.

Chesson, Jr. Eugene; Faustino, Norberto L.; and Munse, William H. 1965. "High-Strength Bolts Subjected to Tension and Shear." *Journal of the Structural Division,* ASCE 91 (no. ST5). 155–180.

Christopher, John E. and Bjorhovde, Reidar 1999. "Semi-Rigid Frame Design Methods for Practicing Engineers." *Engineering Journal,* AISC 36 (no. 1): 12–28.

Cochrane, V. H. 1922. "Rules for Riveted Hole Deduction in Tension Members." *Engineering News Record* (Nov. 16).

Cooper, P. B.; Galambos, T. V.; and Ravindra, M. K. 1978. "LRFD Criteria for Plate Girders." *Journal of the Structural Division,* ASCE 104 (no. ST9): 1389–1407.

Crawford, S. H. and Kulak, G. L. 1971. "Eccentrically Loaded Bolted Connections." *Journal of the Structural Division,* ASCE 97 (no. ST3): 765–783.

Cross, John 2005. "Steel Market Trends." *Modern Steel Construction,* AISC 45 (no. 3): 23–25.

Darwin, D. 1990. *Design of Steel and Composite Beams with Web Openings,* AISC Steel Design Guide Series No. 2. Chicago: AISC.

Disque, Robert O. 1973. "Inelastic K-Factor for Column Design." *Engineering Journal,* AISC 10 (no. 2): 33–35.

Easterling, W. S. and Giroux, L. G. 1993. "Shear Lag Effects in Steel Tension Members." *Engineering Journal,* AISC 30 (no. 3): 77–89.

Federal Emergency Management Agency. 2000. *Recommended Specifications and Quality Assurance Guidelines for Steel Moment-Frame Construction for Seismic Applications.* FEMA-353. Washington.

Fisher, J. W.; Galambos, T. V.; Kulak, G. L.; and Ravindra, M. K. 1978. "Load and Resistance Factor Design Criteria for Connectors." *Journal of the Structural Division,* ASCE 104 (no. ST9): 1427–1441.

Galambos, T. V., ed. 1998. *Guide to Stability Design Criteria for Metal Structures.* 5th ed. Structural Stability Research Council. New York: Wiley–Interscience.

Galambos, T. V. and Ravindra, M. K. 1978. "Properties of Steel for Use in LRFD." *Journal of the Structural Division,* ASCE 104 (no. ST9): 1459–1468.

Gaylord, Edwin H.; Gaylord, Charles N.; and Stallmeyer, James E. 1992. *Design of Steel Structures.* 3rd ed. New York: McGraw-Hill.

Griffis, Lawrence G. 1992. *Load and Resistance Factor Design of W-Shapes Encased in Concrete,* Steel Design Guide Series No. 6. Chicago: AISC.

Hansell, W. C.; Galambos, T. V.; Ravindra, M. K.; and Viest, I. M. 1978. "Composite Beam Criteria in LRFD." *Journal of the Structural Division,* ASCE 104 (no. ST9): 1409–1426.

Hendrick, A, and Murray, T. M. 1984. "Column Web Compression Strength at End-Plate Connections." *Engineering Journal,* AISC 21 (no. 3): 161–169.

Higdon, A.; Ohlsen, E. H.; and Stiles, W. B. 1960. *Mechanics of Materials.* New York: John Wiley and Sons.

International Conference of Building Officials. 1997. *Uniform Building Code.* Whittier, CA.

International Code Council. 2003. *International Building Code.* Falls Church, VA.

Johnston, Bruce G., ed. 1976. *Guide to Stability Design Criteria for Metal Structures.* 3rd ed. Structural Stability Research Council. New York: Wiley-Interscience.

Joint ASCE–AASHO Committee on Flexural Members. 1968. "Design of Hybrid Steel Beams." *Journal of the Structural Division,* ASCE 94 (no. ST6): 1397–1426.

Krishnamurthy, N. 1978. "A Fresh Look at Bolted End-Plate Behavior and Design." *Engineering Journal,* AISC 15 (no. 2): 39–49.

Kulak, G. L. and Timler, P. A. 1984. "Tests on Eccentrically Loaded Fillet Welds." Department of Civil Engineering, University of Alberta, Edmonton. (December).

Kulak, G. L.; Fisher, J. W.; and Struik, J. H. A. 1987. *Guide to Design Criteria for Bolted and Riveted Joints*. 2nd ed. New York: John Wiley and Sons.

Larson, Jay W., and Huzzard, Robert K. 1990. "Economical Use of Cambered Steel Beams." *1990 National Steel Construction Conference Proceedings*. Chicago: AISC.

Lothars, J. E. 1972. *Design in Structural Steel*. 3rd ed. Englewood Cliffs, NJ: Prentice-Hall, Inc.

McGuire, W. 1968. *Steel Structures*. Englewood Cliffs, NJ: Prentice-Hall, Inc.

Munse, W. H. and Chesson, E., Jr. 1963. "Riveted and Bolted Joints: Net Section Design." *Journal of the Structural Division*, ASCE 89 (no. ST1): 107–126.

Murphy, G. 1957. *Properties of Engineering Materials*. 3rd ed. Scranton, PA: International Textbook Co.

Murray, Thomas M. 1983. "Design of Lightly Loaded Steel Column Base Plates." *Engineering Journal*, AISC 20 (no. 4): 143–152.

Murray, Thomas M. 1990. *Extended End-Plate Moment Connections*, Steel Design Guide Series No. 4. Chicago: AISC.

Murray, Thomas M, and Shoemaker, W. Lee 2002. *Flush and Extended Multiple Row Moment End Plate Connections*, Steel Design Guide Series No. 16. Chicago: AISC.

Murray, Thomas M. and Summer, Emmett A. 2003. *Extended End-Plate Moment Connections—Seismic and Wind Applications*. Steel Design Guide Series No. 4., 2nd ed. Chicago: AISC.

Neal, B. G. 1977. *The Plastic Methods of Structural Analysis*. 3rd ed. London: Chapman and Hall Ltd.

Ollgaard, J. G.; Slutter, R. G.; and Fisher, J. W. 1971. "Shear Strength of Stud Connectors in Lightweight and Normal-Weight Concrete." *Engineering Journal*, AISC 8 (no. 2): 55–64.

Ravindra, M. K. and Galambos, T. V. 1978. "Load and Resistance Factor Design for Steel." *Journal of the Structural Division*. ASCE 104 (no. ST9): 1337–1353.

Ravindra, M. K.; Cornell, C. A.; and Galambos, T. V. 1978. "Wind and Snow Load Factors for Use in LRFD." *Journal of the Structural Division*, ASCE 104 (no. ST9): 1443–1457.

Research Council on Structural Connections. 2004. *Specification for Structural Joints Using ASTM A325 or A490 Bolts*. Chicago.

Ricker, David T. 1989. "Cambering Steel Beams." *Engineering Journal*, AISC 26 (no. 4): 136–142.

Ruddy, John L. 1986. "Ponding of Concrete Deck Floors." *Engineering Journal*, AISC 23 (no. 3): 107–115.

Salmon, C. G., and Johnson, J. E. 1996. *Steel Structures, Design and Behavior*. 4th ed. New York: HarperCollins.

Shanley, F. R. 1947. "Inelastic Column Theory." *Journal of Aeronautical Sciences* 14 (no. 5): 261.

Sherman, D. R. 1997. "Designing with Structural Tubing." *Modern Steel Construction*. AISC 37 (no. 2): 36–45.

Southern Building Code Congress International. 1999. *Standard Building Code*. Birmingham, AL.

Steel Joist Institute. 2005. *Standard Specifications, Load Tables, and Weight Tables for Steel Joists and Joist Girders*. Myrtle Beach, SC.

Swanson, J. A. 2002. "Ultimate Strength Prying Models for Boldted T-Stub Connections." *Engineering Journal,* AISC 39 (no. 3): 136–147.

Structural Stability Research Council, Task Group 20. 1979. "A Specification for the Design of Steel-Concrete Composite Columns." *Engineering Journal*, AISC 16 (no. 4): 101–115.

Tall, L., ed. 1964. *Structural Steel Design*. New York: Ronald Press Co.

Thornton, W. A. 1990a. "Design of Small Base Plates for Wide Flange Columns." *Engineering Journal*, AISC 27 (no. 3): 108–110.

Thornton, W. A. 1990b. "Design of Base Plates for Wide Flange Columns—A Concatenation of Methods." *Engineering Journal*, AISC 27 (no. 4): 173–174.

Thornton, W. A. 1992. "Strength and Serviceability of Hanger Connections." *Engineering Journal*, AISC 29 (no. 4): 145–149.

Tide, R. H. R. 2001. "A Technical Note: Derivation of the LRFD Column Design Equations." *Engineering Journal*, AISC 38 (no. 4): 137–139.

Timoshenko, Stephen P. 1953. *History of Strength of Materials*. New York: McGraw–Hill.

Timoshenko, Stephen P., and Gere, James M. 1961. *Theory of Elastic Stability*. 2nd ed. New York: McGraw-Hill.

Viest, I. M.; Colaco, J. P.; Furlong, R. W.; Griffis, L. G.; Leon, R. T.; and Wyllie, Jr. L. A., 1997. *Composite Construction Design for Buildings*. New York: ASCE and McGraw-Hill.

Yura, Joseph A. 1971. "The Effective Length of Columns in Unbraced Frames." *Engineering Journal*, AISC 8 (no. 2): 37–42.

Yura, Joseph A. 2001. "Fundamentals of Beam Bracing." *Engineering Journal*, AISC 38 (no. 1): 11–26.

Yura J. A.; Galambos, T. V.; and Ravindra, M. K. 1978. "The Bending Resistance of Steel Beams." *Journal of the Structural Division*, ASCE 104 (no. ST9): 1355–1370.

Zahn, C. J. 1987. "Plate Girder Design Using LRFD." *Engineering Journal*, AISC 24 (no. 1): 11–20.

Zahn, C. J. and Iwankiw, N. R. 1989. "Flexural-Torsional Buckling and Its Implications for Steel Compression Member Design." *Engineering Journal*, AISC 26 (no. 4): 143–154.

Answers to Selected Problems

Notes 1. Answers are given for all odd-numbered problems except for the following types:

 a. design problems in which there is an element of trial and error in the solution procedure or for which there is more than one acceptable answer

 b. problems for which knowledge of the answer leads directly to the solution

2. All answers, except for approximations, are rounded to three significant figures.

Chapter 1: Introduction

1.5-1 a. 120 ksi b. 13.3% c. 38.9%

1.5-3 c. Approximately 30,100 ksi

1.5-5 c. Approximately 30,000,000 psi d. Approximately 44,000 psi

Chapter 2: Concepts in Structural Steel Design

2-1 a. $R_u = 28.4$ kips (combination 3) b. $\phi R_n = 28.4$ kips c. $R_n = 31.6$ kips d. $R_a = 20.3$ kips (combination 6) e. $R_n = 33.9$ kips

2-3 a. $R_u = 155$ ft-kips (combination 2) b. $R_n = 172$ ft-kips c. $R_a = 108$ ft-kips (combination 2) d. $R_n = 180$ ft-kips

2-5 a. $R_u = 46.8$ psf (combination 3) b. $R_a = 34.5$ kips (combination 1)

Chapter 3: Tension Members

3.2-1 a. 85.1 kips b. 56.6 kips

3.2-3 a. 80.9 kips b. 81.0 kips

3.2-5 a. Unsatisfactory: 102 kips > 98.5 kips b. Unsatisfactory: 70 kips > 65.7 kips

3.3-1 a. 4.14 in.^{2} b. 1.13 in.^{2} c. 3.13 in.^{2} d. 2.31 in.^{2} e. 3.13 in.^{2}

3.3-3 130 kips

3.3-5 a. Not adequate: 220 kips > 172 kips b. Not adequate: 159 kips > 114 kips

3.3-7 a. 341 kips b. 227 kips

3.4-1 222 kips

3.4-3 a. 97.2 kips b. 64.7 kips

3.4-5 a. 318 kips b. 212 kips

3.5-1 86.8 kips

3.5-3 a. 108 kips b. 71.7 kips

3.7-1 a. Required $d = 1.57$ in. b. Required $d = 1.71$ in.

3.7-3 a. Required $d = 0.792$ in. b. Required $d = 1.10$ in.

3.7-5 a. Required $d = 2.06$ in. b. Required $d = 2.00$ in.

3.8-3 6.95 kips

3.8-5 a. Required $d = 0.181$ in. b. Required $d = 0.187$ in.

Chapter 4: Compression Members

4.3-1 a. 1270 kips b. 400 kips

4.3-3 177 kips

4.3-5 a. $\phi_c P_n = 897$ kips, $P_n/\Omega_c = 597$ kips b. $\phi_c P_n = 898$ kips, $P_n/\Omega_c = 597$ kips

4.3-7 148 kips without iteration, 161 kips with iteration

4.3-9 a. 118 kips b. 115 kips

4.7-1 1260 kips

4.7-3 a. Yes: 728 kips < 885 kips b. Yes: 500 kips < 589 kips

4.7-9 a. 1.86 b. 1.86

4.7-11 K_x for $AB = 2.15$, K_x for $BC = 1.57$, K_x for $DE = 1.3$, K_x for $EF = 1.35$

4.7-13 a. 1.7 b. 197 kips

4.8-1   901 kips

4.9-3 $r_x = r_y = 5.07$ in.

4.9-5   23,500 kips

4.9-7   a. 808 kips   b. 212%

4.9-9   a. 1060 kips   b. 707 kips

Chapter 5: Beams

5.2-1   a. $Z = 92.7$ in.^{3}, $M_P = 386$ ft-kips   b. $S = 80.9$ in.^{3}, $M_y = 337$ ft-kips

5.4-3   168 ksi

5.5-1   a. 29.6 kips   b. 31.3 kips

5.5-3   a. Adequate: 1280 ft-kips < 1300 ft-kips
b. Not adequate: 938 ft-kips > 864 ft-kips

5.5-5   a. Not adequate: 505 ft-kips > 461 ft-kips
b. Not adequate: 327 ft-kips > 307 ft-kips

5.5-7   290 ft-kips

5.5-9   a. 1.32   b. 1.32

5.5-11   a. 1.02   b. 1.02

5.5-13   a. Adequate: 169 ft-kips < 213 ft-kips
b. Adequate: 106 ft-kips < 142 ft-kips

5.5-15   a. Not adequate: 1100 ft-kips > 1080 ft-kips
b. Not adequate: 733 ft-kips > 721 ft-kips

5.6-1   a. 0.760 kips/ft   b. 0.758 kips/ft

5.6-3   4480 ft-kips

5.8-1   33.8 kips

5.8-3   a. Shear strength not adequate: 144 kips > 131 kips
b. Shear strength not adequate: 90 kips > 87.5 kips

5.12-1   a. 163 ft-kips   b. 27.5%

5.12-3   a. 221 ft-kips   b. 20.2%

5.15-1   a. Unsatisfactory: Result of interaction equation = 1.11
b. Unsatisfactory: Result of interaction equation = 1.19

5.15-3   a. Unsatisfactory: Result of interaction equation = 1.31
b. Unsatisfactory: Result of interaction equation = 1.23

5.15-5   a. For the loading of Fig. 5.15-5(a), the beam is satisfactory: Result of interaction equation = 0.570. For the loading of Fig. 5.15-5(b), the beam is unsatisfactory: Result of interaction equation = 1.05
b. For the loading of Fig. 5.15-5(a), the beam is satisfactory: Result of interaction equation = 0.535. For the loading of Fig. 5.15-5(b), the beam is satisfactory: Result of interaction equation = 0.983.

Chapter 6: Beam–Columns

6.2-1   a. Satisfactory: Result of interaction equation = 0.868.
b. Satisfactory: Result of interaction equation = 0.868

6.5-1   a. 1.04   b. 1.07

6.5-3   a. Satisfactory: Result of interaction equation = 0.960
b. Satisfactory: Result of interaction equation = 0.985

6.5-5   a. Satisfactory: Result of interaction equation = 0.703   b. Satisfactory: Result of interaction equation = 0.753

6.5-7   a. Not adequate: Result of interaction equation = 1.06   b. Not adequate: Result of interaction equation = 1.08

6.5-9   a. 435 kips   b. 420 kips

6.5-11   a. Not adequate: Result of interaction equation = 1.13
b. Not adequate: Result of interaction equation = 1.16

6.5-13   a. 36.7 kips   b. 38.5 kips

6.6-1   Satisfactory: Result of interaction equation = 0.748

Chapter 7: Simple Connections

7.3-1   a. Satisfactory: $s = 2.75$ in. > 2.33 in.; $L_e = 1.5$ in. = min. L_e b. 118 kips

7.4-1   a. Satisfactory: $s = 3$ in. > 2.33 in.; $L_e = 2$ in. > 1.5 in.
b. For shear, $\phi R_n = 64.9$ kips. For bearing, $\phi R_n = 206$ kips.   c. For shear, $R_n/\Omega = 43.3$ kips. For bearing, $R_n/\Omega = 137$ kips

7.4-3   a. 1.05 bolts: use two.   b. 1.07 bolts: use two.

7.4-5   a. 65.1 kips   b. 63.7 kips

7.6-1   a. 277 kips   b. 231 kips

7.6-3   For both LRFD and ASD: a. 20   b. 10   c. 8   d. A325

7.6-5   a. 58.0 kips   b. 58.0 kips

7.8-1   a. The tee and bolts are adequate: Required $t_f = 0.888$ in. < 0.985 in.   b. The tee and bolts are adequate: Required $t_f = 0.889$ in. < 0.985 in.

7.9-1   a. Adequate: For shear, 12.9 kips < 21.6 kips; For bearing, 12.9 kips < 57.3 kips; For tension, 22.3 kips < 28.6 kips   b. Adequate: For shear, 9.38 kips < 14.4 kips; For bearing, 9.38 kips < 38.2 kips; For tension: 16.2 kips < 17.6 kips

7.9-3   a. 4.78 bolts required   b. 4.78 bolts required

7.9-5   a. required $d = 0.853$ in.   b. Required $d = 0.856$ in.

7.11-1   a. 73.1 kips   b. 72.4 kips

7.11-3   a. 106 kips   b. 104 kips

Chapter 8: Eccentric Connections

8.2-1   22.3 kips

8.2-3   31.8 kips

8.2-5 20.5 kips

8.2-7 a. Required $d = 0.799$ in.   b. Required $d = 0.806$ in.

8.2-9 a. Required $d = 0.697$ in.   b. Required $d = 0.697$ in.

8.2-11 a. 39.5 kips   b. 26.3 kips

8.2-13 a. Six bolts per vertical row
b. Six bolts per vertical row

8.3-1 a. Adequate: For shear, 25.4 kips < 28.3 kips; For tension, 10.9 kips < 21.3 kips
b. Adequate: For shear, 17 kips < 18.9 kips; For tension, 7.29 kips < 14.1 kips

8.3-3 a. Not adequate: For shear, 15 kips < 21.6 kips (OK); For tension: 25 kips > 24.6 kips (N.G.)
b. Not adequate: For shear, 10 kips < 14.4 kips (OK); For tension, 16.7 kips > 16.4 kips (N.G.)

8.3-5 a. Adequate: For shear, 9.69 kips < 21.6 kips; For tension, 29.1 kips < 34.6 kips   b. Adequate: For shear, 6.6 kips < 14.4 kips: For tension, 19.8 kips < 22.8 kips

8.3-7 a. 111 kips   b. 73.8 kips

8.4-1 4.50 kips/in.

8.4-3 13.5 kips/in.

8.4-5 a. 5.34 sixteenths of an inch   b. 5.72 sixteenth of an inch

8.4-7 4.26 kips/in.

8.4-11 a. Adequate: Required size = 1.83 sixteenths of an inch   b. Adequate: Required size = 1.83 sixteenths of an inch

8.4-13 a. 4.37 sixteenths of an inch   b.1.57 sixteenths of an inch

8.5-1 1.18 kips/in.

8.5-3 a. $R_u = 36.5$ kips   b. $R_a = 24.3$ kips

8.5-5 a. 90.5 kips   b. 60.3 kips

8.6-1 a. No: 180 kips > 173 kips   b. No: 120 kips > 115 kips

8.6-3 a. 51.7 kips   b. 122 ft-kips

8.7-1 a. Stiffeners are required   b. Stiffeners are required

8.8-1 a. Adequate: Required number of bolts for tension = 3.03   b. Adequate: Required number of bolts for tension = 3.02

Chapter 9: Composite Construction

9.1-1 a. 1760 in.^{4} b. $f_{top} = 0.926$ ksi (compression), $f_{bot} = 34.4$ ksi, $f_c = 1.10$ ksi

9.1-3 a. 3760 in.^{4} b. $f_{top} = 2.21$ ksi (compression), $f_{bot} = 3.17$ ksi, $f_c = 1.08$ ksi

9.1-5 968 ft-kips

9.2-1 a. Adequate: 146 ft-kips < 251 ft-kips   b. Adequate: 102 ft-kips < 167 ft-kips

9.3-1 a. Satisfactory: Before concrete has cured, $M_u = 51.8$ ft-kips < 75.4 ft-kips; After concrete has cured, $M_u = 155$ ft-kips < 168 ft-kips; $V_u = 24.9$ kips < 79.1 kips   b. Satisfactory: Before concrete has cured, $M_a = 39.5$ ft-kips < 50.1 ft-kips; After concrete has cured , $M_a = 105$ ft-kips < 112 ft-kips; $V_a = 16.8$ kips < 52.8 kips

9.4-1 a. Adequate: Before concrete has cured, $M_u = 233$ ft-kips < 484 ft-kips; After concrete has cured, $M_u = 896$ ft-kips < 966 ft-kips, $V_u = 89.6$ kips < 256 kips
b. Adequate: Before concrete has cured, $M_a = 182$ ft-kips < 322 ft-kips; After concrete has cured, $M_a = 596$ ft-kips < 643 ft-kips, $V_a = 59.6$ kips < 171 kips
c. 98

9.4-3 46

9.6-1 a. 1.73 in.   b. 1.92 in.

9.6-3 a. 1.49 in., 2.02 in. (total)   b. $\Delta_L = 0.840$ in. > $L/360$; select another beam

9.6-5 a. 1.21 in., 1.77 in. (total)   b. $\Delta_L = 0.764$ in. < $L/360$ (OK)

9.7-1 a. 0.539 in.   b. 528 ft-kips

9.7-3 306 ft-kips

9.8-1 306 ft-kips

9.10-1 518 kips

Chapter 10: Plate Girders

10.4-1 2380 ft-kips

10.4-3 3340 ft-kips

10.4-5 a. Not adequate: $M_u = 18{,}700$ ft-kips > $\phi_b M_n = 17{,}700$ ft-kips   b. Not adequate: $M_a = 11{,}900$ ft-kips > $M_n/\Omega_b = 11{,}800$ ft-kips

10.5-1 a. 469 kips   b. 489 kips   c. 235 kips

10.5-3 a Yes: For end panel, $V_u = 95.4$ kips < $\phi_v V_n = 98.2$ kips; For second panel, $V_u = 74.0$ kips < $\phi_v V_n = 352$ kips; For middle panel, $V_u = 51.2$ kips < $\phi_v V_n = 54.7$ kips   b. Yes: For end panel, $V_a = 61.2$ kips < $V_n/\Omega_v = 65.3$ kips; For second panel, $V_a = 47.5$ kips < $V_n/\Omega_v = 234$ kips; For middle panel, $V_a = 32.8$ kips < $V_n/\Omega_v = 36.4$ kips

10.6-1 a. 267 kips   b. 178 kips

Index